NONLINEAR SYSTEM TECHNIQUES AND APPLICATIONS

NONLINEAR SYSTEM TECHNIQUES AND APPLICATIONS

JULIUS S. BENDAT

A Wiley-Interscience Publication
JOHN WILEY & SONS, INC.
New York • Chichester • Weinheim • Brisbane • Singapore • Toronto

This text is printed on acid-free paper. ♾

Published simultaneously in Canada.

Library of Congress Cataloging in Publication Data:

Bendat, Julius S.
Nonlinear system techniques and applications / Julius S. Bendat.
p. cm.
"A Wiley-Interscience publication."
Includes bibliographical references and index.
ISBN 0-471-16576-X (cloth : alk. paper)
1. System analysis. 2. Nonlinear theories. 3. System identification. 4. Stochastic processes. I. Title.
QA402.B4493 1997
003—dc21 97-11446
CIP

Printed in the United States of America

10 9 8 7 6 5 4 3 2 1

CONTENTS

PREFACE

This book is an extensive revision and replacement for my previous Wiley-Interscience book, *Nonlinear System Analysis and Identification from Random Data*, published in 1990. Approximately half of the original material has been rewritten or deleted and replaced by new results obtained after 1990. These changes reflect the technical advances that have taken place in the last eight years as well as an increased awareness of pertinent matters gained through further experience. My main reason for writing this book is to acquaint people with important features and engineering applications of the practical nonlinear system analysis and identification techniques in Chapters 3–7 that were not published in the 1990 book. These new nonlinear system techniques, called Direct and Reverse MI/SO Techniques, can solve nonlinear system problems in many fields because they are valid for random data with arbitrary probability, correlation and spectral properties. I strongly recommend that these techniques be used wherever possible.

In the 1990 book, I gave a prominent role to complicated Volterra methods to analyze and identify bilinear and trilinear systems. These Volterra techniques are now relegated to the final two chapters, Chapters 8 and 9, in this book. They are no longer recommended except as a last resort because they are difficult to compute and interpret, are valid only for Gaussian input data, and can often be replaced by the simpler practical techniques emphasized in this book.

Linear system topics in this book are based on material in the following two Wiley-Interscience books by J. S. Bendat and A. G. Piersol, *Random Data: Analysis and Measurement Procedures, 2nd Edition (1986)* and *Engineering Applications of Correlation and Spectral Analysis, 2nd Edition (1993)*. Knowledge of linear system input/output formulas, statistical error analysis criteria, and digital data processing procedures in these two references are required to understand and apply the nonlinear system techniques in this book.

A brief remark on each chapter. A summary also appears at the beginning of each chapter. Chapter 1 reviews some basic ideas about linear systems and random data, including spectral density functions and input/output relations. Chapter 2 discusses established older techniques for zero-memory nonlinear systems that are useful for many engineering applications. Chapter 3 is a lecture given by me in 1995 where I described how Direct and Reverse MI/SO Techniques can replace many types of single-input/single-output (SI/SO) nonlinear models with equivalent multiple-input/single-output (MI/SO) linear models. Chapter 4 deals with general nonlinear models consisting of linear systems in parallel with arbitrary nonlinear systems; special nonlinear models deal with linear systems in parallel with square-law and/or cubic nonlinear systems. Applications are discussed for oceanographic, automotive and biomedical problems.

The determination of nonlinear system physical parameters with memory is demonstrated in Chapter 5 by computer simulation work on many cases. The identification of nonlinear system response properties of naval vessels from measured ocean engineering data is discussed in great detail in Chapters 6 and 7 based on model basin experimental tests conducted at the Hydromechanics Laboratory of the U.S. Naval Academy (USNA) in June 1994 and July 1995. The ideas and procedures used here are widely applicable to nonlinear problems in many fields. Chapters 8 and 9 review older Volterra techniques to analyze random data passing through bilinear and trilinear systems. These techniques involve complicated multidimensional functions and are included here for possible use only when the simpler practical one-dimensional procedures in earlier chapters are not deemed to be suitable.

Material in this book originated from some of my nonlinear research work that was sponsored from 1977 to 1982 by Shell Internationale Petroleum Maatschappij (SIPM) in The Hague, Netherlands, under Dr. Jan Vugts from SIPM. I was assisted in part of this work for SIPM by Allan Piersol, a colleague for many years who is a co-author of references 1–4. Further nonlinear research studies from 1982 to 1996 were done by me for the Naval Facilities Engineering Service Center (NFESC) in Port Hueneme, California, under Dr. Paul Palo, a senior oceanographic engineer from NFESC, who is a co-author of references 5–7. Support for this work came from NFESC and the Office of Naval Research (ONR) in Washington, DC.

For the two nonlinear system test programs at USNA in 1994 and 1995, the experimental physical data were collected by J. J. Zseleczky, J. K. Waters, and others from USNA. Guidance for this oceanographic work was provided by Dr. Paul Palo. The measured data were digitized and sent to me in Los Angeles for nonlinear system analysis and identification using the new Direct and Reverse MI/SO Techniques in this book. Nonlinear system models were formulated and extensively analyzed with the help of Dr. Robert Coppolino, a principal engineer from Measurement Analysis Corporation in Los Angeles,

California, who is a co-author of references 6 and 7. The nonlinear system models and computer results from analyzing this USNA test data are published for the first time in Chapters 6 and 7 of this book.

I wish to thank all of the above-mentioned people for their important contributions to advance this field, and for their engineering expertise on how to apply the practical techniques in this book to solve nonlinear system problems. I also wish to thank SIPM, NFESC, and ONR for their support in getting to the place where I can now claim that the nonlinear world is blossoming out.

JULIUS S. BENDAT

Los Angeles, California

JULIUS S. BENDAT, the author of *Nonlinear System Techniques and Applications*, is an internationally recognized authority in the field of random process theory and engineering applications. He is a mathematical consultant for a wide range of organizations in both industry and government. Past experience includes President, Measurement Analysis Corporation; Senior Staff, TRW Systems; Research Engineer, Northrop Corporation; Assistant Professor, University of Southern California; Radar Officer, U.S. Navy and Physics Research Assistant, University of California Radiation Laboratory. He received his Ph.D. in Mathematics from the University of Southern California.

Dr. Bendat has presented many short courses in the United States and other countries on the topics discussed in this book and seven previous Wiley books of which he is the author or co-author:

- *Engineering Applications of Correlation and Spectral Analysis, 1st Edition* (1980), *2nd Edition* (1993)
- *Nonlinear System Analysis and Identification from Random Data* (1990)
- *Random Data: Analysis and Measurement Procedures, 1st Edition* (1971), *2nd Edition* (1986)
- *Measurement and Analysis of Random Data* (1966)
- *Principles and Applications of Random Noise Theory* (1958)

NONLINEAR SYSTEM TECHNIQUES AND APPLICATIONS

1

LINEAR SYSTEMS, RANDOM DATA

This first chapter reviews how to classify systems and discusses basic properties for ideal physical systems to be linear, physically realizable, constant-parameter, and stable. The chapter also reviews how to classify random data as stationary, nonstationary, or transient and discusses basic formulas for direct Fourier transform computation of spectral density functions for stationary random data and transient random data. These results represent background material in later chapters that deal with procedures for analyzing and identifying various types of linear and nonlinear systems when stationary random data or transient random data pass through these systems.

1.1 LINEAR SYSTEMS

Physical and engineering systems can be separated into two distinct classifications as linear or nonlinear, with parameters that are either constants or time-varying. This leads to the classifications shown in Figure 1.1. Linear system properties are discussed in references 1 and 2 by Bendat and Piersol. Important types of nonlinear systems are discussed in this book.

A system H is a *linear* system if, for any inputs $x_1 = x_1(t)$ and $x_2 = x_2(t)$, and for any constants c_1, c_2,

$$H[c_1 x_1 + c_2 x_2] = c_1 H[x_1] + c_2 H[x_2] \tag{1.1}$$

This equation contains two properties required of linear systems:

Additive Property

$$H[x_1 + x_2] = H[x_1] + H[x_2] \tag{1.2}$$

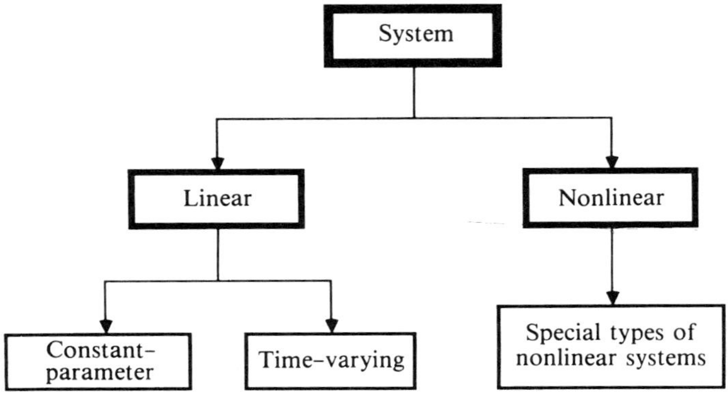

Figure 1.1. Classifications of systems.

Homogeneous Property

$$H[cx] = cH[x] \tag{1.3}$$

By superposition, for linear systems these two properties extend to any finite sum of input terms. The superposition property applies also to integrals that are limiting expressions of finite sums. Failure to satisfy either the additive or the homogeneous property defines a *nonlinear* system and provides a simple, basic way to detect nonlinear system operations.

Systems can be classified as constant-parameter or time-varying. A linear system or a nonlinear system is a *constant-parameter* system if the system response $y(t) = H[x(t)]$ due to an input $x(t)$ is independent of the time when $x(t)$ occurs. Stated another way, if $x(t)$ becomes $x(t + \tau)$, then $y(t)$ becomes $y(t + \tau)$, namely,

$$y(t + \tau) = H[x(t + \tau)] \quad \text{for any } \tau \tag{1.4}$$

A constant-parameter system is governed by a differential-integral equation with constant coefficients. If the coefficients vary with time or if Eq. (1.4) is not satisfied, then the linear or nonlinear system is said to be *time-varying*. Unless stated otherwise, the systems considered in this book are constant-parameter systems.

Systems can be further classified as zero-memory, finite-memory, or infinite-memory. The idea of an *infinite memory* (infinite record lengths) is largely for mathematical convenience and gives negligible error if the system response due to all past inputs is known to be immeasurably small to inputs occurring in the far distant past and if long-time operation is involved. However, if these long-time effects do not decay, or if there is a starting switch in a system so that inputs are present only for a finite period after the switch is closed, then a definite restriction of *finite memory* (finite record lengths) must be imposed to agree with the physical situation. Data obtained experimentally always last only for finite record lengths, leading to statistical errors in estimates of desired theoretical quantities based on infinite record lengths. A *zero-memory* system acts instantaneously on the input data without any memory operations.

For any linear system, random input data with a theoretical Gaussian probability density function describing its amplitude properties will produce random output data that will also have a theoretical Gaussian probability density function. However, for any nonlinear system, Gaussian input data will always produce non-Gaussian output data. Hence, another simple way to detect nonlinear system operations is to measure non-Gaussian output data when these output data are caused by input data that are known, assumed, or verified to be Gaussian. As will be developed in Chapter 2, the particular nature of the non-Gaussian output data can often help to identify the special type of nonlinear operation. Measured data, which can never be truly Gaussian over very wide amplitude ranges, should be judged to be Gaussian or non-Gaussian depending on their properties over amplitude ranges of interest.

It should be noted that different nonlinear transformations may yield approximately similar non-Gaussian outputs so that precise nonlinear system identification usually requires other outside knowledge or additional analyses besides measurement of output probability density functions. Such matters are discussed in this book.

A system H is a *nonlinear* system if, for any inputs $x_1 = x_1(t)$ and $x_2 = x_2(t)$, and for any constant c, the system input/output relations are *not* additive and *not* homogeneous. Specifically,

$$H[x_1 + x_2] \neq H[x_1] + H[x_2] \tag{1.5}$$

$$H[cx] \neq cH[x] \tag{1.6}$$

These nonlinear operations will produce non-Gaussian output data when input data are Gaussian, whereas linear operations preserve the Gaussian structure. Non-Gaussian output data will also occur when input data are non-Gaussian, for example, sine waves. No true sine waves actually exist in nature, but they can be closely approximated.

Any physical system that must operate over very wide dynamic input ranges will ultimately exhibit nonlinear behavior because of saturation and clipping effects. Thus, physical systems should be judged to be linear or nonlinear depending on their specific properties for desired types and levels of inputs. No general theory is available for the analysis of nonlinear systems as exists for linear systems. Instead, different techniques are required for special classifications of nonlinear systems such as those discussed in this book.

1.1.1 Weighting Functions

An ideal constant-parameter linear system can be defined mathematically using a real-valued *unit impulse response function* $h(\tau)$, also call the *weighting function*. This weighting function measures the response of the system at any time to a unit impulse delta function input applied a time τ before. Specifically, $h(\tau)$ measures the output at any time t due to an input applied at time $(t-\tau)$. For a real system to be *physically realizable*, the system cannot respond before the input occurs. Hence, one must have $h(\tau)=0$ for $(t-\tau)>t$, corresponding to

$$h(\tau) = 0 \quad \text{for } \tau < 0 \tag{1.7}$$

This physically realizable weighting function is a special case of a *causal function*, defined as any real-valued function $f(t)$ that is zero for $t \leqslant 0$. Physically realizable systems, whether linear or nonlinear, are always causal.

Consider an arbitrary input $x(\tau)$ to a linear system satisfying Eq. (1.7), where the system operates over the past history of the input from $\tau = -\infty$, the infinite past, to $\tau = t$, the present time. The present value of $x(t)$ can be expressed by the integral

$$x(t) = \int_{-\infty}^{t+} x(\tau)\delta(t-\tau)\,d\tau \tag{1.8}$$

where $t+$ is an infinitesimal value above t and $\delta(\tau)$ is the usual symmetric unit impulse response delta function defined by

$$\begin{aligned} \delta(\tau) &= 0 \quad \text{for } \tau \neq 0, \quad \delta(0) = \infty \\ \int_{-\varepsilon}^{\varepsilon} \delta(\tau) &= 1 \quad \text{for all } \varepsilon > 0 \end{aligned} \tag{1.9}$$

Note that Eq. (1.8), where the integral is the limit of a sum of past delta functions $\delta(t-\tau)$, merely states that $x(t) = x(t)$.

Let the output of the linear system defined by a physically realizable $h(\tau)$ be denoted by $y(t)$. Now, by definition, an input $\delta(t-\tau)$ produces an output $h(t-\tau)$. Hence, by the superposition property for linear systems, the output $y(t)$ due to the input $x(t)$ of Eq. (1.8) becomes

$$y(t) = \int_{-\infty}^{t+} x(\tau)h(t-\tau)\,d\tau = \int_{0}^{\infty} h(\tau)x(t-\tau)\,d\tau \tag{1.10}$$

where the lower limit of zero is understood to be an infinitesimal value below zero. This is the well-known convolution integral expression that is given to describe ideal physically realizable constant-parameter linear system in the time domain. See Figure 1.2.

For linear systems that are not physically realizable, Eq. (1.10) is replaced by

$$y(t) = \int_{-\infty}^{\infty} h(\tau)x(t-\tau)\,d\tau \tag{1.11}$$

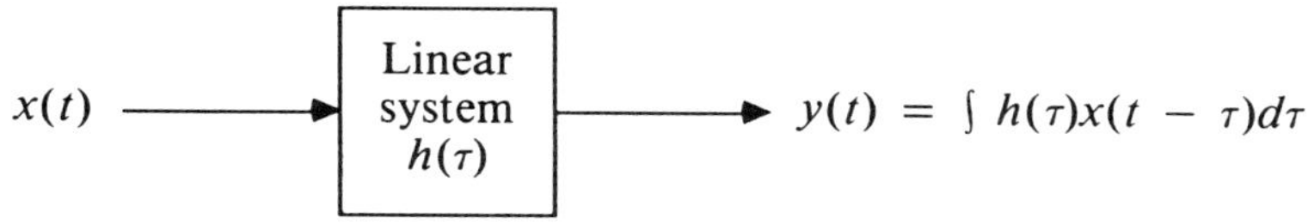

Figure 1.2. Convolution integral relation for ideal constant-parameter linear system.

Equation (1.11) includes Eq. (1.10) as a special case when $h(\tau) = 0$ for $\tau < 0$. It is convenient to use Eq. (1.11) even when Eq. (1.10) is true because of not having to change limits of integration when variables of integration are changed. In later derivations of various theoretical formulas, the theoretical limits of minus infinity to plus infinity will be understood when no limits are shown.

In place of Eq. (1.10), a time-varying linear system is defined by a two-parameter weighting function $h(\tau, t)$ that measures the response of the system at time t to a unit impulse function applied at time $(t - \tau)$. For physical realizability, one must have $h(\tau, t) = 0$ for $\tau < 0$. Thus, physically realizable time-varying linear systems are governed by the time-domain relation

$$y(t) = \int_0^\infty h(\tau, t)x(t - \tau)\, d\tau \tag{1.12}$$

When stationary random data pass through a constant-parameter linear system, the output data will also be stationary random data. However, if the system is time-varying, then the output data become nonstationary random data. To preserve stationarity, one must have a constant-parameter linear system where

$$h(\tau, t) = h(\tau) \quad \text{for all } t \tag{1.13}$$

A physically realizable constant-parameter linear system is said to be *stable* if every possible bounded input produces a bounded output. A necessary and sufficient condition for a linear system to be stable is

$$\int_{-\infty}^{\infty} |h(\tau)|\, d\tau < \infty \tag{1.14}$$

In words, the weighting function $h(\tau)$ must be absolutely integrable over all τ. Stability conditions are assumed for all linear and nonlinear systems considered in this book.

Table 1.1 displays the properties satisfied by ideal physical systems that are linear, physically realizable, constant-parameter, and stable. Many examples of such linear systems are discussed in references 1 and 2.

TABLE 1.1 Properties of Ideal Physical Systems

System Property	Requirement	Benefit
Linear	Additive and homogeneous	Output stays Gaussian
Physically realizable	$h(\tau) = 0 \quad \text{for } \tau < 0$	Output follows input
Constant-parameter	$h(\tau, t) = h(\tau)$	Output stays stationary
Stable	$\int_{-\infty}^{\infty} \lvert h(\tau) \rvert \, d\tau < \infty$	Output stays bounded

1.1.2 Frequency Response Functions

Fourier transforms of both sides of Eq. (1.10) or (1.11) give the familiar relation shown in Figure 1.3 that describes constant-parameter linear systems in the frequency domain, namely,

$$Y(f) = H(f)X(f) \tag{1.15}$$

where $X(f)$, $Y(f)$, and $H(f)$ are Fourier transforms of $x(t)$, $y(t)$, and $h(\tau)$, respectively, as defined by

$$X(f) = \int_{-\infty}^{\infty} x(t)e^{-j2\pi ft} \, dt \tag{1.16}$$

$$Y(f) = \int_{-\infty}^{\infty} y(t)e^{-j2\pi ft} \, dt \tag{1.17}$$

$$H(f) = \int_{-\infty}^{\infty} h(\tau)e^{-j2\pi f\tau} \, d\tau \tag{1.18}$$

A sufficient condition for $H(f)$ to exist is that the linear system be stable. The lower limit in Eq. (1.18) becomes zero for physically realizable (causal) systems where $h(\tau) = 0$ for $\tau < 0$. This complex-valued quantity $H(f)$ is called the *frequency response function* of the system. Note that $H(f)$ is defined for noncausal $h(\tau)$ by Eq. (1.18) when the lower limit can go to minus infinity. For linear systems, Eq. (1.15) shows that $Y(f)$ exists only at frequencies where $X(f)$ exists. This property is not true for nonlinear systems.

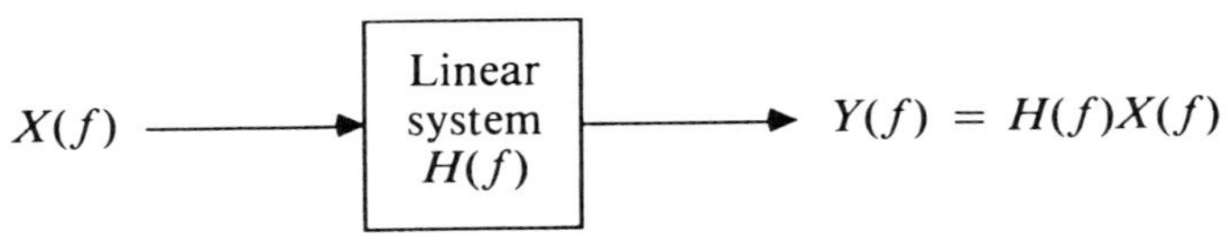

Figure 1.3. Fourier transform relation for ideal constant-parameter linear system.

In terms of the frequency response function $H(f)$, a necessary and sufficient condition that a constant-parameter linear system be physically realizable and stable is that $H(f)$ should be bounded and analytic in the lower-half complex f-plane and on the real f-axis. All poles of $H(f)$ must be in the upper-half complex f-plane. When the complex-valued $H(f)$ is expressed in terms of a real part $H_R(f)$ and an imaginary part $H_I(f)$, a necessary and sufficient condition that a constant-parameter linear system be physically realizable (causal) is that the real part and the imaginary part are Hilbert transform pairs (reference 1).

The frequency response function $H(f)$ can be expressed in terms of a magnitude $|H(f)|$ and an associated phase angle $\phi(f)$ by the equation

$$H(f) = |H(f)|e^{-j\phi(f)} \tag{1.19}$$

The magnitude $|H(f)|$ is called the *gain factor* and the phase angle $\phi(f)$ is called the *phase factor*. The frequency response function can also be expressed in terms of a real part $H_R(f)$ and an imaginary part $H_I(f)$ by the equation

$$H(f) = H_R(f) - jH_I(f) \tag{1.20}$$

Equations (1.19) and (1.20) give the results

$$H_R(f) = |H(f)| \cos \phi(f) \tag{1.21}$$

$$H_I(f) = |H(f)| \sin \phi(f) \tag{1.22}$$

Also,

$$H(f) = [H_R^2(f) + H_I^2(f)]^{1/2} \tag{1.23}$$

$$\phi(f) = \tan^{-1}[H_I(f)/H_R(f)] \tag{1.24}$$

The use of $-j$ rather than $+j$ in Eq. (1.20) is preferred because it yields a $\phi(f)$ in Eq. (1.24) that is a positive instead of a negative arctangent; that is, it makes lag angles positive. This agrees with the physical requirement that it takes a *positive* time-delay for a signal to propagate from an input excitation point to an output response point, not a *negative* time-delay.

Many examples of gain and phase factors from different frequency response functions are discussed and pictured in references 1 and 2. These include theoretical results as well as practical results from measuring experimental data. In succeeding chapters, the notation $H(f)$ or equivalent will always represent a constant-parameter stable finite-memory linear system.

1.2 RANDOM DATA

Random data are any type of data occurring in physical phenomena or in engineering work that does not have an explicit mathematical formula to describe its properties. Instead, any particular time-history record represents

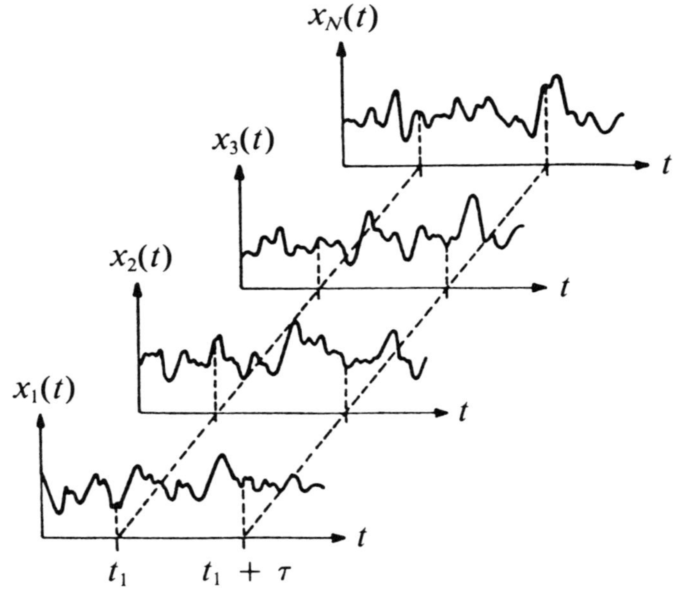

Figure 1.4. Ensemble of time-history records defining a random process.

only one record out of a collection of different time-history records that might have occurred, where the collection of records is called an *ensemble*. This ensemble produces a *random process* that can be defined by analyzing various statistical properties over the collection of records pictured in Figure 1.4. Basic classifications of random data are shown in Figure 1.5. Detailed discussions of each of these classifications and their properties are given in references 1 and 2.

Stationary random data are data whose ensemble-averaged statistical properties are invariant with respect to translations in time. For such data, ensemble-averaged mean values are the same at every time: correlation functions are not functions of two specific times but only of their time difference. *Ergodic data* are stationary random data where long-time averages on any arbitrary time-history record give results that are statistically equivalent to associated ensemble averages over a large collection of records. In practice, stationary random data will automatically be ergodic if there are no sine waves or other deterministic phenomena in the records. Throughout this book, unless stated otherwise, all stationary random data will be assumed to be ergodic. For such data, one long-duration experiment is sufficient to obtain useful information. Theoretically, a stationary random record should never have a beginning or an end.

Nonstationary random data are data whose ensemble-averaged statistical properties change with translations in time. A special class of nonstationary data are called *transient random data* when there is a clearly defined beginning

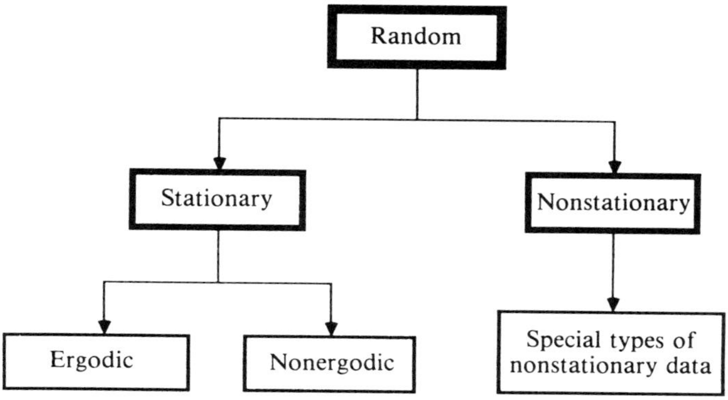

Figure 1.5. Classifications of random data.

and end to the data. In general, techniques for analyzing stationary random data are not appropriate for analyzing nonstationary random data. An exception occurs for analyzing the spectral properties of transient random data, provided there exists an ensemble of similar transient records. Except for a scale factor denoting the time duration of the transient, the same procedures can be followed to obtain the "power" spectral density function of stationary random data versus the "energy" spectral density function of transient random data. With stationary random data, only one long-duration record is required to obtain desired results, since this one record can be subdivided into a collection of suitable shorter records to perform ensemble averages. With transient random data, it is never appropriate to subdivide the transient data into shorter records, since no shorter record is representative of the whole record. Instead, one must repeat the experiment over and over under similar conditions to obtain a collection of suitable records to perform ensemble averages.

When stationary random data pass through constant-parameter linear systems, the output data will also be stationary. Similarly, transient random data through constant-parameter linear systems produce transient output data. However, when stationary or transient random data pass through time-varying linear systems, the output data will be nonstationary. This book concerns only stationary or transient random data at the input and output locations. Special procedures for analyzing different types of nonstationary random data are in Chapter 12 of reference 1.

1.2.1 Basic Properties

For general stationary random data, independent of the time t, basic random data properties from the records in Figure 1.4 are as follows, where $E[\ ]$ denotes an ensemble average over the set of records:

Mean value	$\mu_x = E[x(t)]$	(1.25)
Mean square value	$\psi_x^2 = E[x^2(t)]$	(1.26)
Variance	$\sigma_x^2 = E[\{x(t) - \mu_x\}^2] = \psi_x^2 - \mu_x^2$	(1.27)
Autocorrelation function	$R_{xx}(\tau) = E[x(t)x(t + \tau)]$	(1.28)

The mean value in Eq. (1.25), independent of t, is obtained by summing the amplitude values over the ensemble index k from 1 to N, dividing by N, and theoretically letting N approach infinity as follows:

$$E[x(t)] = \lim_{N \to \infty} \left(\frac{1}{N}\right) \sum_{k=1}^{N} x_k(t) \tag{1.29}$$

Similar computations are used to obtain the other basic properties in Eqs. (1.26) to (1.28). In practice, N will always be finite so that one obtains estimates only of these desired quantities. The fact that N cannot go to infinity leads to random errors in the estimates.

The autocorrelation function $R_{xx}(\tau)$ for stationary random data is a measure of time-related properties in the data that are τ time-units apart. For any value of t, one can easily prove that

$$|R_{xx}(\tau)| \leqslant R_{xx}(0) = E[x^2(t)] = \psi_x^2 \tag{1.30}$$

This shows that the maximum value of $|R_{xx}(\tau)|$ always occurs at $\tau = 0$ and that $R_{xx}(0)$ gives the mean square value of the data.

Consider now any ergodic stationary random process described by a single record $x(t)$. The behavior of $x(t)$ over a long period of time from 0 to T will exhibit the same statistical properties as corresponding ensemble averages over many records at various fixed times. Here, independent of the ensemble index k that denotes a particular ensemble record, basic random data properties using any arbitrary single record $x(t)$ are given as T approaches infinity by the theoretical formulas

Mean value
$$\mu_x = \lim_{T \to \infty} \left(\frac{1}{T}\right) \int_0^T x(t)\, dt \tag{1.31}$$

Mean square value
$$\psi_x^2 = \lim_{T \to \infty} \left(\frac{1}{T}\right) \int_0^T x^2(t)\, dt \tag{1.32}$$

Variance
$$\sigma_x^2 = \lim_{T \to \infty} \left(\frac{1}{T}\right) \int_0^T [x(t) - \mu_x]^2\, dt = \psi_x^2 - \mu_x^2 \tag{1.33}$$

Autocorrelation function
$$R_{xx}(\tau) = \lim_{T \to \infty} \left(\frac{1}{T}\right) \int_0^T x(t)x(t + \tau)\, dt \tag{1.34}$$

In practice, T will always be finite so that one obtains estimates only of these desired results. The fact that T cannot go to infinity leads to random errors in the estimates.

1.2.2 Probability Functions

The amplitude properties of random data can be described using probability density functions. For arbitrary stationary random (ergodic) data, independent of t, let A represent the desired outcome that $x(t)$ has amplitude values that fall inside the amplitude interval $[x_1 < x(t) \leqslant x_2]$. The probability of this outcome A is the fractional time that $x(t)$ falls inside the stated interval relative to the total time T of the record $x(t)$, in the limit as T approaches infinity. For any small interval Δx, let $x_1 = x - (\Delta x/2)$ and let $x_2 = x + (\Delta x/2)$ so as to determine the probability $[x - (\Delta x/2) < x(t) \leqslant x + (\Delta x/2)]$. The *probability density function* $p(x)$ is now defined as the rate of change of this probability with amplitude by dividing by Δx and letting Δx approach zero, namely,

$$p(x) = \lim_{\Delta x \to 0} \text{Probability}\ [x - (\Delta x/2) < x(t) \leqslant x + (\Delta x/2)]/\Delta x \tag{1.35}$$

In practice, Δx will always be finite so that one obtains estimates only of $p(x)$. The fact that Δx cannot go to zero leads to a bias error in the estimate. For small Δx, Eq. (1.35) shows that $p(x)\,\Delta x$ estimates the probability that $x(t)$ falls inside a small interval of width Δx centered at x.

For any probability density function $p(x)$, the *probability distribution function* $P(x)$ determines the probability that the record $x(t)$ at any time will be less than or equal to the amplitude value x, namely,

$$P(x) = \text{Probability}\ [x(t) \leqslant x] = \int_{-\infty}^{x} p(x)\,dx \tag{1.36}$$

Thus

$$\text{Probability}\ [x_1 < x(t) \leqslant x_2] = P(x_2) - P(x_1) = \int_{x_1}^{x_2} p(x)\,dx \tag{1.37}$$

$$\text{Probability}\ [-\infty < x \leqslant \infty] = \int_{-\infty}^{\infty} p(x)\,dx = 1 \tag{1.38}$$

These properties apply to arbitrary probability density functions.

An important class of stationary random data is *Gaussian random data* defined in Section 5.3.3 of reference 1. Basic properties of Gaussian random data are discussed and applied many places in references 1 and 2. Of particular interest are the moment relations stated in Eqs. (9.16) to (9.21) in this book. These relations simplify many input/output formulas for random data passing through linear and nonlinear systems when input data are assumed to be Gaussian.

The first-order Gaussian probability density function $p(x)$ for any stationary random data $x(t)$ can be described completely from knowledge of the mean value μ_x and mean square value ψ_x^2 of the data since these two quantities give the variance and standard deviation of the data. When the mean value is zero, the variance is the same as the mean square value and the Gaussian probability density function takes the familiar form of Eq. (2.4) in this book. Note that this Gaussian $p(x)$ can be defined from knowing only the mean square value of the data. Note also that this mean square value must be finite in order for the Gaussian $p(x)$ to exist. Knowledge of the mean square value is not sufficient to define the $p(x)$ for non-Gaussian data.

Zero mean values will often occur in practice because of using transducers that do not pass DC, or because the first step in data processing is to estimate the mean value and then subtract it off so as to obtain a new variable with zero mean value. It is also customary to divide by the standard deviation of the data so as to obtain a new variable with unity standard deviation. This gives the standardized Gaussian (normal) probability density function that is tabulated in Table A.1 of reference 1, with plots like Figure 4.1 of reference 1 and Figure 2.5 of reference 2.

No real data in nature can be truly Gaussian since, if so, there would be a nonzero probability that any level could be exceeded no matter how large that level might be. However, because of the *central limit theorem*, Section 3.3.1 of reference 1 and Section 2.3.1 of reference 2, many phenomena in nature that are the superposition of many small effects acting together—*if not examined too closely*—appear to follow a Gaussian distribution. The Gaussian assumption thus may be reasonable for input data to a physical system, but should always be questioned with respect to the output data because any data through a nonlinear system will always produce non-Gaussian results.

1.3 SPECTRAL DENSITY FUNCTIONS

Ideas from references 1 and 2 will now be reviewed for direct Fourier transform computation of basic spectral density functions (autospectra and cross-spectra) for either stationary random data or transient random data. Later formulas in this book will be concerned mostly with the spectral analysis of stationary random data passing through various combinations of linear and/or nonlinear systems. Only simple scale factor changes are required to have these results apply to transient random data. Special relations will also be developed for input/output probability density functions and input/output correlation functions, but the emphasis in this book will be devoted to input/output spectral density functions because such frequency-domain formulas are much simpler to compute and interpret than associated time-domain correlation formulas.

For stationary random data, the following expected values of Fourier transform quantities can define the two-sided autospectral density functions

$S_{xx}(f)$ and $S_{yy}(f)$ and the two-sided cross-spectral density function $S_{xy}(f)$:

$$S_{xx}(f) = \frac{1}{T} E[|X(f)|^2] \tag{1.39}$$

$$S_{yy}(f) = \frac{1}{T} E[|Y(f)|^2] \tag{1.40}$$

$$S_{xy}(f) = \frac{1}{T} E[X^*(f)Y(f)] \tag{1.41}$$

In practice, $E[\ ]$ denotes an expected value ensemble average over the quantities inside the brackets. Here, the terms $X(f)$ and $Y(f)$ are finite Fourier transforms of length T for n_d distinct sample records, and the frequency f can theoretically be any value in $-\infty < f < \infty$. However, for digitized data, f is restricted to discrete values spaced $(1/T)$ apart in a bounded frequency range, and Fourier transform formulas become Fourier series formulas. The autospectral density functions $S_{xx}(f)$ and $S_{yy}(f)$ are positive, real-valued, even functions of f; the cross-spectral density function $S_{xy}(f)$ is a complex-valued function of f. If $x(t)$ and $y(t)$ are measured in volts, then $S_{xx}(f)$ and $S_{yy}(f)$ will have units of (volts)^2 per hertz when t has units of seconds. Note that $S_{xy}(f)$ includes $S_{xx}(f)$ and $S_{yy}(f)$ as special cases, and that these quantities always exist for finite length records.

For stationary random data, the one-sided measurable autospectral density functions and cross-spectral density function are zero for $f < 0$ and are defined for $f > 0$ by

$$G_{xx}(f) = \frac{2}{T} E[|X(f)|^2] \tag{1.42}$$

$$G_{yy}(f) = \frac{2}{T} E[|Y(f)|^2] \tag{1.43}$$

$$G_{xy}(f) = \frac{2}{T} E[X^*(f)Y(f)] \tag{1.44}$$

Note that

$$\begin{aligned} G_{xx}(f) &= 2S_{xx}(f) \geqslant 0, & f > 0 \\ &= 0, & f < 0 \end{aligned} \tag{1.45}$$

$$\begin{aligned} G_{yy}(f) &= 2S_{yy}(f) \geqslant 0, & f > 0 \\ &= 0, & f < 0 \end{aligned} \tag{1.46}$$

$$\begin{aligned} G_{xy}(f) &= 2S_{xy}(f), & f > 0 \\ &= 0, & f < 0 \end{aligned} \tag{1.47}$$

General autospectral density functions (autospectra) and cross-spectral density functions (cross-spectra) can have arbitrary spectral properties. The total mean square value of $x(t)$ over all frequencies, independent of its probability nature, is given by the formula

$$\psi_x^2 = E[x^2(t)] = \int_0^\infty G_{xx}(f)\,df = \int_{-\infty}^\infty S_{xx}(f)\,df \tag{1.48}$$

For the special case of zero mean value Gaussian random data, the first-order Gaussian probability density function $p(x)$ of Eq. (2.4) can be determined solely from knowledge of $G_{xx}(f)$ since $G_{xx}(f)$ computes ψ_x^2 and $p(x)$ is known from ψ_x^2. This useful result points out the special importance of autospectral density functions for Gaussian random data.

The data mean square value of $x(t)$ between any two positive frequencies $f_2 > f_1 \geqslant 0$ is given by the formula

$$\psi_x^2(f_1, f_2) = \int_{f_1}^{f_2} G_{xx}(f)\,df \tag{1.49}$$

This last result shows that the autospectral density function $G_{xx}(f)$ represents the rate of change of mean square value with frequency, a property long known from older analog techniques in reference 1 that compute $G_{xx}(f)$ by filtering, squaring, and averaging.

For stationary random data, the *Wiener–Khinchine relation* due to N. Wiener in the United States and A. Khinchine in Russia proves that correlation functions and spectral density functions are Fourier transform pairs, references 1, 2. This does not mean that these two separate time and frequency functions are equally useful for the same applications. Results at specific frequencies of interest are available only from spectral density functions so spectra are always preferred for applications such as determining linear system frequency response functions from measured input/output data. Correlation functions are preferred for other applications such as determining different time delays in multiple path problems. Many engineering cases are discussed and illustrated in reference 2 on where and how to apply correlation functions versus spectral density functions.

The mean value and the mean square value of the data can be computed from either the autocorrelation function or autospectral density function. Hence the probability density function of Gaussian data can be estimated from knowledge of either its autocorrelation function or autospectral density function. This cannot be done for non-Gaussian data. Conversely, knowledge of the probability density function of data does not define its associated autocorrelation function or autospectral density function. For non-Gaussian data, one should always obtain estimates of both its probability density function and autospectral density function since these two functions contain different important amplitude and frequency properties about the data that are not available from either function alone.

Theoretical *white noise* is defined by data where the autospectral density function $G_{xx}(f)$ is a constant over all frequencies from zero to infinity. For such white noise, the total mean square value from Eq. (1.48) would be infinity, which is the reason why true white noise cannot exist in nature. However, data with arbitrary probability density function properties can behave like *bandwidth-limited white noise*, defined as data with a constant autospectrum over some limited frequency range but falling to zero outside this range. A complete contradiction in terms is to assume for any data that the data behave like Gaussian white noise because Gaussian data must have a finite mean square value while white noise has an infinite mean square value. A more reasonable statement is to assume where appropriate that the data behave like Gaussian bandwidth-limited white noise.

True sine waves are data with a fixed amplitude and a fixed frequency that can continue for all time without any amplitude fluctuations or frequency variations. This is impossible for real physical data so no true sine waves can exist in nature. For sine wave data, its autospectral density function is theoretically a delta function with zero bandwidth that approaches infinity at the sine wave frequency, as given by Eq. (3.68) and Figure 3.10 in reference 2. This is another reason why true sine waves cannot occur. However, time-limited sine waves commonly occur due to rotating or reciprocal forcing functions that result in very sharp peaks in autospectrum plots. A narrow-band autospectrum peak can also result from the response of a lightly damped structure to random excitation. These two types of data should not be confused because they produce different effects.

They can be distinguished by performing a simple autospectrum test. One should recompute the autospectrum of the data through each spectral peak with smaller and smaller resolution bandwidths. If the amplitude of the spectral peak increases linearly with the reduction in resolution bandwidth, then the peak can be considered to represent a time-limited sine wave. Otherwise, the peak represents narrow-band random noise. These two types of data can also be distinguished by narrow-band filtering of the data around each spectral peak, followed by computation of their probability density functions to give estimates of the different theoretical results shown in Figures 2.5 and 2.6 in reference 2.

To compute spectral density functions for transient random data, the scale factor of $(1/T)$ should be omitted in Eqs. (1.39) to (1.41) and the expected value operation should be taken over n_d different sample transient records. It is not valid to subdivide any particular transient record, as can be done with stationary random data, since no portion of a transient record would be statistically similar to another portion. New transient records must be obtained to perform the necessary averaging operations. At any selected frequency, the two-sided "energy" spectral density functions for transient random data are defined by

$$\mathscr{S}_{xx}(f) = E[|X(f)|^2] \tag{1.50}$$

$$\mathscr{S}_{yy}(f) = E[|Y(f)|^2] \tag{1.51}$$

$$\mathscr{S}_{xy}(f) = E[X^*(f)Y(f)] \tag{1.52}$$

Note that

$$\mathscr{S}_{xy}(f) = TS_{xy}(f) \tag{1.53}$$

includes $\mathscr{S}_{xx}(f)$ and $\mathscr{S}_{yy}(f)$ as special cases.

For transient random data, the one-sided measurable "energy" spectral density function results are zero for $f < 0$ and are defined for $f > 0$ by

$$\mathscr{G}_{xx}(f) = 2E[|X(f)|^2] \tag{1.54}$$

$$\mathscr{G}_{yy}(f) = 2E[|Y(f)|^2] \tag{1.55}$$

$$\mathscr{G}_{xy}(f) = 2E[X^*(f)Y(f)] \tag{1.56}$$

It follows that $\mathscr{G}_{xy}(f) = TG_{xy}(f)$, where

$$\begin{aligned} \mathscr{G}_{xy}(f) &= 2\mathscr{S}_{xy}(f), \quad f > 0 \\ &= 0, \qquad\qquad f < 0 \end{aligned} \tag{1.57}$$

includes $\mathscr{G}_{xx}(f)$ and $\mathscr{G}_{yy}(f)$ as special cases.

For $f > 0$, these relations prove that

$$\frac{S_{xy}(f)}{S_{xx}(f)} = \frac{\mathscr{S}_{xy}(f)}{\mathscr{S}_{xx}(f)} = \frac{G_{xy}(f)}{G_{xx}(f)} = \frac{\mathscr{G}_{xy}(f)}{\mathscr{G}_{xx}(f)} \tag{1.58}$$

This is the reason why many useful linear system input/output results that involve these functions are valid regardless of whether the data are stationary random data or transient random data, and regardless of whether one computes two-sided or one-sided spectral quantities.

Table 1.2 shows the basic formulas of two- and one-sided spectral density functions for stationary and transient random data. This table demonstrates that only scale factor changes are required to move between these results. The material in this book for the most part will assume stationary random data and use two-sided spectra.

Many examples of autospectral density functions and cross-spectral density functions for both stationary and transient random data are discussed and pictured in references 1 and 2. Applications of these results to solve engineering

TABLE 1.2 Basic Spectral Density Functions

Stationary Random Data

Two-sided spectra

$$S_{xy}(f) = \frac{1}{T} E[X^*(f)Y(f)], \quad -\infty < f < \infty$$

$$S_{xx}(f) = \frac{1}{T} E[X^*(f)X(f)] = \frac{1}{T} E[|X(f)|^2]$$

One-sided spectra

$$G_{xy}(f) = \frac{2}{T} E[X^*(f)Y(f)], \quad f > 0 \text{ only}$$

$$G_{xx}(f) = \frac{2}{T} E[X^*(f)X(f)] = \frac{2}{T} E[|X(f)|^2]$$

Transient Random Data

Two-sided spectra

$$\mathscr{S}_{xy}(f) = E[X^*(f)Y(f)], \quad -\infty < f < \infty$$

$$\mathscr{S}_{xx}(f) = E[X^*(f)X(f)] = E[|X(f)|^2]$$

One-sided spectra

$$\mathscr{G}_{xy}(f) = 2E[X^*(f)Y(f)], \quad f > 0 \text{ only}$$

$$\mathscr{G}_{xx}(f) = 2E[X^*(f)X(f)] = 2E[|X(f)|^2]$$

problems of random data passing through linear systems are treated extensively in these references.

The *coherence function* $\gamma_{xy}^2(f)$ between two stationary random processes $x(t)$ and $y(t)$ is a real-valued function of f defined by

$$\gamma_{xy}^2(f) = \frac{|S_{xy}(f)|^2}{S_{xx}(f)S_{yy}(f)} = \frac{|G_{xy}(f)|^2}{G_{xx}(f)G_{yy}(f)} \tag{1.59}$$

As proved in references 1 and 2, this ordinary coherence function for all f is a nondimensional function that is bounded between zero and one, namely,

$$0 \leqslant \gamma_{xy}^2(f) \leqslant 1 \tag{1.60}$$

The quantity $|\gamma_{xy}(f)|$ represents the positive square root of $\gamma_{xy}^2(f)$.

The coherence function of Eq. (1.59) is a measure of a possible linear relationship between $x(t)$ and $y(t)$. It does not indicate a causal relationship unless one has other physical grounds for knowing that, say, an input $x(t)$ produces an output $y(t)$. If a perfect linear relationship exists between $x(t)$ and $y(t)$ at some frequency f_0, then the coherence function $\gamma_{xy}^2(f_0)$ will be equal to unity at the frequency. If $x(t)$ and $y(t)$ are such that $S_{xy}(f_0) = 0$ at frequency f_0, then the coherence function will be zero at that frequency.

There are four main reasons in practice why a computed coherence function between a measured input $x(t)$ and a measured output $y(t)$ will differ from a value that was expected to be close to unity:

1. Extraneous noise in the input and output measurements.
2. Bias and random errors in spectral density function estimates.
3. The output $y(t)$ due in part to other inputs besides the measured $x(t)$.
4. Nonlinear system operations between $x(t)$ and $y(t)$.

If the first three possibilities are ruled out by good data acquisition, proper data processing, and physical understanding, it is reasonable to conclude that low coherence at particular frequencies is due to nonlinear system effects at these frequencies. Thus, the ordinary coherence function can provide a simple practical way to detect nonlinearity without identifying its precise nature.

1.4 INPUT/OUTPUT SPECTRAL RELATIONS

Input/output spectral relations for many types of linear systems are covered extensively in Chapters 6 and 7 of reference 1 and in Chapters 4–10 of reference 2. The linear systems in these models are assumed to be constant-parameter linear systems. General results are derived when input data in these models are assumed to be stationary random data or transient random data with arbitrary probability and spectral properties. Special results are also developed for Gaussian input data. Thus, these input/output results from references 1 and 2 are very broad in scope and apply to problems in many different fields.

Practical statistical error analysis formulas are derived and discussed in Chapters 8 and 9 of reference 1 and in Chapter 11 of reference 2 that are required to design experiments and evaluate the computed results. Knowledge and use of this material are essential to conduct good engineering work involving these matters.

Detailed spectral procedures are explained in references 1 and 2 to solve engineering and scientific problems for the following models:

1. Single-input/single-output (SI/SO) linear models.
2. Single-input/multiple-output (SI/MO) linear models.
3. Multiple-input/single-output (MI/SO) linear models.
4. Multiple-input/multiple-output (MI/MO) linear models.

TABLE 1.3 Assumptions in Book

1. Systems are constant-parameter stable systems.
2. Memory operations are of finite duration.
3. Data are stationary random ergodic data.
4. Missing integral limits are from $-\infty$ to $+\infty$.
5. Theoretical spectral density functions $S_{xx}(f)$ are two-sided.
6. Measurable spectral density functions $G_{xx}(f)$ are one-sided.
7. Input data have arbitrary probability properties.
8. Input data have arbitrary spectral properties.

Special spectral formulas are listed for two-input/single-output linear models and for three-input/single-output linear models that can be used to analyze nonlinear system engineering problems discussed in this book in Chapters 4–7. Emphasis is placed on models where extraneous output noise can occur since such models are the most important cases in practice.

For an easy reminder, Table 1.3 has various assumptions common to material in later chapters unless stated otherwise.

This concludes Chapter 1.

2

ZERO-MEMORY NONLINEAR SYSTEMS

This chapter begins with a guide to the overall nonlinear material contained in this book. General formulas are derived here to analyze the properties of stationary random data passing through zero-memory nonlinear systems. For any stationary random input data, Gaussian or non-Gaussian, this chapter shows how to determine the output probability density function from the input probability density function, and how this result can help identify different nonlinear sytems. For Gaussian input data with a known autocorrelation function, it also shows how to determine the output autocorrelation function as well as the input/output cross-correlation function. Examples illustrating these results include data through nonlinear two-slope systems, dead-zone systems, clipped systems, smooth-limiter systems, square-law systems, cubic systems, hardening spring systems, and softening spring systems. The chapter concludes with material on how to describe any zero-memory nonlinear system by an optimum third-order polynomial least-squares approximation when input data are assumed to be Gaussian. Practical frequency-domain spectral techniques on how to analyze general nonlinear system models with non-Gaussian input data are developed in Chapters 3–7.

2.1 NONLINEAR MATERIAL IN BOOK

Special nonlinear system relationships and procedures are developed in this chapter and succeeding chapters for many matters that include the following topics:

1. Practical ways to compute and interpret nonlinear system amplitude and frequency results for Gaussian and non-Gaussian data.
2. Input/output probability density functions.
3. Input/output correlation and spectral density functions.

4. Optimum identification of linear and nonlinear system properties from measurements of input/output data.
5. Decomposition of output spectral density functions into linear and nonlinear components.
6. Computation of linear and nonlinear coherence functions.
7. Replacement of single-input/single-output nonlinear models with equivalent multiple-input/single-output linear models.
8. Determination of physical parameters in nonlinear equations of motion from analysis of measured data.
9. Prediction of nonlinear system response properties from measurement of input data and knowledge of nonlinear system properties.
10. Establishment of new standard practical Direct and Reverse MI/SO Techniques for nonlinear system analysis and identification.
11. Applications of the new techniques to many nonlinear engineering problems that occur in different fields.
12. Review of older complicated Volterra functional methods for analysis and identification of bilinear and trilinear systems.

2.2 ZERO-MEMORY NONLINEAR SYSTEMS

A *zero-memory* nonlinear system is a system that acts instantaneously on present inputs in some nonlinear fashion. It does not weight past inputs due to "memory" operations as in convolution integrals for constant-parameter linear systems, where the present value of the output is a function of both present and past values of the input. The initial discussion will be restricted to such nonlinear systems where the output $y(t)$ at any time t is a single-valued nonlinear function of the input $x(t)$ at the *same* time t. Thus, as shown in Figure 2.1,

$$y(t) = g[x(t)] \tag{2.1}$$

where $g(x)$ is a single-valued nonlinear function of x. In succeeding material, the notation $g(x)$ will always represent a zero-memory nonlinear system operation on the input $x = x(t)$, but $g(x)$ is *not* a function of t.

From the definitions of a linear system in Section 1.1, when the system $g(x)$ is *nonlinear* it means that for any constants a_1, a_2 and any inputs $x_1 = x_1(t)$ and $x_2 = x_2(t)$,

$$g(a_1x_1 + a_2x_2) \neq a_1g(x_1) + a_2g(x_2) \tag{2.2}$$

either because $g(x)$ is not additive or because $g(x)$ is not homogeneous. For

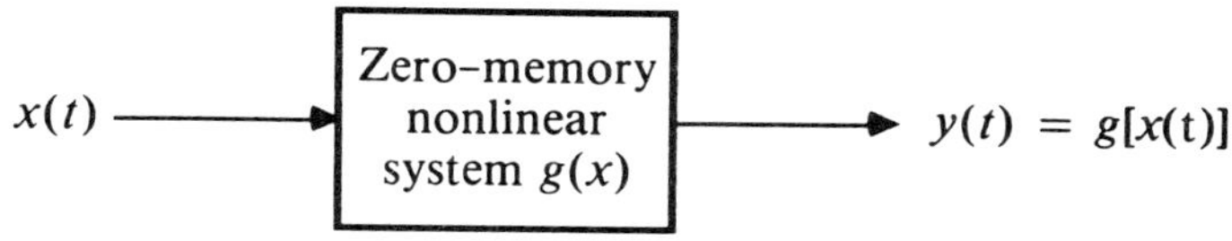

Figure 2.1. Zero-memory nonlinear system.

constant-parameter nonlinear systems:

1. When $x(t)$ is translated to $x(t + \tau)$, then $y(t) = g[x(t)]$ is translated to $y(t + \tau) = g[x(t + \tau)]$.
2. When $x(t)$ is a stationary random process, then $y(t) = g[x(t)]$ is also a stationary random process.

The input/output cross-correlation function $R_{xy}(\tau)$ and the output autocorrelation function $R_{yy}(\tau)$ are here functions only of τ as required for stationary random data, namely,

$$\begin{aligned} R_{xy}(\tau) &= E[x(t)y(t + \tau)] = E[x(t)g\{x(t + \tau)\}] \\ R_{yy}(\tau) &= E[y(t)y(t + \tau)] = E[g\{x(t)\}g\{x(t + \tau)\}] \end{aligned} \tag{2.3}$$

As usual, $E[\ ]$ denotes the expected value operation of the quantity inside the brackets.

Extensions of zero-memory nonlinear systems that occur in practice involve finite-memory operations on past inputs. This can be modeled in simple ways as described in the next section.

2.2.1 Finite-Memory Nonlinear Systems

Finite-memory nonlinear systems are defined by inserting constant-parameter linear systems after and/or before the zero-memory nonlinear systems using appropriate frequency response functions $A(f)$ and $B(f)$ as in Figure 2.2. Special physical cases of interest are when one of the linear systems can be considered to be unity so that memory operations occur only after or before the nonlinear operation, but not in both places as in Figures 2.3 and 2.4.

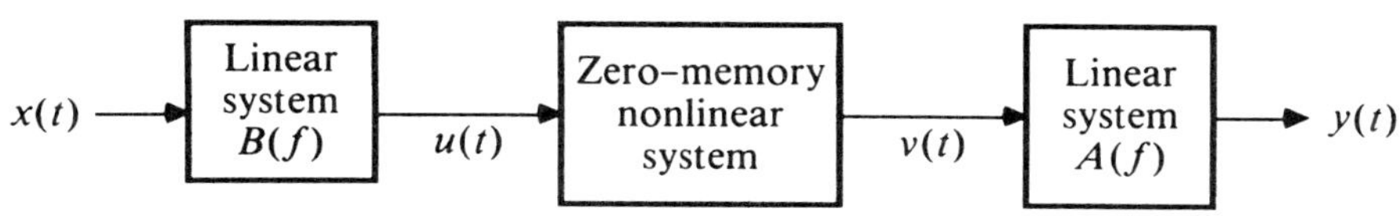

Figure 2.2. Finite-memory nonlinear system with linear systems before and after the zero-memory nonlinear system.

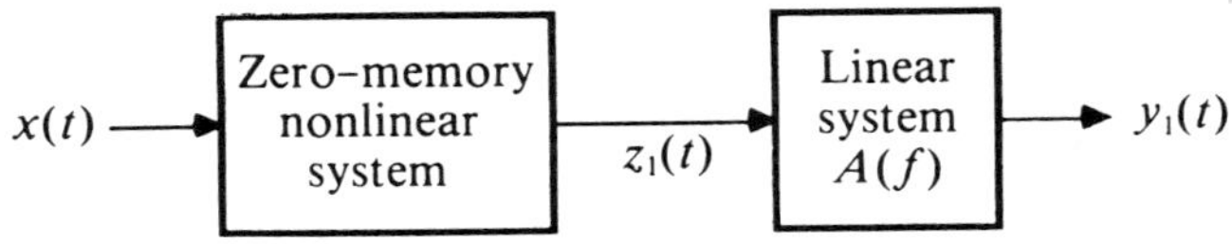

Figure 2.3. Finite-memory nonlinear system with a linear system after the zero-memory nonlinear system.

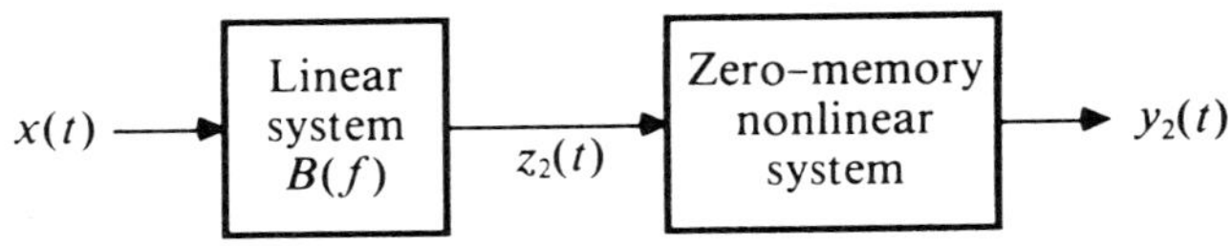

Figure 2.4. Finite-memory nonlinear system with a linear system before the zero-memory nonlinear system.

Linear operations in Figure 2.2, as discussed in references 1 and 2, are appropriate for analyzing the passage of stationary random data from $x(t)$ to $u(t)$ and from $v(t)$ to $y(t)$. New results are required to determine how $v(t)$ is related to $u(t)$ for particular zero-memory nonlinear systems. In Figures 2.3 and 2.4, the passage of stationary random data from $z_1(t)$ to $y_1(t)$ and from $x(t)$ to $z_2(t)$ does not involve any new ideas. It is in the passage of $x(t)$ to $z_1(t)$ and from $z_2(t)$ to $y_2(t)$ through a particular zero-memory nonlinear system that new techniques are needed. Note also that even if $A(f) = B(f)$, and even if the same zero-memory nonlinear system is involved in Figures 2.3 and 2.4, the same input $x(t)$ will in general produce two different outputs $y_1(t) \neq y_2(t)$ that depend on whether the linear systems follow or precede the zero-memory nonlinear system.

Let the quantities $X(f)$, $U(f)$, $V(f)$, $Z(f)$, and $Y(f)$ be the Fourier transforms of $x(t)$, $u(t)$, $v(t)$, $z(t)$, and $y(t)$, respectively. In Figure 2.2, assume that $V(f)$ is some known nonlinear function of $U(f)$ as denoted by

$$V(f) = q[U(f)]$$

Then, the output in Figure 2.2 satisfies

$$Y(f) = A(f)V(f) = A(f)q[U(f)] = A(f)q[B(f)X(f)]$$

In Figure 2.3, assume that $Z_1(f)$ is some known nonlinear function of $X(f)$ as denoted by

$$Z_1(f) = q_1[X(f)]$$

In Figure 2.4, assume that $Y_2(f)$ is some other known nonlinear function of $Z_2(f)$ denoted by

$$Y_2(f) = q_2[Z_2(f)]$$

Now, the outputs in Figures 2.3 and 2.4 satisfy

$$Y_1(f) = A(f)Z_1(f) = A(f)q_1[X(f)]$$
$$Y_2(f) = p_2[Z_2(f)] = q_2[B(f)X(f)]$$

Finite-memory nonlinear systems will be used in Chapters 3–7. The remainder of this chapter concerns only zero-memory nonlinear systems.

2.2.2 Examples of Zero-Memory Nonlinear Systems

Example 1. Two-Slope Systems

$$\begin{aligned} y &= x, & |x| &\leqslant A \\ &= A + b(x - A), & x &\geqslant A \\ &= -A + b(x + A), & x &\leqslant -A \end{aligned}$$

Two cases can occur as sketched in Figure 2.5. See Section 2.6.1.

Example 2. Dead-Zone System

See Figure 2.6 and Section 2.6.2.

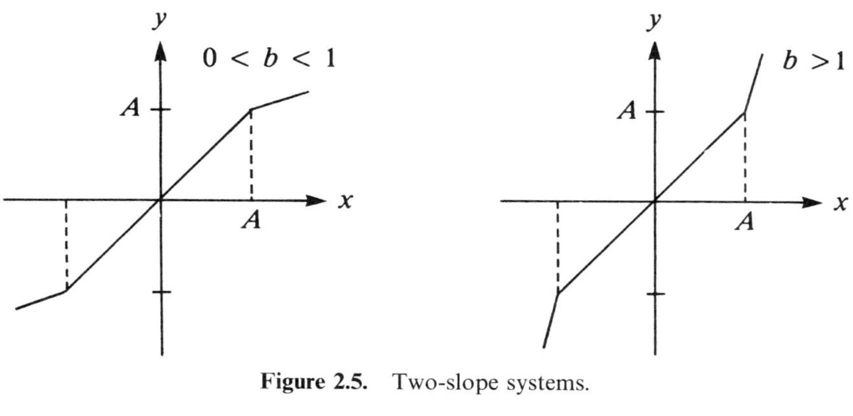

Figure 2.5. Two-slope systems.

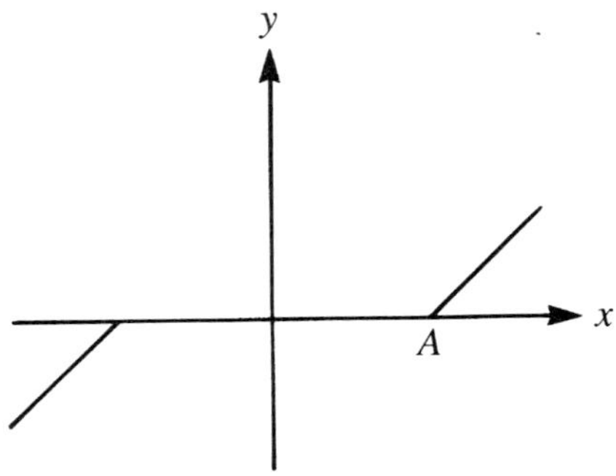

Figure 2.6. Dead-zone system.

Example 3. Clipped System

See Figure 2.7 and Section 2.6.3.

Example 4. Hard-Clipped System

See Figure 2.8 and Section 2.6.4.

Example 5. Square-Law System

See Figure 2.9 and Section 2.7.1.

Example 6. Cubic System

See Figure 2.10 and Section 2.7.2.

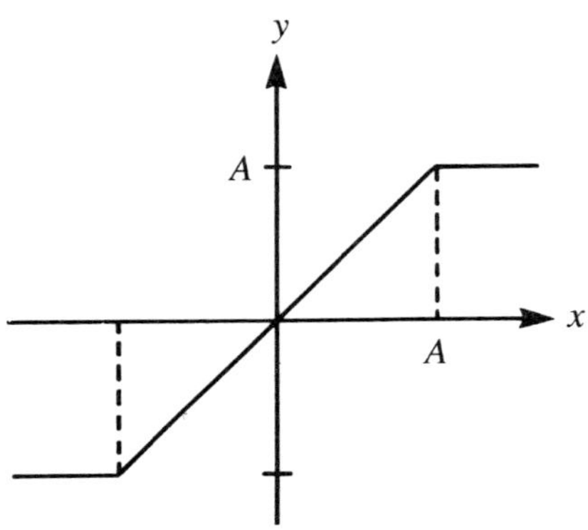

Figure 2.7. Clipped system.

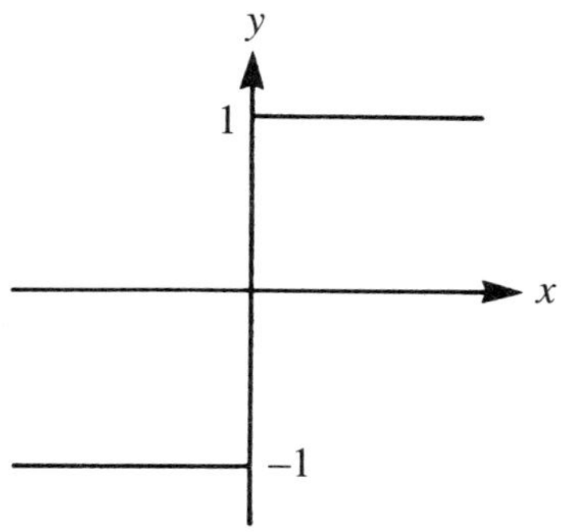

Figure 2.8. Hard-clipped system.

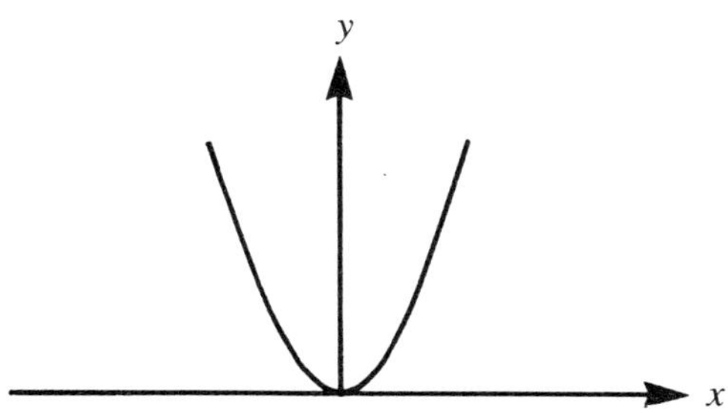

Figure 2.9. Square-law system.

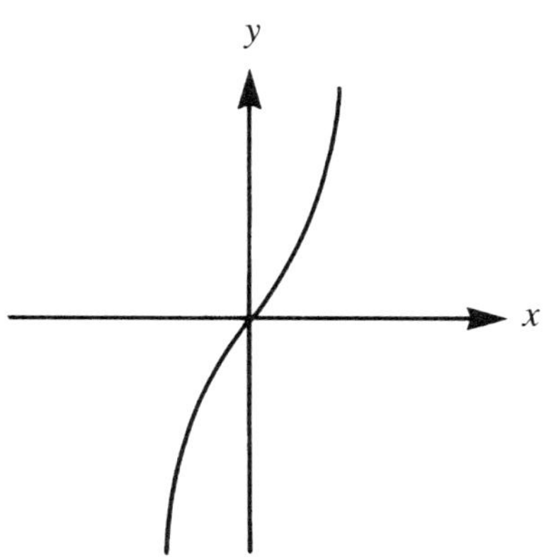

Figure 2.10. Cubic system.

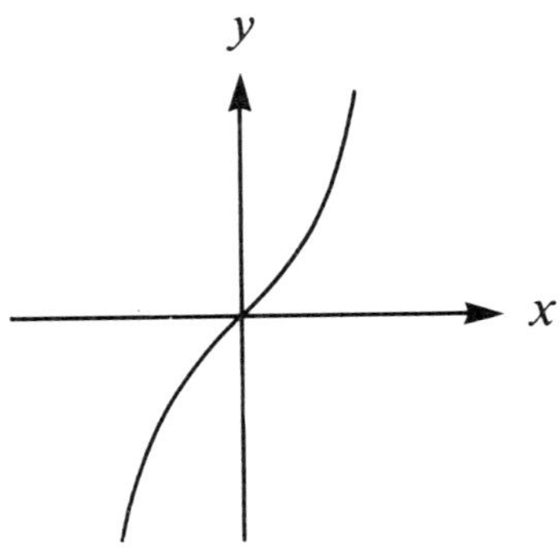

Figure 2.11. Square-law system with sign.

Example 7. Square-Law System with Sign

See Figure 2.11 and Section 2.7.3.

2.3 OUTPUT PROBABILITY DENSITY FUNCTION

An input record $x(t)$ is a representative member of a stationary Gaussian random process with zero mean value if, at any time t, the random variable $x = x(t)$ has the first-order Gaussian probability density function

$$p(x) = \frac{1}{\sigma_x \sqrt{2\pi}} \exp\left(\frac{-x^2}{2\sigma_x^2}\right) \tag{2.4}$$

The quantity

$$\sigma_x^2 = E[x^2] \tag{2.5}$$

is the variance of x when the mean value $\mu_x = E[x] = 0$. For the random variables $x_1 = x(t)$ and $x_2 = x(t + \tau)$ at any times t and τ, their joint probability density function is given by the second-order Gaussian form if

$$p(x_1, x_2) = \frac{1}{2\pi\sigma_x^2\sqrt{1-\rho^2}} \exp\left[\frac{-(x_1^2 - 2\rho x_1 x_2 + x_2^2)}{2\sigma_x^2(1-\rho^2)}\right] \tag{2.6}$$

where $\rho = \rho(\tau)$ is a function of the time-delay τ defined by

$$\rho = \frac{E[x_1 x_2]}{\sqrt{E[x_1^2]E[x_2^2]}} = \frac{E[x_1 x_2]}{\sigma_x^2} \tag{2.7}$$

This quantity ρ is the *correlation coefficient function* between x_1 and x_2. Note that $p(x_1, x_2) = p(x_1)p(x_2)$ when $\rho = 0$ for uncorrelated variables.

The autocorrelation function $R_{xx}(\tau)$ for a stationary random process is defined by

$$R_{xx}(\tau) = E[x(t)x(t + \tau)] = E[x_1 x_2] \tag{2.8}$$

At $\tau = 0$, this gives

$$R_{xx}(0) = E[x(t)x(t)] = E[x_1^2] = E[x_2^2] = \sigma_x^2 \tag{2.9}$$

Hence, the ρ of Eq. (2.7) is equivalent to

$$\rho(\tau) = \rho_{xx}(\tau) = \frac{R_{xx}(\tau)}{R_{xx}(0)} = \frac{R_{xx}(\tau)}{\sigma_x^2} \tag{2.10}$$

Moments for the nonlinear output data from Figure 2.1 can be obtained from knowledge of any $g(x)$ and any $p(x)$ as follows:

$$\begin{aligned} u_y = E[y] = E[g(x)] &= \int g(x)p(x)\, dx \\ E[y^n] = E\{[g(x)]^n\} &= \int [g(x)]^n p(x)\, dx \end{aligned} \tag{2.11}$$

The output data variance is given by

$$\sigma_y^2 = E[y^2] - (E[y])^2 = \psi_y^2 - u_y^2 \tag{2.12}$$

Consider the following important question. Suppose one knows the zero-memory nonlinear relationship $y = g(x)$ of Eq. (2.1). Assume it is not only single-valued but also one-to-one so that a given y corresponds to a unique x. Suppose one also knows the probability density function $p(x)$ for the input $x(t)$, which may or may not be of a Gaussian form. How can one determine the probability density function $p_2(y)$ for the output $y(t)$? The following theorem answers this question.

Theorem 1. Input/Output Probability Density Relation. For any input data $x(t)$ with arbitrary probability density function $p(x)$ where $x(t)$ passes through a zero-memory nonlinear system to produce $y = g(x)$, which is single-valued and one-to-one, the output probability density function $p_2(y)$ for the output $y(t)$ satisfies the relation

$$p_2(y) = \frac{p(x)}{|dy/dx|} = \frac{p(x)}{|g'(x)|} \tag{2.13}$$

This theorem assumes that $(dy/dx) = g'(x)$ exists and is not equal to zero. The absolute value $|dy/dx|$ is required since $p_2(y)$ is nonnegative. The variable x on the right-hand side of Eq. (2.13) should be replaced by its equivalent y from $x = g^{-1}(y)$ to obtain $p_2(y)$ as a function of y.

Corollary. If each value of $y = g(x)$ corresponds to n values of x that are equally likely, then

$$p_2(y) = \frac{np(x)}{|dy/dx|} = \frac{np(x)}{|g'(x)|} \tag{2.14}$$

Proof To prove Theorem 1, observe that for all $y_0 = g(x_0)$, the probability that x lies in $[x_0 - (\Delta x/2),\ x_0 + (\Delta x/2)]$ equals the probability that y lies in $[y_0 - (\Delta y/2),\ y_0 + (\Delta y/2)]$. Thus,

$$\begin{aligned} p(x_0)\Delta x &\approx \text{Prob}[x_0 - (\Delta x/2) < x \leqslant x_0 + (\Delta x/2)] \\ &= \text{Prob}[y_0 - (\Delta y/2) < y \leqslant y_0 + (\Delta y/2)] \approx p_2(y_0)\,\Delta y \end{aligned}$$

In the limit as $\Delta x \to 0$, this gives the desired result

$$p_2(y) = \frac{p(x)}{|dy/dx|}$$

The corollary of Eq. (2.14) comes by using $np(x_0)\,\Delta x \approx p_2(y_0)\,\Delta y$. This completes the proof.

Equations (2.13) and (2.14) are general results for arbitrary $p(x)$. Special results follow by using the Gaussian $p(x)$ of Eq. (2.4), together with the $g(x)$ of particular nonlinear system models. Note that:

1. When $p(x)$ is Gaussian and $g(x)$ is linear, then the resulting $p_2(y)$ will be Gaussian.
2. When $p(x)$ is Gaussian and $g(x)$ is nonlinear, then the resulting $p_2(y)$ will be non-Gaussian.

Hence, if output data are non-Gaussian when input data are Gaussian, then the system must be nonlinear. Analysis of input/output probability density functions thus provides a simple way to detect nonlinear system operations and can also be used to help identify their nature. Examples are discussed in Sections 2.6–2.8.

Consider another important question. Suppose one knows or measures both the input probability density function $p(x)$ and the output probability density function $p_2(y)$. Assume there exists an unknown single-valued one-to-one zero-memory nonlinear transformation $y = g(x)$ that produces $p_2(y)$ from $p(x)$. How can one determine this zero-memory nonlinear function $g(x)$? The following theorem answers this question.

Theorem 2. Identification of Zero-Memory Nonlinear System. When $y = g(x)$ is single-valued and one-to-one, it can be determined from measuring $p(x)$ and $p_2(y)$ by setting the probability distribution function $P_2(y)$ equal to the probability distribution function $P(x)$, namely, for any x_0 and its associated $y_0 = g(x_0)$,

$$P_2(y_0) = \int_{-\infty}^{y_0} p_2(y)\, dy = \int_{-\infty}^{x_0} p(x)\, dx = P(x_0) \tag{2.15}$$

The quantity $P_2(y_0)$ represents the probability that $y(t) \leqslant y_0$ and the quantity $P(x_0)$ represents the probability that $x(t) \leqslant x_0$. For each selected value of x_0 one should determine $P(x_0)$ from the measured $p(x)$, and then find the value of y_0 from the measured $p_2(y)$ such that $P_2(y_0) = P(x_0)$.

This procedure will also identify the form of a single-valued, one-to-one, nonlinear transformation $y = g(x)$ to produce a desired (non-Gaussian) output probability density function $p_2(y)$ from a given (Gaussian) input probability density function $p(x)$. A nonlinear biomedical application of Theorem 2 from reference 19 is in Section 4.8.

2.4 OUTPUT AUTOCORRELATION FUNCTION

Suppose one knows the nonlinear relationship $y = g(x)$ of Eq. (2.1), and suppose $x(t)$ is a Gaussian or non-Gaussian process. What is the predicted output autocorrelation function $R_{yy}(\tau)$? By definition from Eq. (2.3),

$$R_{yy}(\tau) = E[y(t)y(t + \tau)] = E[g\{x(t)\}g\{x(t + \tau)\}] \tag{2.16}$$

Let $x_1 = x(t)$, $x_2 = x(t + \tau)$, $y_1 = g(x_1)$, and $y_2 = g(x_2)$. Then

$$R_{yy}(\tau) = E[y_1 y_2] = \iint g(x_1)g(x_2)p(x_1, x_2)\, dx_1\, dx_2 \tag{2.17}$$

where $p(x_1, x_2)$ is the joint probability density function for x_1 and x_2. This expression is a general result for arbitrary $g(x)$ and arbitrary $p(x_1, x_2)$.

When $y(t)$ is measured, a simpler way to obtain $R_{yy}(\tau)$ is to compute it directly or via inverse Fourier transforms by

$$R_{yy}(\tau) = \mathscr{F}^{-1}[S_{yy}(f)] = \int S_{yy}(f)e^{j2\pi f\tau}\, df \tag{2.18}$$

where $S_{yy}(f)$ is the two-sided output autospectral density function of $y(t)$. With digital processing of data, special steps are required to avoid wraparound errors and triangular weighting effects as explained in reference 1.

The following theorem is useful for predicting output autocorrelation functions after a given zero-memory nonlinear transformation when $y(t)$ is not measured. This theorem requires the input data to be Gaussian.

Theorem 3 (Price). Output Autocorrelation Relation. For Gaussian input data $x(t)$ with known autocorrelation function $R_{xx}(\tau)$, where $x(t)$ passes through any zero-memory nonlinear system $y = g(x)$, the output autocorrelation function $R_{yy}(\tau)$ satisfies the relation

$$\frac{\partial R_{yy}(\tau)}{\partial R_{xx}(\tau)} = E[g'(x_1)g'(x_2)] \tag{2.19}$$

whenever $g'(x) = [dg(x)/dx]$ exists at $x_1 = x(t)$ and $x_2 = x(t + \tau)$.

Corollary. Assuming higher-order derivatives $g^n(x) = [dg^n(x)/dx^n]$ exist,

$$\frac{\partial^n R_{yy}(\tau)}{\partial^n R_{xx}(\tau)} = E[g^n(x_1)g^n(x_2)] \tag{2.20}$$

Proof To prove Theorem 3, assume that $g(x)$ has a Fourier transform $Z(f)$ given by

$$Z(f) = \int g(x)e^{-j2\pi fx}\,dx$$

Then

$$\begin{aligned} g(x_1) &= \int Z(f_1)e^{j2\pi f_1 x_1}\,df_1 \\ g(x_2) &= \int Z(f_2)e^{j2\pi f_2 x_2}\,df_2 \end{aligned} \tag{2.21}$$

where $g(x)$ is the inverse Fourier transform of $Z(f)$. Now, substitute Eq. (2.21) into Eq. (2.16). This gives

$$\begin{aligned} R_{yy}(\tau) &= \iiiint Z(f_1)Z(f_2)e^{j2\pi(f_1x_1 + f_2x_2)}\,p(x_1, x_2)\,dx_1\,dx_2\,df_1\,df_2 \\ &= \iint Z(f_1)Z(f_2)C_{x_1,x_2}(f_1, f_2)\,df_1\,df_2 \end{aligned} \tag{2.22}$$

where

$$C_{x_1,x_2}(f_1, f_2) = \iint e^{j2\pi(f_1x_1+f_2x_2)}p(x_1, x_2)\,dx_1\,dx_2 \tag{2.23}$$

defines the *joint characteristic function* of x_1 and x_2 at f_1 and f_2. Note that

$$C_{x_1,x_2}(f_1, f_2) = E[e^{j2\pi(f_1x_1+f_2x_2)}] \tag{2.24}$$

From sources such as references 1 and 4 or by separate direct proof, it is known for Gaussian data that

$$C_{x_1,x_2}(f_1, f_2) = \exp[-2\pi^2\sigma_x^2(f_1^2 + 2\rho f_1 f_2 + f_2^2)] \tag{2.25}$$

Let the derivative of $g(x)$ with respect to x be denoted by $g'(x)$. Then from Eq. (2.21),

$$\begin{aligned} g'(x_1) &= \int (j2\pi f_1)Z(f_1)e^{j2\pi f_1 x_1}\,df_1 \\ g'(x_2) &= \int (j2\pi f_2)Z(f_2)e^{j2\pi f_2 x_2}\,df_2 \end{aligned} \tag{2.26}$$

Hence,

$$g'(x_1)g'(x_2) = \iint (-4\pi^2 f_1 f_2)Z(f_1)Z(f_2)e^{j2\pi(f_1x_1+f_2x_2)}\,df_1\,df_2$$

Expected values of both sides yield

$$E[g'(x_1)g'(x_2)] = \iint (-4\pi^2 f_1 f_2)Z(f_1)Z(f_2)C_{x_1,x_2}(f_1, f_2)\,df_1\,df_2 \tag{2.27}$$

Consider now Eqs. (2.22) and (2.25). From Eq. (2.25), the partial derivative of $C_{x_1,x_2}(f_1, f_2)$ with respect to ρ yields the result

$$\frac{\partial C_{x_1,x_2}(f_1, f_2)}{\partial \rho} = (-4\pi^2\sigma_x^2 f_1 f_2)C_{x_1,x_2}(f_1,f_2) \tag{2.28}$$

From Eq. (2.22), one obtains the partial derivative

$$\frac{\partial R_{yy}(\tau)}{\partial \rho} = \int Z(f_1)Z(f_2)\,\frac{\partial C_{x_1,x_2}(f_1, f_2)}{\partial \rho}\,df_1\,df_2 \tag{2.29}$$

Substitution of Eq. (2.28) into Eq. (2.29) and comparison with Eq. (2.27) proves that

$$\frac{\partial R_{yy}(\tau)}{\partial \rho} = \sigma_x^2 E[g'(x_1)g'(x_2)] \tag{2.30}$$

From the definition of ρ in Eq. (2.10), one can replace Eq. (2.30) by the equivalent formula that proves the desired result:

$$\frac{\partial R_{yy}(\tau)}{\partial R_{xx}(\tau)} = E(g'(x_1)g'(x_2)] \tag{2.31}$$

The corollary of Eq. (2.20) follows by successive iteration procedures. Equations (2.30) and (2.31) are special cases of Price's theorem (reference 42). This useful result for Gaussian input data will be applied in later sections.

2.5 INPUT/OUTPUT CROSS-CORRELATION FUNCTION

Suppose one knows the nonlinear relationship $y = g(x)$ of Eq. (2.1), and suppose $x(t)$ is a Gaussian or non-Gaussian process. What is the predicted input/output cross-correlation function $R_{xy}(\tau)$? By definition from Eq. (2.3),

$$R_{xy}(\tau) = E[x(t)y(t + \tau)] = E[x(t)g\{x(t + \tau)\}] \tag{2.32}$$

Let $x_1 = x(t)$, $x_2 = x(t + \tau)$, and $y_2 = g(x_2)$. Then

$$R_{xy}(\tau) = E[x_1 y_2] = \iint x_1 g(x_2)p(x_1, x_2)\, dx_1\, dx_2 \tag{2.33}$$

where $p(x_1, x_2)$ is the joint probability density function for x_1 and x_2. This expression is a general result for arbitrary $g(x)$ and arbitrary $p(x_1, x_2)$.

When $x(t)$ and $y(t)$ are both measured, a simpler way to obtain $R_{xy}(\tau)$ is to compute it directly or via inverse Fourier transforms by

$$R_{xy}(\tau) = \mathscr{F}^{-1}[S_{xy}(f)] = \int S_{xy}(f)e^{j2\pi f\tau}\, df \tag{2.34}$$

where $S_{xy}(f)$ is the two-sided input/output cross-spectral density function between $x(t)$ and $y(t)$. As noted after Eq. (2.18), this requires special digital processing steps to avoid errors.

The following theorem is useful for predicting input/output cross-correlation functions after a given zero-memory nonlinear transformation when $y(t)$ is not measured. This theorem requires the input data to be Gaussian.

Theorem 4 (Bussgang). Input/Output Cross-Correlation Relation. For Gaussian input data $x(t)$ with known autocorrelation function $R_{xx}(\tau)$, where $x(t)$ passes through any zero-memory nonlinear system $y = g(x)$, the input/output cross-correlation function $R_{xy}(\tau)$ satisfies the relation

$$R_{xy}(\tau) = \frac{R_{xx}(\tau)}{\sigma_x^2} \int xg(x)p(x)\, dx = \left(\frac{E[xg(x)]}{\sigma_x^2}\right) R_{xx}(\tau) \tag{2.35}$$

Corollary. The input/output cross-spectral density function $S_{xy}(f)$ satisfies the relation

$$S_{xy}(f) = \left(\frac{E[xg(x)]}{\sigma_x^2}\right) S_{xx}(f) \tag{2.36}$$

and the optimum linear system $H_0(f)$ for this zero-memory nonlinear system $g(x)$ is given by

$$H_0(f) = \frac{S_{xy}(f)}{S_{xx}(f)} = \frac{E[xg(x)]}{\sigma_x^2} \tag{2.37}$$

Equation (2.36) follows directly from Eq. (2.35), and Eq. (2.37) from a basic result in references 1 and 2. Note that the optimum linear system here is independent of frequency because the $g(x)$ transformation has no memory. Note also that for nonlinear systems, the optimum linear system is a function of the input variance and will be different for different inputs.

Proof To prove Theorem 4, consider the inside integral of Eq. (2.33) with respect to x_1. This takes the form

$$\int x_1 p(x_1, x_2)\, dx_1 = \frac{1}{2\pi\sigma_x^2\sqrt{1-\rho^2}} \int x_1 \exp\left[\frac{-(x_1^2 - 2\rho x_1 x_2 + x_2^2)}{2\sigma_x^2(1-\rho^2)}\right] dx_1$$

when $p(x_1, x_2)$ is given by Eq. (2.6). The quantity

$$x_1^2 - 2px_1x_2 + x_2^2 = (x_1 - \rho x_2)^2 + (1-\rho^2)x_2^2$$

Now

$$\int x_1 p(x_1, x_2)\, dx_1 = \frac{1}{2\pi\sigma_x^2\sqrt{1-\rho^2}} \int x_1 \exp\left[\frac{-(x_1 - \rho x_2)^2}{2\sigma_x^2(1-\rho^2)} - \frac{x_2^2}{2\sigma_2^2}\right] dx_1$$

Let

$$u = \frac{x_1 - \rho x_2}{\sqrt{1-\rho^2}}; \quad du = \frac{dx_1}{\sqrt{1-\rho^2}}; \quad x_1 = \rho x_2 + u\sqrt{1-\rho^2}$$

Then

$$\begin{aligned}\int x_1 p(x_1, x_2)\, dx_1 &= \frac{1}{2\pi\sigma_x^2}\int (\rho x_2 + u\sqrt{1-\rho^2})\exp\left(\frac{-u^2}{2\sigma_x^2} - \frac{x_2^2}{2\sigma_x^2}\right) du \\ &= \frac{\rho x_2 \exp(-x_2^2/2\sigma_x^2)}{2\pi\sigma_x^2}\int \exp\left(\frac{-u^2}{2\sigma_x^2}\right) du\end{aligned}$$

where the integral involving u vanishes because u is part of an odd function between symmetric limits. The integral

$$\int \exp\left(\frac{-u^2}{2\sigma_x^2}\right) du = \sigma_x\sqrt{2\pi} \tag{2.38}$$

Thus,

$$\int x_1 p(x_1, x_2)\, dx_1 = \frac{\rho x_2 \exp(-x_2^2/2\sigma_x^2)}{\sigma_x\sqrt{2\pi}} \tag{2.39}$$

Finally, by substituting Eq. (2.39) into Eq. (2.33) and replacing ρ by $\rho_{xx}(\tau)$,

$$R_{xy}(\tau) = \frac{\rho_{xx}(\tau)}{\sigma_x\sqrt{2\pi}}\int x_2 g(x_2)\exp\left(\frac{-x_2^2}{2\sigma_x^2}\right) dx_2 \tag{2.40}$$

This proves the desired result

$$R_{xy}(\tau) = \rho_{xx}(\tau)E[xg(x)] = \left(\frac{E[xg(x)]}{\sigma_x^2}\right) R_{xx}(\tau) \tag{2.41}$$

Equations (2.40) and (2.41) are special cases of Bussgang's theorem (reference 17). This useful result for Gaussian input data will be applied in later sections. Equation (2.41) shows that:

1. $R_{xy}(\tau)$ is directly proportional to $R_{xx}(\tau)$ with the factor of proportionality given by $E[xg(x)]/\sigma_x^2$.
2. $R_{xy}(\tau) = 0$ whenever $g(x)$ is an even function of x, since $E[xg(x)]$ will then be the integral of an odd function between symmetric limits. In particular, $R_{xy}(\tau) = 0$ when $g(x) = x^2$, a square-law operation.

3. $S_{xy}(f)$, the two-sided Fourier transform of $R_{xy}(\tau)$, is directly proportional to $S_{xx}(f)$, the two-sided Fourier transform of $R_{xx}(\tau)$ with the factor of proportionality given by $E[xg(x)]/\sigma_x^2$. Hence, $S_{xy}(f)$ will be real-valued whenever $g(x)$ is real-valued, since $S_{xx}(f)$ is always a real-valued function of f. Also, the factor of proportionality here is the optimum linear system for this particular input $x(t)$ only.

2.6 EXAMPLES WITH DISCONTINUOUS DERIVATIVES

Examples of zero-memory nonlinear systems with discontinuous derivatives are discussed in this section. Other examples with continuous derivatives are discussed in Sections 2.7 and 2.8. All of these examples have engineering applications.

2.6.1 Two-Slope Systems

Two-slope systems as per Figure 2.5 are defined by the relation

$$
\begin{aligned}
y = g(x) &= x, && |x| \leqslant A \\
&= A + b(x - A), && x \geqslant A \\
&= -A + b(x + A), && x \leqslant -A
\end{aligned}
\tag{2.42}
$$

Note that

$$
g(-x) = -g(x) \tag{2.43}
$$

Thus, $g(x)$ is an odd function of x. Here,

$$
\begin{aligned}
\frac{dy}{dx} &= 1, \quad |x| \leqslant A \\
&= b, \quad |x| > A
\end{aligned}
\tag{2.44}
$$

with

$$
\begin{aligned}
x = g^{-1}(y) &= y, && |y| \leqslant A \\
&= A + \frac{(y - A)}{b}, && y \geqslant A \\
&= -A + \frac{(y + A)}{b}, && y \leqslant -A
\end{aligned}
\tag{2.45}
$$

One should note that the derivative (dy/dx) is discontinuous at $|x| = A$. From Eq. (2.13), the output probability density function for $b \neq 0$ is now given by

$$\begin{aligned} p_2(y) &= p(x) = p(y), && |y| \leqslant A \\ &= \frac{p(A + [(y - A)/b])}{b}, && y > A \\ &= \frac{p(-A + [(y + A)/b])}{b}, && y < -A \end{aligned} \tag{2.46}$$

Without loss of generality, one can let $\sigma_x = 1$ so that

$$p(x) = \frac{1}{\sqrt{2\pi}} \exp\left(\frac{-x^2}{2}\right) \tag{2.47}$$

Now, the output probability density function

$$\begin{aligned} p_2(y) &= \frac{1}{\sqrt{2\pi}} \exp\left(\frac{-y^2}{2}\right), && |y| \leqslant A \\ &= \frac{1}{b\sqrt{2\pi}} \exp\left[\frac{-(A + [(y - A)/b])^2}{2}\right], && y > A \\ &= \frac{1}{b\sqrt{2\pi}} \exp\left[\frac{-(A - [(y + A)/b])^2}{2}\right], && y < -A \end{aligned} \tag{2.48}$$

Observe that immediately to the left and right of A,

$$\begin{aligned} p_2(A-) &= \frac{1}{\sqrt{2\pi}} \exp\left(\frac{-A^2}{2}\right) \\ p_2(A+) &= \frac{1}{b\sqrt{2\pi}} \exp\left(\frac{-A^2}{2}\right) \end{aligned} \tag{2.49}$$

Two distinct cases occur as follows:

$$\text{If } 0 < b < 1, \quad \text{then } p_2(A+) > p_2(A-)$$
$$\text{If } b > 1, \quad \text{then } p_2(A+) < p_2(A-)$$

Two-slope softening spring systems can be represented by cases where $0 < b < 1$, and two-slope hardening spring systems can be represented by cases where $b > 1$ when x is the displacement and y is the force.

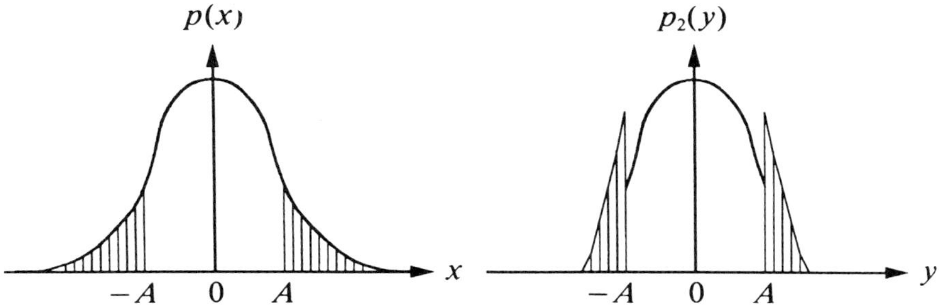

Figure 2.12. Input/output PDF for two-slope system when $0 < b < 1$.

Figures 2.12 and 2.13 show the relationships when $0 < b < 1$ and when $b > 1$ between a Gaussian $p(x)$ input probability density function (PDF) and a non-Gaussian $p_2(y)$ output probability density function for a nonlinear two-slope system. In these figures, the probability value α is defined by the formula

$$\alpha = \int_{-\infty}^{-A} p(x)\, dx = \int_{A}^{\infty} p(x)\, dx = \int_{A}^{\infty} p_2(y)\, dy = \int_{-\infty}^{-A} p_2(y)\, dy \tag{2.50}$$

representing the shaded probability areas. It follows that the probability of x falling inside the range from $-A$ to A is given by

$$1 - 2\alpha = \int_{-A}^{A} p(x)\, dx = \int_{-A}^{A} p_2(y)\, dy \tag{2.51}$$

Figure 2.12 shows that $p_2(y)$ is discontinuous at the value $y = A$ where $p_2(A+) > p_2(A-)$ as y approaches the value A from the right of A versus from the left of A; whereas Figure 2.13 shows that $p_2(A+) < p_2(A-)$ as y ap-

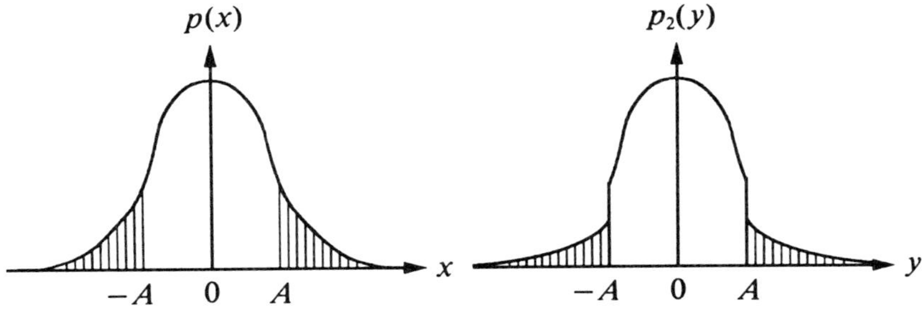

Figure 2.13. Input/output PDF for two-slope system when $b > 1$.

proaches the value A from the right of A versus from the left of A. Since Figures 2.12 and 2.13 both have the same probability of α that $y \geqslant A$, this requires that the tail in Figure 2.12 where $y \geqslant A$ should be much shorter than the tail in Figure 2.13 where $y \geqslant A$. Similar results apply to the tails in Figures 2.12 and 2.13 where $y \leqslant -A$ because of the discontinuity in $p_2(y)$ at the value $y = -A$.

2.6.2 Dead-Zone System

A dead-zone system as per Figure 2.6 is defined by the relation

$$\begin{aligned} y = g(x) &= 0, && |x| \leqslant A \\ &= b(x - A), && x \geqslant A \\ &= b(x + A), && x \leqslant -A \end{aligned} \tag{2.52}$$

Here, the discontinuous derivative

$$\begin{aligned} \frac{dy}{dx} &= 0, \quad |x| \leqslant A \\ &= b, \quad |x| > A \end{aligned} \tag{2.53}$$

with

$$\begin{aligned} &x \text{ undefined}, && y = 0 \\ &x = A + \frac{y}{b}, && y > 0 \\ &\;\;\, = -A + \frac{y}{b}, && y < 0 \end{aligned} \tag{2.54}$$

As before, let

$$p(x) = \frac{1}{\sqrt{2\pi}} \exp\left(\frac{-x^2}{2}\right) \tag{2.55}$$

$$\alpha = \int_A^\infty p(x)\, dx, \quad 1 - 2\alpha = \int_{-A}^{A} p(x)\, dx \tag{2.56}$$

Now, the output probability density function

$$\begin{aligned} p_2(y) &= (1 - 2\alpha)\delta(y), && y = 0 \\ &= \frac{p(A + [y/b])}{b}, && y > 0 \\ &= \frac{p(-A + [y/b])}{b}, && y < 0 \end{aligned} \tag{2.57}$$

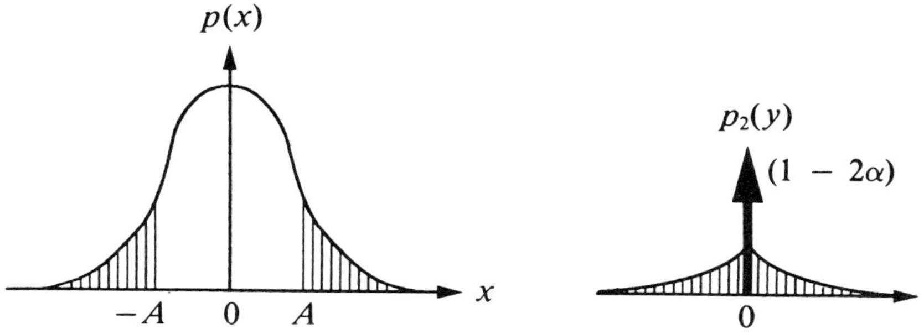

Figure 2.14. Input/output PDF for dead-zone system.

where $\delta(y)$ is the usual delta function. Observe that

$$\int_{0+}^{\infty} p_2(y)\,dy = \int_{-\infty}^{0-} p_2(y)\,dy = \alpha$$
$$\int_{0-}^{0+} p_2(y)\,dy = 1 - 2\alpha \tag{2.58}$$

Figure 2.14 shows the relationship between a Gaussian input probability density function and a non-Gaussian output probability density function for a dead-zone system. The reason for the theoretical delta function at $y = 0$ is that all of the values of x inside the range $|x| \leqslant A$ are restricted to this single value. Thus,

$$\text{probability } [y(t) = 0] = 1 - 2\alpha$$

2.6.3 Clipped System

A clipped system as per Figure 2.7 is defined by the relation

$$\begin{aligned} y = g(x) &= x, && |x| \leqslant A \\ &= A, && x \geqslant A \\ &= -A, && x \leqslant -A \end{aligned} \tag{2.59}$$

Here, the discontinuous derivative

$$\begin{aligned} \frac{dy}{dx} &= 1, && |x| \leqslant A \\ &= 0, && |x| > A \end{aligned} \tag{2.60}$$

with

$$\begin{aligned} x &= y, & |y| &\leqslant A \\ x &\text{ undefined}, & |y| &> A \end{aligned} \tag{2.61}$$

As per Eqs. (2.55) and (2.56), let

$$p(x) = \frac{1}{\sqrt{2\pi}} \exp\left(\frac{-x^2}{2}\right) \tag{2.62}$$

$$\alpha = \int_A^\infty p(x)\, dx, \quad 1 - 2\alpha = \int_{-A}^{A} p(x)\, dx \tag{2.63}$$

The output probability density function becomes here

$$\begin{aligned} p_2(y) &= p(y), & |y| &< A \\ &= \alpha\delta(y - A), & y &= A \\ &= \alpha\delta(y + A), & y &= -A \end{aligned} \tag{2.64}$$

where $p_2(y) = 0$ for $|y| > A$ and $\delta(y)$ is the usual delta function.

$$\begin{aligned} \int_{-A+}^{0} p_2(y)\, dy &= \int_0^{A-} p_2(y)\, dy = \tfrac{1}{2} - \alpha \\ \int_{-\infty}^{-A+} p_2(y)\, dy &= \int_{A-}^{\infty} p_2(y)\, dy = \alpha \end{aligned} \tag{2.65}$$

Figure 2.15 shows the relationship between a Gaussian input probability density function and a non-Gaussian output probability density function for a clipped system. The reason for the theoretical delta functions at $y = A$ and $y = -A$ is that all of the values of x where $x \geqslant A$ are restricted to the single

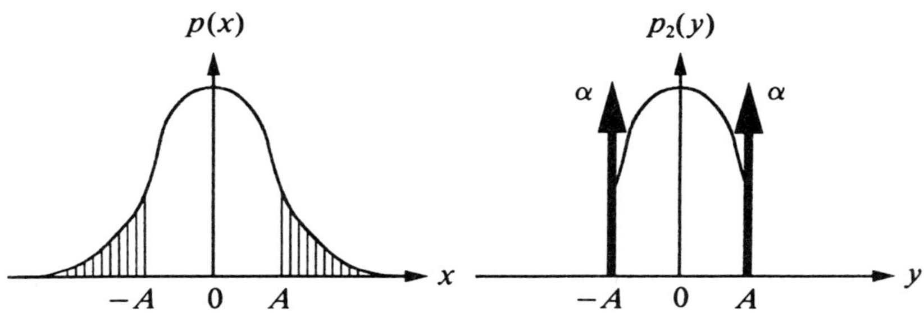

Figure 2.15. Input/output PDF for clipped system.

value $y = A$, and all of the values of x where $x \leqslant -A$ are restricted to the single value $y = -A$. Thus,

$$\text{probability}\,[y(t) = -A] = \text{probability}\,[y(t) = A] = \alpha$$

2.6.4 Hard-Clipped System

A hard-clipped system as per Figure 2.8 is defined as an extreme form of a clipped system where

$$\begin{aligned} y = g(x) &= 1, \quad && x > 0 \\ &= 0, && x = 0 \\ &= -1, && x < 0 \end{aligned} \tag{2.66}$$

The output probability density function becomes

$$\begin{aligned} p_2(y) &= \tfrac{1}{2}\delta(y - 1), \quad && y = 1 \\ &= \tfrac{1}{2}\delta(y + 1), && y = -1 \\ &= 0 && \text{otherwise} \end{aligned} \tag{2.67}$$

where $\delta(y)$ is the usual delta function. The special nature of Eq. (2.67) is sketched in Figure 2.16. Here,

$$\text{probability}\,[y(t) = -1] = \text{probability}\,[y(t) = 1] = 0.50.$$

For the hard-clipped system, the discontinuous derivative

$$\begin{aligned} g'(x) &= 2\delta(x) \quad && \text{at } x = 0 \\ &= 0 && \text{otherwise} \end{aligned} \tag{2.68}$$

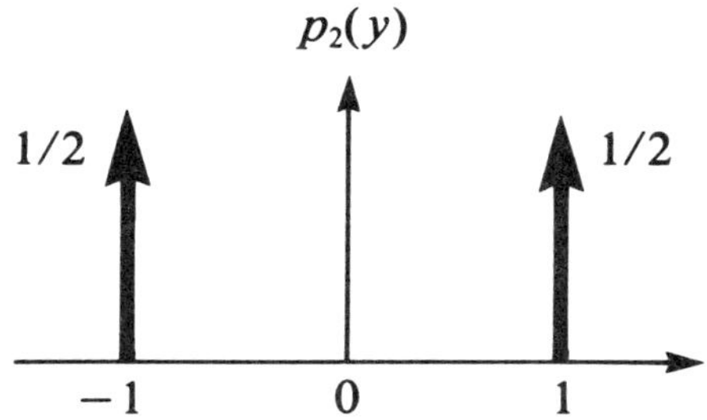

Figure 2.16. Output PDF for hard-clipped system.

The delta function occurs at the value $x = 0$, since there is a discontinuous jump of two units in $g(x)$ when $g(x)$ changes from -1 to $+1$ as x passes through zero.

Price's theorem of Eq. (2.30) and Bussgang's theorem of Eq. (2.35) will now be applied to obtain the output autocorrelation function and the input/output cross-correlation function for the hard-clipped system. Similar results were not carried out for previous systems because of the difficulty of deriving closed-form answers. If desired, however, numerical approximations can be obtained.

Output Autocorrelation Function for Hard-Clipped System. From Eq. (2.30), where $\rho = \rho_{xx}(\tau)$,

$$\begin{aligned}\frac{\partial R_{yy}(\tau)}{\partial \rho} &= \sigma_x^2 E[g'(x_1)g'(x_2)] \\ &= \sigma_x^2 \iint g'(x_1)g'(x_2)p(x_1, x_2)\, dx_1\, dx_2 \qquad (2.69)\end{aligned}$$

Substitutions from Eqs. (2.6) and (2.68) give

$$\begin{aligned}\frac{\partial R_{yy}(\tau)}{\partial \rho} &= \frac{1}{2\pi\sqrt{1-\rho^2}} \iint 4\delta(x_1)\delta(x_2) \exp\left(\frac{-[x_1^2 - 2\rho x_1 x_2 + x_2^2]}{2\sigma_x^2(1-\rho^2)}\right) dx_1\, dx_2 \\ &= \frac{2}{\pi\sqrt{1-\rho^2}} \iint \delta(x_1)\delta(x_2)\, dx_1\, dx_2 = \frac{2}{\pi\sqrt{1-\rho^2}} \qquad (2.70)\end{aligned}$$

Hence,

$$R_{yy}(\tau) = \frac{2}{\pi} \int \frac{d\rho}{\sqrt{1-\rho^2}} + C = \frac{2}{\pi} \sin^{-1}(\rho) + C \qquad (2.71)$$

when C is a constant of integration.

Consider now Eq. (2.17). When $\rho = 0$, this reduces to

$$\begin{aligned}R_{yy}(\tau) &= \iint g(x_1)g(x_2)p(x_1)p(x_2)\, dx_1\, dx_2 \\ &= \left[\int g(x)p(x)\, dx\right]^2 \qquad (2.72)\end{aligned}$$

For cases where $g(x)$ is an odd function of x, as is true for the $g(x)$ of Eq. (2.66), one obtains

$$R_{yy}(\tau) = 0 \quad \text{for } \rho = 0 \qquad (2.73)$$

Thus, the constant $C = 0$ in Eq. (2.71). This proves that

$$R_{yy}(\tau) = \frac{2}{\pi} \sin^{-1}[\rho_{xx}(\tau)] = \frac{2}{\pi} \sin^{-1}\left[\frac{R_{xx}(\tau)}{\sigma_x^2}\right] \tag{2.74}$$

This famous result is known as the *arc-sine law* for the autocorrelation function of hard-clipped Gaussian signals. Measurement of $R_{yy}(\tau)$ enables one to estimate $\rho_{xx}(\tau)$ using the relation

$$\rho_{xx}(\tau) = \sin[(\pi/2)R_{yy}(\tau)] \tag{2.75}$$

One can also determine σ_x^2 from measurement of $x(t)$ to obtain

$$R_{xx}(\tau) = \sigma_x^2 \rho_{xx}(\tau) \tag{2.76}$$

As expected, note also that $R_{yy}(0) = 1$ here.

Input/Output Cross-Correlation Function for Hard-Clipped System. From Eq. (2.35), the $g(x)$ of Eq. (2.66) gives

$$\begin{aligned} R_{xy}(\tau) &= \frac{R_{xx}(\tau)}{\sigma_x^3\sqrt{2\pi}}\left[\int_{-\infty}^{0} (-u)e^{-u^2/2\sigma_x^2}\,du + \int_0^{\infty} ue^{-u^2/2\sigma_x^2}\,du\right] \\ &= \frac{2R_{xx}(\tau)}{\sigma_x^3\sqrt{2\pi}} \int_0^{\infty} ue^{-u^2/2\sigma_x^2}\,du = \left(\frac{\sqrt{2}}{\sigma_x\sqrt{\pi}}\right) R_{xx}(\tau) \end{aligned} \tag{2.77}$$

The associated input/output cross-spectrum formula is

$$S_{xy}(f) = \left(\frac{\sqrt{2}}{\sigma_x\sqrt{\pi}}\right) S_{xx}(f) \tag{2.78}$$

2.6.5 Nonsymmetric Hard-Clipped System

Suppose the hard-clipped system is nonsymmetric as defined by

$$\begin{aligned} y = g(x) &= 0, \qquad x < 0 \\ &= 0.5, \qquad x = 0 \\ &= 1, \qquad x > 0 \end{aligned} \tag{2.79}$$

The discontinuous derivative $g'(x)$ is now

$$\begin{aligned} g'(x) &= \delta(x) \quad \text{at } x = 0 \\ &= 0 \qquad \text{for } x \neq 0 \end{aligned} \tag{2.80}$$

Price's theorem gives

$$\frac{\partial R_{yy}(\tau)}{\partial \rho} = \frac{1}{2\pi\sqrt{1-\rho^2}} \tag{2.81}$$

leading to

$$R_{yy}(\tau) = \left(\frac{1}{2\pi}\right)\sin^{-1}\left[\frac{R_{xx}(\tau)}{\sigma_x^2}\right] + C \tag{2.82}$$

At $\tau = 0$, $R_{xx}(0) = \sigma_x^2$ and $R_{yy}(0) = \frac{1}{2}$ here so that

$$R_{yy}(0) = \left(\frac{1}{2\pi}\right)\sin^{-1}(1) + C = \tfrac{1}{4} + C = \tfrac{1}{2} \tag{2.83}$$

Thus, the constant C in Eq. (2.82) has the value $C = \frac{1}{4}$ for this nonsymmetric hard-clipped system. Results in Eqs. (2.82) and (2.74) should be compared.

2.6.6 Smooth-Limiter System

Consider the nonlinear nonsymmetric smooth-limiter system sketched in Figure 2.17, where

$$\begin{aligned} y = g(x) &= \sqrt{\frac{2}{\pi}} \int_0^x \exp\left(\frac{-t^2}{2}\right) dt, \quad && x \geqslant 0 \\ &= 0, && x \leqslant 0 \end{aligned} \tag{2.84}$$

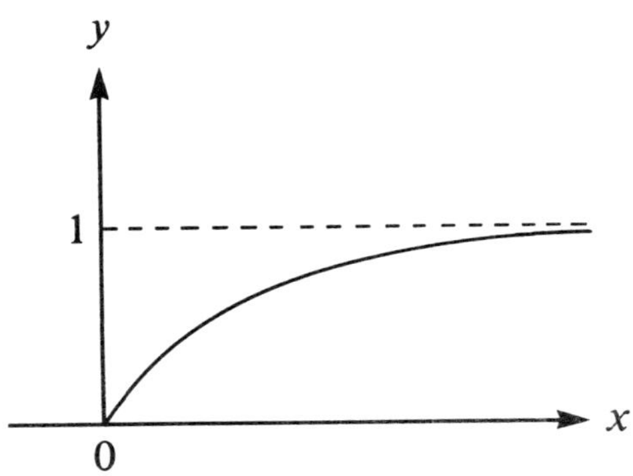

Figure 2.17. Smooth-limiter system.

Equation (2.84) is the equation for the standardized normal probability integral.

For this nonlinear transformation, special values are $g(0) = 0$, $g(1) = 0.683$, $g(2) = 0.954$, $g(3) = 0.997$, and $g(\infty) = 1.000$. Tables are available that give $g(x)$ as a function of x, which also can be used to find x for any $g(x)$ in the range $0 < g(x) < 1$.

The inverse relations are

$$\begin{aligned} x = g^{-1}(y) &= \text{any value of } x \leqslant 0 \text{ for } y = 0 \\ &= \text{unique value of } x \text{ in range } 0 < x < \infty \text{ for } 0 < y < 1 \\ &= \text{undefined for } y < 0 \text{ or } y > 1 \end{aligned}$$

The derivative relations are

$$\begin{aligned} \frac{dy}{dx} = g'(x) &= \sqrt{\frac{2}{\pi}} \exp\left(\frac{-x^2}{2}\right), \quad && x > 0 \\ &= 0, && x < 0 \end{aligned} \tag{2.85}$$

Here, the derivative is discontinuous at $x = 0$.

Assume that $p(x)$, the input probability density function, is Gaussian with mean zero and standard deviation σ, where

$$p(x) = \frac{1}{\sigma\sqrt{2\pi}} \exp\left(\frac{-x^2}{2\sigma^2}\right) \tag{2.86}$$

All of the values of $x \leqslant 0$ are associated with the single value of $y = 0$. Hence, $p_2(y)$, the output probability density function, is such that

$$p_2(y) = \tfrac{1}{2}\delta(y) \quad \text{at } y = 0 \tag{2.87}$$

where the factor $\frac{1}{2}$ occurs because the probability is $\frac{1}{2}$ that $x \leqslant 0$. There are no values of x where $y < 0$ or $y > 1$. Hence,

$$p_2(y) = 0 \quad \text{for } y < 0 \text{ or } y > 1 \tag{2.88}$$

In the range where $0 < y < 1$, the result is

$$p_2(y) = \frac{p(x)}{|dy/dx|} = \sqrt{\frac{\pi}{2}}\, p(x) \exp\left(\frac{x^2}{2}\right) \tag{2.89}$$

where the value of x on the right-hand side should be replaced by its equivalent y from $x = g^{-1}(y)$. As y approaches the value 0 from the inside where $y > 0$,

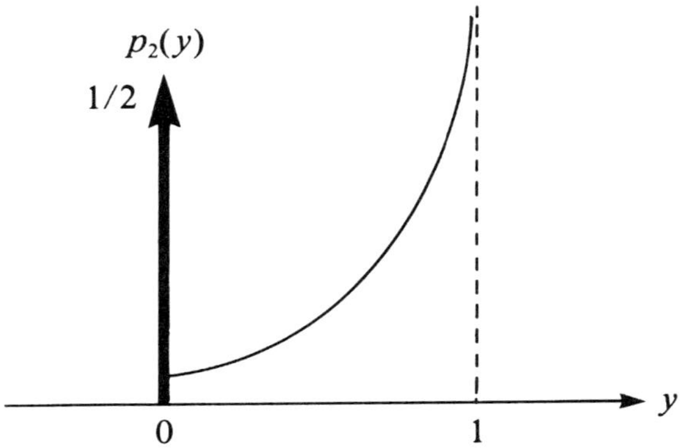

Figure 2.18. Output PDF for smooth-limiter system.

the value of x also approaches zero so that

$$p_2(0+) = \sqrt{\frac{\pi}{2}}\, p(0) = \frac{1}{2\sigma} \tag{2.90}$$

As y approaches the value 1 from the inside where $y < 1$, the value of x approaches infinity so that

$$p_2(1-) = \infty \tag{2.91}$$

A typical shape for $p_2(y)$ is sketched in Figure 2.18. The area under the curve from $0+$ to $1-$ is $\frac{1}{2}$. Here,

$$\text{probability } [y(t) = 0] = 0.50$$
$$\text{probability } [y(t) < 0 \text{ or } y(t) > 1] = 0$$
$$\text{probability } [0 < y(t) < 1] = 0.50$$

Input/Output Cross-Correlation Function for Smooth-Limiter System. By Bussgang's theorem, Eq. (2.35), the input/output cross-correlation function $R_{xy}(\tau)$ for Gaussian input data satisfies the equation

$$R_{xy}(\tau) = \frac{R_{xx}(\tau)}{\sigma^2} \int_0^\infty x g(x) p(x)\, dx \tag{2.92}$$

where, from Eqs. (2.84) and (2.86),

$$g(x) = \sqrt{\frac{2}{\pi}} \int_0^x \exp\left(\frac{-t^2}{2}\right) dt, \quad x \geqslant 0 \tag{2.93}$$

$$p(x) = \frac{1}{\sigma\sqrt{2\pi}} \exp\left(\frac{-x^2}{2\sigma^2}\right) \tag{2.94}$$

Integration by parts formula gives

$$\int u\,dv = uv - \int v\,du$$

Let

$$u = g(x), \quad dv = xp(x)\,dx$$

Then

$$du = g'(x)\,dx = \sqrt{\frac{2}{\pi}} \exp\left(\frac{-x^2}{2}\right) dx, \quad v = -\sigma^2 p(x)$$

At $x = 0$ and $x = \infty$, the product $uv = 0$. Hence,

$$\begin{aligned}\int_0^\infty xg(x)p(x)\,dx &= \sigma^2 \int_0^\infty p(x)g'(x)\,dx \\ &= \frac{\sigma}{\pi} \int_0^\infty \exp\left(\frac{-x^2}{2\sigma^2} - \frac{x^2}{2}\right) dx = \frac{\sigma}{\pi} \int_0^\infty \exp\left(\frac{-x^2}{2\sigma_1^2}\right) dx\end{aligned} \tag{2.95}$$

where σ_1^2 is defined by

$$\sigma_1^2 = \frac{\sigma^2}{1 + \sigma^2} \tag{2.96}$$

The integrand is now in the form of a first-order Gaussian probability density function for $x \geqslant 0$, satisfying

$$\frac{1}{\sigma_1\sqrt{2\pi}} \int_0^\infty \exp\left(\frac{-x^2}{2\sigma_1^2}\right) dx = \frac{1}{2} \tag{2.97}$$

Hence,

$$\int_0^\infty xg(x)p(x)\,dx = \frac{\sigma\sigma_1}{\sqrt{2\pi}} = \frac{\sigma^2}{\sqrt{2\pi(1+\sigma^2)}} \tag{2.98}$$

and

$$R_{xy}(\tau) = \frac{R_{xx}(\tau)}{\sqrt{2\pi(1+\sigma^2)}} \tag{2.99}$$

The associated input/output cross-spectrum formula is

$$S_{xy}(f) = \frac{S_{xx}(f)}{\sqrt{2\pi(1+\sigma^2)}} \tag{2.100}$$

2.6.7 Summary of Results

Sections 2.6.1–2.6.6 produce the following results for the non-Gaussian output probability density function (PDF) compared to a Gaussian PDF, when zero mean value Gaussian input data pass through nonlinear systems.

1. *Two-Slope System* (*Figure 2.12*). For cases where the slope decreases, the output PDF has a discontinuity that produces shorter tails. This also occurs from softening spring and clipped systems.
2. *Two-Slope System* (*Figure 2.13*). For cases where the slope increases, the output PDF has a discontinuity that produces longer tails. This also occurs from hardening spring systems.
3. *Dead-Zone System* (*Figure 2.14*). A delta function effect will appear in the output PDF at output values where the dead-zone effect occurs, with similar tails on both sides of the peak value.
4. *Clipped System* (*Figure 2.15*). A delta function effect will appear in the output PDF at output values where the clipping effect occurs, with a small tail (theoretically zero) for values above a positive clipped value or below a negative clipped value.
5. *Hard-Clipped System* (*Figure 2.16*). Two delta function effects will appear in the output PDF at the two output values where the clipping effect occurs. Similar appearing results can be due to the presence of a strong sine wave in the output data, so one should investigate whether the output data contain such sine waves to distinguish these two possibilities. Sine wave PDFs are discussed in references 1 and 2.
6. *Smooth-Limiter System* (*Figure 2.18*). This output PDF exists only for positive values, with a minimum near $y = 0$ and a very high value where the limit occurs, completely different from a Gaussian PDF and from

symmetric results due to the presence of sine waves. It is similar to one-sided limiter results on sine waves.

7. For Gaussian input data, Sections 2.6.4 and 2.6.6 contain other formulas that predict the output autocorrelation function and input/output cross-correlation functions for the examples of hard-clipped and smooth-limiter systems.

2.7 SQUARE-LAW AND CUBIC SYSTEMS

Three types of zero-memory nonlinear systems will now be discussed representing square-law, cubic, and square-law with sign operations. Such nonlinear systems can occur alone or in parallel with linear operations. From Section 2.9, by least-squares procedures, combinations of linear, square-law, and cubic systems can often be obtained to yield useful approximations to many other types of nonlinear systems that occur in practice. To obtain closed-form answers, input data will be assumed to be a zero mean value Gaussian stationary random process.

2.7.1 Square-Law System

A nonlinear square-law system as per Figure 2.9 is defined by the relation

$$y = g(x) = x^2 \tag{2.101}$$

Here, the continuous derivative

$$\frac{dy}{dx} = g'(x) = 2x \quad \text{and} \quad x = \pm\sqrt{y} \tag{2.102}$$

Thus, the inverse function $x = g^{-1}(y)$ is double-valued. From Eq. (2.14), the output probability density function

$$p_2(y) = \frac{2p(x)}{|2x|} = \frac{p(\sqrt{y})}{\sqrt{y}}, \quad y > 0 \tag{2.103}$$

with $p_2(y) = 0$ for $y < 0$ and $p_2(0) = \infty$. If $p(x)$ has the Gaussian form of Eq. (2.4), then

$$p(\sqrt{y}) = \frac{1}{\sigma_x\sqrt{2\pi}} \exp\left(\frac{-y}{2\sigma_x^2}\right), \quad y > 0 \tag{2.104}$$

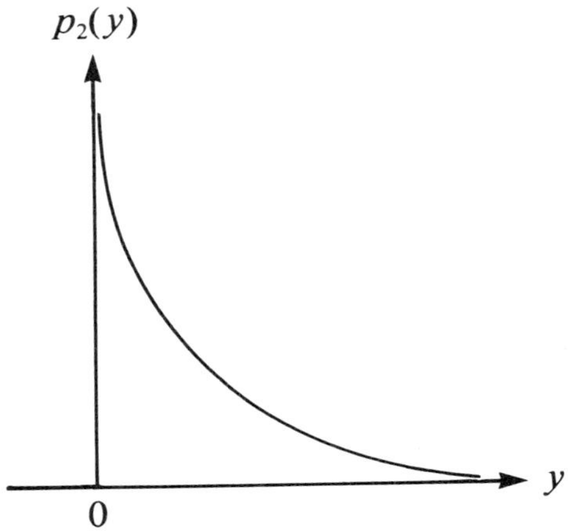

Figure 2.19. Output chi-square PDF for Gaussian data through a square-law system.

Hence,

$$p_2(y) = \frac{1}{\sigma_x\sqrt{2\pi y}} \exp\left(\frac{-y}{2\sigma_x^2}\right), \quad y > 0 \tag{2.105}$$

This is a *chi-square probability density function* with one degree of freedom as sketched in Figure 2.19.

For zero mean value Gaussian input data through a square-law system, the output moments are

$$E[y^n] = E[x^{2n}] = (1)(3)(5) \cdots (2n - 1)\sigma_x^{2n} \tag{2.106}$$

As special cases,

Mean value	$\mu_y = E[y] = E[x^2] = \sigma_x^2$	(2.107)
Mean square value	$\psi_y^2 = R_{yy}(0) = E[y^2] = E[x^4] = 3\sigma_x^4$	(2.108)
Variance	$\sigma_y^2 = E[y^2] - (E[y])^2 = 2\sigma_x^4$	(2.109)

Note that $\mu_y \neq 0$ even though $\mu_x = 0$. Note also that all odd-order moments of the zero mean value Gaussian input data are zero.

The fourth-order moment of the zero mean value Gaussian inputs reduces to the following products of second-order moments, known as the *fourth-order moment relation* (reference 1):

$$E[x_1x_2x_3x_4] = E[x_1x_2]E[x_3x_4] + E[x_1x_3]E[x_2x_4] + E[x_1x_4]E[x_2x_3] \tag{2.110}$$

Higher even-order moments can similarly be reduced.

Output Autocorrelation Function for Square-Law System. From Eq. (2.110), for the zero mean value Gaussian input data, the output autocorrelation function for a square-law system where $y_1 = y(t)$ and $y_2 = y(t + \tau)$ is

$$\begin{aligned} R_{yy}(\tau) &= E[y_1 y_2] = E[x_1^2 x_2^2] = E[x_1 x_1 x_2 x_2] \\ &= E[x_1 x_1]E[x_2 x_2] + 2E[x_1 x_2]E[x_1 x_2] \\ &= \sigma_x^4 + 2R_{xx}^2(\tau) \end{aligned} \tag{2.111}$$

This same result can also be derived from Price's theorem of Eq. (2.19) using

$$\frac{\partial R_{yy}(\tau)}{\partial R_{xx}(\tau)} = E[(2x_1)(2x_2)] = 4R_{xx}(\tau) \tag{2.112}$$

Now

$$R_{yy}(\tau) = 2R_{xx}^2(\tau) + C \tag{2.113}$$

and

$$R_{yy}(0) = 2R_{xx}^2(0) + C = 2\sigma_x^4 + C \tag{2.114}$$

From separate considerations of Eq. (2.85),

$$R_{yy}(0) = E[y^2] = 3\sigma_x^4 \tag{2.115}$$

Hence, $C = \sigma_x^4$ and

$$R_{yy}(\tau) = \sigma_x^4 + 2R_{xx}^2(\tau) \tag{2.116}$$

in agreement with Eq. (2.111).

Input/Output Cross-Correlation Function for Square-Law System. The input/output cross-correlation function for a Gaussian input through a square-law system between $x_1 = x(t)$ and $y_2 = y(t + \tau)$ will be zero, since all odd-order moments are zero for zero mean value Gaussian data. Thus,

$$R_{xy}(\tau) = E[x_1 y_2] = E[x_1 x_2^2] = 0 \tag{2.117}$$

This same result also follows from Bussgang's theorem of Eq. (2.35), since $g(x)$ is an even function of x. The input/output two-sided cross-spectral density function for a Gaussian input through a square-law system is also zero, namely,

$$S_{xy}(f) = 0 \tag{2.118}$$

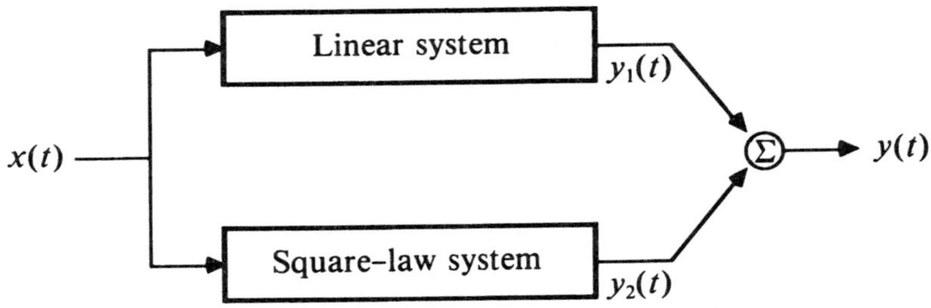

Figure 2.20. Parallel linear and square-law systems.

Thus, usual linear analysis does not recognize square-law effects. This very important result enables one to analyze situations like Figure 2.20 representing a linear system $H(f)$ in parallel with a square-law system. Calculation of $S_{xx}(f)$ and $S_{xy}(f)$ will give $H(f)$ using $S_{xy}(f) = H(f)S_{xx}(f)$, since $S_{xy_2}(f) = 0$. Detection and analysis of the square-law system require new linear techniques or more advanced calculations involving special bispectral density functions to be developed in Chapters 8 and 9. A spectral function of $x(t)$ and $y(t)$ is desired, which will be zero for the linear system and nonzero for the square-law system.

Output Autospectral Density Function for Square-Law System. From Eq. (2.116), the output autospectral density function for the Gaussian input through a square-law system is

$$S_{yy}(f) = \sigma_x^4\delta(f) + 2\int_{-\infty}^{\infty} S_{xx}(u)S_{xx}(f-u)\,du \tag{2.119}$$

where $\delta(f)$ is the usual delta function, while $S_{xx}(f)$ and $S_{yy}(f)$ are two-sided autospectral density functions obtained from the Fourier transforms of $R_{xx}(\tau)$ and $R_{yy}(\tau)$, respectively. For $f \neq 0$, the output spectrum $S_{yy}(f)$ does not include the delta function.

2.7.2 Cubic System

A nonlinear cubic system as per Figure 2.10 is defined by the relation

$$y = g(x) = x^3 \tag{2.120}$$

Here, $x = y^{1/3}$ and the continuous derivative

$$\frac{dy}{dx} = g'(x) = 3x^2 = 3y^{2/3} \tag{2.121}$$

From Eq. (2.13), the output probability density function is

$$p_2(y) = \frac{p(x)}{|3x^2|} = \frac{p(y^{1/3})}{3y^{2/3}} \tag{2.122}$$

If $p(x)$ has the Gaussian form of Eq. (2.4), then

$$p(y^{1/3}) = \frac{1}{\sigma_x\sqrt{2\pi}} \exp\left(\frac{-y^{2/3}}{2\sigma_x^2}\right) \tag{2.123}$$

Hence,

$$p_2(y) = \frac{1}{3\sigma_x y^{2/3}\sqrt{2\pi}} \exp\left(\frac{-y^{2/3}}{2\sigma_x^2}\right) \tag{2.124}$$

Note that $p_2(-y) = p_2(y)$ and $p_2(0) = \infty$ as sketched in Figure 2.21.

For zero mean value Gaussian input data through a cubic system, the output moments are

$$\mu_y = E[y] = E[x^3] = 0 \tag{2.125}$$

$$R_{yy}(0) = E[y^2] = E[x^6] = 15\sigma_x^6 \tag{2.126}$$

Note that $\mu_y = 0$ here when $\mu_x = 0$.

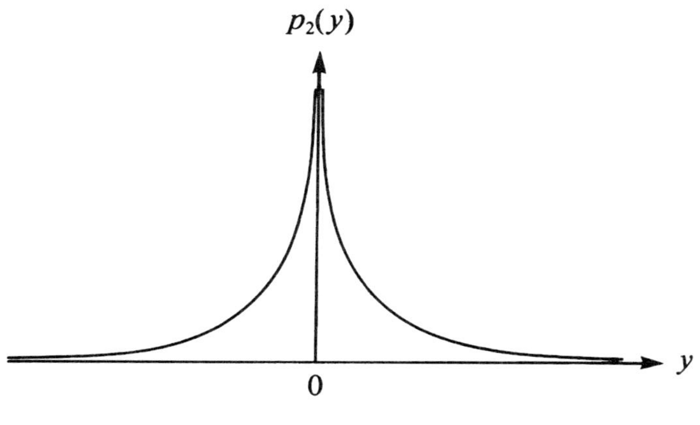

Figure 2.21. Output PDF for Gaussian data through a cubic system.

The sixth-order moment of the zero mean value Gaussian inputs satisfies the relation (reference 1)

$$\begin{aligned} E[x_1x_2x_3x_4x_5x_6] &= E[x_1x_2]E[x_3x_4x_5x_6] \\ &\quad + E[x_1x_3]E[x_2x_4x_5x_6] + E[x_1x_4]E[x_2x_3x_5x_6] \\ &\quad + E[x_1x_5]E[x_2x_3x_4x_6] + E[x_1x_6]E[x_2x_3x_4x_5] \end{aligned} \tag{2.127}$$

Output Autocorrelation Function for Cubic System. The output autocorrelation function for the Gaussian input through a cubic system is given by

$$\begin{aligned} R_{yy}(\tau) &= E[y_1y_2] = E[x_1^3x_2^3] = E[x_1x_1x_1x_2x_2x_2] \\ &= 2E[x_1x_1]E[x_1x_2x_2x_2] + 3E[x_1x_2]E[x_1x_1x_2x_2] \\ &= 2\sigma_x^2[3\sigma_x^2 R_{xx}(\tau)] + 3R_{xx}(\tau)[\sigma_x^4 + 2R_{xx}^2(\tau)] \\ &= 9\sigma_x^4 R_{xx}(\tau) + 6R_{xx}^3(\tau) \end{aligned} \tag{2.128}$$

This same result can also be derived from Price's theorem of Eq. (2.19) using

$$\begin{aligned} \frac{\partial R_{yy}(\tau)}{\partial R_{xx}(\tau)} &= E[(3x_1^2)(3x_2^2)] = 9E[x_1^2x_2^2] \\ &= 9\sigma_x^4 + 18R_{xx}^2(\tau) \end{aligned} \tag{2.129}$$

Now

$$R_{yy}(\tau) = 9\sigma_x^4 R_{xx}(\tau) + 6R_{xx}^3(\tau) + C \tag{2.130}$$

and

$$R_{yy}(0) = 15\sigma_x^6 + C \tag{2.131}$$

But $R_{yy}(0) = 15\sigma_x^6$ from Eq. (2.126). Hence, $C = 0$ and

$$R_{yy}(\tau) = 9\sigma_x^4 R_{xx}(\tau) + 6R_{xx}^3(\tau) \tag{2.132}$$

in agreement with Eq. (2.128).

Input/Output Cross-Correlation Function for Cubic System. The input/output cross-correlation function for the Gaussian input through a cubic system is given by

$$R_{xy}(\tau) = E[x_1y_2] = E[x_1x_2^3] = 3\sigma_x^2 R_{xx}(\tau) \tag{2.133}$$

This same result also follows from Bussgang's theorem of Eq. (2.35) using

$$R_{xy}(\tau) = \frac{R_{xx}(\tau)}{\sigma_x^2} E[x^4] = 3\sigma_x^2 R_{xx}(\tau) \tag{2.134}$$

Equations (2.133) and (2.134) show that the input/output cross-spectral density function for the Gaussian input through a cubic system is

$$S_{xy}(f) = 3\sigma_x^2 S_{xx}(f) \tag{2.135}$$

These relations should be compared with Eqs. (2.117) and (2.118) for the square-law system.

Output Autospectral Density Function for Cubic System. From Eq. (2.128) or (2.132), the output autospectral density function for the Gaussian input through a cubic system is

$$S_{yy}(f) = 9\sigma_x^4 S_{xx}(f) + 6 \iint S_{xx}(u) S_{xx}(v-u) S_{xx}(f-v)\, du\, dv \tag{2.136}$$

In Eq. (2.135), the output $y(t) = x^3(t)$ is correlated with the input $x(t)$. For later applications, it is desirable to have a different output $y_c(t)$ that will be uncorrelated with $x(t)$. This can be done as follows. Let

$$y_c(t) = x^3(t) - 3\sigma_x^2 x(t) \tag{2.137}$$

Then instead of Eqs. (2.135) and (2.136), one will obtain

$$S_{xy_c}(f) = 0 \tag{2.138}$$

$$S_{y_c y_c}(f) = 6 \iint S_{xx}(u) S_{xx}(v-u) S_{xx}(f-v)\, du\, dv \tag{2.139}$$

The corresponding correlation functions become

$$R_{xy_c}(\tau) = 0 \tag{2.140}$$

$$R_{y_c y_c}(\tau) = 6R_{xx}^3(\tau) \tag{2.141}$$

2.7.3 Square-Law System with Sign

Square-law systems with sign as per Figure 2.11 are of great importance in fluid dynamics models involving "dynamic pressure," defined as $q = \frac{1}{2}\rho v|v|$ where ρ is the fluid density and v is the fluid velocity. This quantity q is a key parameter in the description of boundary-layer pressures and drag forces due to the fluid flow over structures.

A nonlinear square-law system with sign is defined by the relation

$$y = g(x) = x|x| \tag{2.142}$$

Hence,

$$\begin{aligned} y &= x^2, \quad & x > 0 \\ &= -x^2, \quad & x < 0 \end{aligned} \tag{2.143}$$

The inverse relation

$$\begin{aligned} x &= +\sqrt{y}, \quad & y > 0 \\ &= -\sqrt{-y}, \quad & y < 0 \end{aligned} \tag{2.144}$$

Also, the continuous derivative

$$\begin{aligned} \frac{dy}{dx} = g'(x) &= 2x, \quad & x > 0 \\ &= -2x, \quad & x < 0 \end{aligned} \tag{2.145}$$

From Eq. (2.13), the output probability density function

$$\begin{aligned} p_2(y) = \frac{p(x)}{|2x|} &= \frac{p(\sqrt{y})}{2\sqrt{y}}, \quad & y > 0 \\ &= \frac{p(\sqrt{-y})}{2\sqrt{-y}}, \quad & y < 0 \end{aligned} \tag{2.146}$$

Thus, $p_2(-y) = p_2(y)$ and $p_2(0) = \infty$ with $p_2(y)$ as sketched in Figure 2.22 if $p(x)$ has the Gaussian form of Eq. (2.4).

For zero mean value Gaussian input data through a square-law system with sign, the output moments

$$\mu_y = E[y] = E[x|x|] = 0 \tag{2.147}$$

$$R_{yy}(0) = E[y^2] = E[x^4] = 3\sigma_x^4 \tag{2.148}$$

Note that the mean value here is the same as for a cubic system, but the mean square value is the same as for a square-law system.

For the square-law system with sign,

$$\begin{aligned} E[xy] &= E[x^2|x|] = E[-x^3]_{x<0} + E[x^3]_{x>0} \\ &= 2E[x^3]_{x>0} = 2\int_0^\infty x^3 p(x)\, dx = (2\sigma_x^3)\sqrt{(2/\pi)} \end{aligned} \tag{2.149}$$

using the $p(x)$ of Eq. (2.4). Hence, from Bussgang's theorem of Eq. (2.35), the

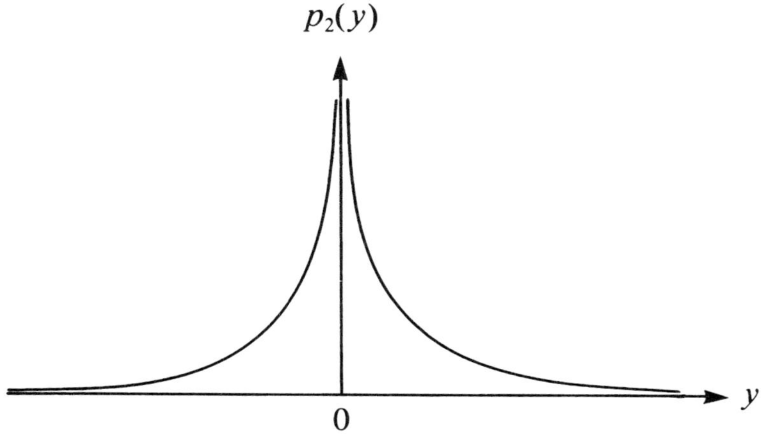

Figure 2.22. Output PDF for Gaussian data through a square-law system with sign.

input/output cross-correlation function for a Gaussian input through a square-law system with sign is

$$R_{xy}(\tau) = 2\sigma_x\sqrt{(2/\pi)}\,R_{xx}(\tau) \tag{2.150}$$

The associated input/output cross-spectrum is

$$S_{xy}(f) = 2\sigma_x\sqrt{(2/\pi)}\,S_{xx}(f) \tag{2.151}$$

The output autocorrelation function for a Gaussian input through a square-law system with sign is given by

$$R_{yy}(\tau) = E[y_1 y_2] = E[(x_1|x_1|)(x_2|x_2|)] \tag{2.152}$$

A useful approximation for this result, derived later in Eq. (2.201), is

$$R_{yy}(\tau) \approx (2.55\sigma_x^2)R_{xx}(\tau) + (0.42/\sigma_x^2)R_{xx}^3(\tau) \tag{2.153}$$

The associated output autospectrum is

$$S_{yy}(f) \approx (2.55\sigma_x^2)S_{xx}(f) + (0.42/\sigma_x^2)\iint S_{xx}(u)S_{xx}(v-u)S_{xx}(f-v)\,du\,dv \tag{2.154}$$

2.7.4 Summary of Results

Sections 2.7.1–2.7.3 produce the following results for the square-law system, the cubic system, and the square-law system with sign when zero mean value Gaussian input data pass through these systems.

TABLE 2.1 Square-Law, Cubic, and Square-Law with Sign Systems

	System		
	Square-Law	Cubic	Square-Law with Sign
Relation	$y = x^2$	$y = x^3$	$y = x\|x\|$
$E[y]$	σ_x^2	0	0
$E[y^2]$	$3\sigma_x^4$	$15\sigma_x^6$	$3\sigma_x^4$
$R_{xy}(\tau)$	0	$3\sigma_x^2 R_{xx}(\tau)$	$2\sigma_x\sqrt{(2/\pi)}R_{xx}(\tau)$
$R_{yy}(\tau)$	$\sigma_x^4 + 2R_{xx}^2(\tau)$	$9\sigma_x^4 R_{xx}(\tau) + 6R_{xx}^3(\tau)$	$(2.55\sigma_x^2)R_{xx}(\tau)$ $+(0.42/\sigma_x^2)R_{xx}^3(\tau)$

1. *Square-Law System.* The output PDF (Figure 2.19) exists only for positive values of y with a very high value near $y = 0$ and with a long tail, as described by a chi-square PDF with one degree of freedom that is completely different from a Gaussian PDF.
2. *Cubic System.* The output PDF (Figure 2.21) will be very sharply peaked near $y = 0$ compared to a Gaussian PDF. Similar measured output PDFs can occur for dead-zone systems (Figure 2.16).
3. *Square-Law System with Sign.* This gives an output PDF (Figure 2.22) that appears the same as for a cubic system output PDF (Figure 2.21). Other analysis is required to distinguish these two situations.

For these three systems, from knowledge of $R_{xx}(\tau)$ for $\tau \geqslant 0$ where $R_{xx}(0) = \sigma_x^2$, the predicted results for $E[y]$, $E[y^2]$, $R_{xy}(\tau)$, and $R_{yy}(\tau)$ are listed in Table 2.1 for easy comparison.

2.8 HARDENING AND SOFTENING SPRING SYSTEMS

Output probability density functions will now be derived for the force output $y(t)$ versus the displacement input $x(t)$ in four models that can represent continuously hardening and softening spring systems. Related alternative output probability density functions follow for the displacement output versus the force input when the roles of $x(t)$ and $y(t)$ are reversed.

2.8.1 Hardening Spring System

For Gaussian input data governed by Eq. (2.4) with $\sigma_x = 1.0$, a continuously hardening spring system can be defined for suitable x and a by the single-

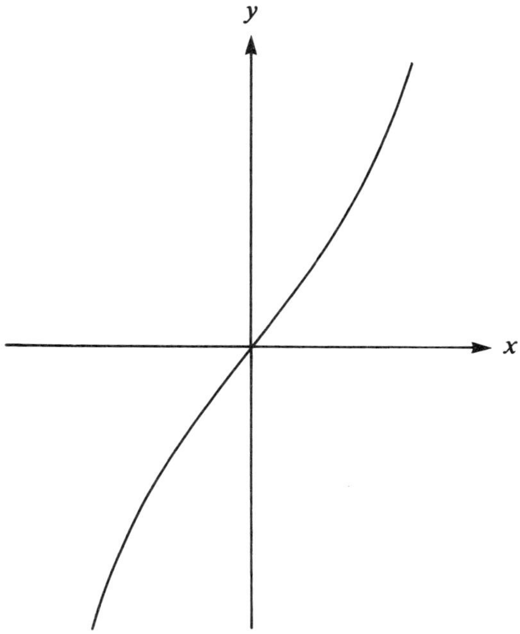

Figure 2.23. Hardening spring system.

valued, one-to-one, zero-memory nonlinear function

$$y = g(x) = \begin{cases} x + ax^2, & x \geqslant 0 \\ x - ax^2, & x \leqslant 0 \end{cases} \tag{2.155}$$

This special $g(x)$ is an odd function of x. Note that the hardening spring system of Eq. (2.155) is equivalent to having a linear system in parallel with a square-law system with sign, since

$$y = x + ax|x| \tag{2.156}$$

Equation (2.155) is sketched in Figure 2.23 for $a = 0.14$ and $|x| \leqslant 3.0$.

For the assumed Gaussian displacement input data,

$$E[x] = 0, \quad E[x^2] = 1.0, \quad E[x^3] = 0, \quad E[x^4] = 3.0 \tag{2.157}$$

For the hardening spring force output data of Eq. (2.155),

$$E[y] = 0, \quad E[y^2] \approx 1.51, \quad E[y^3] = 0, \quad E[y^4] \approx 9.16 \tag{2.158}$$

Thus, the hardening spring causes large kurtosis where $E[y^4] > E[x^4]$. Note also that $\sigma_y \approx 1.23$.

The inverse function $x = g^{-1}(y)$ is given here by

$$\begin{aligned} x &= (-1 + \sqrt{1 + 4ay})/2a, \quad && y > 0 \\ &= (1 - \sqrt{1 - 4ay})/2a, && y < 0 \\ &= 0, && y = 0 \end{aligned} \tag{2.159}$$

The derivative function

$$\begin{aligned} \frac{dy}{dx} &= 1 + 2ax, \quad && x \geqslant 0 \\ &= 1 - 2ax, && x \leqslant 0 \end{aligned} \tag{2.160}$$

Substituting Eq. (2.159) into Eq. (2.160) yields

$$\begin{aligned} \frac{dy}{dx} &= \sqrt{1 + 4ay}, \quad && y > 0 \\ &= \sqrt{1 - 4ay}, && y < 0 \\ &= 1, && y = 0 \end{aligned} \tag{2.161}$$

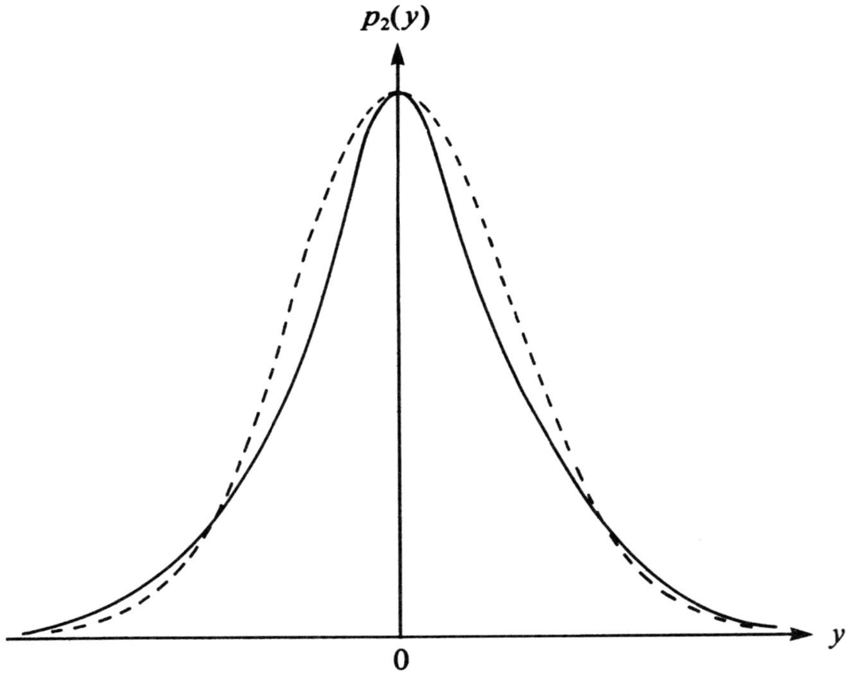

Figure 2.24. Output PDF for hardening spring system.

Hence, by Eq. (2.13), $p_2(0) = p(0)$ with

$$p_2(y) = \frac{p[(-1+\sqrt{1+4ay})/2a]}{\sqrt{1+4ay}}, \quad y > 0$$
$$= \frac{p[(1-\sqrt{1-4ay})/2a]}{\sqrt{1-4ay}}, \quad y < 0 \tag{2.162}$$

This formula shows that the output $p_2(y)$ is an even function of y.

Equation (2.162) is sketched in Figure 2.24 for $a = 0.14$ and $|y| \leqslant 3.5$. Figure 2.24 demonstrates that the PDF for a hardening spring, compared with a Gaussian PDF, will have a sharper peak in the vicinity of $y = 0$ and will have longer tails. The Gaussian PDF is drawn with dashed lines.

2.8.2 Softening Spring System

For Gaussian input data governed by Eq. (2.4) with $\sigma_x = 1.0$, a continuously softening spring system can be defined for suitable x and a by the single-valued one-to-one zero-memory nonlinear function

$$y = g(x) = \begin{cases} x - ax^2, & x \geqslant 0 \\ x + ax^2, & x \leqslant 0 \end{cases} \tag{2.163}$$

As in Eq. (2.155), this different $g(x)$ is an odd function of x. Note that this softening spring system consists of a linear system in parallel with a square-law system with sign, since

$$y = x - ax|x| \tag{2.164}$$

Equation (2.163) is sketched in Figure 2.25 for $a = 0.14$ and $|x| \leqslant 3.0$.

For the assumed Gaussian displacement input data,

$$E[x] = 0, \quad E[x^2] = 1.0, \quad E[x^3] = 0, \quad E[x^4] = 3.0 \tag{2.165}$$

For the softening spring force output data of Eq. (2.163),

$$E[y] = 0, \quad E[y^2] \approx 0.61, \quad E[y^3] = 0, \quad E[y^4] \approx 1.17 \tag{2.166}$$

Thus, the softening spring causes small kurtosis where $E[y^4] < E[x^4]$. Note here that $\sigma_y \approx 0.78$.

The derivative function

$$\frac{dy}{dx} = 1 - 2ax, \quad x \geqslant 0$$
$$= 1 + 2ax, \quad x \leqslant 0 \tag{2.167}$$

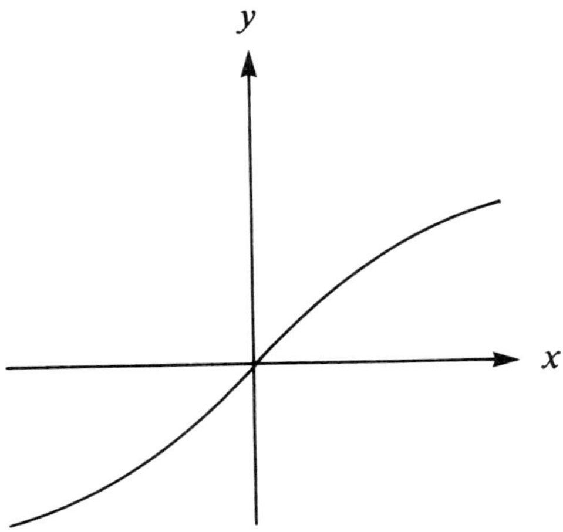

Figure 2.25. Softening spring system.

In particular, for $x > 0$ and $a = 0.14$,

$$\frac{dy}{dx} = 0 \qquad \text{when} \quad x = \frac{1}{2a} \approx 3.57$$

with a maximum value of y given by

$$y_{\max} = \frac{1}{4a} \approx 1.785$$

By restricting $|x| \leqslant 3.5$, it follows that the inverse function $x = g^{-1}(y)$ will be single-valued. This inverse function is given here by

$$\begin{aligned} x &= (1 - \sqrt{1 - 4ay})/2a, & y > 0 \\ &= (-1 + \sqrt{1 + 4ay})/2a, & y < 0 \\ &= 0, & y = 0 \end{aligned} \tag{2.168}$$

Now, by Eq. (2.13), $p_2(0) = p(0)$ with

$$\begin{aligned} p_2(y) &= \frac{p[(1 - \sqrt{1 - 4ay})/2a]}{\sqrt{1 - 4ay}}, & y > 0 \\ &= \frac{p[(-1 + \sqrt{1 + 4ay})/2a]}{\sqrt{1 + 4ay}}, & y < 0 \end{aligned} \tag{2.169}$$

As before, observe that this output $p_2(y)$ is an even function of y.

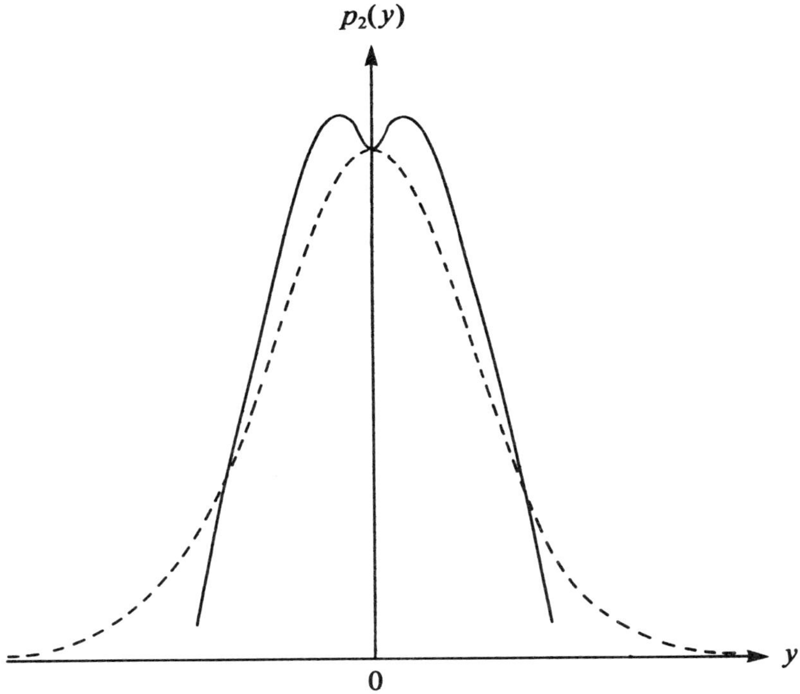

Figure 2.26. Output PDF for softening spring system.

Equation (2.169) is sketched in Figure 2.26 for $a = 0.14$ and $|y| \leqslant 3.5$. Figure 2.26 demonstrates that the PDF for a softening spring, compared with a Gaussian PDF, will have a flatter peak in the vicinity of $y = 0$ and will have shorter tails. Note also the two humps immediately to the right and left of $y = 0$. The Gaussian PDF is drawn with dashed lines.

2.8.3 Hardening/Softening Spring System

A continuously hardening/softening spring system with hardening for $x > 0$ and softening for $x < 0$ can be defined by combining the results from Sections 2.8.1 and 2.8.2 as follows:

$$y = g(x) = x + ax^2 \tag{2.170}$$

This function is neither even nor odd. Equation (2.170) is sketched in Figure 2.27 for $a = 0.14$ and $|x| \leqslant 3.0$. Note that this particular nonlinear transformation has the shape associated with a catenary cable.

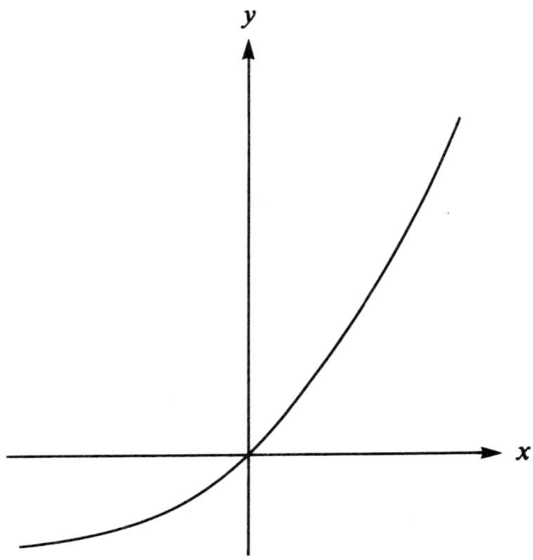

Figure 2.27. Hardening/softening spring system.

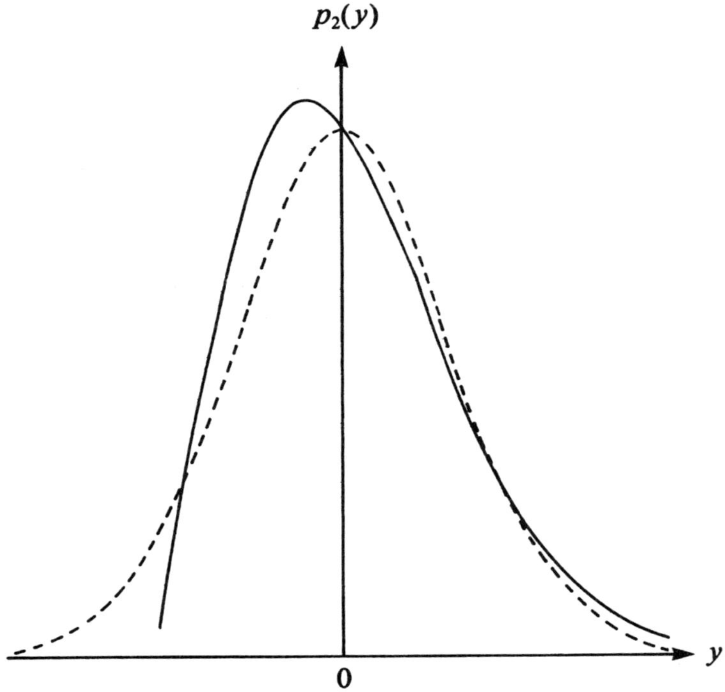

Figure 2.28. Output PDF for hardening/softening spring system.

For this case, the output PDF is given by the relations $p_2(0) = p(0)$ with

$$p_2(y) = \frac{p[(-1 + \sqrt{1 + 4ay})/2a]}{\sqrt{1 + 4ay}} \quad \text{for all } y \neq 0 \tag{2.171}$$

This is sketched in Figure 2.28 for $a = 0.14$ and $|y| \leqslant 3.5$ by combining previous results from Figures 2.24 and 2.26. Note that Figure 2.28 demonstrates that the output PDF for a hardening/softening spring system as defined here, compared with a Gaussian PDF, will be skewed to the left. The Gaussian PDF is drawn with dashed lines.

2.8.4 Softening/Hardening Spring System

A continuously softening/hardening spring system, with softening for $x > 0$ and hardening for $x < 0$, can be defined by combining the results from Sections 2.8.1 and 2.8.2 as follows:

$$y = g(x) = x - ax^2 \tag{2.172}$$

This function is neither even nor odd. Equation (2.172) is sketched in Figure 2.29 for $a = 0.14$ and $|x| \leqslant 3.0$.

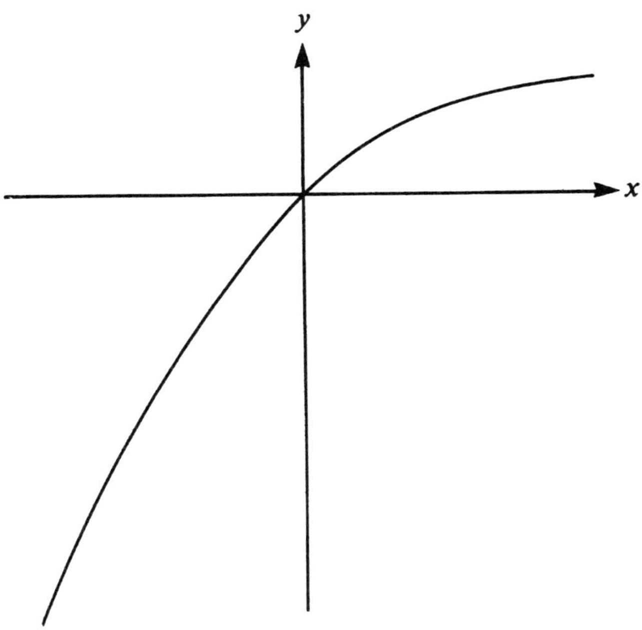

Figure 2.29. Softening/hardening spring system.

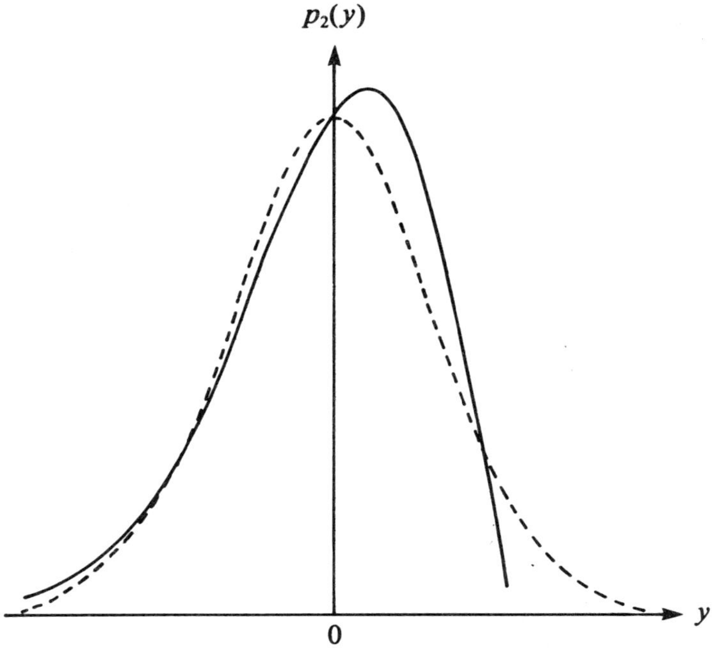

Figure 2.30. Output PDF for softening/hardening spring system.

For this case, the output PDF is now given by the relations $p_2(0) = p(0)$ with

$$p_2(y) = \frac{p[(1 - \sqrt{1 - 4ay})/2a]}{\sqrt{1 - 4ay}} \quad \text{for all } y \neq 0 \tag{2.173}$$

This is sketched in Figure 2.30 for $a = 0.14$ and $|y| \leqslant 3.5$ by combining in the reverse order from Figure 2.28 the previous results of Figures 2.24 and 2.26. In particular, note that Figure 2.30 demonstrates that the output PDF for a softening/hardening spring as defined here, compared with a Gaussian PDF, will be skewed to the right. The Gaussian PDF is drawn with dashed lines.

Figures 2.24, 2.26, 2.28, and 2.30 show the output PDF for unstandardized output variables. However, most computer programs standardize the output variable before computing its PDF. This produces different results as discussed in the next section.

2.8.5 Standardized Output Variable

Suppose the output variable $y(t)$ with probability density function $p_2(y)$, mean value μ_y, and standard deviation σ_y is replaced by a standardized output variable $z(t)$ with probability density function $p_3(z)$, zero mean value $\mu_z = 0$, and unit standard deviation $\sigma_z = 1.0$. The transformation is

$$z(t) = \frac{y(t) - \mu_y}{\sigma_y} \tag{2.174}$$

Now, by the probability density relation of Eqs. (2.13),

$$p_3(z) = \frac{p_2(y)}{|dz/dy|} = \sigma_y p_2(y) = \sigma_y p_2(\mu_y + \sigma_y z) \tag{2.175}$$

This is a general result. When $\mu_y = 0$, one obtains the special formula

$$p_3(z) = \sigma_y p_2(\sigma_y z) \tag{2.176}$$

Consider the continuously hardening and softening spring systems in Figures 2.23 and 2.25 where $\mu_y = 0$. For these cases, at the value $z = 0$,

$$p_3(0) = \sigma_y p_2(0) \tag{2.177}$$

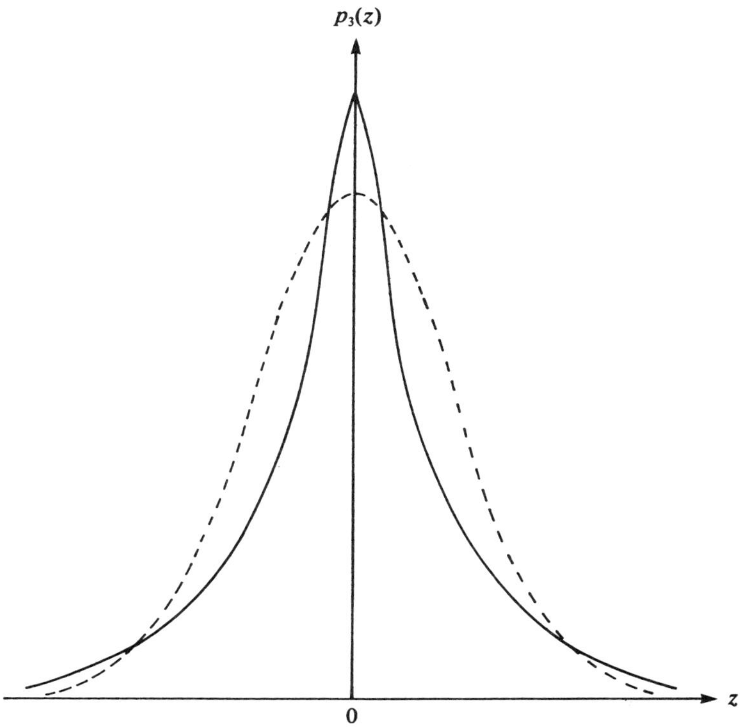

Figure 2.31. Output PDF for hardening spring system with standardized output variable.

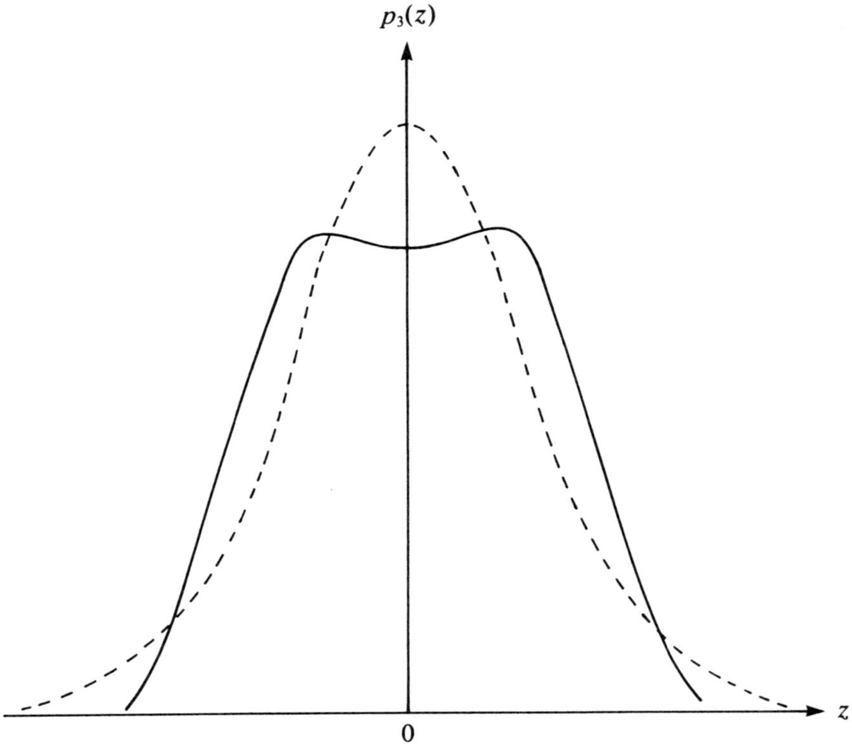

Figure 2.32. Output PDF for softening spring system with standardized output variable.

It follows that

$$\begin{aligned} p_3(0) &> p_2(0) \quad \text{when } \sigma_y > 1 \\ p_3(0) &< p_2(0) \quad \text{when } \sigma_y < 1 \end{aligned} \tag{2.178}$$

Thus, for the hardening spring system where Eq. (2.158) gives $\sigma_y \approx 1.23$, the output PDF of $p_3(z)$ versus z will be more peaked than shown in Figure 2.24 with $p_3(0) \approx 1.23 p_2(0)$. Conversely, for the softening spring system where Eq. (2.166) gives $\sigma_y \approx 0.78$, the output PDF of $p_3(z)$ versus z will be flatter than shown in Figure 2.26 with $p_3(0) \approx 0.78 p_2(0)$. Figures 2.31 and 2.32 based on Eq. (2.176) are plots of $p_3(z)$ versus z for the hardening and softening spring systems where $|z| \leqslant 3.0$. The Gaussian PDF is drawn with dashed lines.

2.8.6 Summary of Results

This chapter proves that, when Gaussian input data pass through different types of zero-memory nonlinear systems, the output probability density function (PDF) will be non-Gaussian and will depend on the type of nonlinearity. The form of the output PDF can help determine the type of nonlinear operation that is occurring.

Sections 2.8.1–2.8.5 produce the following results for the non-Gaussian output PDF compared to a Gaussian PDF when zero mean value Gaussian input data pass through hardening and softening spring systems.

1. *Hardening Spring System* (*Figures 2.23, 2.24, 2.31*). The output PDF will have a sharper peak in the vicinity of $y = 0$, longer tails, and higher kurtosis values.
2. *Softening Spring System* (*Figures 2.25, 2.26, 2.32*). The output PDF will have a flatter peak in the vicinity of $y = 0$, shorter tails, and lower kurtosis values.
3. *Hardening/Softening Spring System* (*Figures 2.27, 2.28*). The output PDF will be skewed to the left.
4. *Softening/Hardening Spring System* (*Figures 2.29, 2.30*). The output PDF will be skewed to the right.
5. The plots shown in Figures 2.23, 2.25, 2.27, and 2.29 are for the force output $y(t)$ versus the displacement input $x(t)$. These same plots can also be used to represent a displacement output versus a force input by interchanging $x(t)$ with $y(t)$ and replacing "hardening" with "softening." The plots shown in Figures 2.24, 2.26, 2.28, and 2.30 are output probability density functions for the force output to a Gaussian displacement input. When the roles of $x(t)$ and $y(t)$ are reversed, then these same particular plots can represent the output probability density functions for the displacement output to a Gaussian force input for systems where "hardening" and "softening" are reversed. For example, for a Gaussian force input $x(t)$, Figure 2.24 is the plot of the output probability density function $p_2(y)$ for the displacement output $y(t)$ in a "softening" spring system, and Figure 2.26 is the plot of the output probability density function $p_2(y)$ for the displacement output $y(t)$ in a "hardening" spring system.

2.9 THIRD-ORDER POLYNOMIAL LEAST-SQUARES APPROXIMATION

The replacement of zero-memory nonlinear systems by an optimum third-order polynomial least-squares approximation will now be derived. Suppose $y = g(x)$ represents a zero-memory nonlinear system. What is the least-squares approximation to $y = g(x)$ by the *third-order* polynomial

$$v = a_1 x + a_2 x^2 + a_3 x^3 \tag{2.179}$$

under the assumption that the input data x follow a zero mean value Gaussian distribution? To be specific, what should be the choices of the coefficient a_1, a_2, a_3 so as to minimize the mean square error Q defined by

$$Q = E[(y - v)^2] = E[(y - a_1 x - a_2 x^2 - a_3 x^3)^2] \tag{2.180}$$

over all possible choices of these coefficients?

For any y, the assumptions on x give

$$\begin{aligned} Q = E[y^2] - 2a_1 E[xy] - 2a_2 E[x^2 y] - 2a_3 E[x^3 y] \\ + a_1^2 E[x^2] + 2a_1 a_3 E[x^4] + a_2^2 E[x^4] + a_3^2 E[x^6] \end{aligned} \tag{2.181}$$

using the fact that $E[x] = E[x^3] = E[x^5] = 0$. The setting of partial derivatives of Q with respect to a_1, a_2, and a_3 equal to zero shows

$$\begin{aligned} \frac{\partial Q}{\partial a_1} &= -2E[xy] + 2a_1 E[x^2] + 2a_3 E[x^4] = 0 \\ \frac{\partial Q}{\partial a_2} &= -2E[x^2 y] + 2a_2 E[x^4] = 0 \\ \frac{\partial Q}{\partial a_3} &= -2E[x^3 y] + 2a_1 E[x^4] + 2a_3 E[x^6] = 0 \end{aligned} \tag{2.182}$$

For zero mean value Gaussian data, the expected values $E[x^2] = \sigma_x^2$, $E[x^4] = 3\sigma_x^4$, and $E[x^6] = 15\sigma_x^6$ now give the following set of equations to be solved for a_1, a_2, and a_3:

$$\begin{aligned} a_1 \sigma_x^2 + 3a_3 \sigma_x^4 &= E[xy] \\ 3a_2 \sigma_x^4 &= E[x^2 y] \\ 3a_1 \sigma_x^4 + 15a_3 \sigma_x^6 &= E[x^3 y] \end{aligned} \tag{2.183}$$

Hence, for any $y = g(x)$, solving Eq. (2.183) shows that one obtains the v of Eq. (2.179) by choosing

$$a_1 = \frac{15\sigma_x^6 E[xy] - 3\sigma_x^4 E[x^3 y]}{6\sigma_x^8} \tag{2.184}$$

$$a_2 = \frac{E[x^2 y]}{3\sigma_x^4} \tag{2.185}$$

$$a_3 = \frac{\sigma_x^2 E[x^3 y] - 3\sigma_x^4 E[xy]}{6\sigma_x^8} \tag{2.186}$$

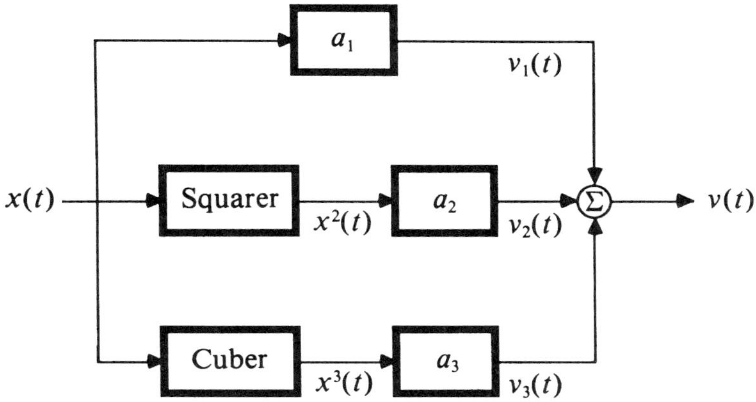

Figure 2.33. Optimum third-order polynomial least-squares model for a zero-memory nonlinear system.

These three equations are important results for many engineering applications. When substituted into Eq. (2.179), they show how an arbitrary zero-memory nonlinear system can be replaced by the optimum third-order polynomial of Figure 2.33, consisting of a linear operation x in parallel with a squaring operation x^2 and a cubing operation x^3. This model will be extended in Section 4.6.

It is instructive to compare the optimum third-order results with a *linear* least-squares approximation to y given by

$$v = b_1 x \tag{2.187}$$

Here, one minimizes the mean square error Q defined by

$$\begin{aligned} Q &= E[(y - v)^2] = E[(y - b_1 x)^2] \\ &= E[y^2] - 2b_1 E[xy] + b_1^2 E[x^2] \end{aligned}$$

Now

$$\frac{\partial Q}{\partial b_1} = -2E[xy] + 2b_1 E[x^2] = 0$$

shows that, for any $y = g(x)$,

$$b_1 = \frac{E[xy]}{E[x^2]} = \frac{E[xy]}{\sigma_x^2} \tag{2.188}$$

This value of b_1 is *not* the same as the a_1 in Eq. (2.184), and Eq. (2.187) is inferior to Eq. (2.179). Note that b_1 is the same as the factor of proportionality in Bussgang's theorem, Eq. (2.35). Similarly, other approximations involving subsets of the terms in Eq. (2.179) will, in general, give poorer results than Eq. (2.179).

Application to Square-Law System with Sign. Consider the application of these matters to the case of a *square-law system with sign* where

$$y = x|x| \tag{2.189}$$

For this particular y, the Gaussian assumptions on x yield

$$\begin{aligned} E[xy] &= E[x^2|x|] = 2E[x^3]_{x>0} = 2\sigma_x^3\sqrt{(2/\pi)} \\ E[x^2y] &= E[x^3|x|] = 0 \\ E[x^3y] &= E[x^4|x|] = 2E[x^5]_{x>0} = 8\sigma_x^5\sqrt{(2/\pi)} \end{aligned} \tag{2.190}$$

Hence, from Eq. (2.188),

$$b_1 = 2\sigma_x\sqrt{(2/\pi)} \tag{2.191}$$

so that the *linear* least-squares approximation to $x|x|$ by Eq. (2.187) is

$$v = 2\sigma_x\sqrt{(2/\pi)}\,x \tag{2.192}$$

However, from Eqs. (2.184)–(2.186), one obtains

$$a_1 = \sigma_x\sqrt{(2/\pi)} \tag{2.193}$$

$$a_2 = 0 \tag{2.194}$$

$$a_3 = \sqrt{(2/\pi)}/3\sigma_x \tag{2.195}$$

Hence, the *third-order* polynomial least-squares approximation to $x|x|$ by Eq. (2.179) is

$$v = [\sigma_x\sqrt{(2/\pi)}]x + [\sqrt{(2/\pi)}/3\sigma_x]x^3 \tag{2.196}$$

This useful result shows how to approximate $x|x|$ by a combination of a linear system in parallel with a cubic system. This gives a much better representation for $x|x|$ than the simpler linear result of Eq. (2.192) as will now be demonstrated by computing $R_{xv}(\tau)$ and $R_{vv}(\tau)$.

From Eq. (2.179) for third-order approximations, the input/output cross-correlation function for the assumed x, using $x_1 = x(t)$ and $x_2 = x(t + \tau)$, is, in general,

$$\begin{aligned} R_{xv}(\tau) &= a_1 E[x_1 x_2] + a_3 E[x_1 x_2^3] \\ &= a_1 R_{xx}(\tau) + 3a_3 \sigma_x^2 R_{xx}(\tau) \end{aligned} \tag{2.197}$$

For the example of a square-law system with sign, where a_1 and a_3 satisfy Eqs. (2.193) and (2.195), the special result is

$$R_{xv}(\tau) = 2\sigma_x \sqrt{(2/\pi)}\, R_{xx}(\tau) \tag{2.198}$$

in agreement with the known correct result of Eq. (2.150). The linear approximations of Eqs. (2.187) and (2.192) also give

$$R_{xv}(\tau) = b_1 R_{xx}(\tau) = 2\sigma_x \sqrt{(2/\pi)}\, R_{xx}(\tau) \tag{2.199}$$

Thus, for the square-law system with sign, $R_{xv}(\tau)$ is the same for third-order and linear approximations. This is the expected result from Bussgang's theorem of Eq. (2.35).

From Eq. (2.179) for third-order approximations, the output autocorrelation function for the assumed x, using $x_1 = x(t)$ and $x_2 = x(t + \tau)$, is, in general,

$$\begin{aligned} R_{vv}(\tau) &= a_1^2 E[x_1 x_1] + a_1 a_3 E[x_1 x_2^3] + a_1 a_3 E[x_1^3 x_2] \\ &\quad + a_2^2 E[x_1^2 x_2^2] + a_3^2 E[x_1^3 x_2^3] \\ &= a_1^2 R_{xx}(\tau) + 3a_1 a_3 \sigma_x^2 R_{xx}(\tau) + 3a_1 a_3 \sigma_x^2 R_{xx}(\tau) \\ &\quad + a_2^2 [\sigma_x^4 + 2R_{xx}^2(\tau)] + a_3^2 [9\sigma_x^4 R_{xx}(\tau) + 6R_{xx}^3(\tau)] \end{aligned} \tag{2.200}$$

For the example of a square-law system with sign, from Eqs. (2.193)–(2.195) one obtains

$$R_{vv}(\tau) = [8\sigma_x^2/\pi] R_{xx}(\tau) + [4/3\pi\sigma_x^2] R_{xx}^3(\tau) \tag{2.201}$$

with

$$R_{vv}(0) = [28/3\pi]\sigma_x^4 \approx 2.971\sigma_x^4 \tag{2.202}$$

This differs by less than 1% from the known correct result of Eq. (2.148) that $R_{yy}(0) = 3\sigma_x^4$. For the linear approximation of Eqs. (2.187) and (2.192) one obtains

$$R_{vv}(\tau) = b_1^2 R_{xx}(\tau) = [8\sigma_x^2/\pi] R_{xx}(\tau) \tag{2.203}$$

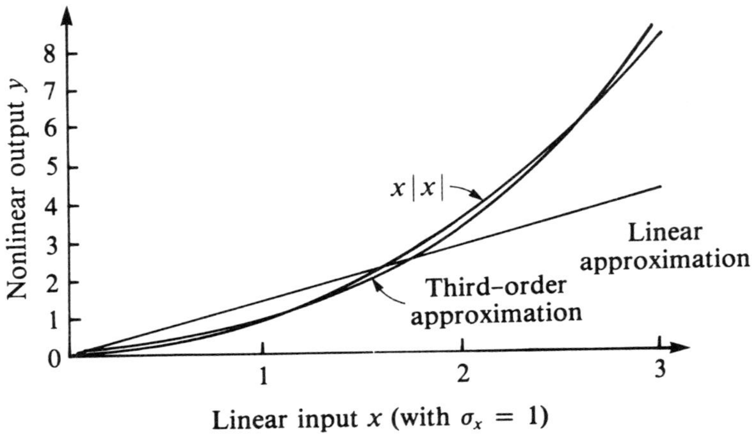

Figure 2.34. Least-squares approximation of $x|x|$.

with

$$R_{vv}(0) = [8/\pi]\sigma_x^4 \approx 2.546\sigma_x^4 \tag{2.204}$$

This result is significantly different from the correct $R_{yy}(0) = 3\sigma_x^4$. The conclusion is that one should use the third-order polynomial approximation of Eq. (2.196) in preference to the linear approximation of Eq. (2.192) if one desires to analyze a square-law system with sign by such least-squares approximations. Figure 2.34 shows graphically the differences between these two approximations and the desired curve of $x|x|$ for values of x in the range $0 \leqslant x \leqslant 3\sigma_x$.

This concludes Chapter 2.

3

DIRECT AND REVERSE MI/SO TECHNIQUES FOR ANALYSIS AND IDENTIFICATION OF NONLINEAR SYSTEMS

This chapter is based on a keynote lecture that I delivered at the Third International Conference on Stochastic Structural Dynamics held in San Juan, Puerto Rico, on January 15–18, 1995. In writing this book, I have retained the flavor of this presentation by leaving this chapter in the lecture form. The figures in this chapter were slides in the original lecture. These figures are shown here together with words that I said when each figure was displayed. My objective in this chapter is to motivate the reader to apply wherever possible the Direct and Reverse MI/SO techniques that I discussed in this lecture, which are covered in this book in greater detail in Chapter 4–7. These new practical MI/SO techniques are applicable to any nonlinear system that can reasonably be modeled by means of a nonlinear differential or integrodifferential equation of motion. SDOF and MDOF nonlinear systems are included, where the coefficients can be constants or frequency-dependent. Simulated or measured excitation data and response data can have arbitrary probability and spectral properties. Each of the identified nonlinear components can be evaluated at any desired frequency with separate coherence functions. Thus these techniques represent a significant advance in using real-world data to help improve the design and understanding of nonlinear physical systems.

Figure 3.1 states some of the main features of the Direct and Reverse MI/SO Techniques that are not available in other nonlinear methods, such as least-squares time-domain procedures, or in higher-order Volterra series. The first key result in this work is the recognition that single-input/single-output (SI/SO) nonlinear models without feedback can be converted into equivalent direct multiple-input/single-output (MI/SO) linear models without feedback. The second key result in this work is the recognition that single-input/single-output (SI/SO) nonlinear models with feedback can be converted into equivalent reverse multiple-input/single-output (MI/SO) linear models without feedback. Each of the other features are also important such as requiring only functions of one frequency and removing restrictions regarding the probability or spectral natures of the excitation and response data. Thus this technique applies to Gaussian or to non-Gaussian data with arbitrary spectral density functions. Also, confidence limits on the linear and nonlinear terms can be obtained easily at each frequency using ordinary coherence functions.

FEATURES OF DIRECT AND REVERSE MI/SO TECHNIQUES

1. Single-input/single-output (SI/SO) nonlinear models without feedback can be converted into equivalent direct dynamic multiple-input/single-output (MI/SO) linear models without feedback.

2. Single-input/single-output (SI/SO) nonlinear models with feedback can be converted into equivalent reverse dynamic multiple-input/single-output (MI/SO) linear models without feedback.

3. An exact nonlinear representation is obtained using a linear frequency response function of one frequency for each nonlinear component.

4. The nonlinear system amplitude properties can be identified as well as the physical parameters with frequency-dependent coefficients.

5. There are no restrictions on the probability or spectral natures of the excitation or response data.

6. All results are easy to compute, simple to interpret, and can be evaluated at every frequency for both linear and nonlinear terms using appropriate coherence functions.

7. The Direct and Reverse MI/SO Techniques are superior to other nonlinear methods such as least-squares time-domain procedures or higher-order Volterra series that do not have these features.

Figure 3.1.

A zero-memory nonlinear system (Chapter 2) is a system that acts on an input $x(t)$ in a nonlinear fashion $g[x(t)]$ so as to produce an output $y(t)$ at the same instant of time. There is no weighting of past inputs to give the present output $y(t)$. A simple example is a square-law device where $y(t) = x^2(t)$. A finite-memory nonlinear system is a zero-memory nonlinear system that is followed or preceded by a linear system. In Figure 3.2, the zero-memory nonlinear system is followed by a linear system defined by a frequency response function $A(f)$ using A for "after." One could also have a zero-memory nonlinear system preceded by a linear system defined by a $B(f)$ using B for "before." It turns out that general results for many nonlinear physical problems can be obtained only for cases where the linear systems follow the zero-memory nonlinear operations, so these cases will be discussed here.

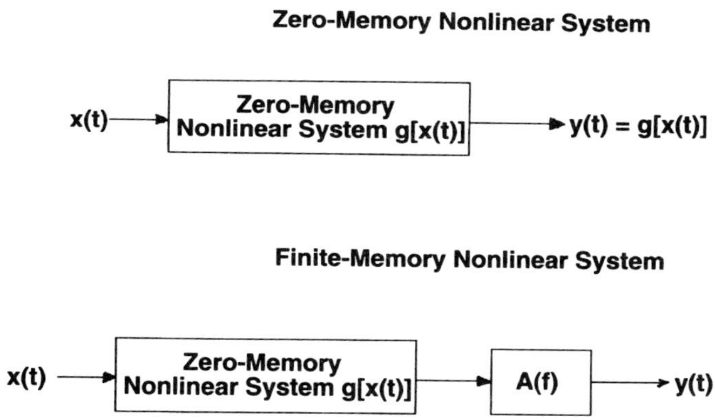

Figure 3.2.

A Volterra series (Chapter 8), also called a "power series with memory," is a functional extension of linear (first-order) systems to bilinear (second-order), trilinear (third-order), and higher-order systems. These extensions in Figure 3.3 require knowledge of multidimensional frequency response functions $H(f, g)$ to describe the bilinear system and $H(f, g, h)$ to describe the trilinear system in place of the simple one-dimensional frequency response function $H(f)$ for the linear system. Given the input excitation data and the output response data, general input/output formulas to identify the linear, bilinear, and trilinear frequency response functions are known only for Gaussian input data (Chapter 9). These higher-order functions are difficult to obtain in practice because they require large amounts of data and are complicated to interpret. Also, practical statistical error analysis criteria are not available to design experiments or to evaluate calculated results.

Volterra Series of Linear, Bilinear, and Trilinear Systems

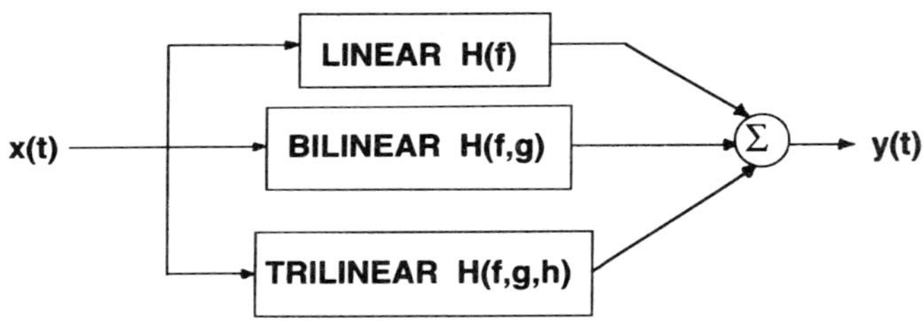

Figure 3.3.

A special Volterra series is shown in Figure 3.4 consisting of a linear system $A_1(f)$ in parallel with two nonlinear systems: a squarer followed by a different linear system $A_2(f)$, and a cuber followed by another linear system $A_3(f)$. The squarer path represents a special bilinear system and the cuber path represents a special trilinear system. The output $x_2(t)$ from the squarer system is $x^2(t)$ and the output $x_3(t)$ from the cuber system is $x^3(t)$. This model can be used as a third-order approximation for various types of nonlinear transformations such as $x(t)|x(t)|$ (Chapter 2). For non-Gaussian input data, the three outputs from the linear, squarer, and cuber paths will usually be correlated. For Gaussian input data, only the linear and cuber system outputs will be correlated.

Nonlinear System of Linear, Squarer, and Cuber Systems

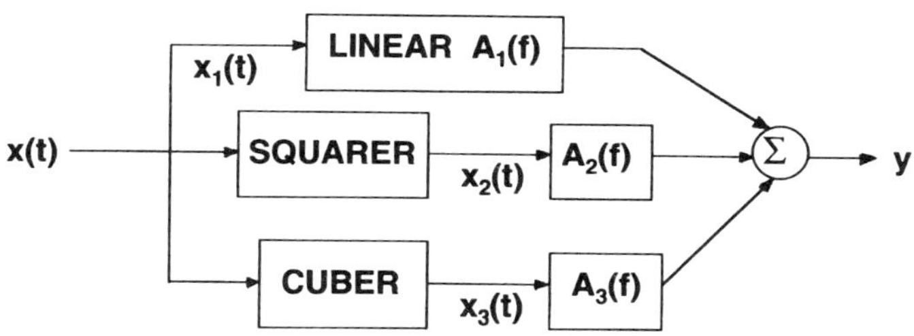

Figure 3.4.

A more general single-input/single-output (SI/SO) nonlinear system without feedback is shown in Figure 3.5, where in place of the squarer there is an arbitrary known or assumed zero-memory nonlinear system $g_2[x(t)]$ with output $x_2(t)$, and in place of the cuber there is an arbitrary known or assumed zero-memory nonlinear system $g_3[x(t)]$ with output $x_3(t)$. This nonlinear model in Figure 3.5 will fit many more physical nonlinear problems than the nonlinear model in Figure 3.4 and is the one recommended to use in practice. Given the input excitation data $x(t) = x_1(t)$ and the output response data $y(t)$, the system identification problem here is to identify the three linear frequency response functions $A_1(f)$, $A_2(f)$, and $A_3(f)$. This is easy to solve by well-established multiple-input/single-output (MI/SO) linear procedures if one recognizes that this general nonlinear model can be replaced by an equivalent three-input/single-output linear model where the three inputs are the known $x_1(t)$, $x_2(t)$, and $x_3(t)$ and the output is the known $y(t)$. ***This is the basis of the Direct MI/SO Technique for analysis and identification of nonlinear systems!***

General single-input/single-output nonlinear system

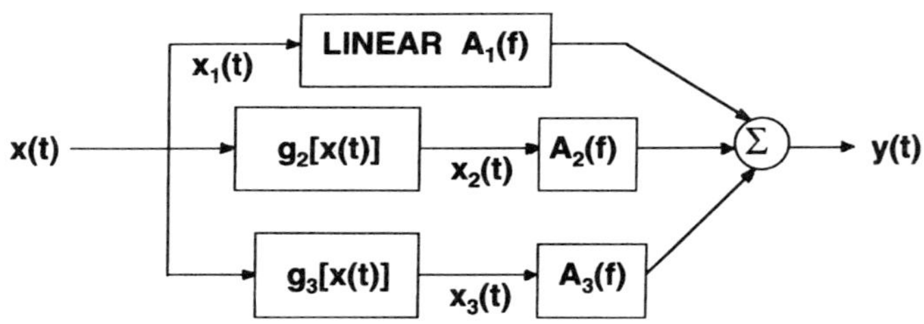

Figure 3.5.

Figure 3.6 shows the three-input/single-output linear system with possible correlated input records that is absolutely 100% equivalent to the general single-input/single-output nonlinear system in Figure 3.5. The identification of the three linear frequency response functions $\{A_i(f)\}$ requires one to replace the three correlated input records $\{x_i(t)\}$ in Figure 3.6 with three uncorrelated input records $\{u_i(t)\}$ as shown in Figure 3.7, where the notation "u" is used here to indicate "uncorrelated." Also, the three original linear frequency response functions in Figure 3.6 have to be replaced by new linear frequency response functions $\{L_i(f)\}$ as shown in Figure 3.7 so as to preserve the same output response data $y(t)$ and the same output noise data $n(t)$ in both Figures 3.6 and 3.7.

It is straightforward to compute the uncorrelated input records from the original input records. The first uncorrelated input record $u_1(t)$ in Figure 3.7 is the same as $x_1(t)$ in Figure 3.6. The second uncorrelated input record $u_2(t)$ is the result obtained by removing the linear effects of $x_1(t)$ from $x_2(t)$. The third uncorrelated input record $u_3(t)$ is the result obtained by removing the linear effects of both $x_1(t)$ and $x_2(t)$ from $x_3(t)$. The $\{L_i(f)\}$ functions can be computed by three separate simple single-input/single-output spectral calculations, after which it is easy to obtain the $\{A_i(f)\}$ functions by algebraic equations. These Direct MI/SO Techniques apply to data with arbitrary probability and spectral properties.

The coherence function between $u_1(t)$ and $y(t)$ in Figure 3.7 states the percentage of the output spectrum of $y(t)$ that is due to $u_1(t)$ passing through $L_1(f)$. Similarly, the coherence functions between $u_2(t)$ and $y(t)$ and between $u_3(t)$ and $y(t)$ state the percentage of the output spectrum of $y(t)$ that is due to $u_2(t)$ and $u_3(t)$ passing through $L_2(f)$ and $L_3(f)$, respectively. The sum of these three ordinary coherence functions is the cumulative coherence function for Figure 3.7. These results occur because these three inputs are uncorrelated.

All of these matters for MI/SO linear models like Figures 3.6 and 3.7 are developed fully in two books by Bendat and Piersol (references 1 and 2).

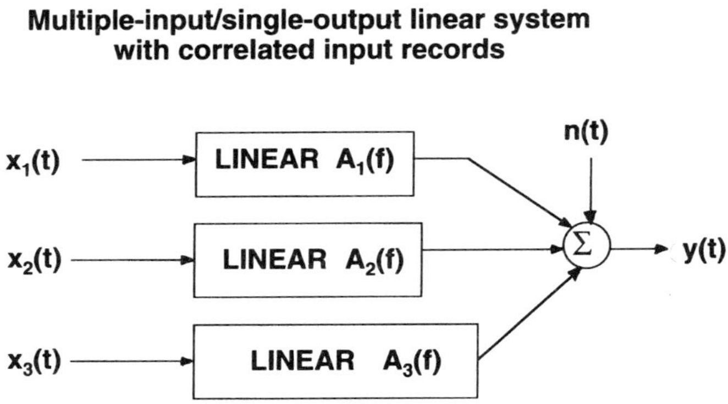

Figure 3.6.

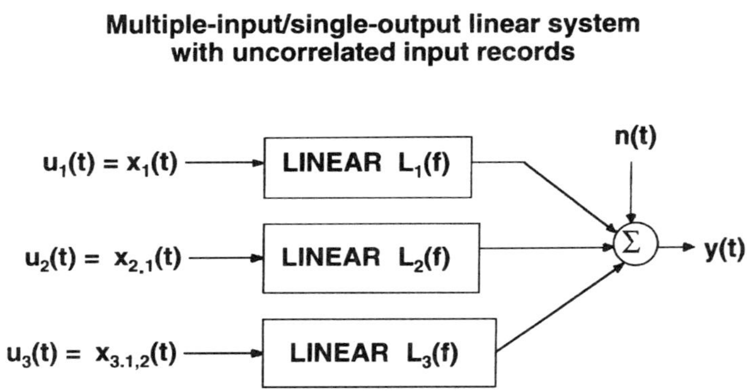

Figure 3.7.

Figure 3.8 shows the formula for a mechanical system defined by a linear SDOF differential equation of motion where the mass, damping, and stiffness coefficients are constants. Figure 3.9 shows the formula for a linear SDOF integrodifferential equation of motion where the damping coefficient is frequency-dependent. These equations are well-known and often assumed to govern many physical problems. The frequency-domain relations are always simpler than the associated time-domain relations because convolution integrals in the time domain can be replaced by product functions in the frequency domain. The particular example of a frequency-dependent damping coefficient shows that the constant damping coefficient is just a special case. Similarly, one can have frequency-dependent stiffness coefficients in place of constant stiffness coefficients. The analysis to follow identifies nonlinear system properties where the coefficients for the physical parameters can be constants or frequency-dependent.

SINGLE-INPUT/SINGLE-OUTPUT LINEAR SYSTEM WITH CONSTANT COEFFICIENTS

SDOF differential equation of motion with constant coefficients

$$m\ddot{y}(t) + c\dot{y}(t) + ky(t) = x(t)$$

where

$x(t)$ = excitation, $y(t)$ = response, m,c,k = constants

TIME-DOMAIN RELATION

$x(t) \rightarrow$ [$h(\tau)$] $\rightarrow$ $y(t) = x(t) * h(t) = \int h(\tau)\, x(t-\tau)\, d(\tau)$

$H(f)$ = Fourier transform of $h(\tau) = 1/[k - (2\pi f)^2 m + j(2\pi f)c]$

FREQUENCY-DOMAIN RELATION

$X(f) \rightarrow$ [$H(f)$] $\rightarrow$ $Y(f) = H(f)\, X(f)$

The frequency-domain relation is simpler than the time-domain relation.

Figure 3.8.

SINGLE-INPUT/SINGLE-OUTPUT LINEAR SYSTEM WITH FREQUENCY-DEPENDENT COEFFICIENT

SDOF integrodifferential equation of motion with a frequency-dependent damping coefficient is given by

$$m\ddot{y}(t) + \int c(\tau)\, \dot{y}(t-\tau)\, d\tau + ky(t) = x(t)$$

where

$x(t)$ = excitation, $y(t)$ = response, m, k = constants

$$H(f) = 1/[k - (2\pi f)^2 m + j(2\pi f)C(f)]$$

$c(\tau)$ = Inverse Fourier transform of $C(f)$ represents a physical parameter with memory. The constant damping case is just a special case where $C(f) = c$ and $c(\tau) = c\delta(\tau)$.

There is no requirement to restrict the analysis with a priori assumptions that the physical parameters are constants.

Figure 3.9.

Figure 3.10 shows equivalent "forward" and "reverse" linear systems that can represent single-input/single-output problems. In the reverse linear system, the mathematical input is the real measured physical response record, and the mathematical output is the real measured physical excitation record. The linear frequency response function $A_1(f)$ in the reverse system is the reciprocal of the linear frequency response function $H(f)$ in the forward system.

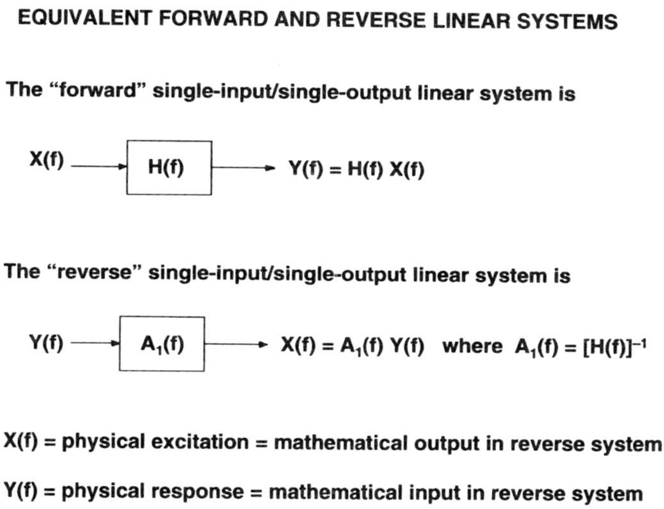

Figure 3.10.

Figure 3.11 shows the formula for a mechanical system defined by a nonlinear SDOF differential equation of motion where the nonlinear restoring force $p(u, \dot{u}, t)$ can be a function of u, $\dot{u}$, and t. Here, $F(t)$ is a physical force input and $u(t)$ is a physical displacement output. The total restoring force is defined by $F_r(u, \dot{u}, t)$. Two physical examples of nonlinear systems are given by

$$p(u, \dot{u}, t) = u^3(t), \qquad \text{a cubic stiffness force}$$

$$p(u, \dot{u}, t) = \dot{u}(t)|\dot{u}(t)|, \qquad \text{a quadratic viscous damping force}$$

SDOF NONLINEAR SYSTEM AND EQUATIONS

$$m\ddot{u}(t) + c\dot{u}(t) + ku(t) + p(u,\dot{u},t) = F(t)$$

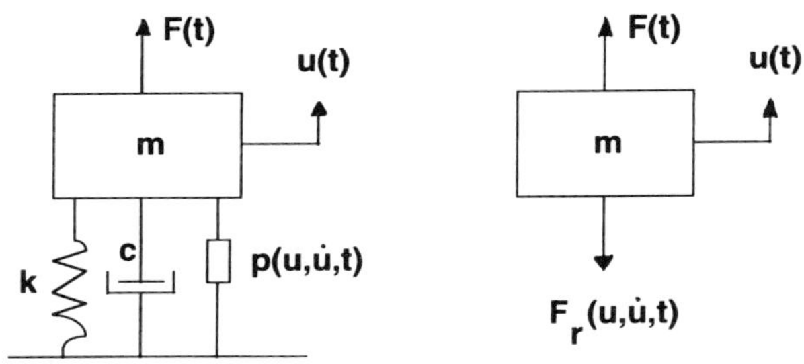

$$F_r(u,\dot{u},t) = F(t) - m\ddot{u}(t) = ku(t) + c\dot{u}(t) + p(u, \dot{u}, t)$$

Figure 3.11.

The single-input/single-output nonlinear model with feedback in Figure 3.12 is the same as the mechanical system shown in Figure 3.11. Such a nonlinear model with feedback is quite difficult to analyze for either of the following problems:

1. Prediction of the output response data from knowledge of the input force data and the linear and nonlinear system properties.
2. Identification of the linear and nonlinear system properties from knowledge of the input force data and output response data.

Complicated time-domain iterative procedures are usually employed that give restricted applications due to assumptions required on either the data or the systems. For any nonlinear system, the physical output $u(t)$ in Figure 3.12 will always be non-Gaussian independent of the probability nature of the physical input $F(t)$.

SINGLE-INPUT/SINGLE-OUTPUT NONLINEAR MODEL WITH FEEDBACK

$$m\ddot{u}(t) + c\dot{u}(t) + ku(t) + p(u, \dot{u}, t) = F(t)$$

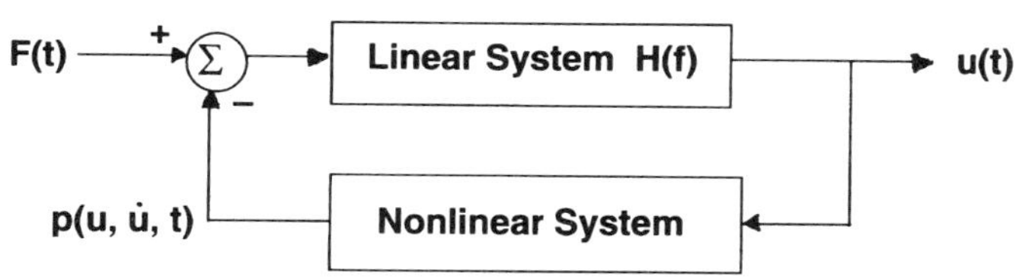

F(t) = force input

u(t) = displacement output

p(u, u̇, t) = nonlinear term

Figure 3.12.

The reverse single-input/single-output nonlinear model without feedback is illustrated in Figure 3.13 where the previous physical force input $F(t)$ is now considered to be the mathematical output, and the previous physical displacement output $u(t)$ is now considered to be the mathematical input. The same nonlinear SDOF differential equation of motion applies to this mathematical model as before. Now, however, it is simple to solve this model because the nonlinear system shown here can often be replaced by a general SI/SO nonlinear model such as Figure 3.5 and then solved by the standard MI/SO linear procedures outlined in Figures 3.6 and 3.7. ***This is the basis of the Reverse MI/SO Technique for analysis and identification of nonlinear systems!***

The mathematical output $F(t)$ in Figure 3.13 can theoretically be Gaussian if that is assumed for the original physical force input $F(t)$. There is no claim here that the non-Gaussian mathematical input $u(t)$ can produce a Gaussian output $F(t)$. This mathematical model is used only to determine the nature of nonlinear system properties for the physical parameters that can exist between the real physical input and the real physical output in the previous feedback model of Figure 3.12.

REVERSE SINGLE-INPUT/SINGLE-OUTPUT NONLINEAR MODEL WITHOUT FEEDBACK

$$m\ddot{u}(t) + c\dot{u}(t) + ku(t) + p(u, \dot{u}, t) = F(t)$$

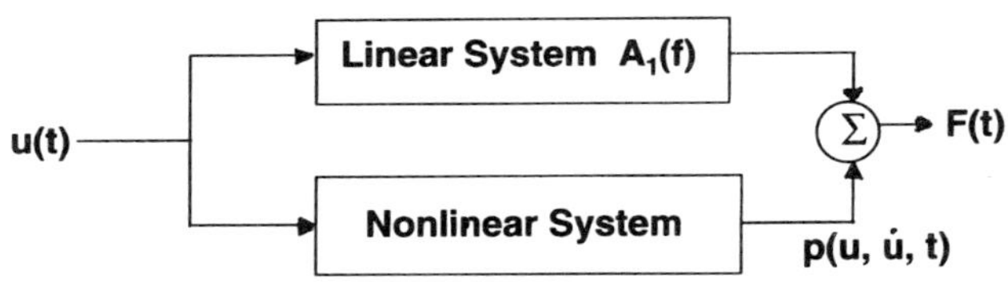

u(t) = displacement input

F(t) = force output

p(u, u̇, t) = nonlinear term

Figure 3.13.

The Reverse MI/SO Technique for nonlinear system identification is illustrated in Figure 3.14 for the Duffing single-input/single-output nonlinear system where the nonlinear restoring force is given by $p(u, \dot{u}, t) = k_3 u^3(t)$, when k_3 is a constant coefficient. This involves the reversal of the physical input $F(t)$ and the physical excitation $u(t)$ to create the equivalent two-input/single-output linear mathematical model shown, where $F(t)$ is now the mathematical output with $u(t)$ as the first mathematical input and $u^3(t)$ as the second mathematical input.

One should note here that the mathematical input $u(t)$ will always be non-Gaussian since it represents the physical response from a nonlinear system, and that the two mathematical inputs $u(t)$ and $u^3(t)$ will always be correlated. These two inputs should be replaced by the notation $x(t)$ and $x^3(t)$ when using the model shown in Figure 3.6 and should not be confused with the later computed uncorrelated inputs that would occur in Figure 3.7. From knowledge of $u(t)$ and $F(t)$, without any restrictions on their probability or spectral properties, it is simple to solve this two-input/single-output linear system to identify the two frequency response functions $A_1(f)$ and $A_2(f)$.

REVERSE MI/SO TECHNIQUE FOR NONLINEAR SYSTEM IDENTIFICATION

The Duffing single-input/single-output nonlinear system is

$$m\ddot{u}(t) + c\dot{u}(t) + ku(t) + k_3u^3(t) = F(t)$$

where

F(t) = physical excitation = mathematical output
u(t) = physical response = first mathematical input
u³(t) = nonlinear term = second mathematical input

The equivalent two-input/single-output linear system is here

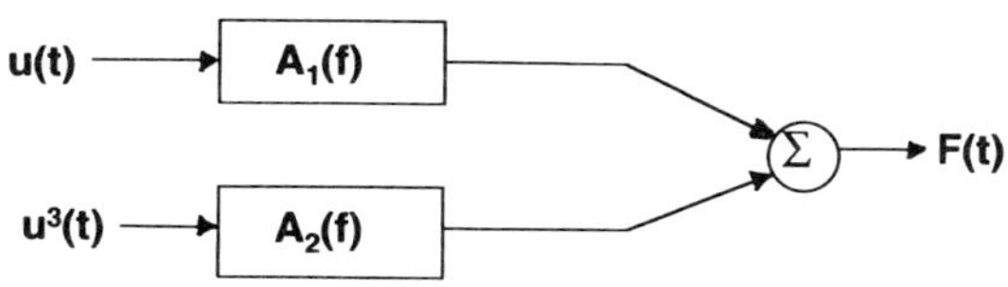

This Reverse MI/SO linear system can be solved easily.

Figure 3.14.

The $A_1(f)$ results from some computer simulation studies of a Duffing nonlinear system are shown in Figures 3.15 and 3.16. Spectral analysis of the data was performed here using only 16 averages. The Reverse SI/SO results (indicated by the dashed lines) are obtained by doing a standard single-input/single-output linear analysis between the input $u(t)$ and the output $F(t)$. This gives the wrong values for the magnitude and phase of $A_1(f)$, as shown on these two figures, as well as no information about $A_2(f)$. The Reverse MI/SO results (indicated by the solid lines) are obtained by doing the correct two-input/single-output linear analysis for this case. This gives the correct values for the magnitude and phase of $A_1(f)$, as shown on these two figures, as well as the correct value for $A_2(f)$, which is not shown. The differences between these wrong results from a standard SI/SO analysis and the correct results from Reverse MI/SO analysis are quite striking and show that conventional modal analysis is subject to serious errors when nonlinear effects occur. These strong differences have also been found in other computer simulation studies of various nonlinear systems with constant coefficients and with frequency-dependent coefficients.

Note that the magnitude of $A_1(f)$ in Figure 3.15 is the reciprocal of the magnitude of $H(f)$ in Figure 3.8 for a SDOF linear system. Hence, the magnitude of $A_1(f)$ displays a minimum value rather than a peak value at the undamped natural frequency $f_n = 3.0\,\text{Hz}$ here, and rises rather than falls for frequencies above f_n.

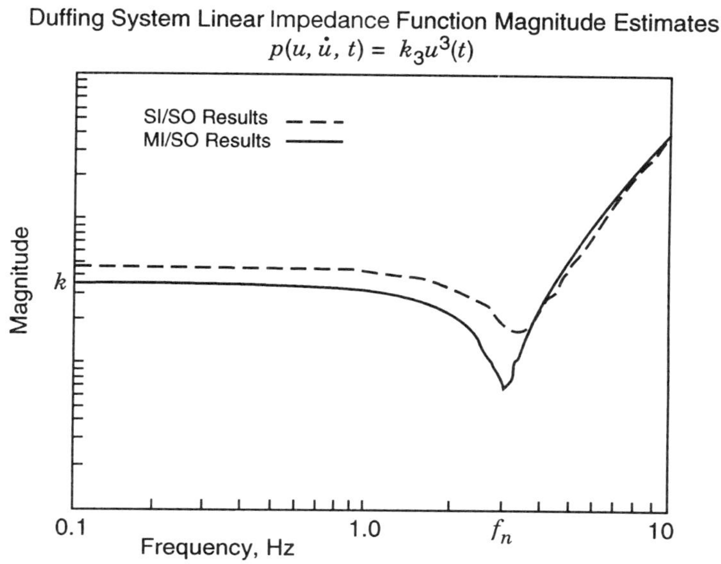

Figure 3.15.

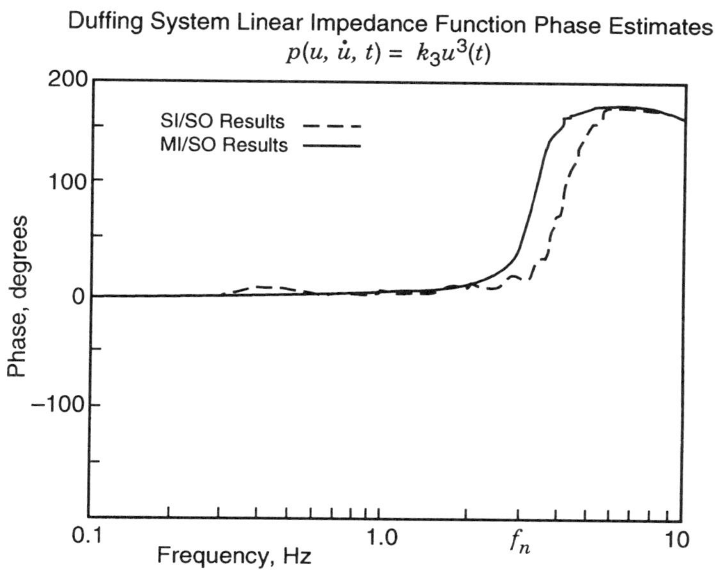

Figure 3.16.

Figure 3.17 shows the Duffing system cumulative coherence function estimates for the correct two-input/single-output reverse dynamics linear model. The lower curve gives the coherence function between the first uncorrelated input $u_1(t) = u(t)$ and the output $y(t) = F(t)$ via the path defined by $L_1(f)$. The upper curve adds on the coherence function between the second uncorrelated input $u_2(t)$ and the output $y(t)$ via the path defined by $L_2(f)$. The sum of these two coherence functions is the upper curve called the cumulative coherence function. These coherence functions provide practical quantitative results to predict exactly what happens at every frequency of interest. Neither of these curves represents the ordinary coherence function that would be obtained if a conventional single-input/single-output spectral analysis were conducted between the input $u(t)$ and the output $F(t)$ through the path defined by the frequency response function $A_1(f)$ in Figure 3.14 since $A_1(f) \neq L_1(f)$ for the Duffing system.

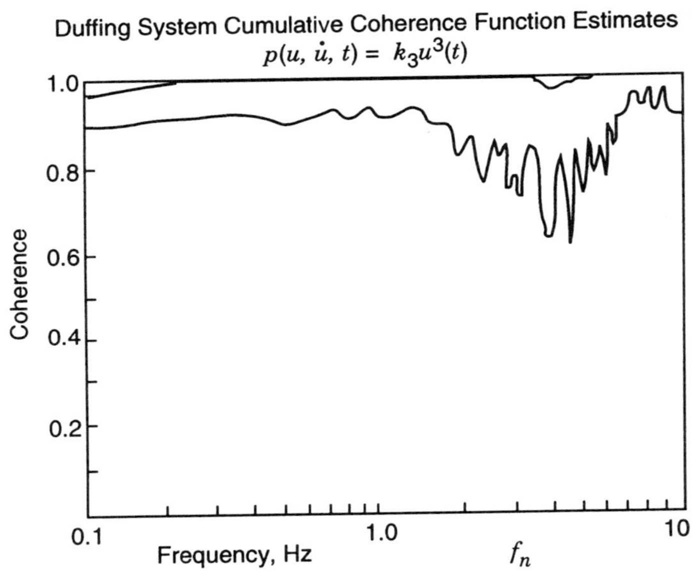

Figure 3.17.

Some of the main conclusions from computer studies with simulated data and from laboratory test programs with real data are listed on Figure 3.18. One should state them loudly because of their importance. These results prove that the Direct and Reverse MI/SO Techniques are two of the most significant recent developments in this field that can and should be applied by engineers and scientists to analyze and identify nonlinear systems. This concludes Chapter 3.

CONCLUSIONS FROM COMPUTER STUDIES AND FROM LABORATORY TEST PROGRAMS

1. Direct and Reverse MI/SO Techniques can be implemented using established procedures and computer programs by changing SI/SO nonlinear models into equivalent MI/SO linear models. The method does not require any assumptions regarding the probability functions or spectral shapes of input/output data.

2. These MI/SO techniques can be used with simulated or measured data to identify each linear and nonlinear term in proposed nonlinear integro-differential equations of motion.

3. Nonlinear system amplitude properties can be determined as well as the frequency properties of coefficients for the linear and nonlinear system physical parameters. The spectral contributions from each linear and nonlinear term can be evaluated using appropriate coherence functions.

4. Computed cumulative coherence functions with simulated or measured data give improved results over wider frequency ranges when nonlinear terms are included and show the particular frequencies where the nonlinear terms are important. The computed coefficients with measured data are usually frequency-dependent.

5. These features make the Direct and Reverse MI/SO Techniques the most accurate practical methods available today to analyze and identify the dynamic properties of nonlinear physical systems based on simulated or measured data.

Figure 3.18.

4

PARALLEL LINEAR AND NONLINEAR SYSTEMS

This chapter presents a general methodology for analyzing the spectral properties of stationary random data passing through SI/SO nonlinear systems with parallel linear and nonlinear systems. From measurements of input/output data, this methodology shows how to decompose the output spectral density function into its uncorrelated linear and nonlinear components. A large class of SI/SO nonlinear models are made to be equivalent to direct MI/SO linear models that are either two-input/single-output linear models or three-input/single-output linear models. Important special cases are linear systems in parallel with nonlinear square-law and cubic systems, called Case 1 and Case 2 nonlinear models. Formulas are stated on how to evaluate statistical errors in estimates. Engineering applications in this chapter include nonlinear system models for (1) wave forces on structures, (2) slowly varying drift forces, (3) an automative application, and (4) a biomedical study. The chapter concludes with connections between reverse MI/SO Techniques for nonlinear system analysis and identification from random data, and the determination of physical parameters in nonlinear differential equations of motion.

4.1 ANALYSIS METHODOLOGY

A general physical situation is pictured in Figure 4.1 of a single-input/single-output nonlinear model with parallel linear and nonlinear systems of arbitrary type where the outputs can be correlated. As shown, the total output $y(t)$ due to the input $x(t)$ is the sum of a linear output $y_h(t)$, a nonlinear output $y_v(t)$, and an extraneous noise term $n(t)$ that will be assumed uncorrelated with $y_h(t)$ and $y_v(t)$. The input data can be Gaussian or non-Gaussian. The nonlinear output $y_v(t)$ can here be any desired function of $x(t)$. Also, there is no

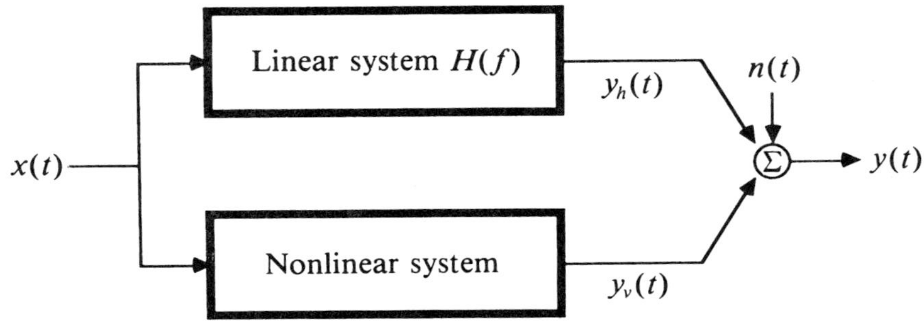

Figure 4.1. Single-input/single-output nonlinear model with parallel linear and nonlinear systems where outputs can be correlated.

requirement that $y_h(t)$ and $y_v(t)$ be uncorrelated. Thus,

$$y(t) = y_h(t) + y_v(t) + n(t) \tag{4.1}$$

and the output autospectral density function becomes

$$S_{yy}(f) = S_{y_h y_h}(f) + S_{y_v y_v}(f) + S_{y_h y_v}(f) + S^*_{y_h y_v}(f) + S_{nn}(f) \tag{4.2}$$

where, in general, the cross-spectrum term

$$S_{y_h y_v}(f) \neq 0 \tag{4.3}$$

From measurements only of $x(t)$ and $y(t)$, in practice, it is difficult to decompose $S_{yy}(f)$ into its linear and nonlinear components because $S_{y_h y_v}(f) \neq 0$. A recommended analysis methodology will now be outlined to help solve such decomposition problems by showing how to change this nonlinear model of Figure 4.1 so that $x(t)$ will give a different optimum linear output $y_o(t)$ and a revised uncorrelated nonlinear output $y_u(t)$ as pictured in Figure 4.2. Specifically, it is required that

$$y(t) = y_o(t) + y_u(t) + n(t) \tag{4.4}$$

with the cross-spectrum term

$$S_{y_o y_u}(f) = 0 \tag{4.5}$$

For this new model, it will be much simpler to decompose $S_{yy}(f)$ into its linear and nonlinear components. The desired $S_{yy}(f)$ will be

$$S_{yy}(f) = S_{y_o y_o}(f) + S_{y_u y_u}(f) + S_{nn}(f) \tag{4.6}$$

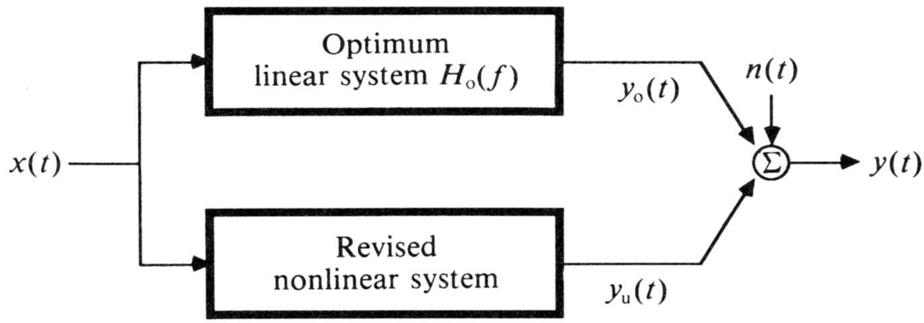

Figure 4.2. Nonlinear model with parallel optimum linear system and revised nonlinear system where outputs are uncorrelated.

Equations (4.1) and (4.4) have the same $y(t)$ and $n(t)$. Hence, it is required that the two models satisfy the time and Fourier transform equations

$$y_h(t) + y_v(t) = y_o(t) + y_u(t) \tag{4.7}$$

$$Y_h(f) + Y_v(f) = Y_o(f) + Y_u(f) \tag{4.8}$$

The optimum linear sytem $H_o(f)$ for Figure 4.2 may or may not be the same as the linear system $H(f)$ shown in Figure 4.1, depending on the correlation between $x(t)$ and $y_v(t)$. To be specific, by taking appropriate Fourier transforms of Eq. (4.1), one obtains

$$Y(f) = Y_h(f) + Y_v(f) + N(f) \tag{4.9}$$

where

$$Y_h(f) = H(f)X(f) \tag{4.10}$$

and $Y_v(f)$ is the Fourier transform of the nonlinear output $y_v(t)$. Now, assuming $S_{xn}(f) = 0$, the input/output cross-spectrum $S_{xy}(f)$ between $x(t)$ and the total output $y(t)$ is

$$S_{xy}(f) = S_{xy_h}(f) + S_{xy_v}(f) \tag{4.11}$$

Here, $S_{xy_h}(f)$ is the cross-spectrum between $x(t)$ and the linear output $y_h(t)$ given by

$$S_{xy_h}(f) = H(f)S_{xx}(f) \tag{4.12}$$

and $S_{xy_v}(f)$ is the cross-spectrum between $x(t)$ and the nonlinear output $y_v(t)$.

The optimum linear system $H_o(f)$ is known from the usual formula

$$H_o(f) = \frac{S_{xy}(f)}{S_{xx}(f)} \tag{4.13}$$

Hence, Eqs. (4.11)–(4.13) prove that

$$H_o(f) = H(f) + \frac{S_{xy_v}(f)}{S_{xx}(f)} \tag{4.14}$$

Thus, $H_o(f) = H(f)$ if and only if $S_{xy_v}(f) = 0$. In general, $S_{xy_v}(f) \neq 0$ so that $H_o(f) \neq H(f)$.

If $H(f)$ is replaced by $H_o(f)$ as shown in Figure 4.2, then the output $Y(f)$ can be separated into a new optimum linear component $Y_o(f)$ and a revised nonlinear component $Y_u(f)$ where Fourier transforms of Eq. (4.4) give

$$Y(f) = Y_o(f) + Y_u(f) + N(f) \tag{4.15}$$

with

$$Y_o(f) = H_o(f)X(f) \tag{4.16}$$

From Eq. (4.8), $Y_u(f)$ must satisfy

$$Y_u(f) = Y_h(f) + Y_v(f) - Y_o(f) \tag{4.17}$$

Hence, from Eqs. (4.10), (4.14), and (4.16),

$$Y_o(f) = Y_h(f) + \left[\frac{S_{xy_v}(f)}{S_{xx}(f)}\right] X(f) \tag{4.18}$$

and from Eq. (4.17),

$$Y_u(f) = Y_v(f) - \left[\frac{S_{xy_v}(f)}{S_{xx}(f)}\right] X(f) \tag{4.19}$$

It follows immediately that

$$S_{xy_u}(f) = \frac{1}{T} E[X^*(f)Y_u(f)] = 0 \tag{4.20}$$

where expected value operations are over an ensemble of records of length T. Thus, the input $x(t)$ is uncorrelated with the output $y_u(t)$ in Figure 4.2.

Equation (4.16) then shows that

$$S_{y_o y_u} = \frac{1}{T} E[Y_o^*(f)Y_u(f)] = H_o^*(f)S_{xy_u}(f) = 0 \tag{4.21}$$

This proves that the outputs $y_o(t)$ and $y_u(t)$ in Figure 4.2 are uncorrelated as stated in Eq. (4.5). From Eq. (4.15), one now obtains the desired result of Eq. (4.6), namely,

$$S_{yy}(f) = S_{y_o y_o}(f) + S_{y_u y_u}(f) + S_{nn}(f) \tag{4.22}$$

with

$$S_{y_o y_o}(f) = |H_o(f)|^2 S_{xx}(f) = \frac{|S_{xy}(f)|^2}{S_{xx}(f)} \tag{4.23}$$

The formula for $S_{y_u y_u}(f)$ depends on the particular type of nonlinear operations that may be involved. Equation (4.19) shows that $S_{y_u y_u}(f)$ is a function of $S_{y_v y_v}(f)$ and $S_{xy_v}(f)$ using the $y_v(t)$ of Figure 4.1. Specifically,

$$S_{y_u y_u}(f) = \frac{1}{T} E[Y_u^*(f)Y_u(f)] = S_{y_v y_v}(f) - \frac{|S_{xy_v}(f)|^2}{S_{xx}(f)} \tag{4.24}$$

Note that $S_{y_u y_u}(f) = S_{y_v y_v}(f)$ if and only if $S_{xy_v}(f) = 0$. In general, $S_{xy_v}(f) \neq 0$ so that $S_{y_u y_u}(f) \neq S_{y_v y_v}(f)$.

The ordinary linear coherence function $\gamma_{xy}^2(f)$ from the optimum linear output $y_o(t)$ is

$$\gamma_{xy}^2(f) = \frac{S_{y_o y_o}(f)}{S_{yy}(f)} = \frac{|S_{xy}(f)|^2}{S_{xx}(f)S_{yy}(f)} \tag{4.25}$$

The associated nonlinear coherence function $q_{xy}^2(f)$ from the uncorrelated nonlinear output $y_u(t)$ is

$$q_{xy}^2(f) = \frac{S_{y_u y_u}(f)}{S_{yy}(f)} \tag{4.26}$$

where $S_{y_u y_u}(f)$ is computed by Eq. (4.24). In terms of these coherence functions, the output noise spectrum $S_{nn}(f)$ becomes

$$S_{nn}(f) = [1 - \gamma_{xy}^2(f) - q_{xy}^2(f)]S_{yy}(f) \tag{4.27}$$

This shows the importance of obtaining $q_{xy}^2(f)$ to reduce $S_{nn}(f)$. For good models, $S_{nn}(f)$ should be small compared to $S_{yy}(f)$ at selected frequencies of interest. These frequencies are where the peak values occur in the output spectrum of $S_{yy}(f)$.

4.2 TWO-INPUT/SINGLE-OUTPUT LINEAR MODELS

Suppose the nonlinear system in Figure 4.1 is a general finite-memory nonlinear system consisting of a zero-memory nonlinear system with output $v(t)$, which is followed by a linear system $A(f)$ with output $y_v(t)$. Figure 4.1 then becomes Figure 4.3, where the linear output $y_h(t)$ and the nonlinear output $y_v(t)$ can be correlated. By applying the methodology in Section 4.1, Figure 4.3 can be replaced by Figure 4.4 consisting of a revised zero-memory nonlinear system with uncorrelated output $u(t)$, which is followed by the same linear system $A(f)$ as before but with output $y_{\mathrm{u}}(t)$. The previous linear system $H(f)$ is replaced by the optimum linear system $H_{\mathrm{o}}(f)$. Now, in Figure 4.4, the linear output $y_{\mathrm{o}}(t)$ and the nonlinear output $y_{\mathrm{u}}(t)$ are uncorrelated.

Relations between quantities shown in Figures 4.3 and 4.4 can be stated as special cases of results in Section 4.1. These relations can also be derived independently and include the following formulas:

$$C(f) = [S_{xy_v}(f)/S_{xx}(f)] = [S_{xv}(f)/S_{xx}(f)]A(f) \tag{4.28}$$

$$H_{\mathrm{o}}(f) = [S_{xy}(f)/S_{xx}(f)] = H(f) + C(f) \tag{4.29}$$

$$Y_{\mathrm{o}}(f) = Y_h(f) + C(f)X(f) \tag{4.30}$$

$$Y_{\mathrm{u}}(f) = Y_v(f) - C(f)X(f) \tag{4.31}$$

$$Y_{\mathrm{o}}(f) = H_{\mathrm{o}}(f)X(f), \quad Y_h(f) = H(f)X(f) \tag{4.32}$$

$$Y_{\mathrm{u}}(f) = A(f)U(f), \quad Y_v(f) = A(f)V(f) \tag{4.33}$$

$$U(f) = V(f) - [S_{xv}(f)/S_{xx}(f)]X(f) \tag{4.34}$$

$$S_{x\mathrm{u}}(f) = 0 \tag{4.35}$$

$$S_{\mathrm{uu}}(f) = S_{vv}(f)[1 - \gamma_{xv}^2(f)] \tag{4.36}$$

$$S_{\mathrm{u}y}(f) = S_{vy}(f) - [S_{vx}(f)/S_{xx}(f)]S_{xy}(f) \tag{4.37}$$

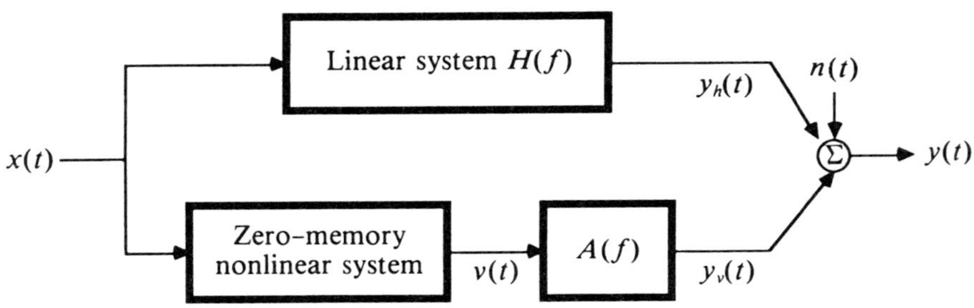

Figure 4.3. Single-input/single-output nonlinear model with parallel linear and zero-memory nonlinear systems where outputs can be correlated.

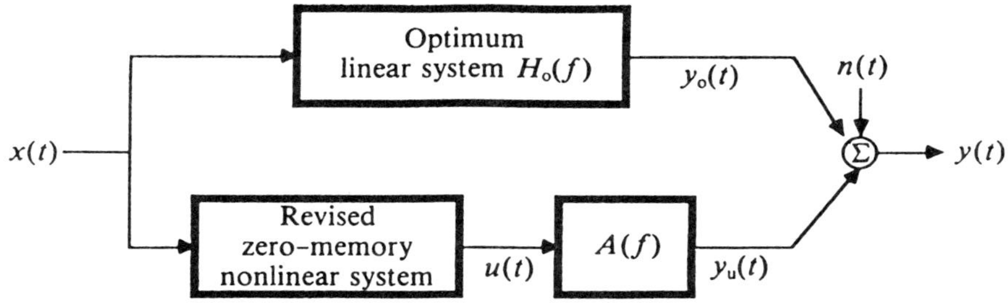

Figure 4.4. Nonlinear model with parallel optimum linear system and revised zero-memory nonlinear system where outputs are uncorrelated.

For the single-input/single-output nonlinear models of Figures 4.3 and 4.4, assume that:

1. $x(t)$ and $y(t)$ can be measured, and
2. $v(t)$ can be calculated from $x(t)$.

The single-input/single-output nonlinear models of Figures 4.3 and 4.4 can then be replaced by the equivalent two-input/single-output linear models of Figures 4.5 and 4.6, respectively. From knowledge of $x(t)$, $v(t)$, and $y(t)$, one can compute their individual autospectral density functions as well as their various cross-spectral density functions. One can then compute $S_{uu}(f)$ and $S_{uy}(f)$ from Eqs. (4.36) and (4.37), respectively.

It is straightforward to compute various desired quantities in Figures 4.5 and 4.6. When $C(f) \neq 0$, Eqs. (4.28) and (4.29) show that $H_o(f) \neq H(f)$ and is dependent on the particular input excitation. Thus, assuming a conventional

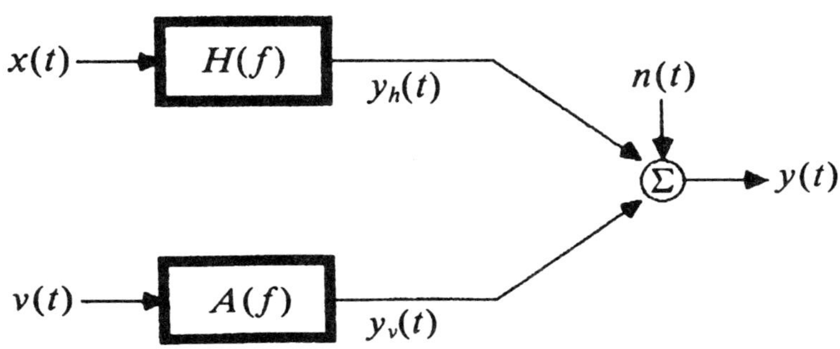

Figure 4.5. Two-input/single-output linear model equivalent to Figure 4.3 where inputs can be correlated.

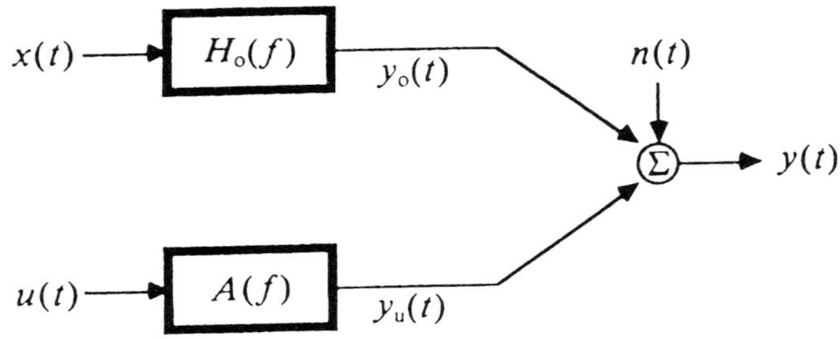

Figure 4.6. Two-input/single-output linear model equivalent to Figure 4.4 where inputs are uncorrelated.

single-input/single-output linear model in place of the correct nonlinear model of Figure 4.1 and computing only $H_o(f)$ leads to an erroneous estimate of $H(f)$ and no estimate of $A(f)$. Using Fourier transforms, the two correlated inputs in Figure 4.5 are $X_1(f) = X(f)$ and $X_2(f) = V(f)$. The two uncorrelated inputs in Figure 4.6 are $U_1(f) = X(f)$ and $U_2(f) = U(f)$, where $U(f)$ is computed by Eq. (4.34).

Some important equations for Figures 4.3–4.6 are

$$S_{yy}(f) = S_{y_o y_o}(f) + S_{y_u y_u}(f) + S_{nn}(f) \tag{4.38}$$

$$H_o(f) = \frac{S_{xy}(f)}{S_{xx}(f)} = |H_o(f)|e^{-j\phi_o(f)} \tag{4.39}$$

$$S_{y_o y_o}(f) = |H_o(f)|^2 S_{xx}(f) = \gamma_{xy}^2(f) S_{yy}(f) \tag{4.40}$$

$$A(f) = \frac{S_{uy}(f)}{S_{uu}(f)} = |A(f)|e^{-j\phi_a(f)} \tag{4.41}$$

$$S_{y_u y_u}(f) = |A(f)|^2 S_{uu}(f) = q_{xy}^2(f) S_{yy}(f) \tag{4.42}$$

$$H(f) = H_o(f) - [S_{xy}(f)/S_{xx}(f)]A(f) \tag{4.43}$$

$$S_{y_h y_h}(f) = |H(f)|^2 S_{xx}(f) \tag{4.44}$$

$$S_{y_v y_v}(f) = |A(f)|^2 S_{vv}(f) \tag{4.45}$$

$$S_{y_h y_v}(f) = H^*(f) A(f) S_{xv}(f) \tag{4.46}$$

The meaning of these equations is clear. Equation (4.40) gives the spectral density function for the optimum linear output term $y_o(t)$; Eq. (4.42) gives the spectral density function for the uncorrelated nonlinear output term $y_u(t)$. Equations (4.44) and (4.45) give the spectral density functions for the correlated linear and nonlinear output terms $y_h(t)$ and $y_v(t)$, respectively. The frequency response functions $H_o(f)$, $A(f)$, and $H(f)$ are computed by Eqs. (4.39), (4.41), and (4.43), respectively. Linear and nonlinear coherence functions are defined by Eqs. (4.40) and (4.42). Note that only first-order autospectral and cross-spectral density functions are required in these equations, and the results apply to Gaussian or non-Gaussian data.

There are two main problems of concern in all applications that involve parallel linear and nonlinear systems like Figure 4.3.

1. *Response Prediction Problem.* Given the frequency response functions $H(f)$ and $A(f)$, plus measurement only of the input record $x(t)$, assuming $v(t)$ can be calculated from $x(t)$, determine the spectral properties of the linear response record $y_h(t)$ and the nonlinear response record $y_v(t)$. This problem is solved for Figure 4.3 using Eqs. (4.44) and (4.45).
2. *System Identification Problem.* From simultaneous measurements of both the input record $x(t)$ and the total output record $y(t)$, assuming $v(t)$ can be calculated from $x(t)$, identify the optimum frequency function properties of both $H(f)$ and $A(f)$ to minimize the autospectrum of the extraneous output noise $n(t)$. This problem is solved for Figure 4.3 using Eqs. (4.43) and (4.41).

4.3 STATISTICAL ERRORS IN ESTIMATES

Statistical normalized random error and bias error formulas will now be stated for estimates of quantities in Figures 4.4 and 4.6. These formulas are derived in reference 1 and summarized in reference 2. The basis for these results is that $x(t)$ and $u(t)$ are uncorrelated inputs to $H_o(f)$ and $A(f)$, respectively. For these error formulas to be strictly valid, both $x(t)$ and $u(t)$ should be Gaussian. This can often be assumed for the input data $x(t)$ but will not be true for $u(t)$, since $u(t)$ comes from a nonlinear operation on $x(t)$. Thus, error formulas for $A(f)$, $S_{y_u y_u}(f)$, and $q_{xy}^2(f)$ should be regarded only as guidelines.

The normalized random errors for the two output spectra estimates $S_{y_o y_o}(f)$ and $S_{y_u y_u}(f)$ are approximated by

$$\varepsilon[\hat{S}_{y_o y_o}(f)] \approx \frac{\{2[1 - \hat{q}_{xy}^2(f)] - \hat{\gamma}_{xy}^2(f)\}^{1/2}}{|\hat{\gamma}_{xy}(f)|\sqrt{n_d}} \tag{4.47}$$

$$\varepsilon[\hat{S}_{y_u y_u}(f)] \approx \frac{\{2[1 - \hat{\gamma}_{xy}^2(f)] - \hat{q}_{xy}^2(f)\}^{1/2}}{|\hat{q}_{xy}(f)|\sqrt{n_d}} \tag{4.48}$$

where n_d is the number of independent averages used to compute the original spectral density quantities. Required values of n_d to achieve desired values of $\varepsilon[\hat{S}_{y_o y_o}(f)]$ or $\varepsilon[\hat{S}_{y_u y_u}(f)]$ are obtained by solving these equations for n_d. Bias errors for these spectra estimates will be minimized when there is good bandwidth resolution.

The normalized random errors for the two coherence function estimates $\hat{\gamma}_{xy}^2(f)$ and $\hat{q}_{xy}^2(f)$ are approximated by

$$\varepsilon[\hat{\gamma}_{xy}^2(f)] \approx \frac{\sqrt{2}\{[1 - 2\hat{\gamma}_{xy}^2(f)][1 - \hat{q}_{xy}^2(f)] + \hat{\gamma}_{xy}^4(f)\}^{1/2}}{|\hat{\gamma}_{xy}(f)|\sqrt{n_d}} \tag{4.49}$$

$$\varepsilon[\hat{q}_{xy}^2(f)] \approx \frac{\sqrt{2}\{[1 - 2\hat{q}_{xy}^2(f)][1 - \hat{\gamma}_{xy}^2(f)] + \hat{q}_{xy}^4(f)\}^{1/2}}{|\hat{q}_{xy}(f)|\sqrt{n_d}} \tag{4.50}$$

Bias errors for these coherence estimates are

$$b[\hat{\gamma}_{xy}^2(f)] \approx \frac{[1 - \hat{\gamma}_{xy}^2(f)]^2}{n_d} \tag{4.51}$$

$$b[\hat{q}_{xy}^2(f)] \approx \frac{[1 - \hat{q}_{xy}^2(f)]^2}{n_d - 1} \tag{4.52}$$

Results in Eqs. (4.47)–(4.52) apply to Figure 4.2 as well as to Figures 4.4 and 4.6 when $\hat{\gamma}_{xy}^2(f)$ and $\hat{q}_{xy}^2(f)$ can be computed.

Normalized random errors for the frequency response function estimates $\hat{H}_o(f)$ and $\hat{A}(f)$ are required for both gain factor and phase factor estimates. For gain factor estimates $|\hat{H}_o(f)|$ and $|A(f)|$, the normalized random errors are approximated by

$$\varepsilon[|\hat{H}_o(f)|] \approx \frac{[1 - \hat{\gamma}_{xy}^2(f)]^{1/2}}{|\hat{\gamma}_{xy}(f)|\sqrt{2n_d}} \tag{4.53}$$

$$\varepsilon[|\hat{A}(f)|] \approx \frac{[1 - \hat{\gamma}_{xy}^2(f) - \hat{q}_{xy}^2(f)]^{1/2}}{|\hat{q}_{xy}(f)|\sqrt{2(n_d - 1)}} \tag{4.54}$$

The standard deviations (in radians) of the two associated phase factor estimates $\hat{\phi}_o(f)$ and $\hat{\phi}_a(f)$ (in radians) are approximated by

$$\sigma[\hat{\phi}_o(f)] \approx \sin^{-1}\{\varepsilon[|\hat{H}_o(f)|]\} \tag{4.55}$$

$$\sigma[\hat{\phi}_a(f)] \approx \sin^{-1}\{\varepsilon[|\hat{A}(f)|]\} \tag{4.56}$$

For small values of $\varepsilon[|\hat{H}_o|]$ and $\varepsilon[|\hat{A}|]$, these equations can further be simplified to

$$\sigma[\hat{\phi}_o(f)] \approx \varepsilon[|\hat{H}_o(f)|] \tag{4.57}$$

$$\sigma[\hat{\phi}_a(f)] \approx \varepsilon[|\hat{A}(f)|] \tag{4.58}$$

Various bias errors in frequency response function estimates that should be minimized as much as possible in practice are also covered in reference 1.

These random error formulas from Section 4.3 are suitable for estimates from nonlinear models of the engineering applications in Sections 4.4 to 4.8. The same random error formulas apply to theoretical two-sided spectral quantities involving $S(f)$ functions and to measurable one-sided spectral quantities involving $G(f)$ functions as listed in Table 1.2. The sensitivity of the wave force results in Section 4.4 to the Gaussian assumption on input data is discussed in reference 4.

Example of Experimental Design. Suppose that stationary random data $x(t)$ and $y(t)$ are to be measured for the nonlinear model of Figure 4.2 or 4.4. Suppose that the desired bandwidth resolution is $\Delta f = 0.10$ Hz and that the desired normalized random errors in estimates of $\hat{G}_{y_o y_o}(f)$ and $\hat{G}_{y_u y_u}(f)$ are 10% each. What should be the total record length T_{total} for the experiment when the expected coherence function estimates are $\hat{\gamma}_{xy}^2(f) = 0.60$ and $\hat{q}_{xy}^2(f) = 0.20$?

The total record length $T_{\text{total}} = n_d T$ where $T = (1/\Delta f)$ and n_d is the number of distinct subrecords of length T each. From $\Delta f = 0.10$ Hz, it follows that each subrecord should be of length

$$T = (1/\Delta f) = 10 \text{ s}$$

From Eq. (4.47), the solution for n_d to estimate $G_{y_o y_o}(f)$ is

$$n_d \approx \frac{\{2[1 - \hat{q}_{xy}^2(f)] - \hat{\gamma}_{xy}^2(f)\}}{\hat{\gamma}_{xy}^2(f)\varepsilon^2[\hat{G}_{y_o y_o}(f)]}$$

For $\hat{\gamma}_{xy}^2(f) = 0.60$, $q_{xy}^2(f) = 0.20$, and $\varepsilon = 0.10$, this gives

$$n_d \approx 167$$

Hence, for estimates of $G_{y_o y_o}$, the experiment should be designed so that the total record length is

$$T_{\text{total}} = n_d T \approx 28 \text{ min}$$

From Eq. (4.48), the resolution for n_d to estimate $G_{y_u y_u}(f)$ is

$$n_d \approx \frac{\{2[1 - \hat{\gamma}_{xy}^2(f)] - \hat{q}_{xy}^2(f)\}}{\hat{q}_{xy}^2(f)\varepsilon^2[\hat{G}_{y_u y_u}(f)]}$$

For $\gamma_{xy}^2(f) = 0.60$, $\hat{q}_{xy}^2(f) = 0.20$, and $\varepsilon = 0.10$, this gives

$$n_d \approx 300$$

Hence, for estimates of $G_{y_u y_u}(f)$, the experiment should be designed so that the total record length is

$$T_{\text{total}} = n_{\text{d}} T \approx 50 \text{ min}$$

This numerical example indicates the nature of increased numbers of averages and longer total record lengths to obtain equally valid estimates of nonlinear output spectra compared to linear output spectra for situations where $\hat{q}^2_{xy}(f) < \hat{\gamma}^2_{xy}(f)$. Similar requirements apply to other parallel linear and nonlinear systems where the nonlinear coherence functions are smaller than the linear coherence functions.

4.4 NONLINEAR WAVE FORCE MODELS

Two ocean engineering applications will now be discussed that involve Direct MI/SO Techniques for analyzing parallel linear and nonlinear systems. These applications concern nonlinear models for the following:

1. Wave forces on structures.
2. Slowly varying drift forces on structures.

Oceanographic research and development work consider systems such as vertical piles, ships, or offshore oil and gas platforms. Through changes in terminology and interpretation, similar models and procedures are applicable to the analysis of linear and nonlinear properties for electrical, mechanical, and propulsion systems in airplanes, aerospace vehicles, automotive vehicles, power plants, bridges, tall buildings, and other engineering fields.

The general problem to be analyzed for wave forces on small-diameter fixed structures is shown in Figure 4.7. The structure is assumed to be subject to stationary random ocean waves represented by the wave velocity input $x(t)$ at a specified depth. This input record produces a wave force output at the same depth denoted by the output record $y(t)$.

Morison's equation is widely used to study such wave forces on structures (references 4, 38, 51, and 58) where the output wave force $y(t)$ due to an input

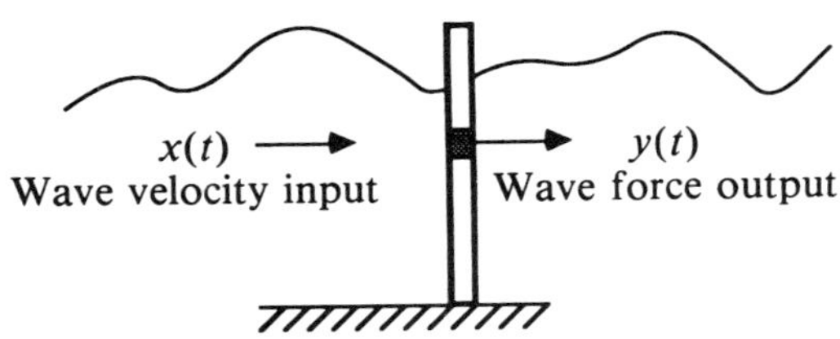

Figure 4.7. Illustration of wave force problem.

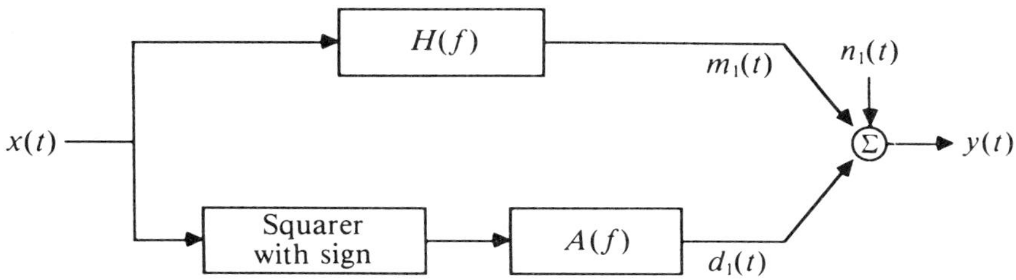

Figure 4.8. Case 1 nonlinear wave force model with parallel linear and nonlinear systems.

wave velocity $x(t)$ consists of two main components:

1. A *linear* inertial force $m(t)$ proportional to the derivative $\dot{x}(t)$ of $x(t)$.
2. A *nonlinear* drag force $d(t)$ proportional to $x(t)|x(t)|$, a squarer with sign.

Morison's equation is thus of the form

$$y(t) = m(t) + d(t) = C_1\dot{x}(t) + C_2x(t)|x(t)| \tag{4.59}$$

where the coefficients C_1 and C_2 are assumed to be constants, independent of frequency. Experimental work shows that these coefficients can vary with frequency so that Morison's equation needs to be extended.

Generalizations of Morison's equation are drawn in the Case 1 and Case 2 nonlinear wave force models of Figures 4.8 and 4.9, respectively, where C_1 and C_2 are replaced by linear frequency response functions $H(f)$ and $A(f)$ or $H(f)$ and $B(f)$. The problem of concern is to determine the spectral properties of the linear inertial component $m(t)$ and the nonlinear drag component $d(t)$ from measurements of the input wave velocity $x(t)$ and the total wave force output

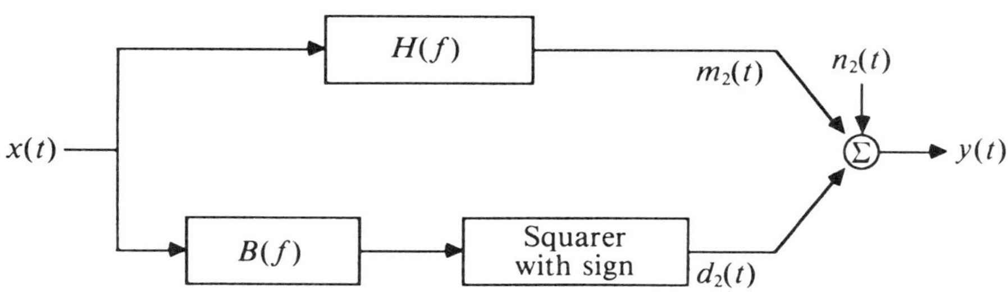

Figure 4.9. Case 2 nonlinear wave force model with parallel linear and nonlinear systems.

$y(t)$. This problem is difficult for two main reasons:

1. $d(t)$ has a nonlinear relationship involving a squarer with sign.
2. The parallel outputs $m(t)$ and $d(t)$ are correlated.

Note also that Figures 4.8 and 4.9 show different possible values for outputs $m_1(t)$ versus $m_2(t)$ and $d_1(t)$ versus $d_2(t)$, as well as for the noise terms $n_1(t)$ versus $n_2(t)$, so that it is not obvious in advance which of these models will best fit the measured data.

Formulas to analyze the Case 1 nonlinear wave force model of Figure 4.8 for general non-Gaussian input data with arbitrary spectral properties are stated in the previous Section 4.2. To obtain desired results, it is sufficient merely to recognize the appropriate equivalent terms in Figures 4.3 and 4.8 as follows:

$$v(t) = \text{output from squarer with sign}$$

$$y_h(t) = m_1(t) \quad \text{and} \quad y_v(t) = d_1(t)$$

The Response Prediction Problem and the System Identification Problem can both be solved for the Case 1 nonlinear wave force problem by using the general equations listed in Section 4.2.

For Gaussian input data, the two models in Figures 4.8 and 4.9 were studied extensively in past work (references 4 and 58), where the squarer with sign, $x|x|$, was replaced by its third-order polynomial least-squares approximation of Eq. (2.196). This gives a nonlinear system where a linear operation is in parallel with a cubic operation. Ocean engineering experiments were performed at the Netherlands Ship Model Basin in The Hague, where the data were analyzed using formulas based on Gaussian ideas. This experimental work showed that the Case 1 nonlinear wave force model of Figure 4.8 is superior to the Case 2 nonlinear wave force model of Figure 4.9 in producing smaller spectral values for the extraneous output noise. This is fortunate because the analysis required for the Case 1 nonlinear model is much simpler than the analysis required for the Case 2 nonlinear model.

4.5 NONLINEAR DRIFT FORCE MODELS

The general problem to be analyzed for slowly varying drift forces on structures is shown in Figure 4.10. Here, $x(t)$ and $y(t)$ have different interpretations than the $x(t)$ and $y(t)$ in Figure 4.7. The input $x(t)$ represents in Figure 4.10 the wave elevation (height) rather than the wave velocity of Figure 4.7. This input record produces a force on the ship that causes the ship to move. The output $y(t)$ represents in Figure 4.10 the ship motion output rather than the wave force output of Figure 4.7.

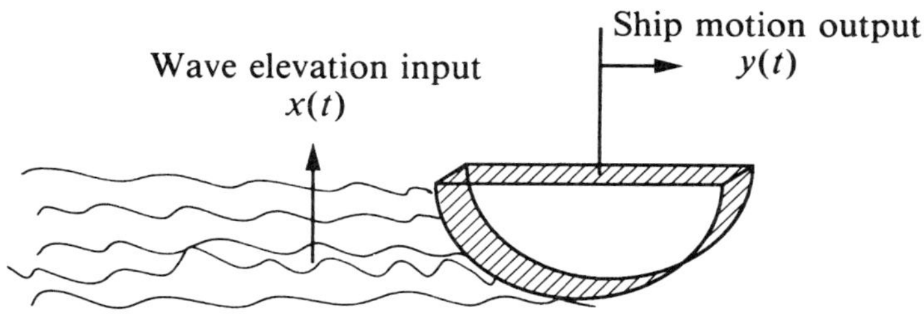

Figure 4.10. Illustration of drift force problem.

In Figure 4.10, the force acting on the ship is assumed to consist of two components:

1. A *linear* term proportional to $x(t)$.
2. A *nonlinear* term proportional to the squared envelope signal $u(t)$ of $x(t)$.

This nonlinear term is called the *slowly varying drift force*. The resulting ship motion output $y(t)$ can be defined by

$$y(t) = k_1 x(t) + k_2 u(t) \tag{4.60}$$

where the coefficients k_1 and k_2 are assumed to be constants, independent of frequency. The squared envelope signal $u(t)$ is a zero-memory nonlinear function of $x(t)$ where frequencies in $u(t)$ are lower than frequencies in $x(t)$ so models like Figure 2.33 are not appropriate.

A generalization of Eq. (4.60) is drawn in Figure 4.11, where the coefficients k_1 and k_2 are replaced by linear frequency response functions $H_1(f)$ and $H_2(f)$. The problem of concern is to determine the spectral properties of the linear component $y_1(t)$ and the nonlinear component $y_2(t)$ from measurements of the input wave elevation input $x(t)$ and the total ship motion output $y(t)$.

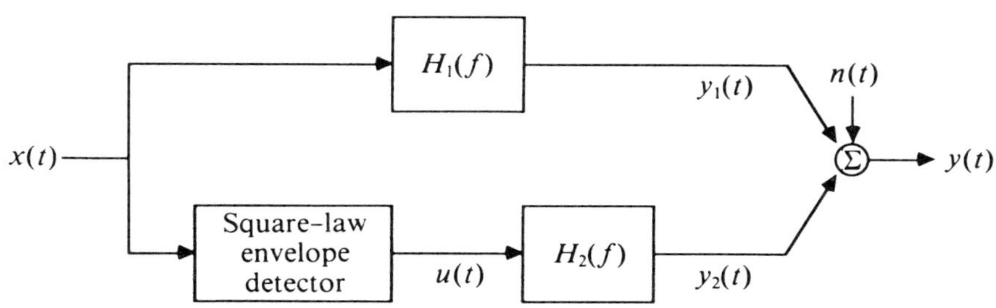

Figure 4.11. Nonlinear drift force model with parallel linear and square-law envelope detector systems.

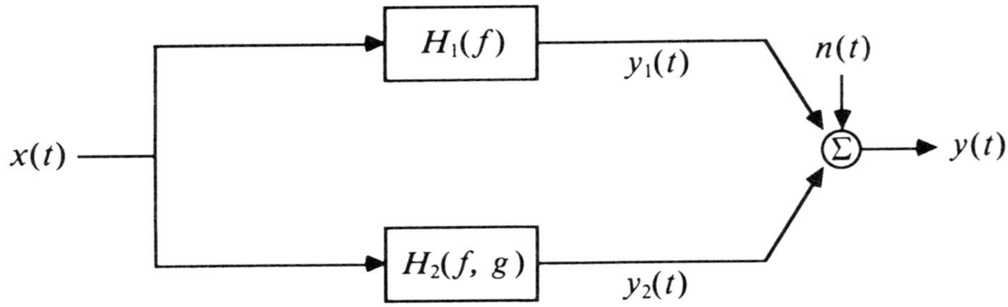

Figure 4.12. Nonlinear drift force model with parallel linear and bilinear systems.

For Gaussian input data, the outputs $y_1(t)$ and $y_2(t)$ will be uncorrelated. This makes Figure 4.11 a special case of Figure 4.4 in Section 4.1, where $y_1(t)$ is the same as $y_o(t)$ and $y_2(t)$ is the same as $y_u(t)$. It is not necessary here to replace Figure 4.3 by Figure 4.4. Note also that Figure 4.11 is equivalent to a two-input/single-output linear model with uncorrelated inputs $x(t)$ and $u(t)$.

For arbitrary non-Gaussian input data, the outputs $y_1(t)$ and $y_2(t)$ will be correlated. The square-law envelope detector output should be designated by $v(t)$ instead of $u(t)$ as shown in Figure 4.3. It is then required to apply the procedure discussed in Section 4.2 that changes Figure 4.3 into Figure 4.4. The Response Prediction Problem and the System Identification Problem can both be solved for the nonlinear drift force model by using the general equations listed in Section 4.2.

Previous nonlinear drift force models such as Figure 4.12 (references 20, 32, and 48) are more complicated than the proposed nonlinear model of Figure 4.11 because they employ a bilinear frequency response function $H_2(f, g)$ to represent the nonlinear system. Solutions involve two-dimensional bispectral density functions that are hard to compute and interpret, and the input/output formulas apply only to Gaussian input data. Techniques to identify bilinear and trilinear models are developed in Chapters 8 and 9.

4.6 CASE 1 AND CASE 2 NONLINEAR MODELS

Two broad cases of great physical interest will now be examined that are called Case 1 and Case 2 nonlinear models. A strong mathematical basis for the importance of these two cases is that they are nonlinear system extensions of the optimum third-order polynomial least-squares model displayed in Figure 2.33.

4.6.1 Case 1 Square-Law and Cubic Models

Case 1 single-input/single-output nonlinear models with correlated outputs are pictured in Figure 4.13 where the squarer and cuber nonlinear systems are

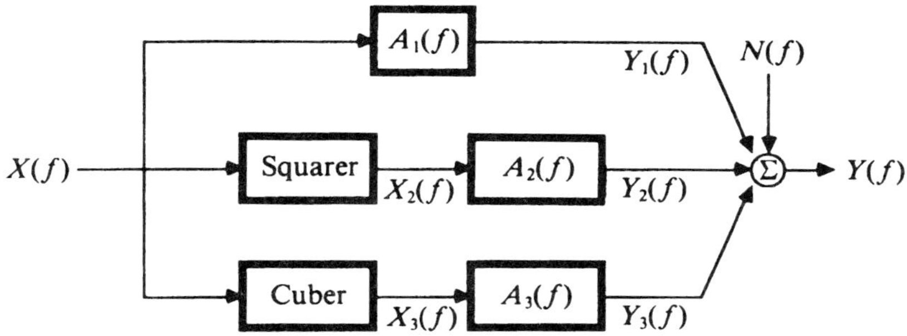

Figure 4.13. Case 1 nonlinear model with correlated outputs.

followed by the $A_2(f)$ and $A_3(f)$ linear systems, respectively, using the symbol "A" for "after." In Figure 4.13, the following Fourier transforms exist:

$$X(f) = X_1(f) = \text{Fourier transform of input } x(t) \tag{4.61}$$

$$X_2(f) = \text{Fourier transform of squarer output } x_2(t) = x^2(t) \tag{4.62}$$

$$X_3(f) = \text{Fourier transform of cuber output } x_3(t) = x^3(t) \tag{4.63}$$

$$Y(f) = \text{Fourier transform of total output } y(t) \tag{4.64}$$

$$N(f) = \text{Fourier transform of output noise } n(t) \tag{4.65}$$

The input/output Fourier transform relation for Figure 4.13 is

$$Y(f) = Y_1(f) + Y_2(f) + Y_3(f) + N(f) \tag{4.66}$$

where the output terms are

$$\begin{aligned} Y_1(f) &= A_1(f)X_1(f) \\ Y_2(f) &= A_2(f)X_2(f) \\ Y_3(f) &= A_3(f)X_3(f) \end{aligned} \tag{4.67}$$

Equations (4.61)–(4.67) apply to arbitrary input/output data.

The Case 1 single-input/single-output nonlinear model of Figure 4.13 is equivalent to the three-input/single-output linear model of Figure 4.14 where the inputs $X(f)$, $X_2(f)$, and $X_3(f)$ are correlated. Recognition of this equivalence in references 3 and 4 is significant because it allows for the use of well-established MI/SO linear techniques to analyze the original Case 1 nonlinear model. These techniques are developed in great detail in references 1 and 2 from earlier work on these matters in references 10–13 and 24. The basis for these MI/SO linear techniques is the computation of conditioned spectral density functions where the linear effects of preceding input records are removed from succeeding input records. Thus, using the notation in reference

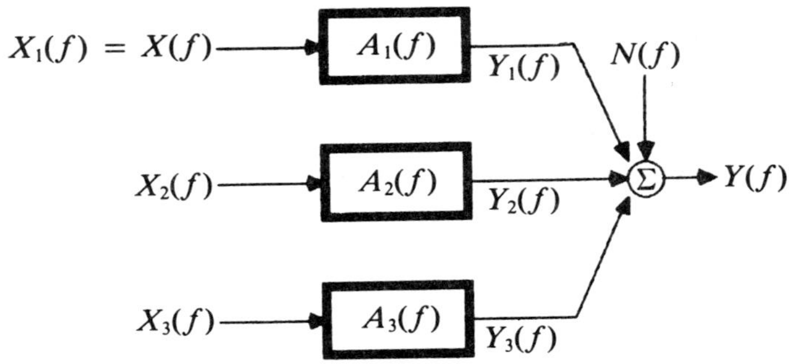

Figure 4.14. Three-input/single-output linear model with correlated inputs equivalent to Figure 4.13.

1, the first input record $X_1(f) = X(f)$ is left alone, the second input record $X_2(f)$ is replaced by $X_{2\cdot 1}(f)$, and the third input record is replaced by $X_{3\cdot 2!}(f)$, where these new conditoned input records are mutually uncorrelated. There is no requirement that the input data $X(f)$ be Gaussian to identify the linear frequency response functions $A_1(f)$, $A_2(f)$, and $A_3(f)$ in Figures 4.13 and 4.14 from measurement of $X(f)$ and the total output $Y(f)$.

4.6.2 Extended Case 1 Nonlinear Models

The Case 1 nonlinear models of Figures 4.13 and 4.14 can be generalized further as shown in Figure 4.15 by letting $X_2(f)$ be the Fourier transform of any zero-memory nonlinear function $g_2[x(t)]$ and by letting $X_3(f)$ be the Fourier transform of any different zero-memory nonlinear function $g_3[x(t)]$. Figures 4.1 and 4.2 are important special cases where $g_2[x(t)] = x^2(t)$ and $g_3[x(t)] = x^3(t)$. Similarly, single-input/single-output nonlinear models with an arbitrary number of parallel paths can be analyzed when $X_2(f)$ and $X_3(f)$ extend to a sum of different $X_i(f)$ that are Fourier transforms of different $g_i[x(t)]$, $i = 2, 3, \ldots, q$, with each of these nonlinear systems then followed by different linear systems $A_i(f)$, $i = 2, 3, \ldots, q$. This extended Case 1 type

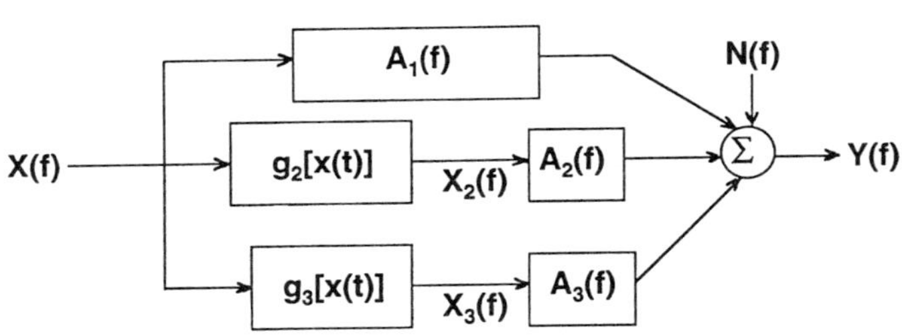

Figure 4.15. Extended Case 1 nonlinear model.

nonlinear model can be replaced by a multiple-input/single-output linear model with the correlated inputs $X(f)$ and $X_i(f)$, $i = 2, \ldots, q$, and the $A_i(f)$ can be identified by the conditioned spectral density techniques in references 1 and 2. Nonlinear system zero-memory (amplitude-domain) operations are defined by the $g_i[x(t)]$ separately from the nonlinear system frequency-domain operations that are defined by the $A_i(f)$.

A complete discussion on how to analyze extended Case 1 nonlinear models of a linear system in parallel with two general nonlinear systems like Figure 4.15 is contained in Section 13.3 of reference 2. Detailed formulas are listed for system identification with non-Gaussian input data. For the special situation of a Case 1 square-law and cubic model, special formulas are listed for system identification with Gaussian input data. A step-by-step procedure is given in Section 13.3.4 of reference 2 on how to determine the physical parameters in extended Case 1 nonlinear models. This material is useful to everyone who works in this field. ***It shows how to apply the Direct and Reverse MI/SO Techniques for analysis and identification of nonlinear systems.***

4.6.3 Case 2 Square-Law and Cubic Models

Case 2 single-input/single-output nonlinear models with correlated outputs are pictured in Figure 4.16 where the zero-memory squarer and cuber nonlinear systems are preceded by the $B_2(f)$ and $B_3(f)$ linear systems, respectively, using the symbol "B" for "before." In Figure 4.16, the following Fourier transforms exist:

$$X(f) = \text{Fourier transform of input } x(t) \tag{4.68}$$

$$U_2(f) = B_2(f)X(f) = \text{Fourier transform of } u_2(t) \tag{4.69}$$

$$U_3(f) = B_3(f)X(f) = \text{Fourier transform of } u_3(t) \tag{4.70}$$

$$Y(f) = \text{Fourier transform of total output } y(t) \tag{4.71}$$

$$N(f) = \text{Fourier transform of output noise } n(t) \tag{4.72}$$

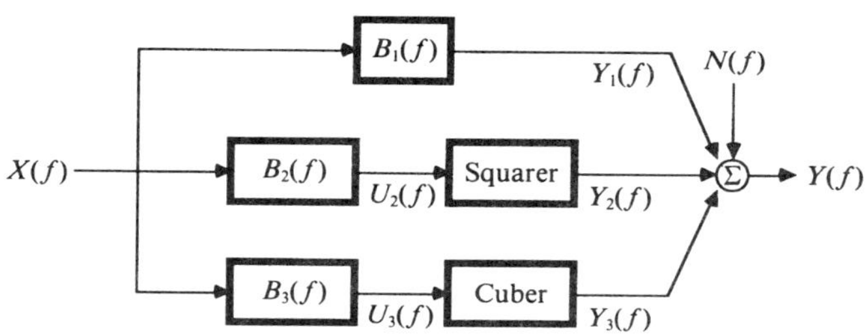

Figure 4.16. Case 2 nonlinear model with correlated outputs.

The input/output Fourier transform relation for Figure 4.16 is

$$Y(f) = Y_1(f) + Y_2(f) + Y_3(f) + N(f) \tag{4.73}$$

where the output terms are

$$\begin{aligned} Y_1(f) &= B_1(f)X(f) \\ Y_2(f) &= \text{function of } U_2(f) \\ Y_3(f) &= \text{function of } U_3(f) \end{aligned} \tag{4.74}$$

Equations (4.68)–(4.74) apply to arbitrary input/output data.

The quantities $U_2(f)$ and $U_3(f)$ cannot be determined from knowledge of only the input $x(t)$ since they also require knowledge of the frequency response functions $B_2(f)$ and $B_3(f)$. This is in contrast to Case 1 nonlinear models where the quantities $X_2(f)$ and $X_3(f)$ can be determined from knowledge of only $x(t)$ without knowing the frequency response functions $A_2(f)$ and $A_3(f)$. Thus, in general, Case 2 nonlinear models like Figure 4.16 cannot be replaced by equivalent three-input/single-output linear models as done previously for Case 1 nonlinear models. It follows that conventional spectral analysis techniques cannot identify the systems defined by $B_1(f)$, $B_2(f)$, and $B_3(f)$ in Case 2 nonlinear models. No practical procedures are known to solve such models for non-Gaussian input data.

For Gaussian input data, special bispectral and trispectral density functions that are functions of a single frequency are defined in Sections 8.4.4 and 9.4.4 that can be used to identify the $B(f)$ systems in Figure 4.16. These special functions are simpler than general bispectral density functions of two frequencies or general trispectral density functions of three frequencies, but they are still much more complicated to compute than usual spectral density functions. These advanced matters are discussed further in Chapters 8 and 9.

4.7 NONLINEAR AUTOMOTIVE APPLICATION

A Direct MI/SO automotive application of extended Case 1 nonlinear models like Figure 4.15 will now be discussed, where a single-input/single-output (SI/SO) nonlinear model is converted into an equivalent three-input/single-output (MI/SO) linear model. This application deals with data collected from an automobile shock subsystem (MacPherson strut) during a road simulator test program. The analysis was done by Dr. Robert Coppolino from Measurement Analysis Corporation in Los Angeles, California.

The input data $x(t)$ collected during this automotive test program are time histories of measured displacement excitations; the output data $F(t)$ are time histories of measured resulting force response. These experimental data were analyzed to try to identify linear and nonlinear system properties of the automobile shock subsystem. The linear system properties were obtained using conventional linear system SI/SO spectral techniques. The nonlinear system properties were obtained using nonlinear system MI/SO spectral techniques.

Results from the nonlinear MI/SO spectral analysis of this shock subsystem input/output data indicate that there are two main nonlinear force terms:

1. A restoring force that acts as a piecewise nonlinear stiffness function.
2. A dissipative friction force that acts as a nonlinear damping function to oppose the motion.

Results also show that conventional SI/SO linear analysis gives the wrong linear system term and, of course, fails completely to identify either of the two nonlinear system terms.

The schematic of a shock subsystem drawn below illustrates the role of each stiffness and friction force contribution, the physical input motion $x(t)$, and the physical output force $F(t)$.

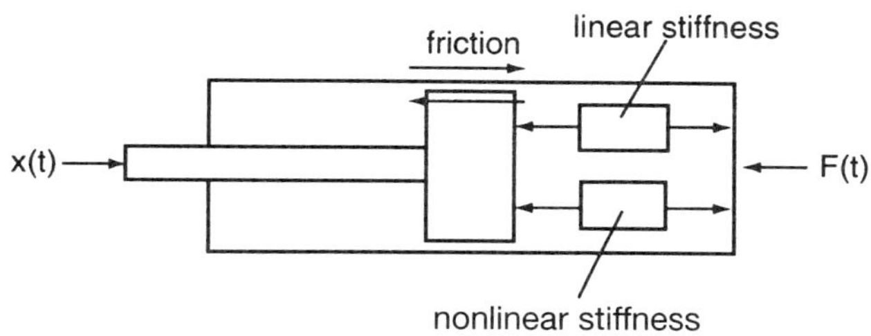

Shock Subsystem Model

4.7.1 Analysis of Shock Subsystem Data

The analysis of shock subsystem data from this automotive study was conducted in three parts:

Preliminary examination of time history records

Linear system SI/SO spectral analysis

Nonlinear system MI/SO spectral analysis

Preliminary examination of the displacement input data $x(t)$ and the force output data $F(t)$ consists of checks on the appearance of their time history plots (Figure 4.17), examination of a force versus displacement plot (Figure 4.18), computation of their probability density functions (Figure 4.19), and computation of their autospectral density functions (Figure 4.20).

The linear system SI/SO spectral analysis is based on the model of Figure 1.3 where the single-input/single-output Fourier transform formula with extraneous output noise is

$$Y(f) = H(f)X(f) + N(f) \tag{4.75}$$

Here,

$$\begin{aligned}
\text{X} = X(f) &= \text{Fourier transform of displacement input } x(t)\\
\text{Y} = Y(f) &= \text{Fourier transform of force output } F(t)\\
\text{H} = H(f) &= \text{Frequency response function from } x(t) \text{ to } F(t)\\
\text{N} = N(f) &= \text{Fourier transform of extraneous output noise}
\end{aligned}$$

From the SI/SO analysis, plots are obtained of the ordinary coherence function between input and output (Figure 4.21), together with the magnitude and phase of the SI/SO estimated frequency response function H (Figures 4.22 and 4.23).

The nonlinear system MI/SO spectral analysis is based on the extended Case 1 nonlinear model of Figure 4.15 where the three-input/single-output Fourier transform formula with extraneous output noise is

$$Y(f) = A_1(f)X_1(f) + A_2(f)X_2(f) + A_3(f)X_3(f) + N(f) \tag{4.76}$$

Here,

$$\begin{aligned}
\text{X1} = X_1(f) &= \text{Fourier transform of displacement input } x(t) = x_1(t)\\
\text{X2} = X_2(f) &= \text{Fourier transform of nonlinear input } x_2(t)\\
\text{X}_3 = X_3(f) &= \text{Fourier transform of nonlinear input } x_3(t)\\
\text{Y} = Y(f) &= \text{Fourier transform of force output } F(t)\\
\text{A1} = A_1(f) &= \text{Frequency response function from } x_1(t) \text{ to } F(t)\\
\text{A2} = A_2(f) &= \text{Frequency response function from } x_2(t) \text{ to } F(t)\\
\text{A3} = A_3(f) &= \text{Frequency response function from } x_3(t) \text{ to } F(t)\\
\text{N} = N(f) &= \text{Fourier transform of extraneous output noise}
\end{aligned}$$

The nonlinear input $x_2(t)$ is a restoring force that acts like a piecewise nonlinear stiffness term defined by

$$\begin{aligned}
x_2(t) &= 0 && \text{if } |x(t)| \leqslant 0.20\\
&= x(t) - 0.20 && \text{if } x(t) \geqslant 0.20\\
&= x(t) + 0.20 && \text{if } x(t) \leqslant -0.20
\end{aligned} \tag{4.77}$$

Selection of the displacement "break" value $|x(t)| = 0.20$ was made based on examination of the force versus displacement plot in Figure 4.18.

The nonlinear input $x_3(t)$ is a dissipative friction force that acts like a nonlinear damping term defined by

$$\begin{aligned} x_3(t) &= 1 \quad \text{if } \dot{x}(t) > 0 \\ &= -1 \quad \text{if } \dot{x}(t) < 0 \end{aligned} \tag{4.78}$$

From the MI/SO analysis, plots are obtained of the cumulative coherence function from the three inputs to the output (Figure 4.24), the magnitude and phase of the first MI/SO estimated frequency response function (Figures 4.25 and 4.26), the magnitude and phase of the second MI/SO estimated frequency response function (Figures 4.27 and 4.28), and the magnitude and phase of the third MI/SO estimated frequency response function (Figures 4.29 and 4.30).

4.7.2 Computed Results from Measured Data

Digital data processing parameters used in this automotive study are listed in Table 4.1. The data were bandpass filtered from 5 to 25 Hz. Figures from the data analysis are listed in Table 4.2. Note that the total test duration was only 32 seconds. These results will now be discussed.

TABLE 4.1 Digital Data Processing Parameters

f_2 = sampling frequency	= 256 Hz
$\Delta t = 1/f_s$ = time interval between samples	= 0.0039 s
$f_c = f_s/2$ = Nyquist cutoff frequency	= 128 Hz
N = number of sample values per subrecord	= 512
$T = N\,\Delta t$ = time of each subrecord	= 2.0 s
n_d = number of distinct (different) subrecords	= 16
$N_{total} = n_d N$ = total number of sample values	= 8192
$T_{total} = n_d T$ = total test duration	= 32 s
$\Delta f = 1/T$ = bandwidth resolution	= 0.50 Hz

TABLE 4.2 Figures from Analysis of Shock Subsystem Data

4.17	Time history records of displacement input and force output
4.18	Force versus displacement plot
4.19	Probability density functions of input and output
4.20	Autospectral density functions of input and output
4.21	Ordinary coherence function between input and output
4.22	Magnitude of SI/SO frequency response function
4.23	Phase of SI/SO frequency response function
4.24	Cumulative coherence function from three inputs to output
4.25	Magnitude of first MI/SO frequency response function
4.26	Phase of first MI/SO frequency response function
4.27	Magnitude of second MI/SO frequency response function
4.28	Phase of second MI/SO frequency response function
4.29	Magnitude of third MI/SO frequency response function
4.30	Phase of third MI/SO frequency response function

Preliminary Data Analysis. The displacement input time history record in Figure 4.17 shows that most data samples are associated with a low-amplitude displacement input. This fact is confirmed in the small rms value for the displacement input probability density function in Figure 4.19. The solid curves plotted in Figure 4.19 are the computed standardized probability density functions; the dashed curves are standardized Gaussian probability density functions with zero mean values and unity variances. Note that the displacement input and force output both appear to be definitely non-Gaussian. Had the input been Gaussian and the output non-Gaussian, this would have indicated a possible system nonlinearity. Note also that the force output probabilty density function has the appearance for the passage of Gaussian data through a nonlinear softening spring system as displayed in Figure 2.32. Insight into the nature of the nonlinearities in this system is gained from study of the force versus displacement plot in Figure 4.18. The "break" in apparent slope for displacements of magnitude greater than 0.20 suggests the presence of a piecewise nonlinear stiffness. The autospectral density functions (autospectra) in Figure 4.20 for the input data G_{XX} and output data G_{FF} both show a peak value in the vicinity of 12 Hz with the output spectrum approximately 10,000 times that of the input spectrum.

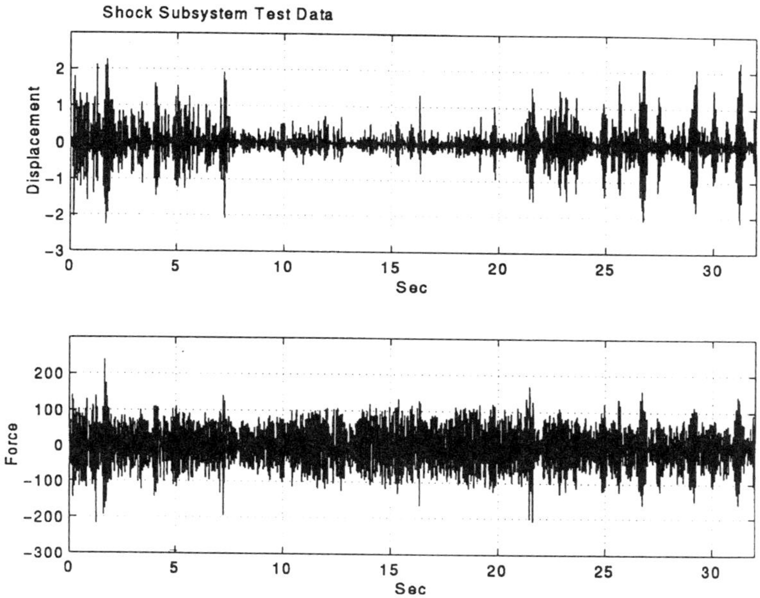

Figure 4.17.

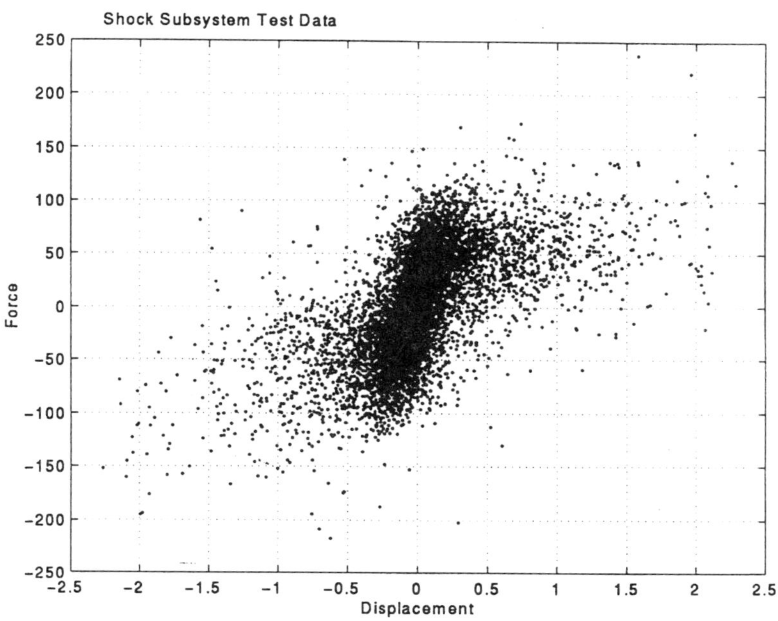

Figure 4.18.

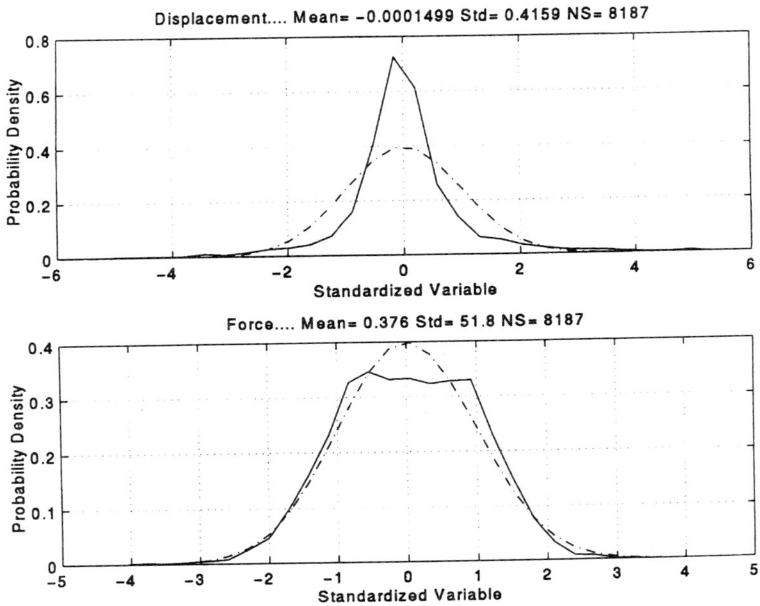

Figure 4.19.

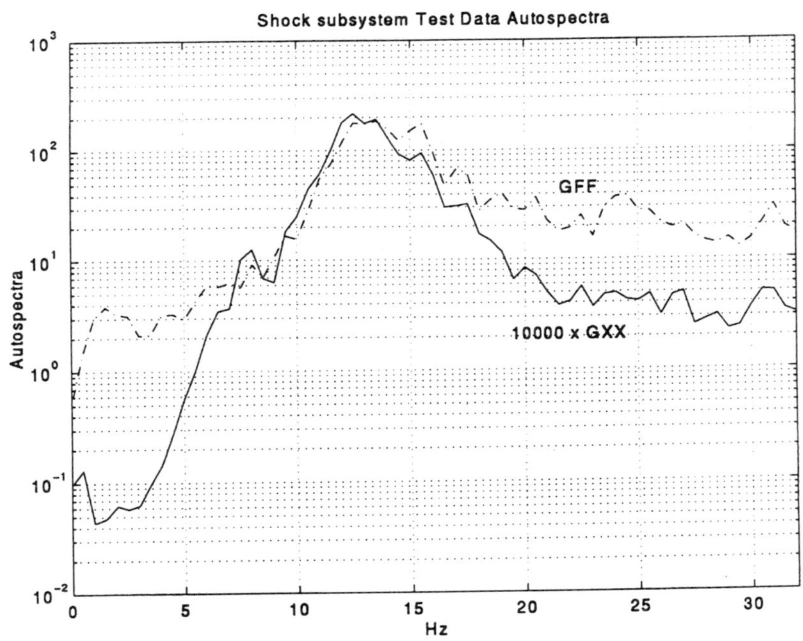

Figure 4.20.

Linear SI/SO Spectral Analysis. A conventional linear system SI/SO spectral analysis was performed on the test data that was collected for the output force data and the input displacement data. Figure 4.21 illustrates the ordinary coherence function between this input and output. Coherence values below, say, 0.80 in ranges of significant input power are due to either noise, other unmeasured inputs, or system nonlinearities. Note the limited frequency range from about 12 to 13.5 Hz where this ordinary coherence function is above 0.80. Figures 4.22 and 4.23 illustrate the magnitude and phase of the estimated SI/SO frequency response function, denoted here by H. This result is the optimum linear estimate of the frequency response function that minimizes any output extraneous noise. It is absolutely wrong here since it was known in advance that the shock subsystem is nonlinear. Better estimates of this desired linear frequency reponse function are obtained by the nonlinear system Direct MI/SO spectral analysis that was performed next.

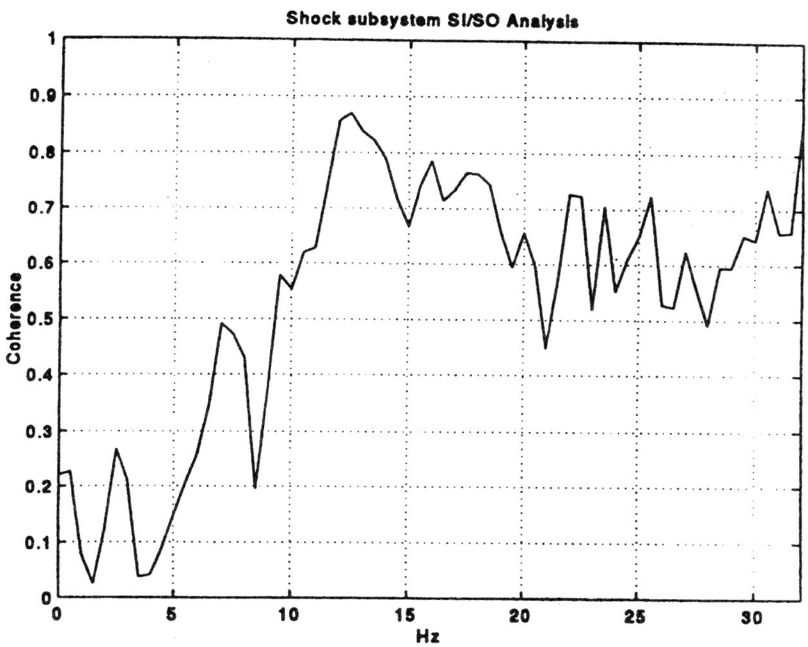

Figure 4.21.

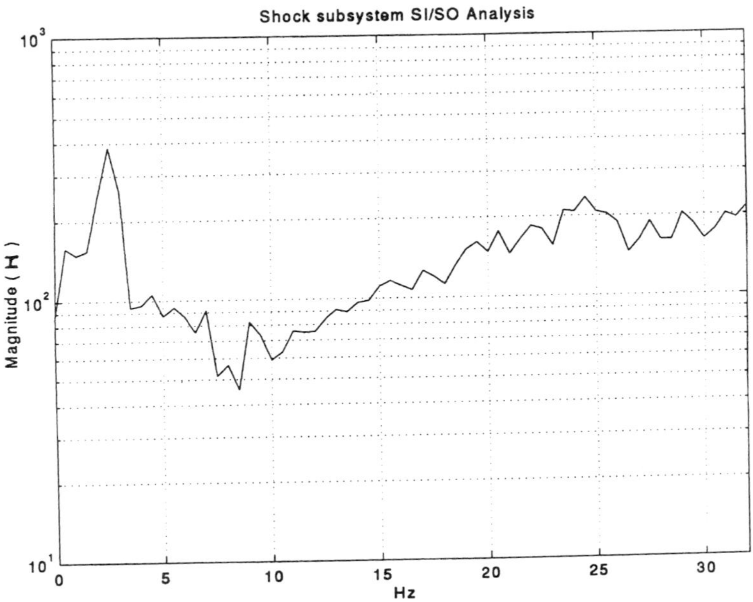

Figure 4.22.

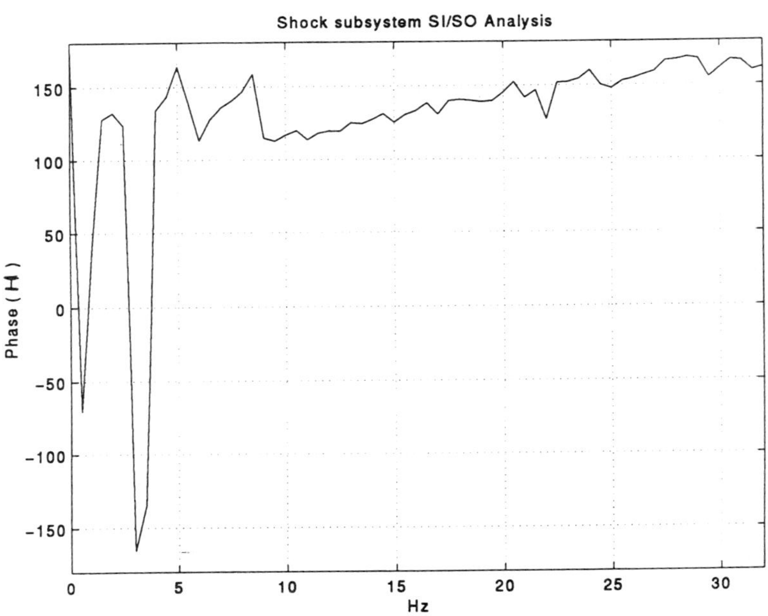

Figure 4.23.

Nonlinear MI/SO Spectral Analysis. The cumulative coherence function for this analysis from all three input functions is shown in the top solid curve in Figure 4.24. Note the greatly improved coherence values resulting from inclusion of the two nonlinear input terms. The cumulative coherence function is now above 0.80 in the extended frequency range from 10 to 30 Hz, with the upper two curves showing how the two nonlinear terms contribute to the overall result. The lowest dashed curve is the same as the ordinary coherence function in Figure 4.21 and states the percentage of the output spectrum that is due to the first input. The middle curve in Figure 4.24 states the percentage of the output spectrum that is due to the first two inputs.

Estimates of the magnitude and phase of the first MI/SO frequency response function, denoted here by A1, are illustrated in Figures 4.25 and 4.26. This gives better estimates of the correct linear frequency response function in place of the wrong estimates shown in Figures 4.22 and 4.23. The near uniform magnitude and phase functions in Figures 4.25 and 4.26 indicate that this linear frequency response function term is primarily stiffness and damping.

Estimates of the magnitude and phase of the second MI/SO frequency response function, denoted here by A2, are illustrated in Figures 4.27 and 4.28. Estimates of the magnitude and phase of the third MI/SO frequency response function, denoted here by A3, are illustrated in Figures 4.29 and 4.30. These results are the frequency response functions associated with the two nonlinear input functions.

Large random errors in these estimates are due mainly to the fact that (1) the displacement input time history records were limited by low amplitudes and (2) the input/output time history records for the MI/SO spectral analysis were relatively short. Future tests should be conducted with higher level displacement inputs and with much longer input/output records to try to reduce the random errors to acceptable values.

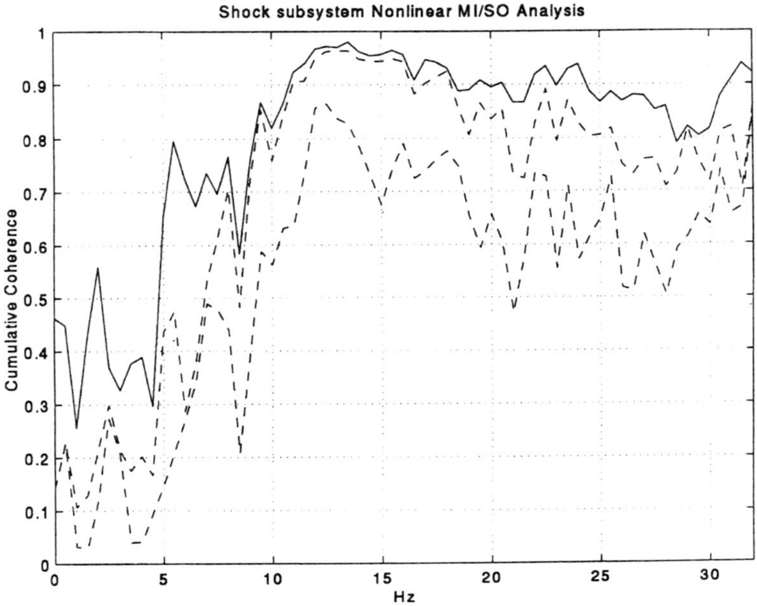

Figure 4.24.

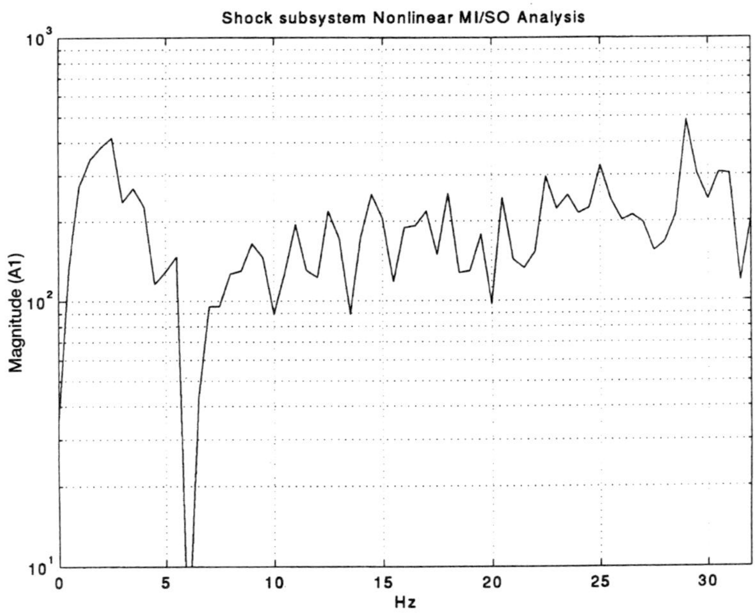

Figure 4.25.

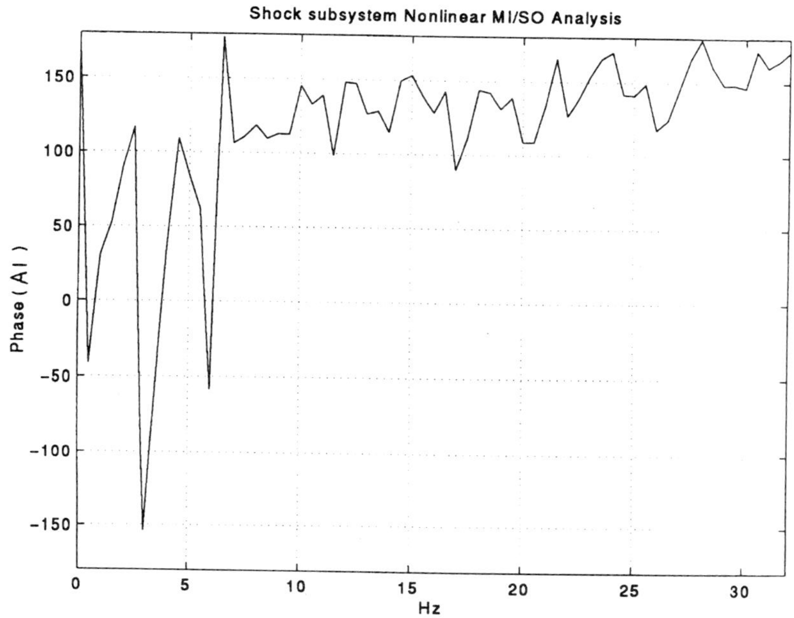

Figure 4.26.

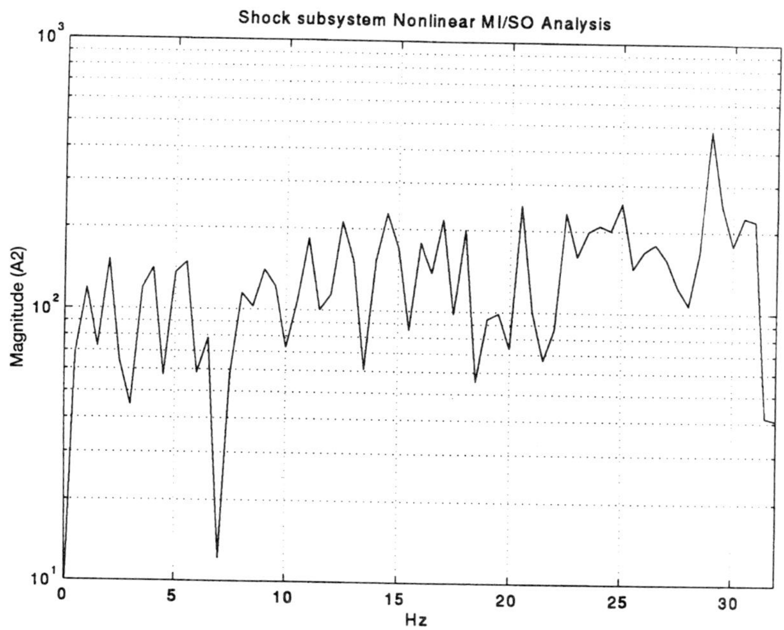

Figure 4.27.

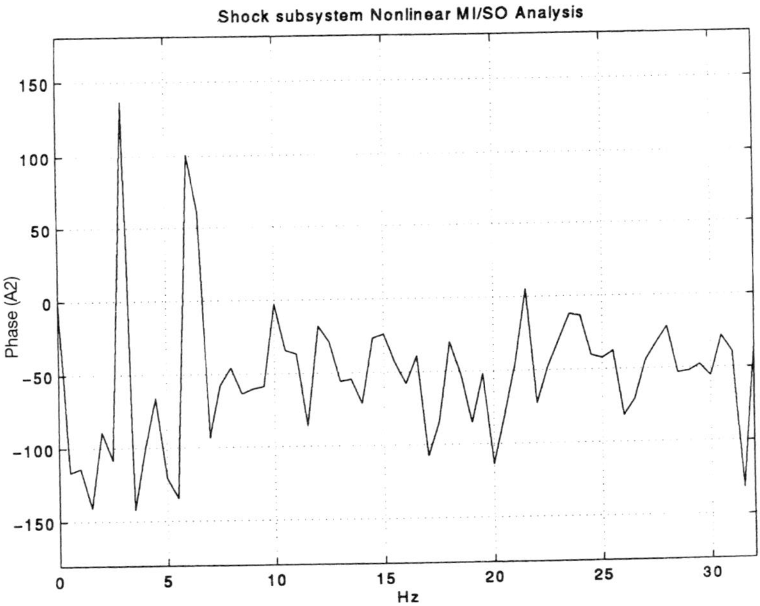

Figure 4.28.

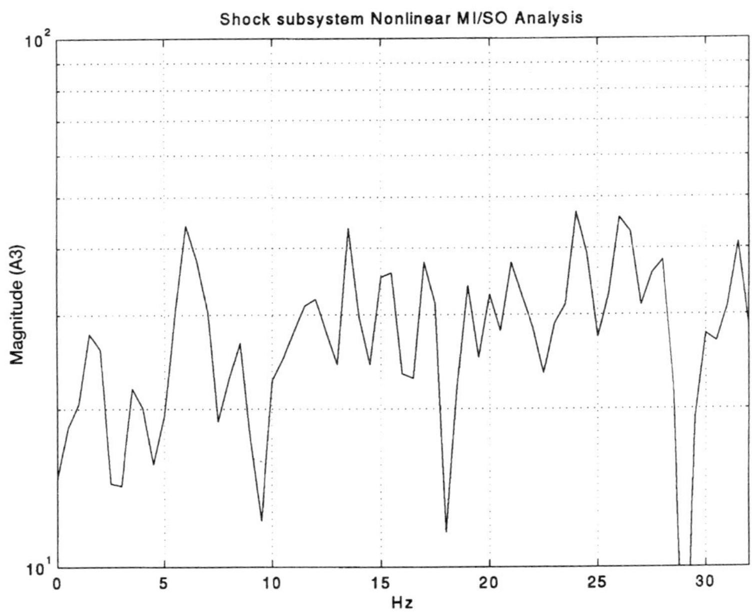

Figure 4.29.

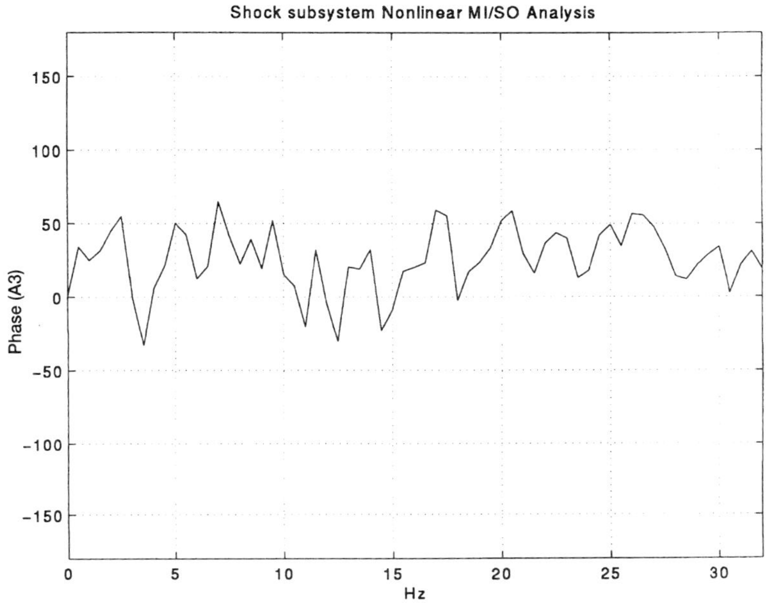

Figure 4.30.

4.8 NONLINEAR BIOMEDICAL STUDY

A biomedical application of the special Case 1 nonlinear model of Figure 4.13 is discussed and illustrated in reference 19. This research was performed by Dr. Mark Chertoff and others at the Department of Hearing and Speech, Kansas City Medical Center in Kansas City, Kansas, as part of a contract from the National Institutes of Health in Washington, D.C. The study involves a nonlinear mechano-electric transduction (MET) process in the cochlea of an ear of a Mongolian gerbil where cochlear partition motion inputs produces nonlinear electrical signal voltage outputs from deflection of hair cells. Tests were conducted using Gaussian input noise on six Mongolian gerbils weighing between 45 and 70 grams.

A main objective of this biomedical study was to determine the nonlinear cochlear hearing transformations in both normal and noise-damaged ears of Mongolian gerbils. New nonlinear results were obtained using the Case 1 nonlinear model of Figure 4.13 that provided the parameters of a third-order polynomial equation to describe the desired nonlinear transformation. Coherence functions were calculated to assess the validity of the overall nonlinear model and the relative importance of the linear, square-law, and cubic terms. The linear term was found to account for the largest magnitude response, followed by the cubic and square-law terms. Results of this study were encouraging towards quantifying frequency-dependent changes in cochlear physiology as a result of cochlear damage. This biomedical research work is still underway with different extended Case 1 nonlinear models like Figure 4.15 being investigated to obtain further results.

An initial part of this bimedical study involved the application of Theorem 2 in Chapter 2, Eq. (2.15), to identify the form of a zero-memory nonlinear function between the input and output records. For this analysis, histograms were computed of both the input sound pressure recordings and the output electrical signal recordings.

Figure 4.31 shows the standardized histograms for the Gaussian input data and for the non-Gaussian output data. Figure 4.32 shows exactly how the calculation procedure of Theorem 2 is applied to find the associated values of x_0 and y_0 where their probability distribution functions are equal. The particular value shown here is where $P(x_0) = P(y_0) = 0.15$. Figure 4.33 displays the complete nonlinear function $y = g(x)$ for this data over the range $-1.50 \leqslant x \leqslant 1.50$ where y varies from $-700 \leqslant y \leqslant 500$. Note that this nonlinear function is sigmoidal with non-symmetric saturation. For input pressures larger than 1.0 pascals, the positive output voltages are smaller in magnitude than the corresponding negative output voltages. The form of the zero-memory nonlinear function in Figure 4.33 justifies consideration of the Case 1 nonlinear model in this biomedical study.

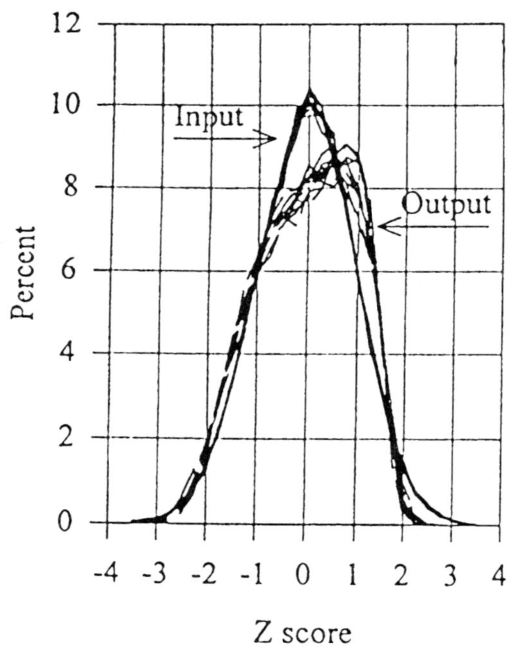

Figure 4.31. Input and output histograms.

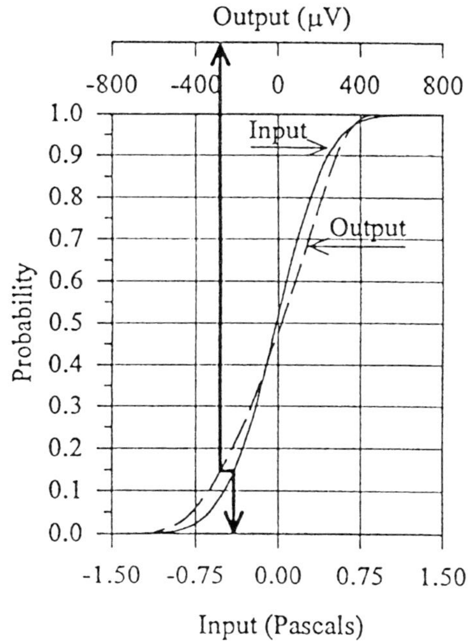

Figure 4.32. Calculation procedure.

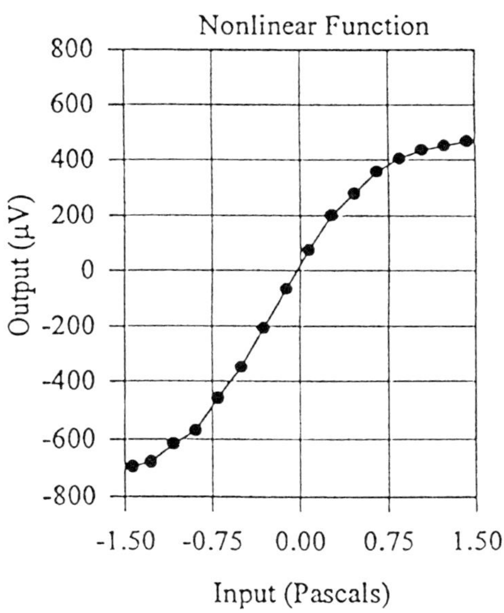

Figure 4.33. Complete nonlinear function.

4.9 NONLINEAR DIFFERENTIAL EQUATIONS OF MOTION

Random data analysis of the nonlinear systems in Section 4.2 converts a single-input/single-output nonlinear model with parallel linear and finite-memory nonlinear systems into an equivalent Direct MI/SO two-input/single-output linear model. Similarly, as discussed in Section 4.6.2, a more general single-input/single-output nonlinear model consisting of a linear system in parallel with an arbitrary number of different finite-memory nonlinear systems can be converted into an equivalent Direct MI/SO multiple-input/single-output linear model. Conversion of these Case 1 type single-input/single-output nonlinear models allows them to be analyzed by known practical linear techniques from references 1 and 2.

Important connections exist between this new approach for random data analysis of nonlinear systems and nonlinear differential equations of motion. The connections will be shown here for nonlinear differential equations of motion representing single-degree-of-freedom (SDOF) systems with nonlinear stiffness terms, where the physical parameters can be determined from input/output measurements in an experimental program. These techniques also apply to determining nonlinear damping terms or other types of nonlinear properties in proposed nonlinear differential equations of motion, when experimental measurements are made only of the input data and the total output data from the system under investigation.

4.9.1 Single-Degree-of-Freedom Linear Systems

If $x(t)$ is a force input to a typical SDOF linear system with mass m, damping coefficient c, and stiffness coefficient k, and if $y(t)$ is the resulting displacement output, then the linear differential equation of motion in the time domain is

$$m\ddot{y}(t) + c\dot{y}(t) + ky(t) = x(t) \tag{4.79}$$

Fourier transforms of both sides give the associated frequency-domain relation

$$[k - (2\pi f)^2 m + j(2\pi f)c]Y(f) = X(f) \tag{4.80}$$

where $X(f)$ and $Y(f)$ are Fourier transforms of $x(t)$ and $y(t)$, respectively. Equation (4.80) is the same as

$$Y(f) = H(f)X(f) \tag{4.81}$$

where $H(f)$ is the linear frequency response function given by

$$H(f) = [k - (2\pi f)^2 m + j(2\pi f)c]^{-1} \tag{4.82}$$

Equation (4.82) can also be expressed by

$$H(f) = (1/k)[1 - (f/f_n)^2 + j2\zeta(f/f_n)]^{-1} \tag{4.83}$$

where the undamped natural frequency f_n and the damping ratio ζ are defined by

$$f_n = (1/2\pi)\sqrt{k/m} \tag{4.84}$$

$$\zeta = c/2\sqrt{km} \tag{4.85}$$

Assume that input data are stationary random data. Then output data will also be stationary random data. From Eq. (4.81), it follows by spectral analysis that the desired linear frequency response function can be identified by the familiar formula

$$H(f) = \frac{S_{xy}(f)}{S_{xx}(f)} \tag{4.86}$$

Here, $S_{xy}(f)$ is the cross-spectral density function between the force input $x(t)$ and the displacement output $y(t)$, and $S_{xx}(f)$ is the autospectral density function of the force input $x(t)$. Computation of $H(f)$ from measured $x(t)$ and $y(t)$ by means of Eq. (4.86) enables one to determine the physical parameters m, c, k, f_n, and ζ that appear in Eqs. (4.84) and (4.85). This then gives the associated linear differential equation of motion (4.79). Here, Gaussian force input data produces Gaussian displacement output data.

The connections between the measured $H(f)$ and the physical parameters can be shown in the following way. From calculation of $H(f)$, one can find the frequency f_n where the peak value of $|H(f)|$ occurs and one can determine the values of $H(0)$ and $|H(f_n)|$. See Figure 8.10. From Equation (4.83) these values will be

$$H(0) = (1/k) \tag{4.87}$$

$$|H(f_n)| = (1/2\zeta k) \tag{4.88}$$

This knowledge of f_n, $H(0)$, and $|H(f_n)|$ will then give the physical parameters k, m, ζ, and c by using the relations of Eqs. (4.84) – (4.88) to obtain

$$k = 1/H(0) \tag{4.89}$$

$$m = k/(2\pi f_n)^2 = 1/(2\pi f_n)^2 H(0) \tag{4.90}$$

$$\zeta = 1/2k|H(f_n)| = H(0)/2|H(f_n)| \tag{4.91}$$

$$c = 2\zeta\sqrt{km} = 1/(2\pi f_n)|H(f_n)| \tag{4.92}$$

These formulas are well-known results that show the connection between measurement of $H(f)$ and the physical parameters in its associated SDOF linear differential equation of motion.

4.9.2 Reversal of Input and Output

Consider an experimental program where the force and displacement are measured. To apply the new approach for random data analysis of nonlinear systems to nonlinear differential equations of motion, as in Section 2.8, the roles of the force and displacement should be reversed so that the displacement becomes the input $x(t)$ and the force becomes the output $y(t)$. The linear differential equation of motion in the time domain instead of Eq. (4.79) is now

$$y(t) = m\ddot{x}(t) + c\dot{x}(t) + kx(t) \tag{4.93}$$

where $x(t)$ is the displacement input and $y(t)$ is the force output. The associated Fourier transform relation for Eq. (4.93) in the frequency domain is

$$Y(f) = A_1(f)X(f) \tag{4.94}$$

where

$$A_1(f) = [k - (2\pi f)^2 m + j(2\pi f)c] \tag{4.95}$$

Note that $A_1(f)$ is the inverse of the $H(f)$ of Eq. (4.82), namely,

$$A_1(f) = [H(f)]^{-1} \tag{4.96}$$

From Eq. (4.94), by ordinary spectral analysis of stationary random data, the frequency response function $A_1(f)$ can be identified by the formula

$$A_1(f) = \frac{S_{xy}(f)}{S_{xx}(f)} \tag{4.97}$$

Here, $S_{xy}(f)$ is now the cross-spectral density function between the displacement input $x(t)$ and the force output $y(t)$, and $S_{xx}(f)$ is the autospectral density function of the displacement input $x(t)$. Computation of $A_1(f)$ from measured $x(t)$ and $y(t)$ by means of Eq. (4.97) enables one to determine the parameters $m, c,$ and k in Eq. (4.95), which then gives the reverse linear differential equation of motion (4.93). Instead of Eqs. (4.87) and (4.88), one will now obtain the results that the minimum value of $|A_1(f)|$ occurs at f_n with

$$A_1(0) = k \tag{4.98}$$

$$|A_1(f_n)| = 2\zeta k \tag{4.99}$$

Then

$$m = k/(2\pi f_n)^2 = A_1(0)/(2\pi f_n)^2 \tag{4.100}$$

$$c = 2\zeta\sqrt{km} = |A_1(f_n)|/(2\pi f_n) \tag{4.101}$$

These computed parameters apply to both Eq. (4.93) and Eq. (4.79).

4.9.3 Single-Degree-of-Freedom Nonlinear Systems

When nonlinear stiffness terms are involved with SDOF systems, a nonlinear differential equation of motion is typically

$$m\ddot{y}(t) + c\dot{y}(t) + ky(t) + k_2y^2(t) + k_3y^3(t) = x(t) \tag{4.102}$$

where $x(t)$ is the physical force input and $y(t)$ is the resulting physical displacement output. Here $y^2(t)$ and $y^3(t)$ represent nonlinear stiffness terms multiplied by constants k_2 and k_3, respectively. The associated Fourier transform relation using the $H(f)$ of Eq. (4.82) is

$$[H(f)]^{-1}Y(f) + k_2Y_2(f) + k_3Y_3(f) = X(f) \tag{4.103}$$

where $Y_2(f)$ and $Y_3(f)$ are Fourier transforms of $y^2(t)$ and $y^3(t)$, respectively. Equation (4.103) is the same as

$$Y(f) = H(f)[X(f) - k_2Y_2(f) - k_3Y_3(f)] \tag{4.104}$$

This result represents a feedback system. From knowledge of the physical parameters m, c, and k in $H(f)$, and the physical parameters k_2 and k_3, simulation studies with stationary random data can determine the displacement output $Y(f)$ and $y(t)$ as a function of the force input $X(f)$ and $x(t)$. Here, Gaussian force input data produces non-Gaussian displacement output data.

With measured experimental data, none of this simulation work is required and none of the physical parameters need to be specified. Instead, the physical parameters can be determined by applying the new Reverse MI/SO Techniques discussed in this book. The roles of $x(t)$ and $y(t)$ should be reversed so that Eq. (4.102) becomes the reverse nonlinear relation

$$y(t) = m\ddot{x}(t) + c\dot{x}(t) + kx(t) + k_2x^2(t) + k_3x^3(t) \tag{4.105}$$

Here, $x(t)$ is now the measured non-Gaussian mathematical displacement input and $y(t)$ is the measured mathematical force output. The associated Fourier transform relation to Eq. (4.105), using the $A_1(f)$ of Eq. (4.95), is

$$Y(f) = A_1(f)X(f) + k_2X_2(f) + k_3X_3(f) \tag{4.106}$$

where $X_2(f)$ and $X_3(f)$, the Fourier transforms of $x^2(t)$ and $x^3(t)$, respectively, can be computed from the measured $x(t)$.

Equation (4.106) represents a Case 1 nonlinear model displayed in Figure 4.13 where $A_2(f) = k_2$ and $A_3(f) = k_3$. For non-Gaussian displacement input data, the conditioned spectral density techniques from references 1 and 2 can be used to identify the systems $A_1(f)$, $A_2(f)$, and $A_3(f)$ because Figure 4.13 is always equivalent to a three-input/single-output linear model with correlated inputs $X(f)$, $X_2(f)$, and $X_3(f)$, as shown in Figure 4.14.

When the constants k_2 and k_3 in Eq. (4.106) are replaced by more general linear systems described by $A_2(f)$ and $A_3(f)$, respectively, Eq. (4.106) takes the form

$$Y(f) = A_1(f)X(f) + A_2(f)X_2(f) + A_3(f)X_3(f) + N(f) \tag{4.107}$$

where $N(f)$ is the Fourier transform of extraneous uncorrelated output noise $n(t)$ representing all possible deviations from an ideal model. More complicated nonlinear stiffness and damping terms will be obtained if these new procedures show that the linear systems $A_2(f)$ and $A_3(f)$ are not constants but instead are functions of frequency.

For example, suppose one finds that $A_1(f)$ satisfies Eq. (4.95), $N(f) = 0$, and

$$A_2(f) = 0 \tag{4.108}$$

$$A_3(f) = k_3 + j(2\pi f)c_3 \tag{4.109}$$

where k_3 and c_3 are constants. The corresponding reverse nonlinear differential equation of motion in place of Eq. (4.105) will then be

$$\begin{aligned} y(t) &= m\ddot{x}(t) + c\dot{x}(t) + kx(t) + k_3x^3(t) + c_3 \frac{d}{dt}[x^3(t)] \\ &= m\ddot{x}(t) + c\dot{x}(t) + kx(t) + k_3x^3(t) + 3c_3x^2(t)\dot{x}(t) \end{aligned} \tag{4.110}$$

Special cases of this equation should be noted. When $c_3 = 0$, Eq. (4.110) takes the form of Duffing's equation. When $k_3 = 0$, Eq. (4.110) takes the form of Van der Pol's equation. These particular nonlinear equations are important in the field of nonlinear vibrations in electrical and mechanical systems.

A SDOF nonlinear differential equation of motion with arbitrary nonlinear terms can be expressed as

$$m\ddot{y}(t) + c\dot{y}(t) + ky(t) + g[y(t)] = x(t) \tag{4.111}$$

where

$x(t)$ = physical force input

$y(t)$ = physical displacement output

$g[y(t)]$ = arbitrary nonlinear transformation of $y(t)$

Assume that the quantity $g[y(t)]$ is a sum of $(q - 1)$ different zero-memory nonlinear functions $g_i[y(t)]$ with different constant coefficients a_i such that

$$g[y(t)] = \sum_{i=2}^{q} a_i g_i[y(t)] \tag{4.112}$$

Then, the proposed nonlinear differential equation of motion is

$$m\ddot{y}(t) + c\dot{y}(t) + ky(t) + \sum_{i=2}^{q} a_i g_i[y(t)] = x(t) \tag{4.113}$$

Verification of terms in Eq. (4.113) can be obtained from measured experimental data by applying the techniques discussed here where (1) the roles of $x(t)$ and $y(t)$ are reversed, and (2) the resulting single-input/single-output nonlinear model is replaced by an equivalent Reverse MI/SO multiple-input/single-output linear model. Specifically, by reversing the meaning of $x(t)$ and $y(t)$, Eq. (4.113) becomes the reverse nonlinear relation

$$y(t) = m\ddot{x}(t) + c\dot{x}(t) + kx(t) + \sum_{i=2}^{q} a_i g_i[x(t)] \tag{4.114}$$

where

$$x(t) = \text{mathematical displacement input}$$

$$y(t) = \text{mathematical force output}$$

Then, from Eq. (4.114), Fourier transforms of both sides give the associated frequency-domain formula

$$Y(f) = A_1(f)X(f) + \sum_{i=2}^{q} a_i X_i(f) \tag{4.115}$$

where $A_1(f)$ is given by Eq. (4.95) and where the terms $X_i(f)$ are the Fourier transforms of $g_i[x(t)]$ for $i = 2, 3, \ldots, q$. A further extension of Eq. (4.115) occurs by replacing the constants a_i with linear frequency response functions $A_i(f)$, where $i = 2, 3, \ldots, q$. Thus, as per Eq. (4.107), Eq. (4.115) with the addition of extraneous uncorrelated output noise becomes an extended Case 1 type nonlinear model represented by

$$Y(f) = A_1(f)X(f) + \sum_{i=2}^{q} A_i(f)X_i(f) + N(f) \tag{4.116}$$

From measurement of $x(t)$ and $y(t)$, one can compute all of the proposed $g_i[x(t)]$ to obtain the various $X_i(f)$ terms in Eq. (4.116). One can also compute $X(f)$ and $Y(f)$. The solution of the $A_i(f)$ in Eq. (4.116) is now a straightforward matter using the multiple-input/single-output linear techniques in references 1 and 2. The particular reverse nonlinear differential equation of motion of Eq. (4.114) will be verified if $A_1(f)$ agrees with Eq. (4.95), if the $A_i(f)$ for $i = 2, 3, \ldots, q$ are constants, and if the output noise $N(f)$ is negligible. Thus, physical parameters in proposed SDOF nonlinear differential equations of motion such as Eq. (4.113) can be determined from measured random data.

If the $A_i(f)$ for $i = 2, 3, \ldots, q$ vary with frequency, one obtains more general SDOF nonlinear differential equations of motion than Eq. (4.113). The $g_i[x(t)]$ for $i = 2, 3, \ldots, q$ identify the different proposed nonlinear system amplitude-domain properties and the linear frequency response functions $A_i(f)$ identify the proposed nonlinear system frequency-domain properties.

This concludes Chapter 4.

5

DETERMINATION OF PHYSICAL PARAMETERS WITH MEMORY IN NONLINEAR SYSTEMS

This chapter illustrates application of the Reverse MI/SO Technique for SDOF nonlinear systems with frequency-dependent linear and nonlinear damping coefficients that are defined by memory functions of an exponential or exponential-cosine form. Seven computer simulation cases are developed and discussed in detail to identify these physical parameters using both the Reverse MI/SO Technique and the usual optimum linear SI/SO technique. The SI/SO technique gives erroneous estimates compared to the correct estimates from the Reverse MI/SO Technique. The last section discusses how to formulate six-DOF nonlinear system models for ocean engineering applications, like those covered in Chapters 6 and 7, and shows how to replace these nonlinear physical models with equivalent mathematical reverse dynamics linear models. Similar nonlinear system models apply to problems in aerospace, automotive, biomedical, and other fields.

5.1 REVERSE MI/SO TECHNIQUE

The Reverse MI/SO Technique was developed from applied research work performed for the Naval Civil Engineering Laboratory (now the Naval Facilities Engineering Service Center) in Port Hueneme, California. Some of these results are discussed and illustrated in Sections 13.4 and 13.5 of reference 2. Three publications on these matters are references 5–7. The responses from various SDOF nonlinear systems such as Duffing, Van der Pol, Mathieu, and Deadband were simulated to numerically evaluate the technique. The system dynamics were described by nonlinear second-order integrodifferential equations of motion with constant coefficients (i.e., zero memory) or with frequency-dependent coefficients of an exponential or exponential-cosine form. In all cases, Reverse MI/SO successfully identified the known system coefficients.

Direct MI/SO Techniques and Reverse MI/SO Techniques are both very practical analysis and identification techniques because they are frequency-domain techniques that are applicable to many general types of nonlinear systems. They do not represent an "optimum" linear approximation for a nonlinear problem. They require only slightly more computations than a conventional linear SI/SO spectral analysis, and, most importantly, they yield coherence functions (goodness-of-fit) for each linear and nonlinear term. These coherence functions allow for unambiguous assessment of the validity of the proposed nonlinear model.

The most useful application of Reverse MI/SO is in finding an optimum low-order nonlinear integrodifferential equation of motion from measured or simulated data that minimizes the extraneous output noise. There are practically no restrictions on the proposed form of the nonlinear terms. No error is introduced by including extra possible nonlinear terms. Reverse MI/SO is also capable of identifying the frequency dependence of all linear and nonlinear paths through a nonlinear system. In fact, Reverse MI/SO automatically identifies the best nonlinear integrodifferential equation of motion rather than simply the best nonlinear differential equation of motion describing the system dynamics. This is a valuable tool for nonlinear stability studies.

These features make Reverse MI/SO a very practical technique for identifying phenomena such as rigid-body dynamics of ships on a free surface, where the roll of the ship is known to be highly nonlinear even for moderate excitations, and where the added mass and damping are known to be frequency-dependent. Similar nonlinear properties can be identified for different nonlinear problems in many other fields.

The fact that Reverse MI/SO naturally returns an integrodifferential equation of motion is significant because it allows for physical interpretations by engineers and scientists regarding specific effects and their behavior (e.g., that the coefficient associated with the first derivative of the response representing a damping effect is frequency-dependent). This makes it useful for fundamental studies into the physics of various phenomena. Such results are not available from other higher-order statistical techniques such as Volterra series models or time-domain iteration studies.

For these reasons, Reverse MI/SO is considered to be an important new practical tool for nonlinear systems analysis and identification that offers significant advantages compared to existing techniques. This technique will now be demonstrated for nonlinear systems where its physical parameters have memory. Direct MI/SO Techniques are applied in Chapters 6 and 7.

5.2 PHYSICAL PARAMETERS WITH MEMORY

Consider first a simple differential equation of motion for a nonlinear SDOF system with nonlinear viscous damping whose physical parameters are constants with zero memory as defined by

$$m\ddot{u}(t) + c\dot{u}(t) + ku(t) + dz(t) = F(t) \tag{5.1}$$

where

$$F(t) = \text{force input} = \text{physical excitation} \tag{5.2}$$

$$u(t) = \text{displacement output} = \text{physical response} \tag{5.3}$$

$$z(t) = \dot{u}(t)|\dot{u}(t)| = \text{nonlinear viscous damping term} \tag{5.4}$$

$$m = \text{mass coefficient} \tag{5.5}$$

$$c = \text{linear damping coefficient} \tag{5.6}$$

$$k = \text{linear stiffness coefficient} \tag{5.7}$$

$$d = \text{nonlinear viscous damping coefficient} \tag{5.8}$$

Taking the Fourier transforms of both sides of Eq. (5.1) yields the following reverse dynamic system that is described by Eq. (5.9), namely,

$$A_1(f)X_1(f) + A_2(f)X_2(f) = Y(f) \tag{5.9}$$

where

$$Y(f) = \text{Fourier transform of } F(t) = \text{mathematical output} \tag{5.10}$$

$$X_1(f) = X(f) = \text{Fourier transform of } u(t) = \text{first mathematical input} \tag{5.11}$$

$$X_2(f) = \text{Fourier transform of } z(t) = \text{second mathematical input} \tag{5.12}$$

$$A_1(f) = k - (2\pi f)^2 m + j(2\pi f)c \tag{5.13}$$

$$A_2(f) = d \tag{5.14}$$

Equation (5.9) is equivalent to a two-input/single-output linear model with correlated input records as shown in Figure 5.1. These input records $X_1(f)$ and $X_2(f)$ produce correlated output records that, together with an uncorrelated

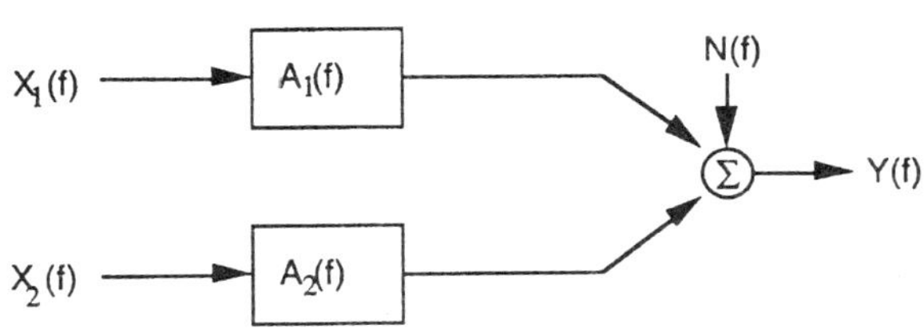

Figure 5.1. Two-input/single-output lnear model with correlated input records.

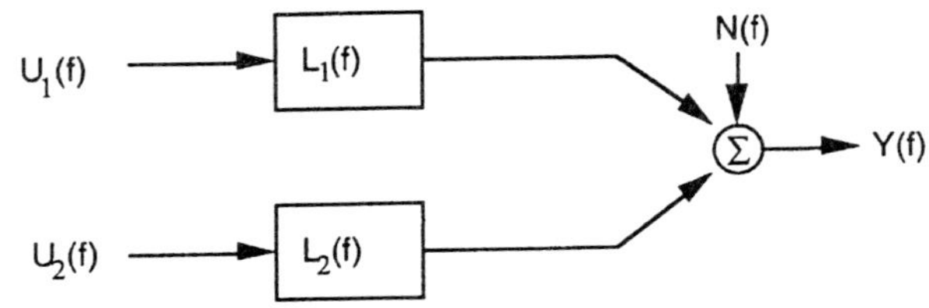

Figure 5.2. Two-input/single-output linear model with uncorrelated input records.

noise output record $N(f)$, sum to the total output record $Y(f)$. Practical procedures to identify the frequency response functions $A_1(f)$ and $A_2(f)$ from measured input/output random data are developed and applied in Chapter 4. These procedures show how to replace any given pair of correlated input records with a new pair of mathematically computed uncorrelated input records $U_1(f)$ and $U_2(f)$ that are calculated using conditioned spectral density functions as shown in Figure 5.2. Note also that the linear systems $A_1(f)$ and $A_2(f)$ in Figure 5.1 are replaced by different linear systems $L_1(f)$ and $L_2(f)$ in Figure 5.2. This is required to have the same $Y(f)$ and $N(f)$.

The constant physical parameters k, m, and c can be identified from calculation of the $A_1(f)$ of Eq. (5.13) by the relations

$$\text{Real}\,[A_1(f)] = k - (2\pi f)^2 m \tag{5.15}$$

$$\text{Imag}\,[A_1(f)] = (2\pi f)c \tag{5.16}$$

$$\text{Real}\,[A_1(f)] = k \quad \text{at} \quad f = 0 \tag{5.17}$$

$$\{\text{Real}\,[A_1(f)]\}/(2\pi f)^2 \to -m \quad \text{as}\, f \to \infty \tag{5.18}$$

$$\{\text{Imag}\,[A_1(f)]\}/(2\pi f) = c \quad \text{at all}\ f \tag{5.19}$$

The constant physical parameter d can be identified from inspection of the calculated $A_2(f)$ of Eq. (5.14).

Seven computer simulation cases are illustrated and discussed in this chapter. Equations (5.1) to (5.19) represent Case 1, namely:

Case 1. Linear and nonlinear damping, each without memory.

Consider next analytical and numerical extensions of Eq. (5.1) when the physical parameters of the linear and nonlinear damping terms have coefficients with memory as described by the following general SDOF nonlinear integrodifferential equation of motion:

$$m\ddot{u}(t) + \int_0^t c(\tau)\dot{u}(t-\tau)\,d\tau + ku(t) + \int_0^t d(\tau)z(t-\tau)\,d\tau = F(t) \tag{5.20}$$

where $F(t)$, $u(t)$, and $z(t)$ are the same as before in Eqs. (5.2)–(5.4), but the previous constant damping coefficients c and d are replaced by memory terms $c(\tau)$ and $d(\tau)$, respectively. Thus, the linear and nonlinear damping terms are now influenced by prior as well as by current motions. To be specific,

$$c(\tau) = \text{linear damping memory term} \tag{5.21}$$

$$d(\tau) = \text{nonlinear damping memory term} \tag{5.22}$$

Case 1 with constant linear and nonlinear damping coefficients occurs when $c(\tau) = c\delta(\tau)$ and $d(\tau) = d\delta(\tau)$, where $\delta(\tau)$ is the usual delta function.

Taking the Fourier transforms of both sides of Eq. (5.20) yields the result

$$A_1(f)X_1(f) + A_2(f)X_2(f) = Y(f) \tag{5.23}$$

where $X_1(f)$ and $X_2(f)$ are Eqs. (5.11) and (5.12), but now

$$A_1(f) = k - (2\pi f)^2 m + j(2\pi f)C(f) \tag{5.24}$$

$$A_2(f) = D(f) \tag{5.25}$$

The quantities $C(f)$ and $D(f)$ are

$$C(f) = \text{Fourier transform of } c(\tau) \tag{5.26}$$

$$D(f) = \text{Fourier transform of } d(\tau) \tag{5.27}$$

Equations (5.23)–(5.27) are general results for any forms of $C(f)$ and $D(f)$. For the nonmemory Case 1, $C(f) = c$ and $D(f) = d$.

Two important special cases of physical parameters with memory are treated in the next two sections, namely:

1. Exponential memory terms.
2. Exponential-cosine memory terms.

5.2.1 Exponential Memory Terms

Exponential memory terms for Eqs. (5.21) and (5.22) can be represented by

$$c(\tau) = ca\exp(-a\tau) \tag{5.28}$$

$$d(\tau) = dp\exp(-p\tau) \tag{5.29}$$

where c and d are the linear and nonlinear damping constants from Eqs. (5.6) and (5.8), respectively, and where

$$a = \text{exponential memory term in } c(\tau) \tag{5.30}$$

$$p = \text{exponential memory term in } d(\tau) \tag{5.31}$$

The Fourier transforms of Eqs. (5.28) and (5.29) are

$$C(f) = ca/[a + j(2\pi f)] \tag{5.32}$$

$$D(f) = dp/[p + j(2\pi f)] \tag{5.33}$$

The nonmemory Case 1 occurs when a and p approach infinity.

For this exponential memory case, the integrodifferential Eq. (5.20) can be expressed using the following relations:

$$m\ddot{u}(t) + u_c(t) + ku(t) + u_d(t) = F(t) \tag{5.34}$$

$$\dot{u}_c(t) + au_c(t) = ca\dot{u}(t) \tag{5.35}$$

$$\dot{u}_d(t) + pu_d(t) = dpz(t) \tag{5.36}$$

where Eqs. (5.35) and (5.36) result from taking Fourier transforms of Eqs. (5.37) and (5.38) and converting back to the time domain since $u_c(t)$ and $u_d(t)$ are given by

$$u_c(t) = \int_0^t ca\,[\exp(-a\tau)]\dot{u}(t - \tau)\,d\tau \tag{5.37}$$

$$u_d(t) = \int_0^t dp\,[\exp(-p\tau)]z(t - \tau)\,d\tau \tag{5.38}$$

with

$$z(t) = \begin{cases} [\dot{u}(t)]^2 & \text{if } \dot{u}(t) > 0 \quad (5.39) \\ -[\dot{u}(t)]^2 & \text{if } \dot{u}(t) < 0 \quad (5.40) \end{cases}$$

Equations (5.24) and (5.25) become here

$$A_1(f) = k - (2\pi f)^2 m + j(2\pi f)ca/[a + j(2\pi f)] \tag{5.41}$$

$$A_2(f) = dp/[p + j(2\pi f)] \tag{5.42}$$

The real and imaginary parts of Eqs. (5.41) and (5.42) are

$$\text{Real}\,[A_1(f)] = k - (2\pi f)^2 m + ca(2\pi f)^2/[a^2 + (2\pi f)^2] \tag{5.43}$$

$$\text{Imag}\,[A_1(f)] = ca^2(2\pi f)/[a^2 + (2\pi f)^2] \tag{5.44}$$

$$\text{Real}\,[A_2(f)] = dp^2/[p^2 + (2\pi f)^2] \tag{5.45}$$

$$\text{Imag}\,[A_2(f)] = -dp(2\pi f)/[p^2 + (2\pi f)^2] \tag{5.46}$$

The physical parameters k, m, c, and a can be identified from calculation of

Eqs. (5.43) and (5.44) by the relations

$$\text{Real}\,[A_1(f)] = k \quad \text{at } f = 0 \tag{5.47}$$

$$\{\text{Real}\,[A_1(f)]\}/(2\pi f)^2 \to -m \quad \text{as } f \to \infty \tag{5.48}$$

$$\{\text{Real}\,[A_1(f)] - k + (2\pi f)^2 m\}/(2\pi f)^2 = c/a \quad \text{at } f = 0 \tag{5.49}$$

$$\{\text{Imag}\,[A_1(f)]\}/(2\pi f) = ca^2/[a^2 + (2\pi f)^2] \tag{5.50}$$

$$(2\pi f)\,\text{Imag}\,[A_1(f)] \to ca^2 \quad \text{as } f \to \infty \tag{5.51}$$

The physical parameters d and p can be identified from calculation of Eqs. (5.45) and (5.46) by the relations

$$\text{Real}\,[A_2(f)] = d \quad \text{at } f = 0 \tag{5.52}$$

$$\{\text{Imag}\,[A_2(f)]\}/(2\pi f) = -(d/p) \quad \text{at } f = 0 \tag{5.53}$$

$$(2\pi f)\,\text{Imag}\,[A_2(f)] \to -dp \quad \text{as } f \to \infty \tag{5.54}$$

Equations (5.28)–(5.54) represent Case 4 in this chapter, namely:

Case 4. Linear and nonlinear damping, each with exponential memory. Special situations of Case 4 are:

Case 2. Linear damping with exponential memory.

Case 3. Nonlinear damping with exponential memory.

For Case 2, the appropriate relations for Eqs. (5.21) and (5.22) are

$$c(\tau) = ca\exp(-a\tau) \qquad d(\tau) = d\delta(\tau) \tag{5.55}$$

where $\delta(\tau)$ is the usual delta function. Here, the Fourier transform of $C(f)$ is given by Eq. (5.32) but the Fourier transform of $D(f)$ is

$$D(f) = d \tag{5.56}$$

which is the result of Eq. (5.33) when $p \to \infty$.

For Case 3, the appropriate relations for Eqs. (5.21) and (5.22) are

$$c(\tau) = c\delta(\tau) \qquad d(\tau) = dp\exp(-p\tau) \tag{5.57}$$

where $\delta(\tau)$ is the usual delta function. Here, the Fourier transform of $D(f)$ is given by Eq. (5.33) but the Fourier transform of $C(f)$ is

$$C(f) = c \tag{5.58}$$

which is the result of Eq. (5.32) when $a \to \infty$.

5.2.2 Exponential-Cosine Memory Terms

This case is an extension of the analytical relations in the previous section where exponential-cosine memory terms for Eqs. (5.21) and (5.22) can be represented by

$$c(\tau) = ca\exp(-a\tau)\cos b\tau \tag{5.59}$$

$$d(\tau) = dp\exp(-p\tau)\cos q\tau \tag{5.60}$$

Here, c and d are the linear and nonlinear damping constants from Eqs. (5.6) and (5.8), respectively; a and p are the exponential memory terms from Eqs. (5.30) and (5.31), respectively; and

$$b = 2\pi f_b = \text{cosine memory term in } c(\tau) \tag{5.61}$$

$$q = 2\pi f_q = \text{cosine memory term in } d(\tau) \tag{5.62}$$

These results reduce to the previous exponential case when the cosine memory terms b and q are both zero. The Fourier transforms of Eqs. (5.59) and (5.60) are

$$C(f) = ca(a + j2\pi f)/[(a + j2\pi f)^2 + b^2] \tag{5.63}$$

$$D(f) = dp(p + j2\pi f)/[(p + j2\pi f)^2 + q^2] \tag{5.64}$$

For this exponential-cosine memory case, the integrodifferential Eq. (5.20) can be expressed using the following relations:

$$m\ddot{u}(t) + u_c(t) + ku(t) + u_d(t) = F(t) \tag{5.65}$$

$$\ddot{u}_c(t) + 2a\dot{u}_c(t) + (a^2 + b^2)u_c(t) = ca[\ddot{u}(t) + a\dot{u}(t)] \tag{5.66}$$

$$\ddot{u}_d(t) + 2p\dot{u}_d(t) + (p^2 + q^2)u_d(t) = dp[\dot{z}(t) + pz(t)] \tag{5.67}$$

where Eqs. (5.66) and (5.67) result from taking Fourier transforms of Eqs. (5.68) and (5.69) and converting back to the time domain since $u_c(t)$ and $u_d(t)$ are given here by

$$u_c(t) = \int_0^t ca[\exp(-a\tau)](\cos b\tau)\dot{u}(t - \tau)\,d\tau \tag{5.68}$$

$$u_d(t) = \int_0^t dp[\exp(-p\tau)](\cos q\tau)z(t - \tau)\,d\tau \tag{5.69}$$

The quantity $z(t)$ is given in Eqs. (5.39) and (5.40) with

$$\dot{z}(t) = \begin{cases} 2\dot{u}(t)\ddot{u}(t) & \text{if } \dot{u}(t) > 0 \quad (5.70) \\ -2\dot{u}(t)\ddot{u}(t) & \text{if } \dot{u}(t) < 0 \quad (5.71) \end{cases}$$

Equations (5.24) and (5.25) become here

$$A_1(f) = k - (2\pi f)^2 m + j(2\pi f)ca(a + j2\pi f)/[(a + j2\pi f)^2 + b^2] \tag{5.72}$$

$$A_2(f) = dp(p + j2\pi f)/[(p + j2\pi f)^2 + q^2] \tag{5.73}$$

The real and imaginary parts of Eqs. (5.72) and (5.73) are

$$\text{Real}[A_1(f)] = k - (2\pi f)^2 m + ca(2\pi f)^2(a^2 - b^2 + 4\pi^2 f^2)/D_1(f) \tag{5.74}$$

$$\text{Imag}[A_1(f)] = ca^2(2\pi f)(a^2 + b^2 + 4\pi^2 f^2)/D_1(f) \tag{5.75}$$

with

$$D_1(f) = (a^2 + b^2 - 4\pi^2 f^2)^2 + 16a^2\pi^2 f^2 \tag{5.76}$$

$$\text{Real}[A_2(f)] = dp^2(p^2 + q^2 + 4\pi^2 f^2)/D_2(f) \tag{5.77}$$

$$\text{Imag}[A_2(f)] = -dp(2\pi f)(p^2 - q^2 + 4\pi^2 f^2)/D_2(f) \tag{5.78}$$

and

$$D_2(f) = (p^2 + q^2 - 4\pi^2 f^2)^2 + 16p^2\pi^2 f^2 \tag{5.79}$$

The physical parameters k, m, c, a, and b can be identified from calculation of Eqs. (5.74) and (5.75) by the relations

$$\text{Real}[A_1(f)] = k \quad \text{at } f = 0 \tag{5.80}$$

$$\{\text{Real}[A_1(f)]\}/(2\pi f)^2 \to -m \quad \text{as } f \to \infty \tag{5.81}$$

$$\text{Real}[A_1(f)] \to ca \quad \text{as } f \to \infty \tag{5.82}$$

$$\{\text{Imag}[A_1(f)]\}/(2\pi f) = ca^2(a^2 + b^2 + 4\pi^2 f^2)/D_1(f) \tag{5.83}$$

$$(2\pi f)\,\text{Imag}[A_1(f)] \to ca^2 \quad \text{as } f \to \infty \tag{5.84}$$

The physical parameters d, p, and q can be identified from calculation of Eqs. (5.77) and (5.78) by the relations

$$\text{Real}[A_2(f)] = dp^2/(p^2 + q^2) \quad \text{at } f = 0 \tag{5.85}$$

$$\{\text{Imag}[A_2(f)]\}/(2\pi f) = -dp(p^2 - q^2)/(p^2 + q^2)^2 \quad \text{at } f = 0 \tag{5.86}$$

$$(2\pi f)\,\text{Imag}[A_2(f)] \to -dp \quad \text{as } f \to \infty \tag{5.87}$$

Equations (5.59)–(5.87) represent Case 7 in this chapter, namely:

Case 7. Linear and nonlinear damping, each with exponential-cosine memory.

Special situations of Case 7 are:

Case 5. Linear damping with exponential-cosine memory.
Case 6. Nonlinear damping with exponential-cosine memory.

For Case 5, the appropriate relations for Eqs. (5.21) and (5.22) are

$$c(\tau) = ca\exp(-a\tau)\cos b\tau \qquad d(\tau) = d\delta(\tau) \tag{5.88}$$

where $\delta(\tau)$ is the usual delta function. Here, the Fourier transform of $C(f)$ is given by Eq. (5.65) but the Fourier transform of $D(f)$ is

$$D(f) = d \tag{5.89}$$

which is the result of Eq. (5.64) when $p \to \infty$.

For Case 6, the appropriate relations for Eqs. (5.21) and (5.22) are

$$c(\tau) = c\delta(\tau) \qquad d(\tau) = dp\exp(-p\tau)\cos q\tau \tag{5.90}$$

where $\delta(\tau)$ is the usual delta function. Here, the Fourier transform of $D(f)$ is given by Eq. (5.66) but the Fourier transform of $C(f)$ is

$$C(f) = c \tag{5.91}$$

which is the result of Eq. (5.63) when $a \to \infty$.

5.3 COMPUTER SIMULATION STUDIES

Computer algorithms that perform multiple-input/multiple-output (MI/MO) linear system analysis, as well as probability and spectral calculations, were initially designed as MATLAB .m files (reference 35). The .m files employ routines contained in the MATLAB Signal Processing Toolbox (reference 36). These advanced capabilities have been extended and are now part of the MATLAB Integrated Test Analysis Processor (ITAP) TOOLBOX, under development by Dr. Robert Coppolino in Los Angeles, California. The computer data processing work on the simulated data in this chapter, and on the measured data in Chapters 4, 6, and 7 was done by Dr. Coppolino with the MATLAB/ITAP programs.

Key routines available in MATLAB/ITAP include the following:

1. **Probl.m** Preliminary evaluation of individual time history records that compute and display (a) a selected time history segment, (b) probability density with comparison to a standardized Gaussian probability density, (c) autospectrum, and (d) key numerical properties.
2. **Mimogen.m** General purpose MI/MO analysis for an arbitrary number of input and output time history records. This routine computes and displays estimated frequency response functions and estimated cumulative coherence functions. Special cases are automatically treated by MI/SO, SI/MO, and SI/SO analysis.
3. MATLAB/ITAP also contains other routines to perform dynamic modeling, response simulation, analysis of measured data, and experimental modal identification.

Material in this chapter presents MATLAB/ITAP results from analyzing simulated data for seven SDOF damping cases:

One case of a nonlinear system without memory

Three cases of nonlinear systems with exponential memory

Three cases of nonlinear systems with exponential-cosine memory

Numerical solutions for these seven SDOF cases were obtained using the Adams–Bashford predictor–corrector procedure (reference 30), when the systems were subjected to broadband random excitation. The excitation signals were defined using (1) the random number generator in MATLAB (normal distribution option) and (2) an eight-pole Butterworth digital filter set to yield random signals with roughly uniform spectral content from 0 to 25 Hz. Required MI/SO spectral analyses were performed on the simulated excitation and response time history records to estimate the reverse dynamic system frequency response functions. The spectral analyses were conducted with a Hanning window to reduce leakage and with 50% overlapping to recover "lost" data.

The seven SDOF damping cases are as follows:

Case 1. Linear and nonlinear damping, each without memory.

Case 2. Linear damping with exponential memory.

Case 3. Nonlinear damping with exponential memory.

Case 4. Linear and nonlinear damping, each with exponential memory.

Case 5. Linear damping with exponential-cosine memory.

Case 6. Nonlinear damping with exponential-cosine memory.

Case 7. Linear and nonlinear damping, each with exponential-cosine memory.

TABLE 5.1 Notaton for SDOF Equations

m = mass coefficient
c = linear damping constant coefficient
k = linear stiffness constant coefficient
$c(\tau) = ca\exp(-a\tau)\cos b\tau$ = linear damping memory term
a = linear damping exponential term
$b = 2\pi f_b$ = linear damping periodic term
d = nonlinear damping constant coefficient
$d(\tau) = dp\exp(-p\tau)\cos q\tau$ = nonlinear damping memory term
p = nonlinear damping exponential term
$q = 2\pi f_q$ = nonlinear damping periodic term
RMS_F = root mean square value for the random excitation level of the force input data

TABLE 5.2 Simulated System Parameters

Parameter	Case 1	Case 2	Case 3	Case 4	Case 5	Case 6	Case 7
m	1.0	1.0	1.0	1.0	1.0	1.0	1.0
c	3.77	3.77	3.77	3.77	3.77	3.77	3.77
k	355.3	355.3	355.3	355.3	355.3	355.3	355.3
a	∞	40	∞	40	40	∞	40
b	0	0	0	0	10π	0	10π
d	100	100	100	100	100	100	100
p	∞	∞	20	20	∞	20	20
q	0	0	0	0	0	5π	5π
RMS_F	5.0	5.0	5.0	5.0	5.0	5.0	5.0

TABLE 5.3 Digital Data Processing Parameters

f_s = sampling frequency = 100 Hz
$\Delta t = 1/f_s$ = time interval between samples = 0.01 s
$f_c = f_s/2$ = Nyquist cutoff frequency = 50 Hz
N = number of sample values per subrecord = 1024
$T = N\,\Delta t$ = time of each subrecord = 10.24 s
n_d = number of distinct (disjoint, different) subrecords = 32
$N_{\text{total}} = n_d N$ = total number of sample values = 32,768
$T_{\text{total}} = n_d T$ = total time for all subrecords = 327.68 s
$\Delta f = 1/T$ = bandwidth resolution = 0.097 Hz

TABLE 5.4 Figures from Computer Simulations

	Case 1	Case 2	Case 3	Case 4	Case 5	Case 6	Case 7
$A_1(f)$	5.5	5.13	5.21	5.29	5.37	5.45	5.53
$A_2(f)$	5.6	5.14	5.22	5.30	5.38	5.46	5.54
Cumulative coherence	5.7	5.15	5.23	5.31	5.39	5.47	5.55
Real $[A_1(f)/(2\pi f)^2]$	5.8	5.16	5.24	5.32	5.40	5.48	5.56
Imag $[A_1(f)/(2\pi f)]$	5.9	5.17	5.25	5.33	5.41	5.49	5.57
$(2\pi f)$ Imag $[A_1(f)]$	5.10	5.18	5.26	5.34	5.42	5.50	5.58
Real $[A_2(f)]$	5.11	5.19	5.27	5.35	5.43	5.51	5.59
SI/SO vs MI/SO	5.12	5.20	5.28	5.36	5.44	5.52	5.60

The notation for the various SDOF linear and nonlinear equations with memory coefficients is given in Table 5.1. Simulated system parameters are shown in Table 5.2. From Eqs. (4.84) and (4.85), these system parameters simulate a SDOF linear system where the undamped natural frequency $f_n = 3.0$ Hz and the damping ratio $\zeta = 0.10$. The digital data processing parameters are listed in Table 5.3. Table 5.4 cross-lists the graphical results according to the case type and the particular function computed in each figure. This table is included to allow for quick inspection and comparison of the various results found in this investigation.

5.4 DISCUSSION OF SIMULATION RESULTS

Random excitation and response time history records from all seven cases were analyzed using Reverse MI/SO spectral analysis techniques. As shown in Figure 5.1, the two reverse dynamic mathematical inputs consist of the displacement response $u(t)$ as the first mathematical input, and the nonlinear viscous damping term $\dot{u}(t)|\dot{u}(t)|$ as the second mathematical input. The single reverse dynamic mathematical output term is the excitation force $F(t)$.

Throughout the seven cases studied, moderate statistical bias and random errors occur in the estimates for the linear damping functions. Relatively small statistical bias and random errors occur in the estimates of the nonlinear damping functions. These statistical errors do not compromise the fidelity of the overall identified linear and nonlinear systems as indicated by the high values of the cumulative coherence functions shown for the seven cases in Figures 5.7, 5.15, 5.23, 5.31, 5.39, 5.47, and 5.55, respectively. The results for all seven cases will now be discussed.

Time history segments for $F(t)$ and $u(t)$ are illustrated in Figure 5.3 and the corresponding autospectra are given in Figure 5.4. The same data were used for Cases 1–7.

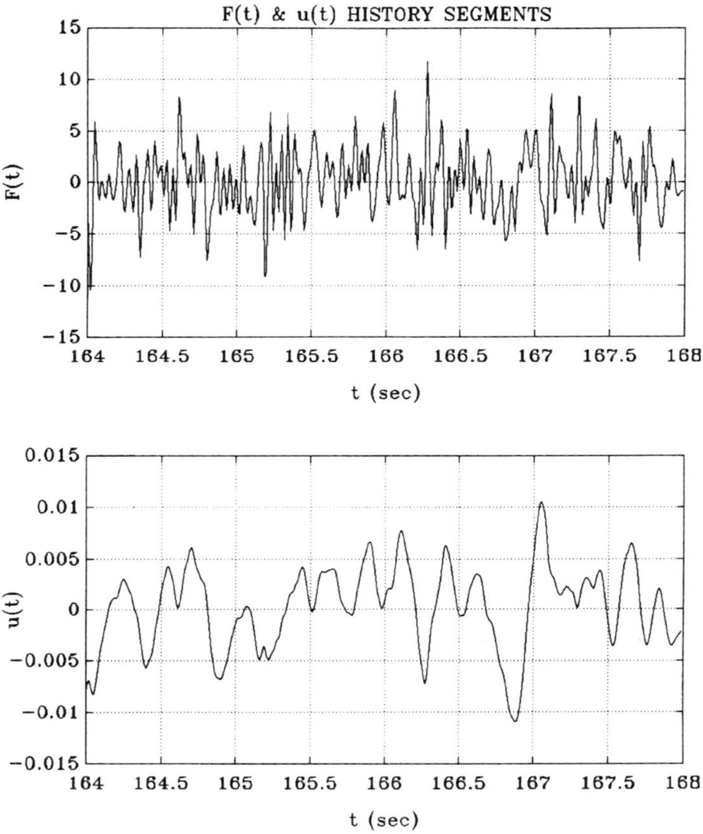
F(t) & u(t) HISTORY SEGMENTS
F(t)
15
10
5
0
−5
−10
−15
164
164.5
165
165.5
166
166.5
167
167.5
168
t (sec)
u(t)
0.015
0.01
0.005
0
−0.005
−0.01
−0.015
164
164.5
165
165.5
166
166.5
167
167.5
168
t (sec)

Figure 5.3.

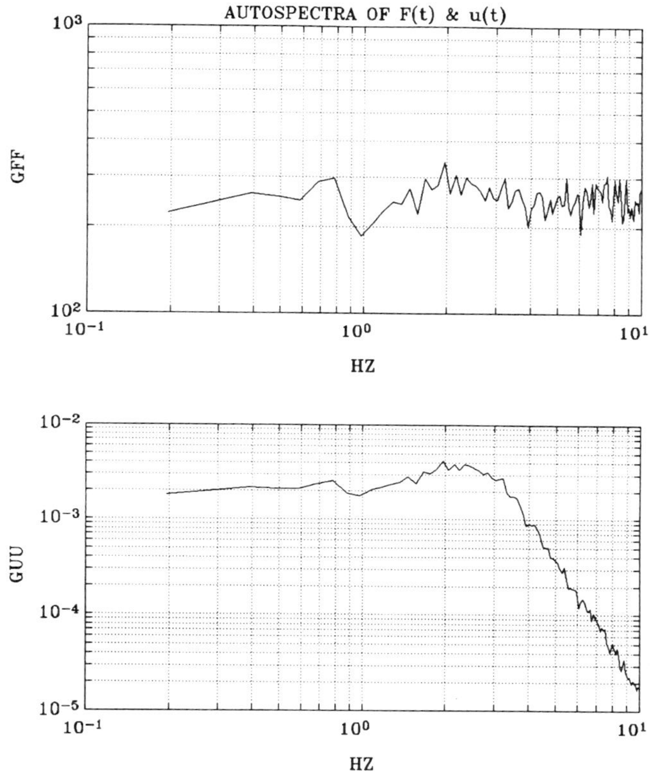

Figure 5.4.

Case 1. Linear and Nonlinear Damping, Each Without Memory. The calculated frequency response functions associated with the linear and nonlinear system components in Case 1 are displayed in Figures 5.5 and 5.6, respectively. In each of these figures, the Reverse MI/SO spectral analysis result is depicted by the circles at discrete frequencies, and the exact analytical reference result is shown as a solid line curve. This convention is used for all seven example cases to compare the MI/SO results with the exact reference results.

The close comparison of MI/SO functions with the true frequency response functions is confirmed here by the cumulative coherence functions shown in Figure 5.7. The near unity value of the lower coherence component indicates that most of the output autospectrum can be accounted for by an optimum linear approximation. However, as shown later in Figure 5.12 and in related figures for the other cases, such optimum linear results can be completely wrong and will produce erroneous physical parameter values for modal analysis work when the nonlinear effects are ignored. This fact is a significant result from these numerical computer simulation studies!

Asymptotic properties of the linear frequency response function, $A_1(f)$, are described by the plots shown in Figures 5.8–5.10. Equations (5.15)–(5.19) can be used to estimate the physical parameters k, m, and c. The result in Figure 5.9 estimates with a small positive bias error the simulated value c for the constant linear damping coefficient, in agreement with Eq. (5.19). No bias error appears in Figure 5.8 and only a slight bias error in Figure 5.10. Properties of the nonlinear frequency response function, $A_2(f)$, illustrated in Figure 5.11, give the correct simulated value d for the constant nonlinear damping coefficient without any bias error, in agreement with Eq. (5.14).

Figure 5.12 compares the estimates for $A_1(f)$ from the appropriate MI/SO based reverse dynamic system analysis and from the wrong SI/SO analysis, where it is implicitly assumed that the subject system is linear. Circles represent the properly computed MI/SO estimates, and asterisks represent the erroneously computed SI/SO estimates. Note the substantial difference in the two estimates even though the coherence functions in Figure 5.7 show that the system nonlinearity is small.

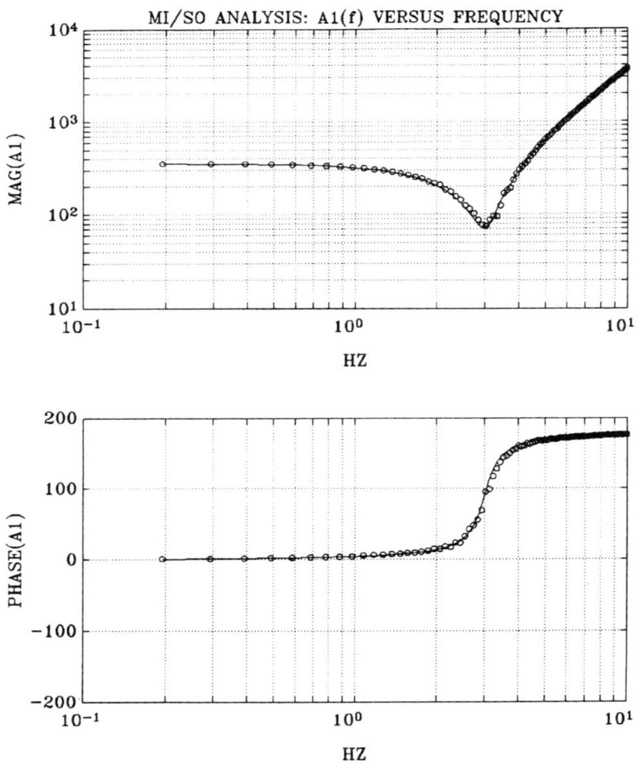

Figure 5.5.

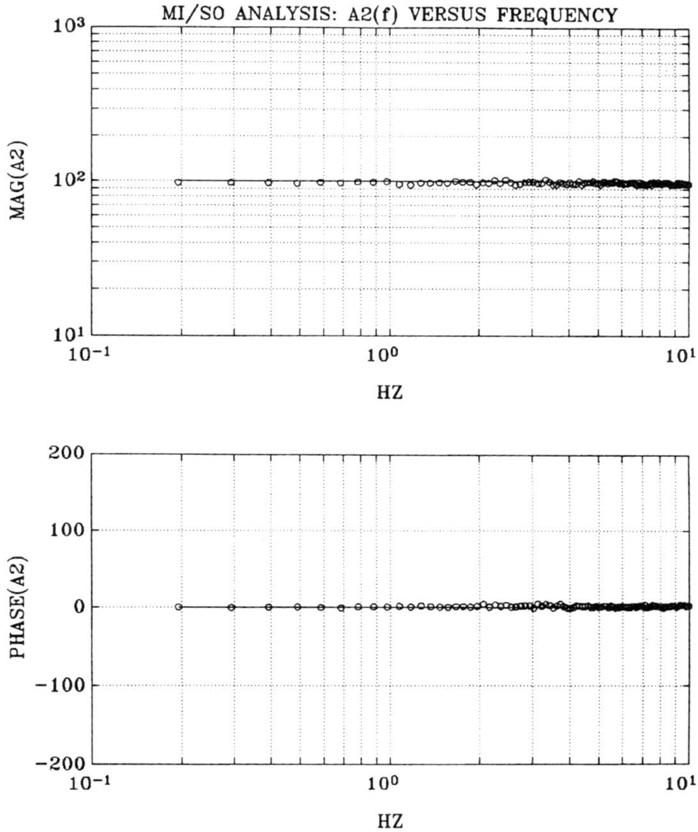
MI/SO ANALYSIS: A2(f) VERSUS FREQUENCY
MAG(A2)
10³
10²
10¹
10⁻¹
10⁰
10¹
HZ
PHASE(A2)
200
100
0
−100
−200
10⁻¹
10⁰
10¹
HZ

Figure 5.6.

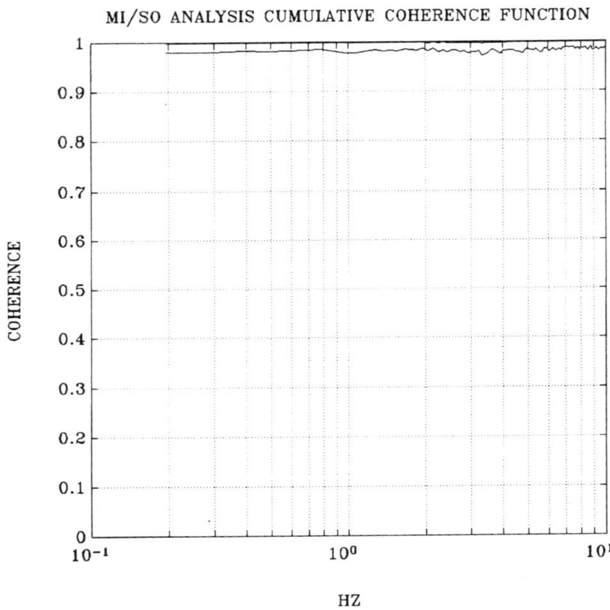

Figure 5.7.

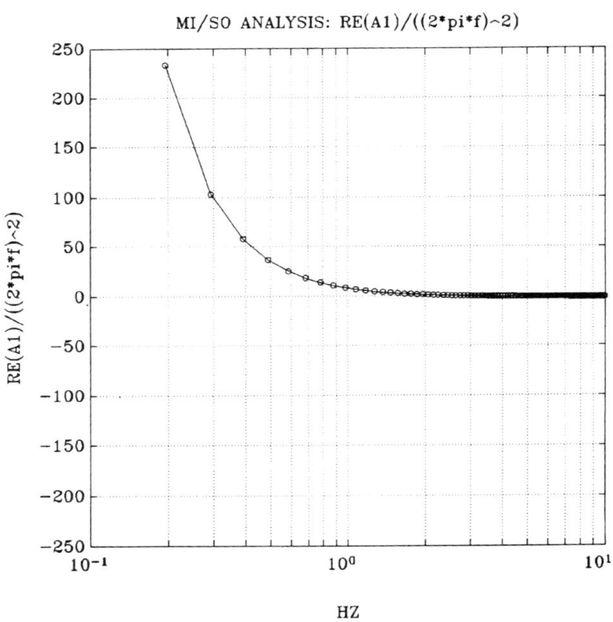

Figure 5.8.

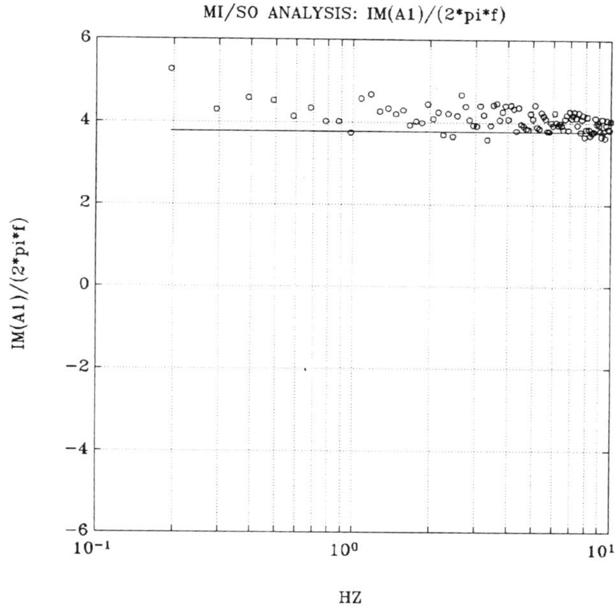

Figure 5.9.

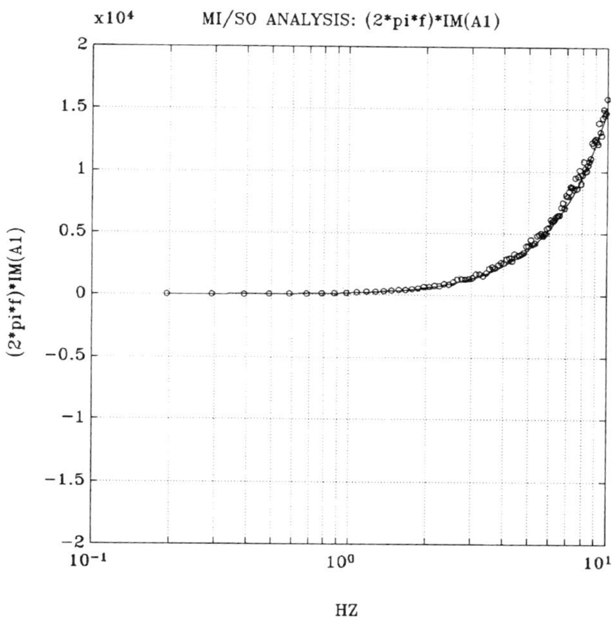

Figure 5.10.

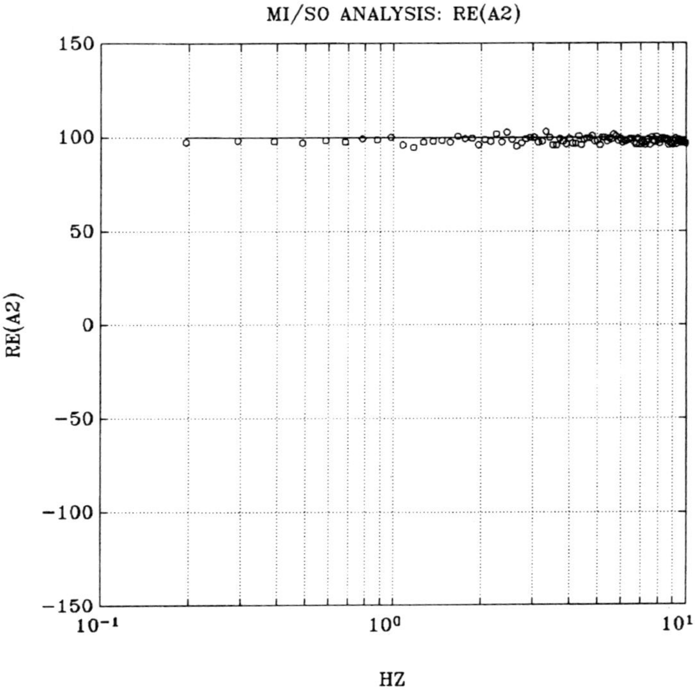

Figure 5.11.

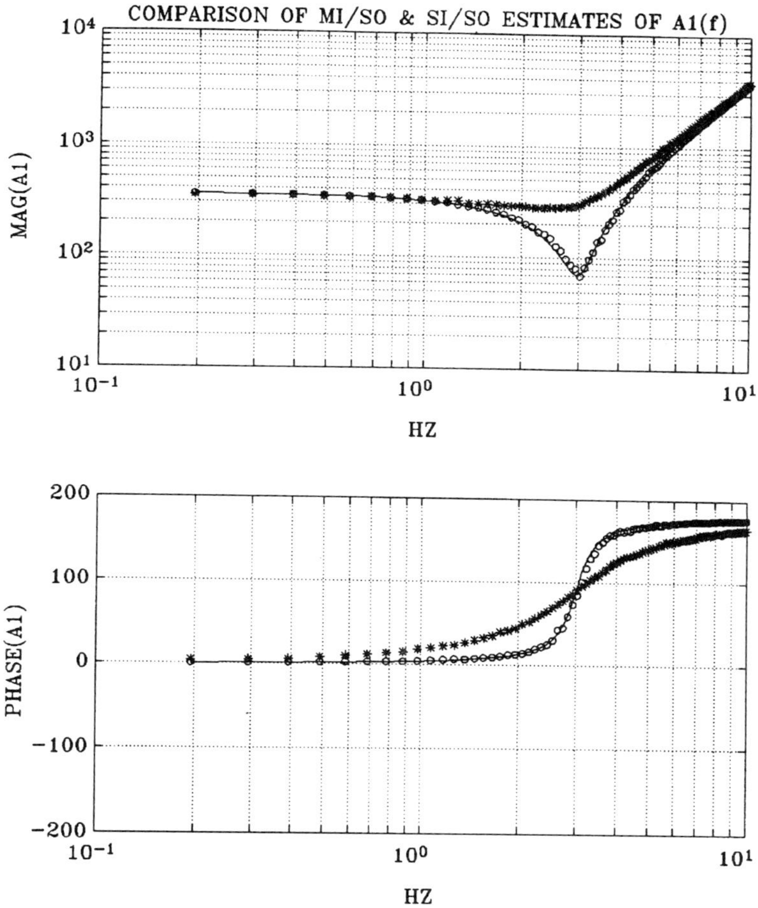

Figure 5.12.

Case 2. Linear Damping with Exponential Memory. The calculated frequency response functions associated with the linear and nonlinear system components in Case 2 are displayed in Figures 5.13 and 5.14. Note that Figure 5.13 by itself does not indicate any linear damping with memory since it is very similar to the previous Figure 5.5 for the nonmemory case. The close comparison of MI/SO functions with the exact frequency response functions is confirmed here by the cumulative coherence functions shown in Figure 5.15. The near unity value of the lower coherence component indicates that most of the output autospectrum can be accounted for by an optimum linear approximation that may not give correct physical parameter values.

Asymptotic properties of the linear frequency response function, $A_1(f)$, are described by the plots shown in Figures 5.16–5.18. Equations (5.47)–(5.51) can be used to estimate the physical parameters k, m, c, and a. The result in Figure 5.17 has small bias error and indicates linear damping here with exponential memory, in agreement with Eq. (5.44). Figure 5.17 is quite different from the previous Case 1 nonmemory result in Figure 5.9. Properties of the nonlinear frequency response function, $A_2(f)$, illustrated in Figure 5.19, give the correct value here for the constant nonlinear damping coefficient d with negligible bias error, in agreement with Eq. (5.56).

Figure 5.20 compares the estimates for $A_1(f)$ from the appropriate MI/SO based reverse dynamic system analysis and from the wrong SI/SO analysis, where it is implicitly assumed that the subject system is linear. Circles represent the properly computed MI/SO estimates, and asterisks represent the erroneously computed SI/SO estimates. There are substantial differences in the two estimates even though the coherence functions in Figure 5.15 show that the system nonlinearity is not strong.

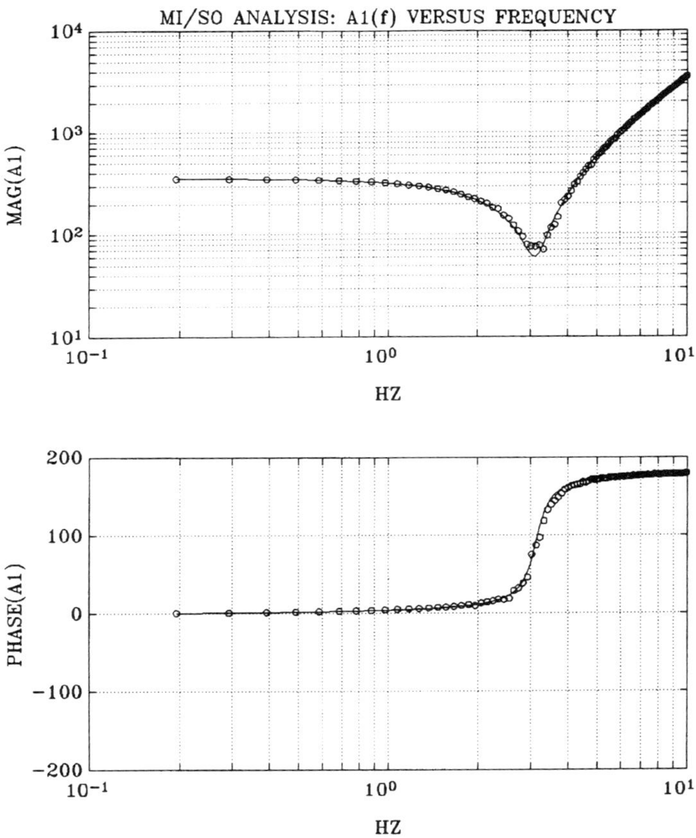

Figure 5.13.

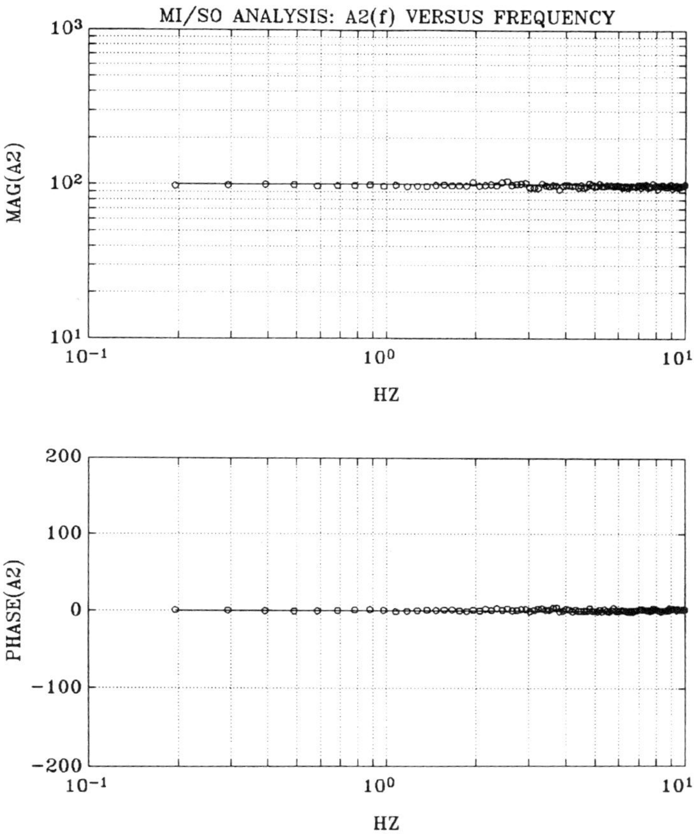
MI/SO ANALYSIS: A2(f) VERSUS FREQUENCY
MAG(A2)
10^3
10^2
10^1
10^-1
10^0
10^1
HZ
PHASE(A2)
200
100
0
-100
-200
10^-1
10^0
10^1
HZ

Figure 5.14.

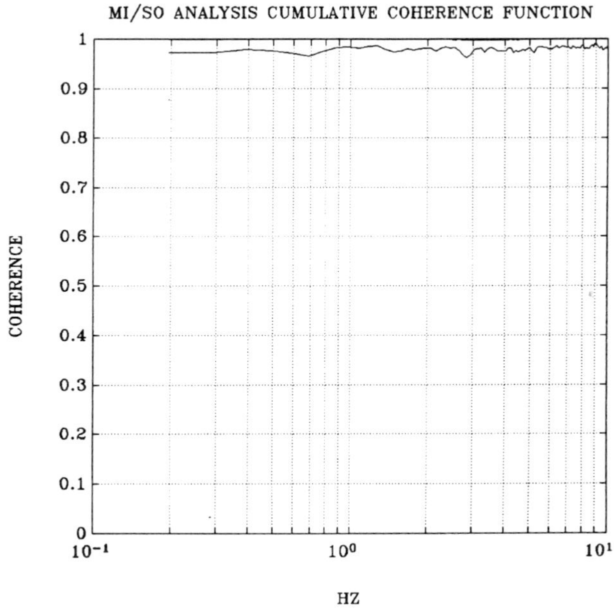

Figure 5.15.

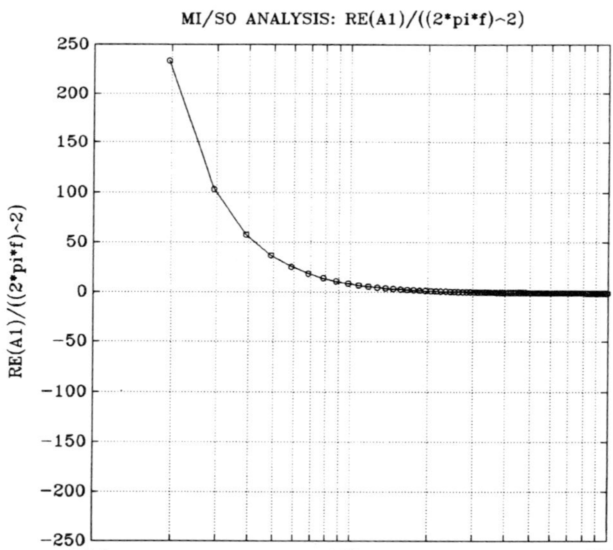

Figure 5.16.

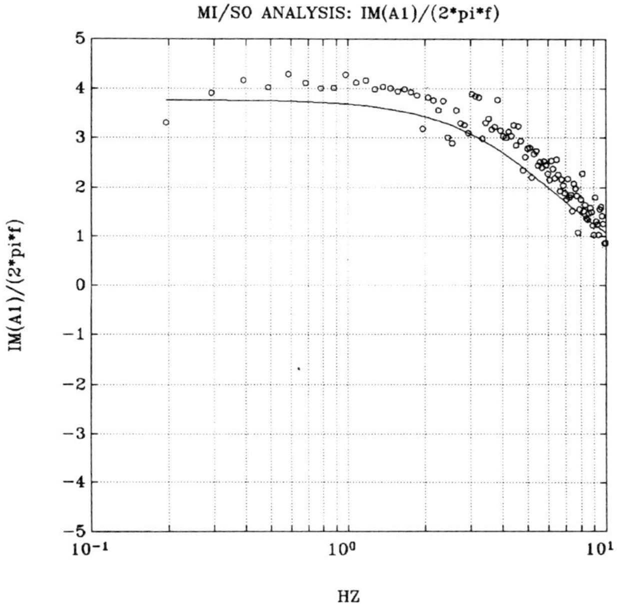

Figure 5.17.

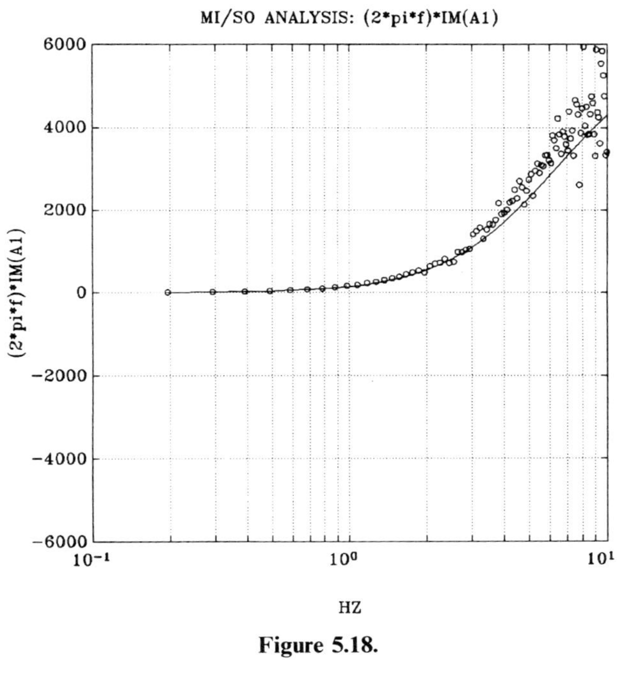

Figure 5.18.

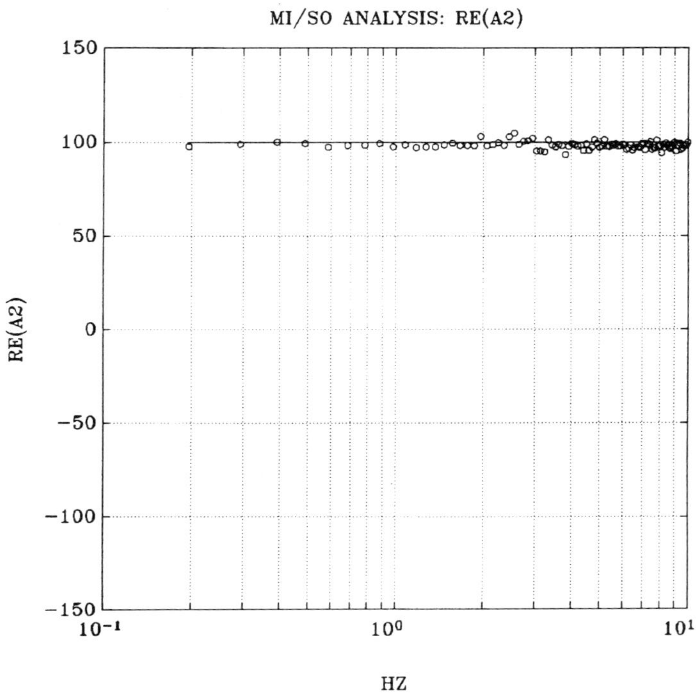

Figure 5.19.

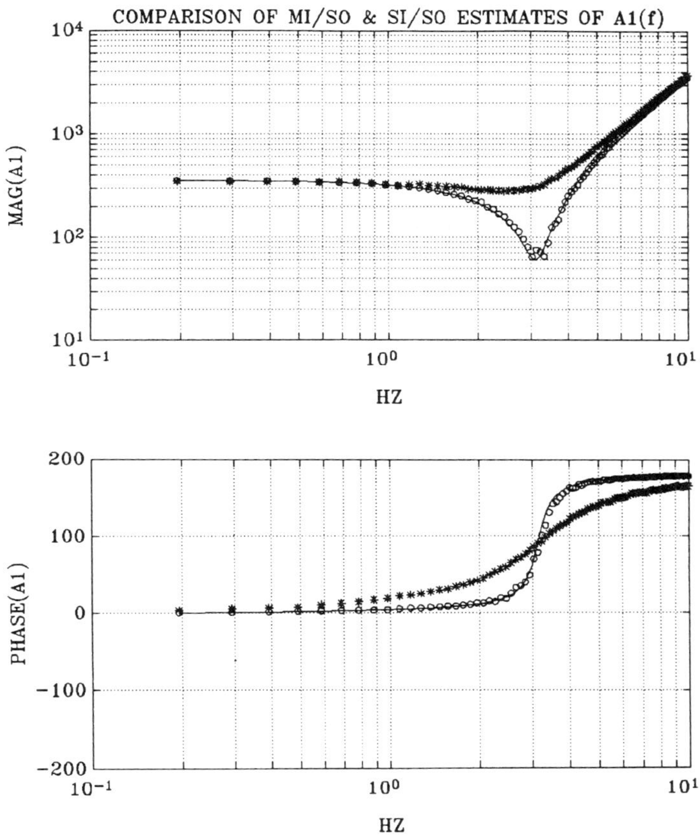

Figure 5.20.

Case 3. Nonlinear Damping with Exponential Memory. The calculated frequency response functions associated with the linear and nonlinear system components in Case 3 are displayed in Figures 5.21 and 5.22. Note that Figure 5.22, in agreement with Eq. (5.33), indicates nonlinear damping with memory since it is quite different from the previous Figures 5.6 and 5.14. The close comparison of MI/SO functions with the exact frequency response functions is confirmed here by the cumulative coherence functions shown in Figure 5.23. The near unity value of the lower coherence component indicates that most of the output autospectrum can be accounted for by an optimum linear approximation that may not give correct physical parameter values.

Asymptotic properties of the linear frequency response function, $A_1(f)$, are described by the plots shown in Figures 5.24–5.26. The result in Figure 5.25 has some bias error but still indicates a constant linear damping coefficient c, in agreement with Eq. (5.58). Properties of the nonlinear frequency response function, $A_2(f)$, illustrated in Figure 5.27, indicate nonlinear damping here with exponential memory, in agreement with Eq. (5.45). Note how Figure 5.27 differs from the previous results in Figure 5.11 or 5.19. Equations (5.52) and (5.53) can be used to estimate the physical parameters d and p.

Figure 5.28 compares the estimates for $A_1(f)$ from the appropriate MI/SO based reverse dynamic system analysis and from the wrong SI/SO analysis, where it is implicitly assumed that the subject system is linear. Circles represent the properly computed MI/SO estimates, and asterisks represent the erroneously computed SI/SO estimates. As before, the two estimates are quite different even though the coherence functions in Figure 5.23 show that the system nonlinearity is not strong.

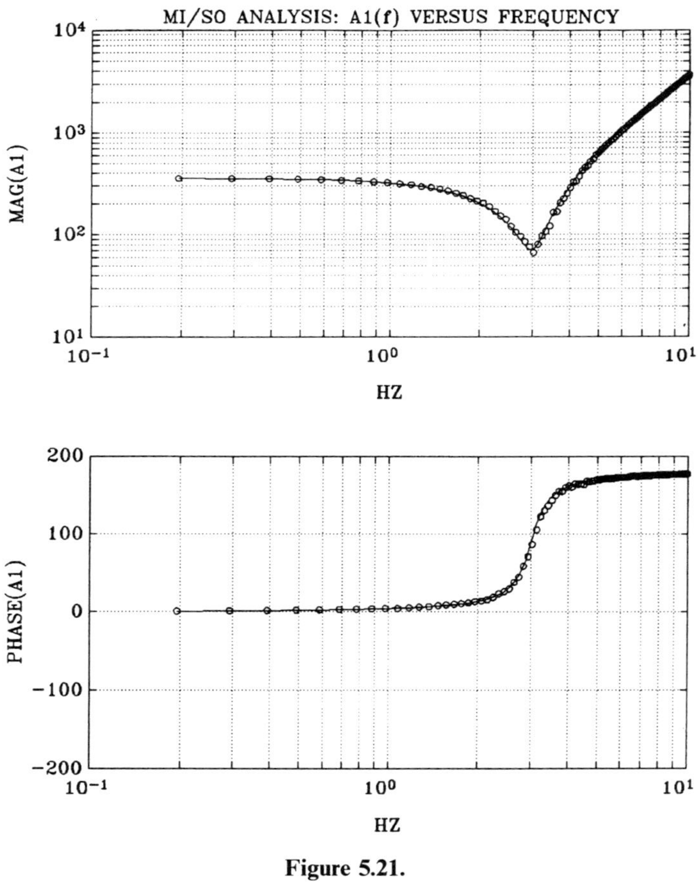

Figure 5.21.

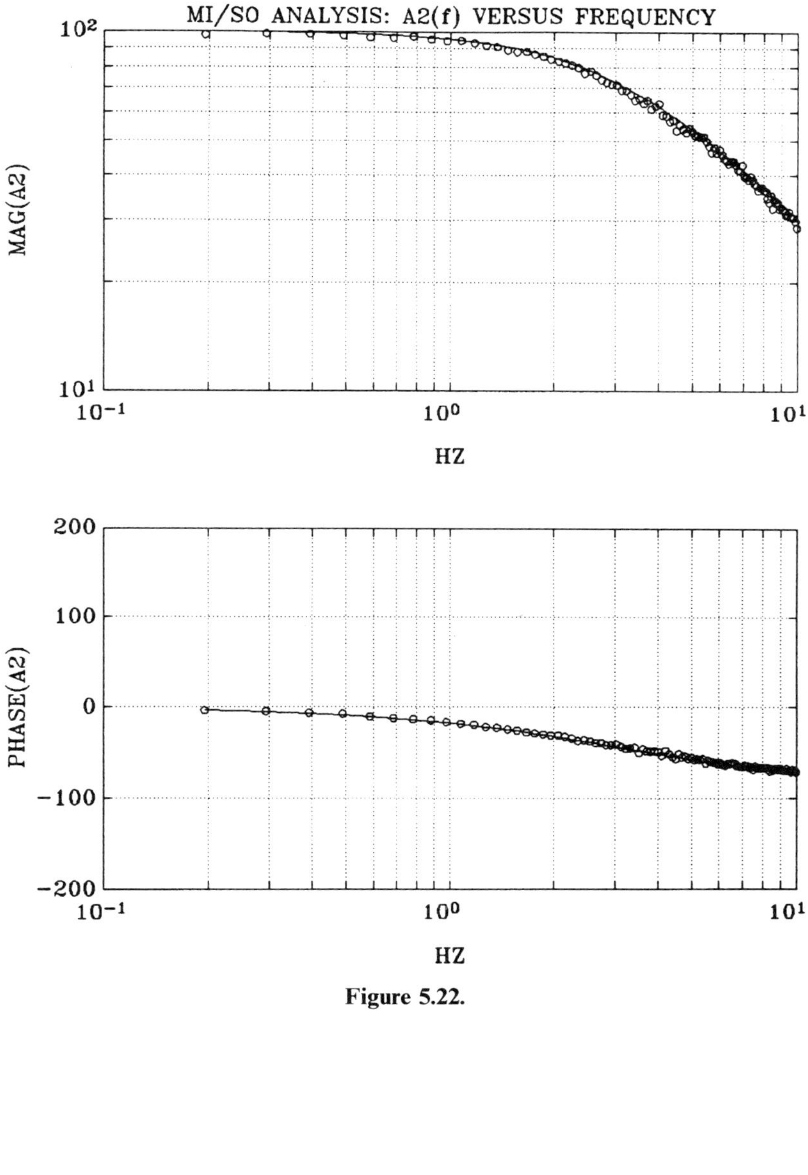
MI/SO ANALYSIS: A2(f) VERSUS FREQUENCY
MAG(A2)
10^2
10^1
10^{-1}
10^0
10^1
HZ
PHASE(A2)
200
100
0
−100
−200
10^{-1}
10^0
10^1
HZ

Figure 5.22.

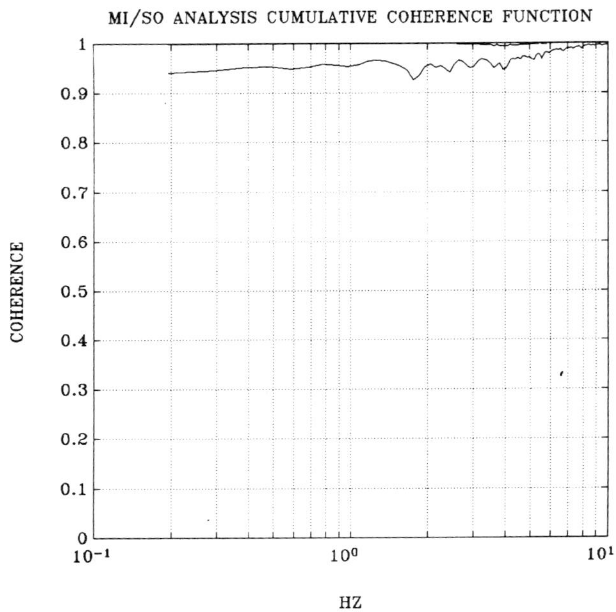

Figure 5.23.

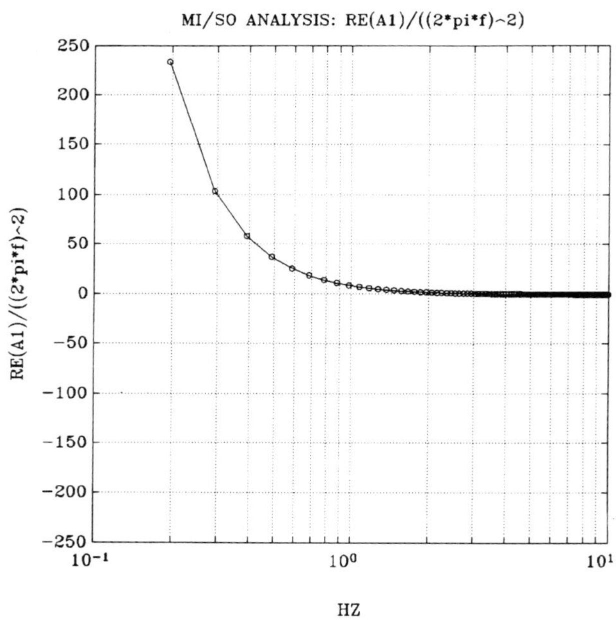

Figure 5.24.

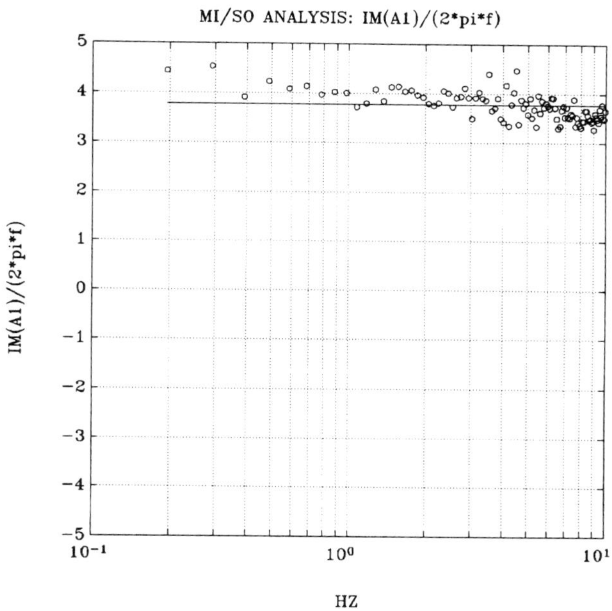

Figure 5.25.

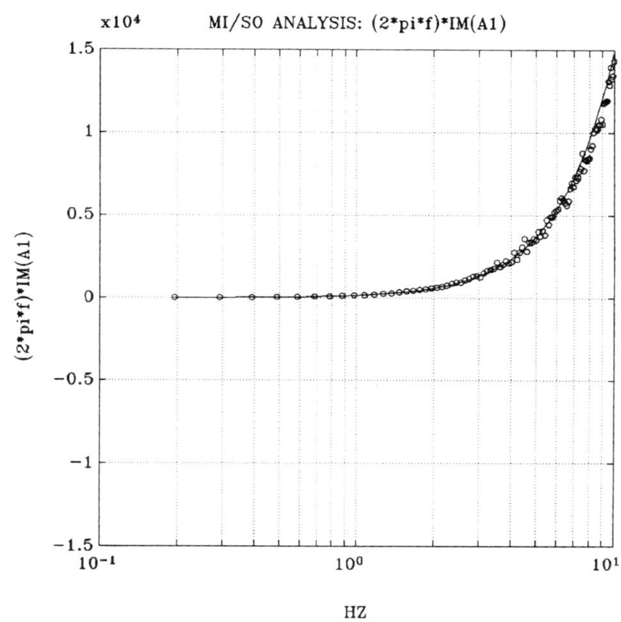

Figure 5.26.

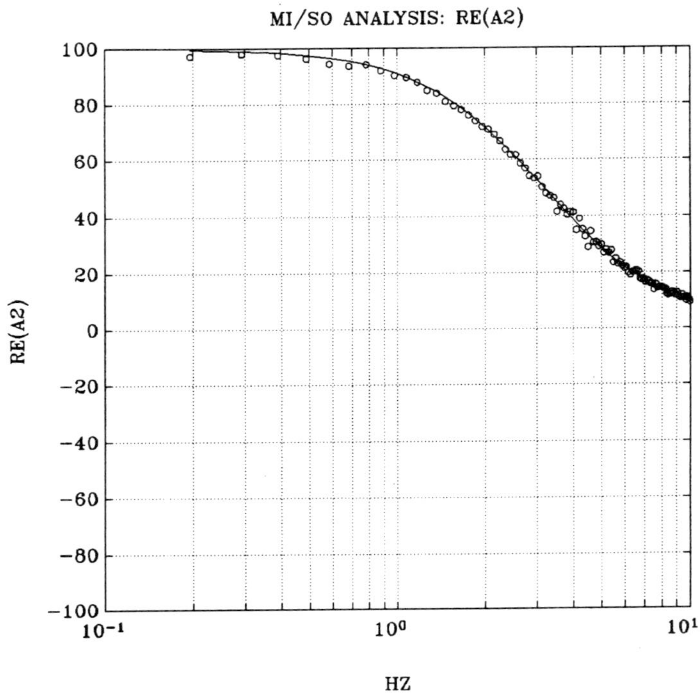
MI/SO ANALYSIS: RE(A2)
RE(A2)
100
80
60
40
20
0
−20
−40
−60
−80
−100
10⁻¹
10⁰
10¹
HZ

Figure 5.27.

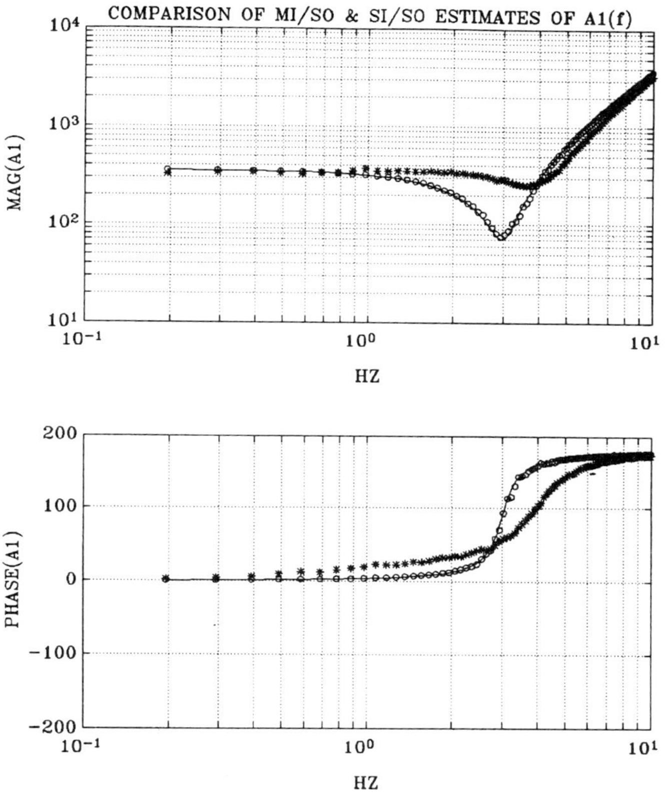

Figure 5.28.

Case 4. Linear and Nonlinear Damping, Each with Exponential Memory. The calculated frequency response functions associated with the linear and nonlinear system components in Case 4 are displayed in Figures 5.29 and 5.30, respectively. Figure 5.29 is quite similar in appearance to the previous Figures 5.5, 5.13, and 5.21, but Figure 5.30 like Figure 5.22 is quite different from the previous Figures 5.6 and 5.14. The close comparison of MI/SO functions with the exact frequency response functions is confirmed here by the cumulative coherence functions shown in Figure 5.31, which sum to near unity values at all frequencies.

Asymptotic properties of the linear frequency response function, $A_1(f)$, are described by the plots shown in Figures 5.32–5.34. Equations (5.47)–(5.51) can be used to estimate the physical parameters k, m, c, and a. The result in Figure 5.33 shows some bias error but still clearly indicates linear damping with exponential memory, in agreement with Eq. (5.44). Properties of the nonlinear frequency response function, $A_2(f)$, illustrated in Figure 5.35, have negligible bias error to indicate nonlinear damping with exponential memory, in agreement with Eq. (5.45). Equations (5.52) and (5.53) can be used to estimate the physical parameters d and p.

Figure 5.36 compares the estimates for $A_1(f)$ from the appropriate MI/SO based reverse dynamic system analysis and from the wrong SI/SO analysis, where it is implicitly assumed that the subject system is linear. Circles represent the properly computed MI/SO estimates, and asterisks represent the erroneously computed SI/SO estimates. Note the substantial difference in the two estimates even though the coherence functions in Figure 5.31 show that the system nonlinearity is not strong.

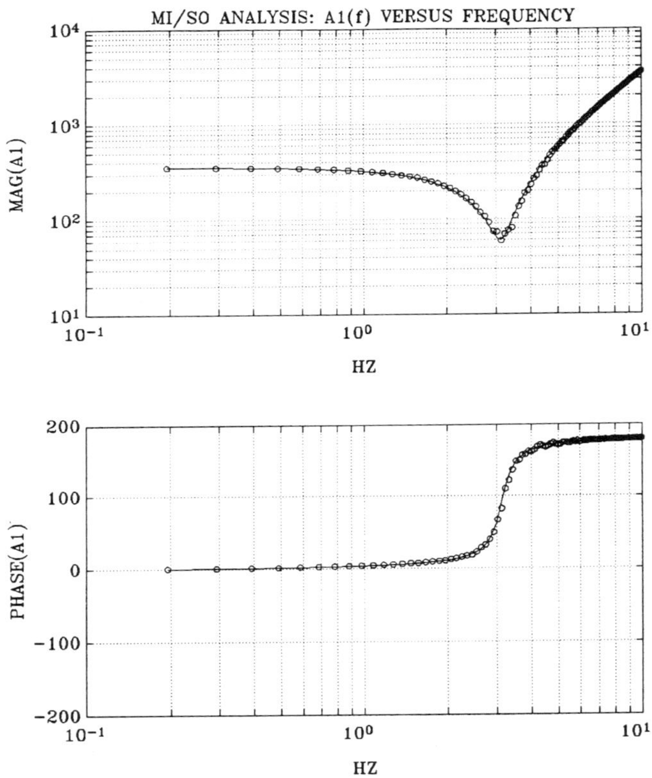

Figure 5.29.

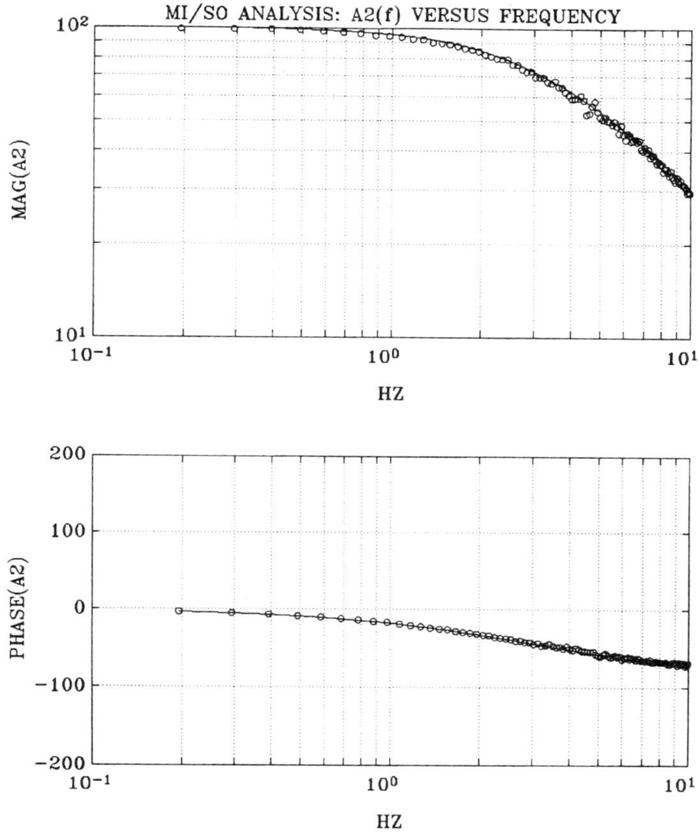
MI/SO ANALYSIS: A2(f) VERSUS FREQUENCY
MAG(A2)
10^2
10^1
10^{-1}
10^0
10^1
HZ
PHASE(A2)
200
100
0
−100
−200
10^{-1}
10^0
10^1
HZ

Figure 5.30.

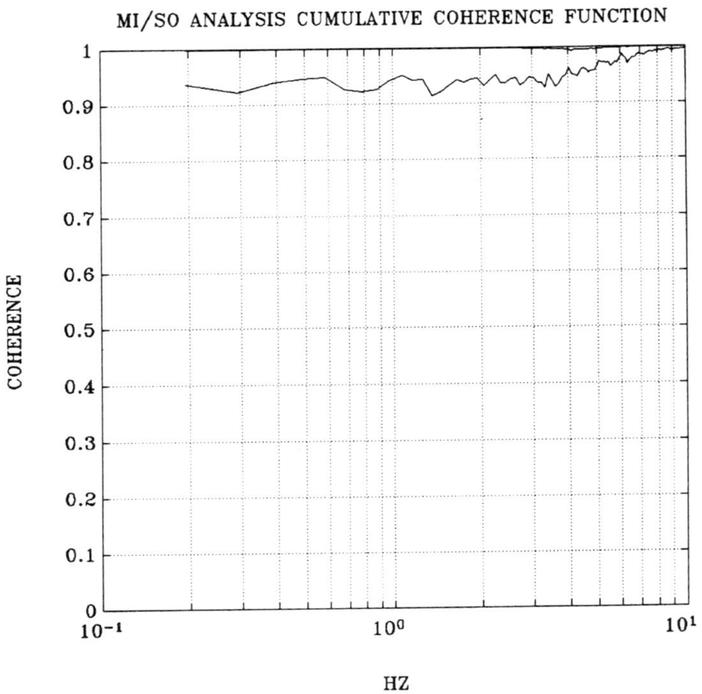
MI/SO ANALYSIS CUMULATIVE COHERENCE FUNCTION
1
0.9
0.8
0.7
0.6
0.5
0.4
0.3
0.2
0.1
0
COHERENCE
10⁻¹
10⁰
10¹
HZ

Figure 5.31.

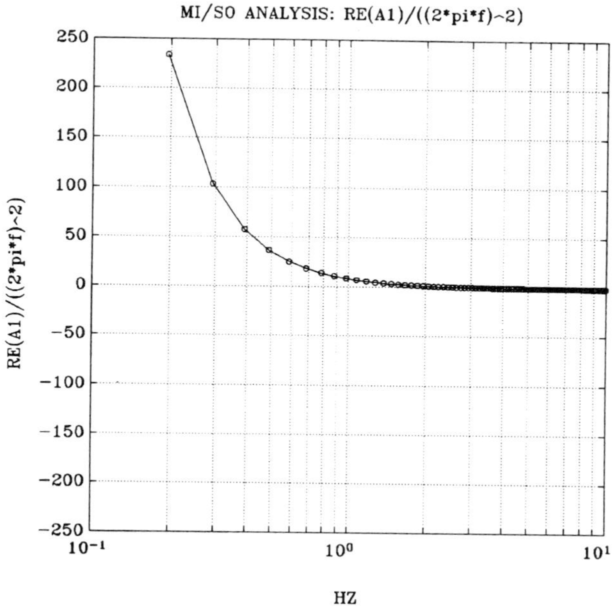

Figure 5.32.

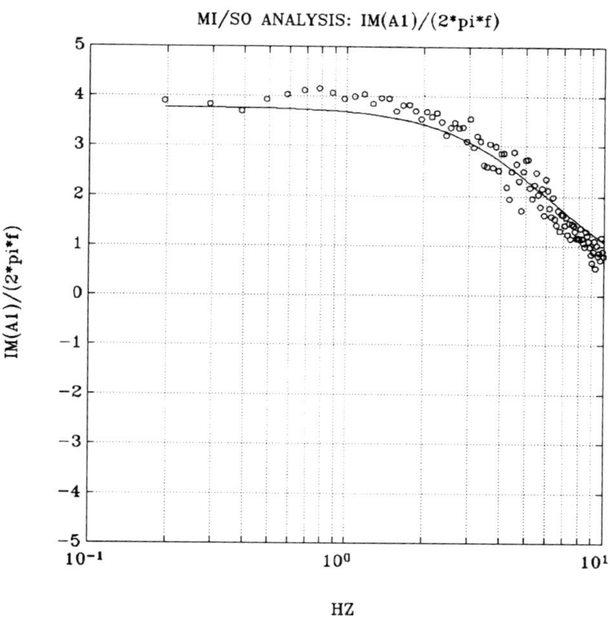

Figure 5.33.

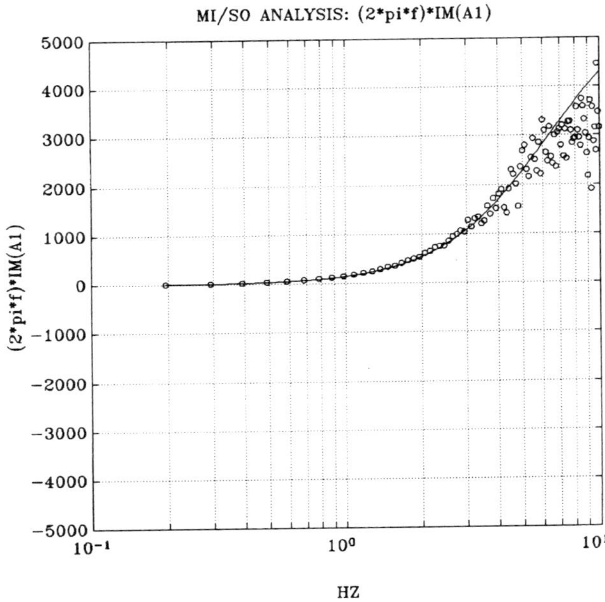

Figure 5.34.

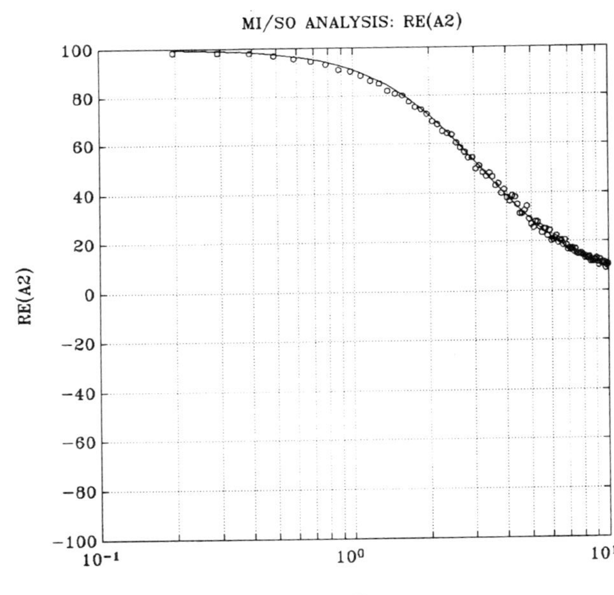

Figure 5.35.

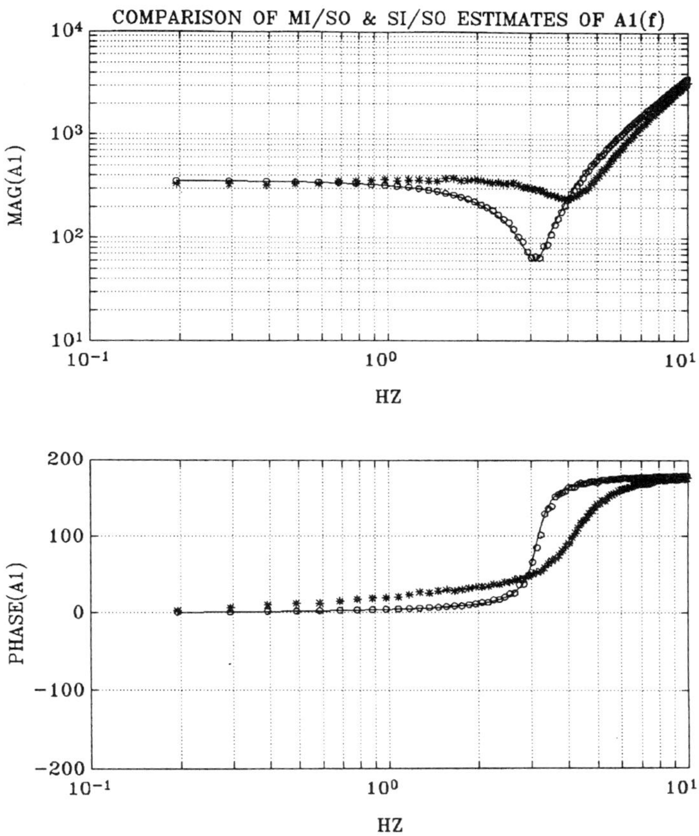

Figure 5.36.

Case 5. Linear Damping with Exponential-Cosine Memory. The calculated frequency response functions associated with the linear and nonlinear system components in Case 5 are displayed in Figures 5.37 and 5.38. Note that Figure 5.37 by itself does not indicate any linear damping with memory since it is very similar to the previous Figure 5.5 for the nonmemory case. The close comparison of MI/SO functions with the exact frequency response functions is confirmed here by the cumulative coherence functions shown in Figure 5.39. The near unity value of the lower coherence component indicates that most of the output autospectrum can be accounted for by an optimum linear approximation that may not give correct physical parameter values.

Asymptotic properties of the linear frequency response function, $A_1(f)$, are described by the plots shown in Figures 5.40–5.42. Equations (5.80)–(5.84) can be used to estimate the physical parameters k, m, c, a, and b. The result in Figure 5.41 has some bias error but still indicates linear damping here with exponential-cosine memory, in agreement with Eq. (5.75). To distinguish linear damping exponential-cosine cases in Figure 5.41 from the previous linear damping exponential cases in Figures 5.17 and 5.33, special curve-fitting techniques should be used based on Eqs. (5.80)–(5.87). Properties of the nonlinear frequency response function, $A_2(f)$, illustrated in Figure 5.43, give the correct value here for the constant nonlinear damping coefficient d with negligible bias error, in agreement with Eq. (5.89).

Figure 5.44 compares the estimates for $A_1(f)$ from the appropriate MI/SO based reverse dynamic system analysis and from the wrong SI/SO analysis, where it is implicitly assumed that the subject system is linear. Circles represent the properly computed MI/SO estimates, and asterisks represent the erroneously computed SI/SO estimates. There are substantial differences in the two estimates even though the coherence functions in Figure 5.39 show that the system nonlinearity is quite small.

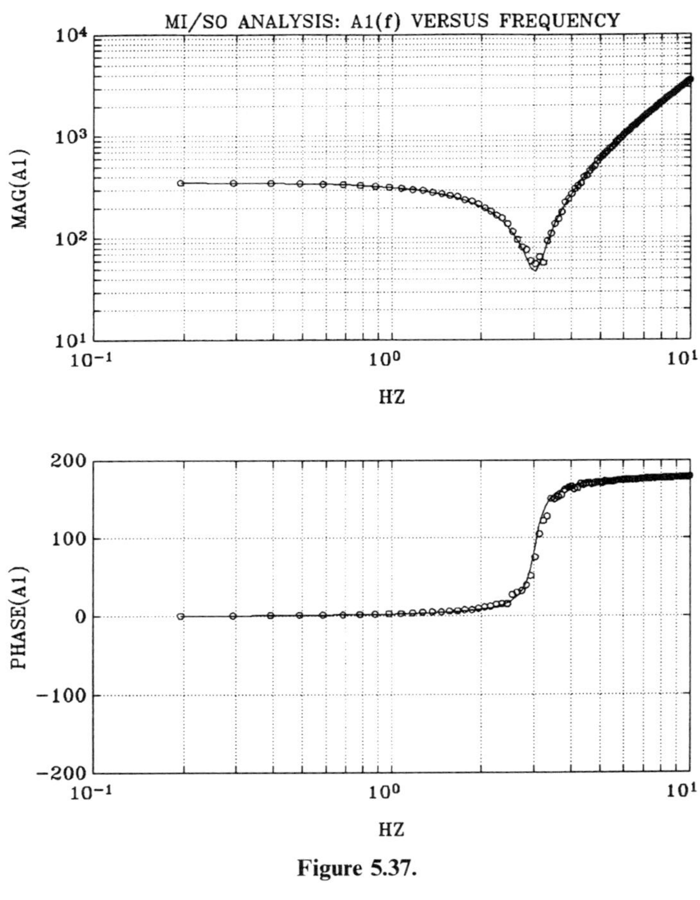

Figure 5.37.

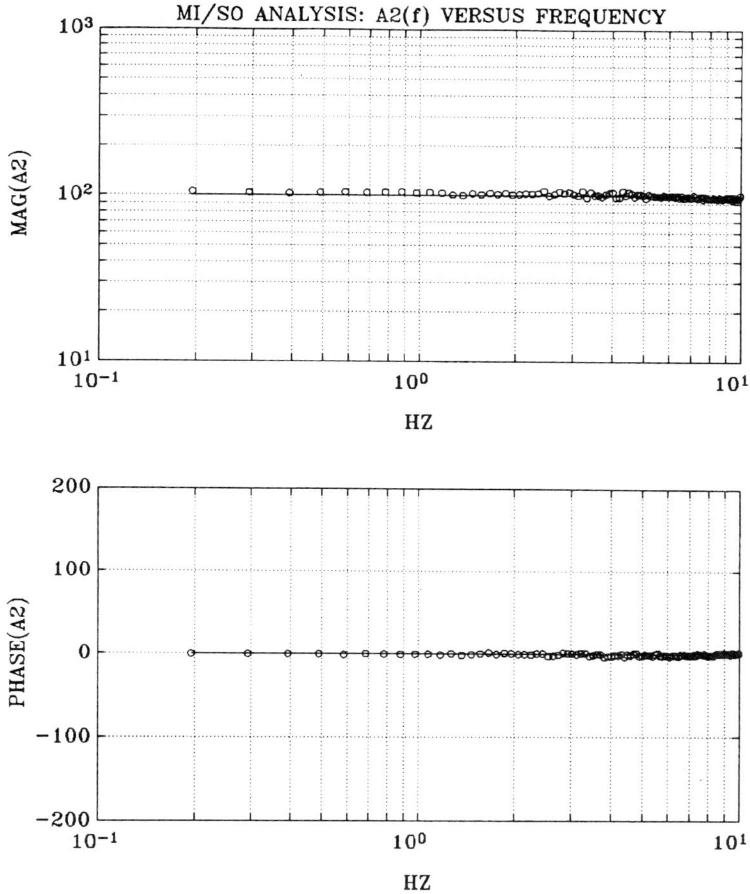
MI/SO ANALYSIS: A2(f) VERSUS FREQUENCY
MAG(A2)
10^3
10^2
10^1
10^{-1}
10^0
10^1
HZ
PHASE(A2)
200
100
0
−100
−200
10^{-1}
10^0
10^1
HZ

Figure 5.38.

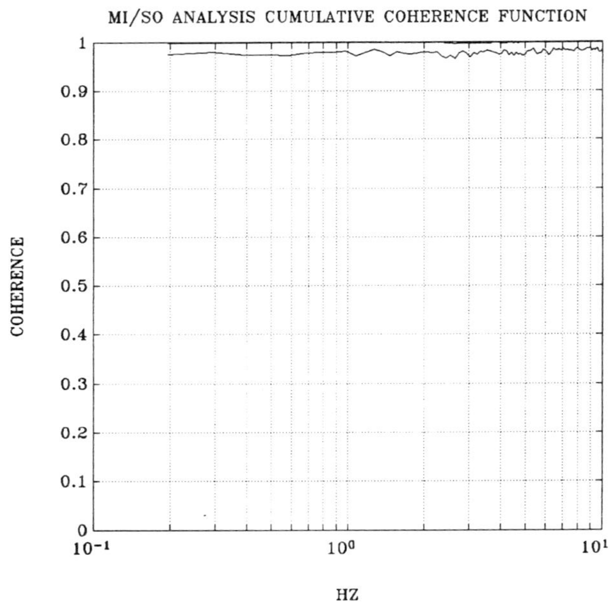

Figure 5.39.

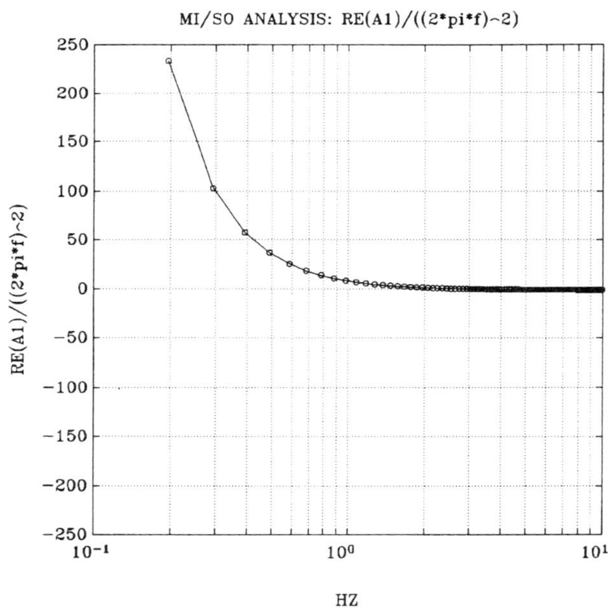

Figure 5.40.

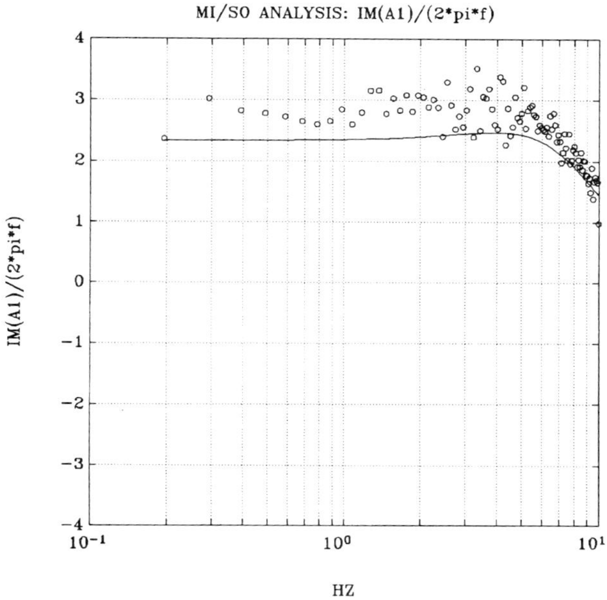

Figure 5.41.

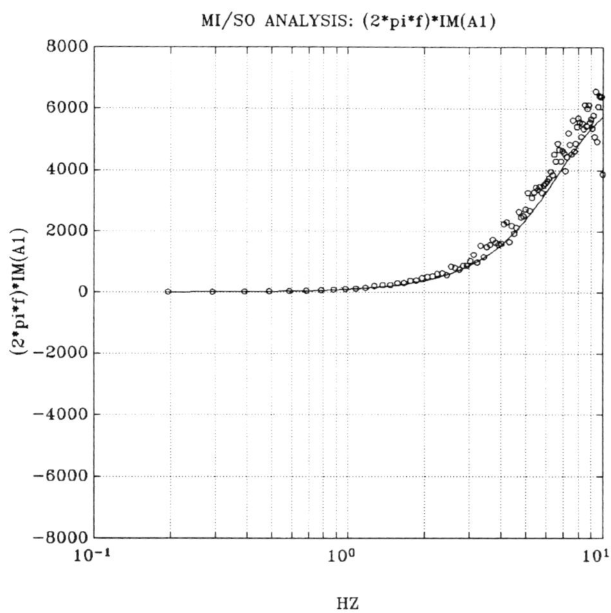

Figure 5.42.

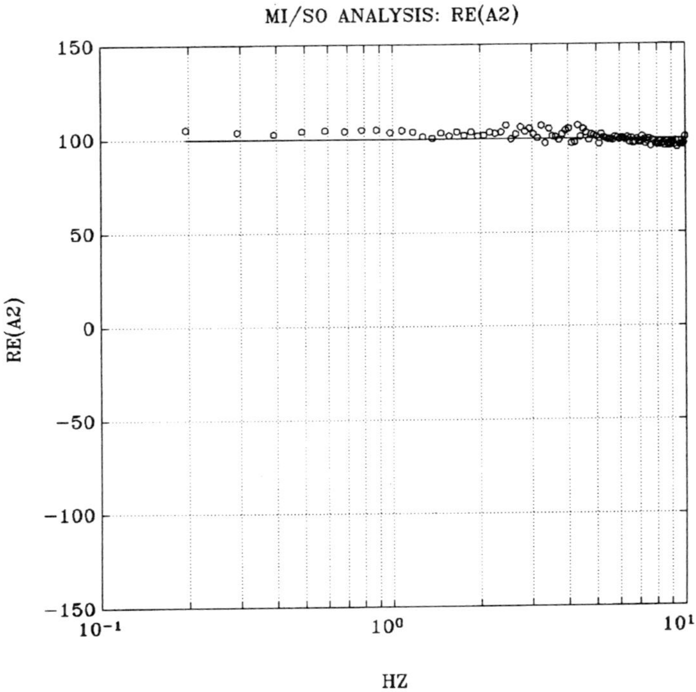

Figure 5.43.

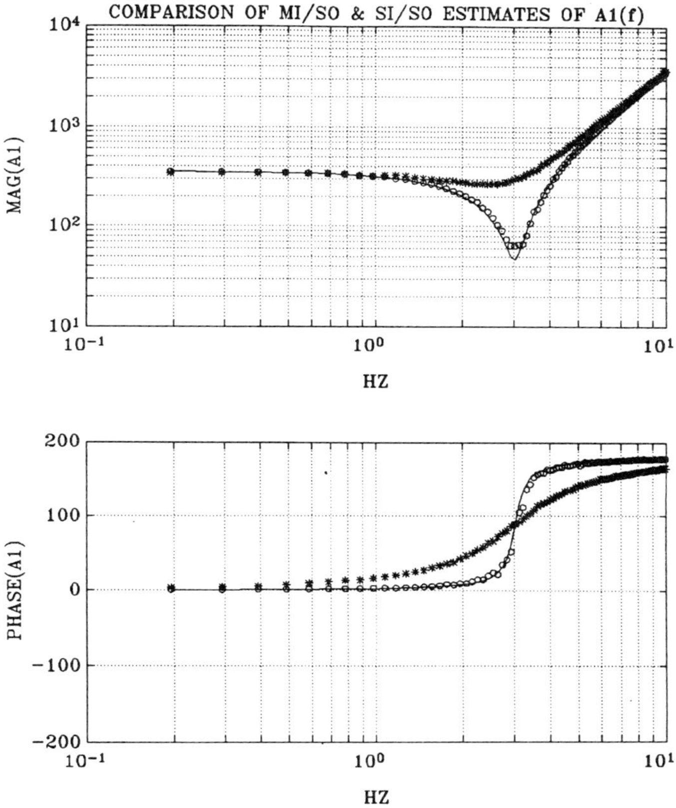

Figure 5.44.

Case 6. Nonlinear Damping with Exponential-Cosine Memory. The calculated frequency response functions associated with the linear and nonlinear system components in Case 6 are displayed in Figures 5.45 and 5.46, respectively. Note that Figure 5.46, in agreement with Eq. (5.73), indicates nonlinear damping with memory since it is quite different from the previous Figures 5.6 and 5.14. The close comparison of MI/SO functions with the exact frequency response functions is confirmed here by the cumulative coherence functions shown in Figure 5.47. The near unity value of the lower coherence component indicates that most of the output autospectrum can be accounted for by an optimum linear approximation that may not give correct physical parameter values.

Asymptotic properties of the linear frequency response function, $A_1(f)$, are described by the plots shown in Figures 5.48–5.50. The result in Figure 5.49 has large bias error but still indicates a constant linear damping coefficient c, in agreement with Eq. (5.91). Properties of the nonlinear frequency response function, $A_2(f)$, illustrated in Figure 5.51, indicate nonlinear damping here with expontenial-cosine memory, in agreement with Eq. (5.77). To distinguish nonlinear damping exponential-cosine cases in Figure 5.51 from the previous nonlinear damping exponential cases in Figures 5.27 and 5.35, special curve-fitting techniques should be used based on Eqs. (5.80)–(5.87). Note how Figure 5.51 differs from the previous results in Figure 5.27 or 5.35. Equations (5.85)–(5.87) can be used to estimate the physical parameters d, p, and q.

Figure 5.52 compares the estimates for $A_1(f)$ from the appropriate MI/SO based reverse dynamic system analysis and from the wrong SI/SO analysis, where it is implicitly assumed that the subject system is linear. Circles represent the properly computed MI/SO estimates, and asterisks represent the erroneously computed SI/SO estimates. Note again the substantial difference in the two estimates even though the coherence functions in Figure 5.47 show that the system nonlinearity is not strong.

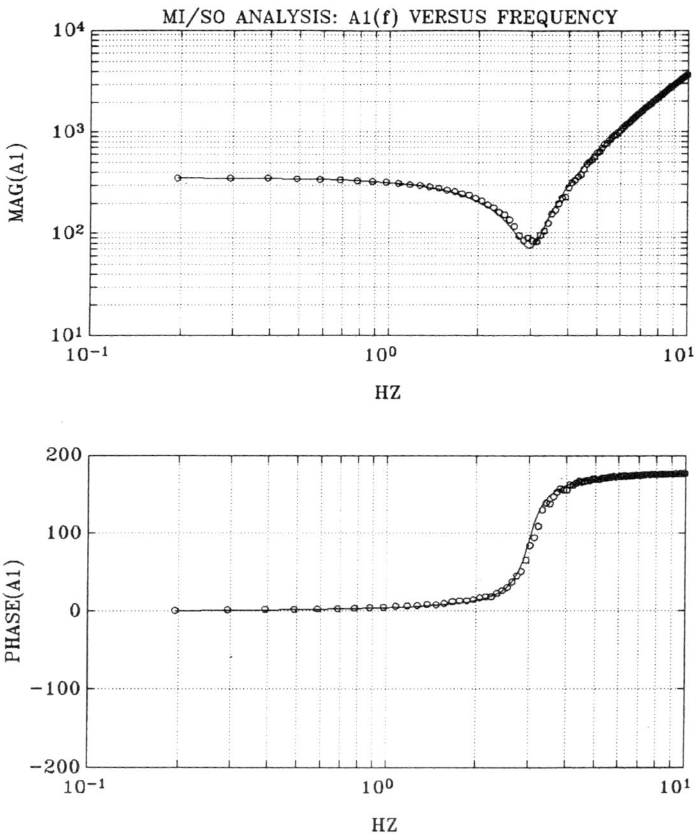

Figure 5.45.

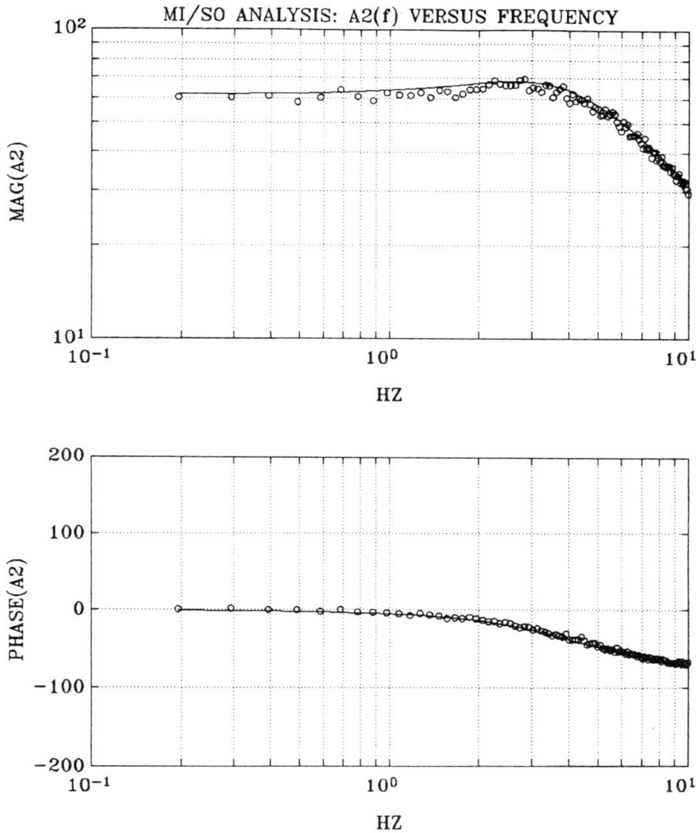
MI/SO ANALYSIS: A2(f) VERSUS FREQUENCY
MAG(A2)
10²
10¹
10⁻¹
10⁰
10¹
HZ
PHASE(A2)
200
100
0
-100
-200
10⁻¹
10⁰
10¹
HZ

Figure 5.46.

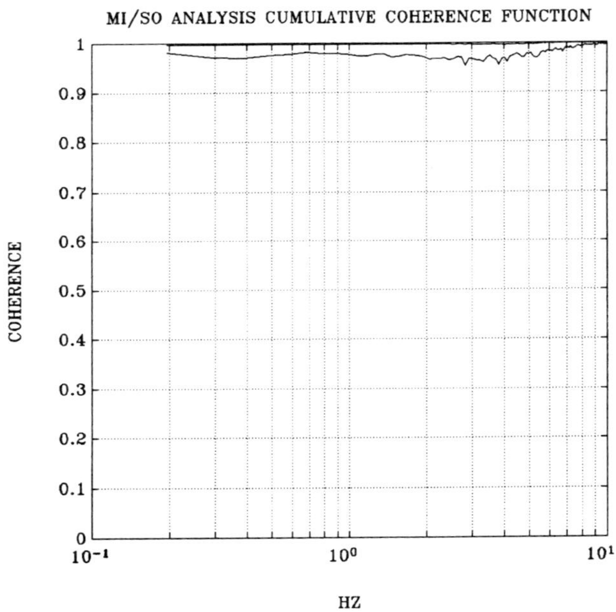

Figure 5.47.

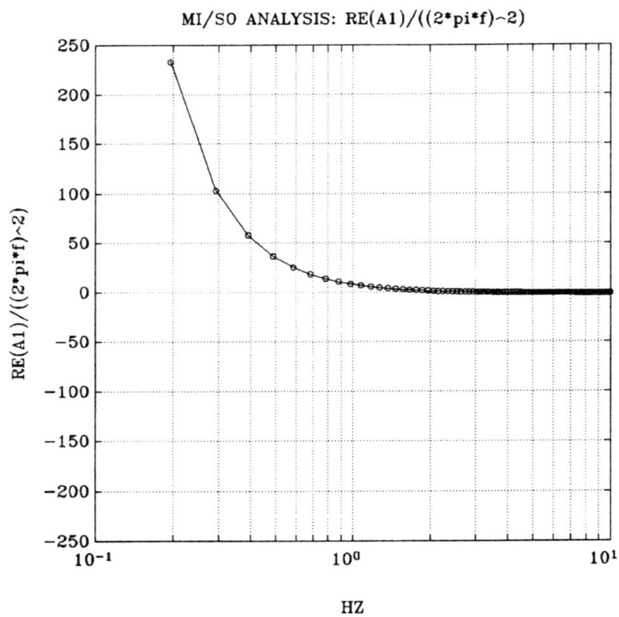

Figure 5.48.

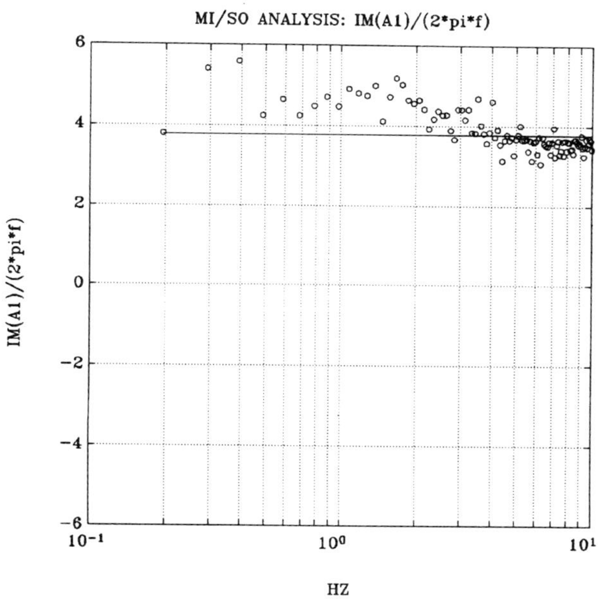

Figure 5.49.

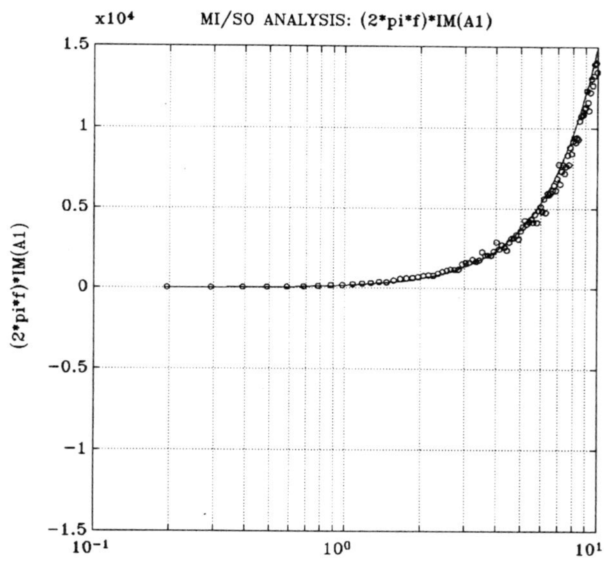

Figure 5.50.

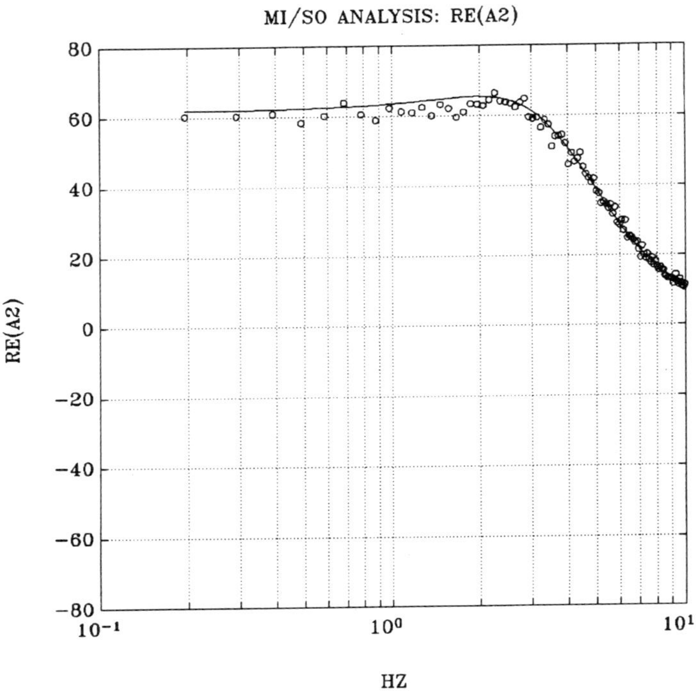

Figure 5.51.

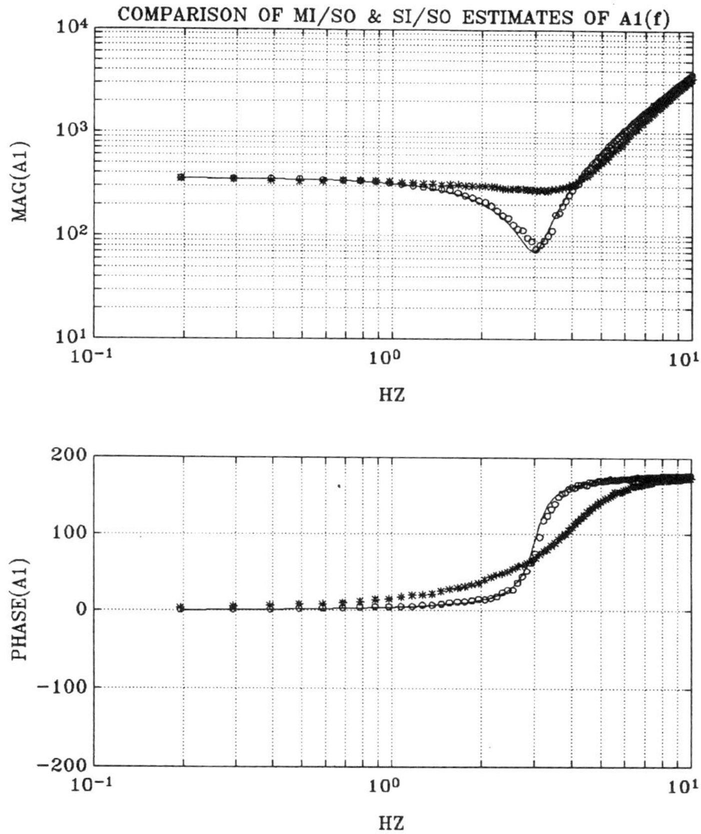

Figure 5.52.

Case 7. Linear and Nonlinear Damping, Each with Exponential-Cosine Memory. The calculated frequency response functions associated with the linear and nonlinear system components in Case 7 are displayed in Figures 5.53 and 5.54. Figure 5.53 is quite similar in appearance to the previous Figures 5.5, 5.13, 5.21, 5.29, 5.37, and 5.45, but Figure 5.54, like Figures 5.30 and 5.46, is quite different from the previous Figures 5.6, 5.14, and 5.38. The close comparison of MI/SO functions with the exact frequency response functions is confirmed here by the cumulative coherence functions shown in Figure 5.55 that sum to near unity values at all frequencies.

Asymptotic properties of the linear frequency response function, $A_1(f)$, are described by the plots shown in Figures 5.56–5.58. Equations (5.80)–(5.84) can be used to estimate the physical parameters k, m, c, a, and b. The result in Figure 5.57, even with considerable bias error, indicates linear damping with exponential-cosine memory, in agreement with Eq. (5.75). Properties of the nonlinear frequency response function, $A_2(f)$, illustrated in Figure 5.59, with negligible bias error indicate nonlinear damping with exponential-cosine memory, in agreement with Eq. (5.77). Equations (5.85)–(5.87) can be used to estimate the physical parameters d, p, and q.

Figure 5.60 compares the estimates for $A_1(f)$ from the appropriate MI/SO based reverse dynamic system analysis and from the wrong SI/SO analysis, where it is implicitly assumed that the subject system is linear. Circles represent the properly computed MI/SO estimates, and asterisks represent the erroneously computed SI/SO estimates. Note as in all of the other cases the substantial difference in the two estimates even though the coherence functions in Figure 5.55 show that the system nonlinearity is not strong.

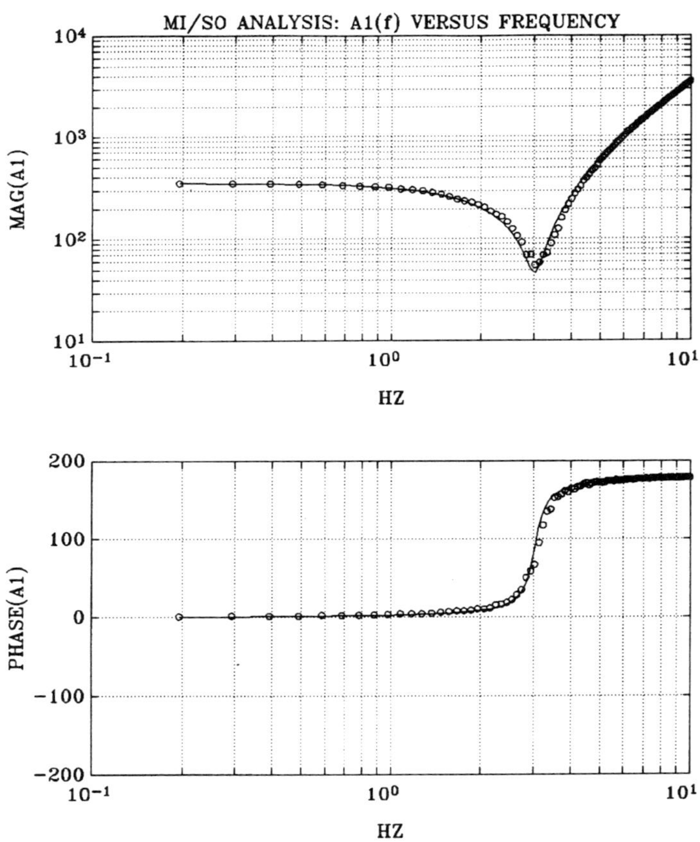

Figure 5.53.

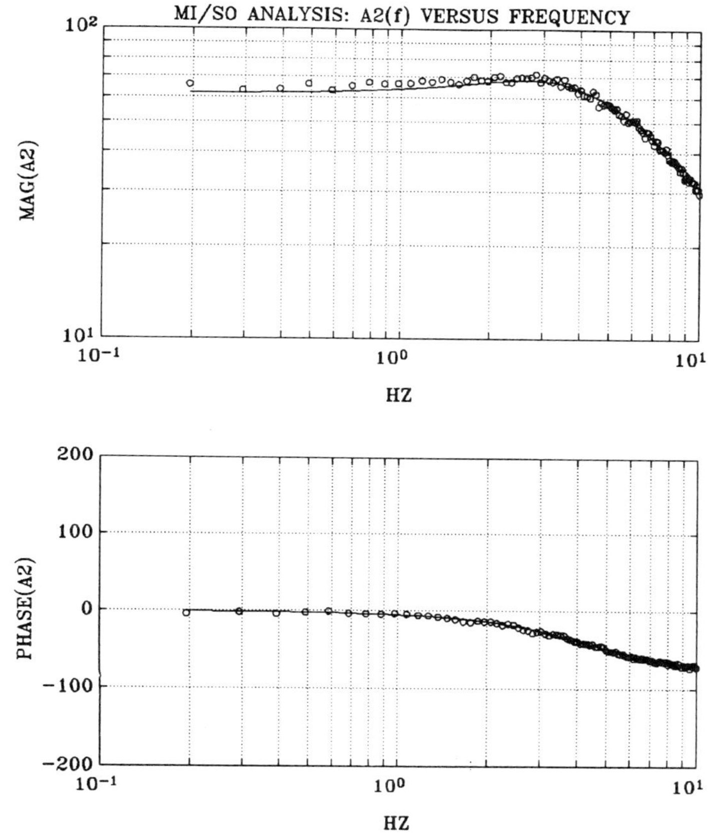
MI/SO ANALYSIS: A2(f) VERSUS FREQUENCY
MAG(A2)
10^2
10^1
10^{-1}
10^0
10^1
HZ
PHASE(A2)
200
100
0
−100
−200
10^{-1}
10^0
10^1
HZ

Figure 5.54.

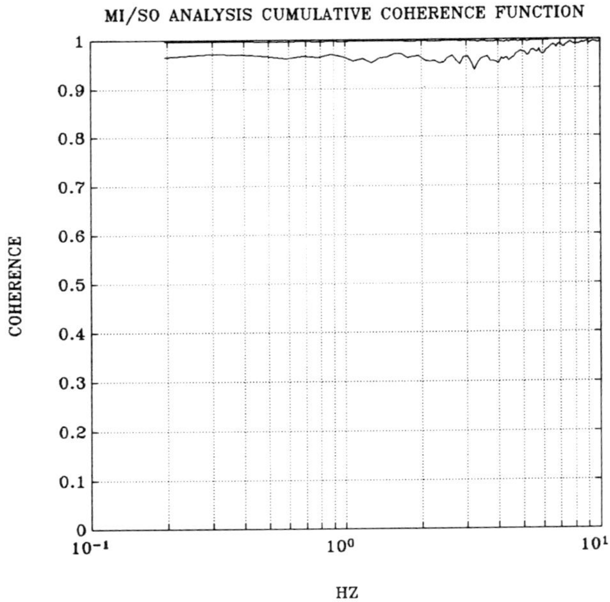

Figure 5.55.

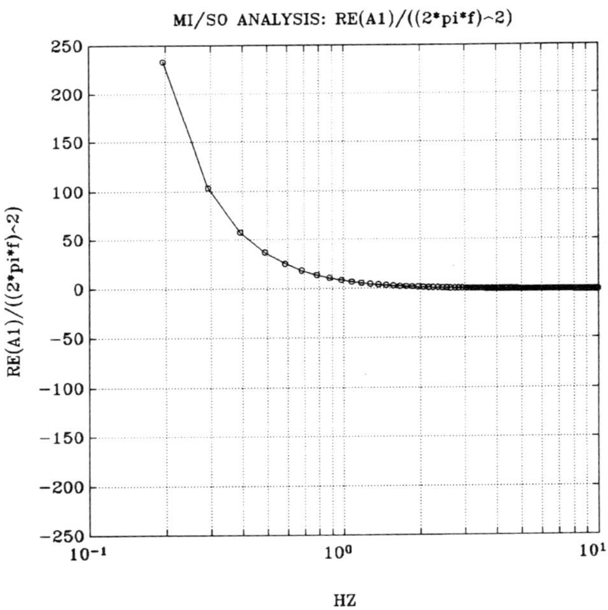

Figure 5.56.

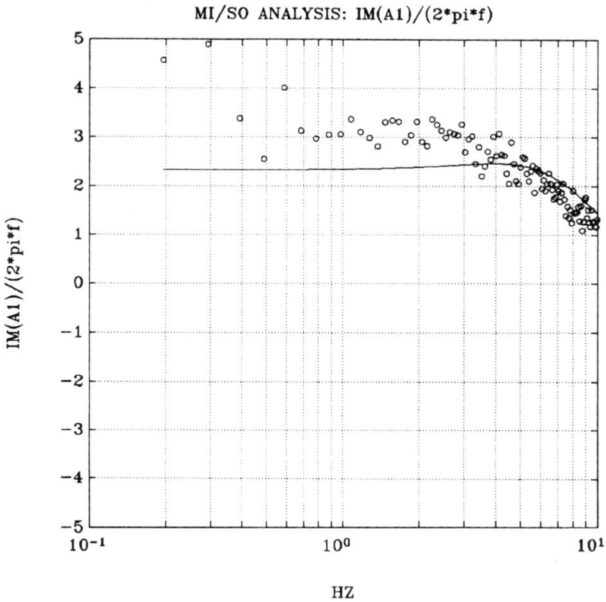

Figure 5.57.

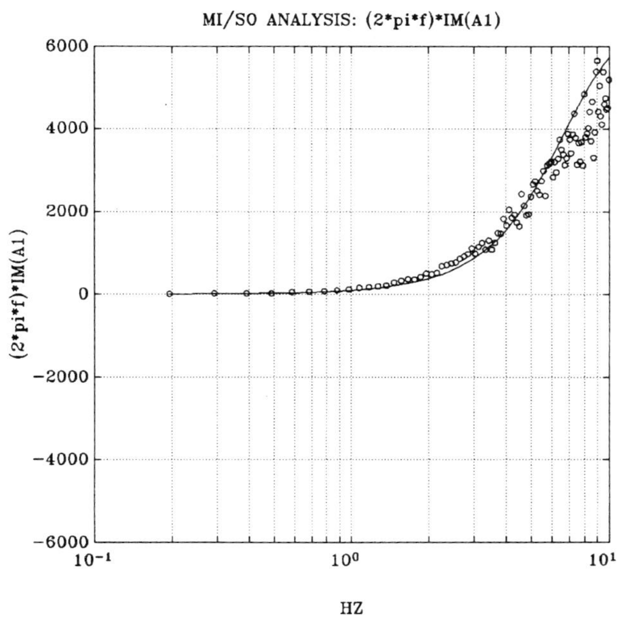

Figure 5.58.

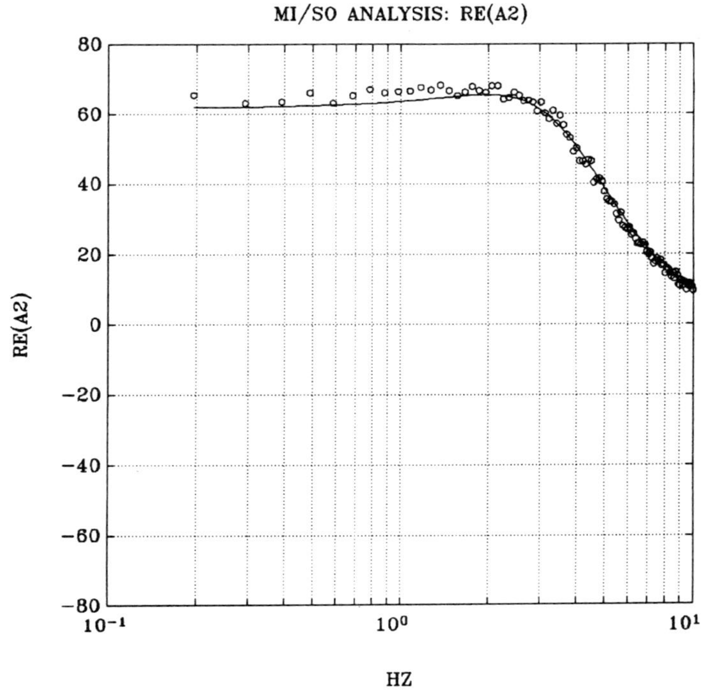

Figure 5.59.

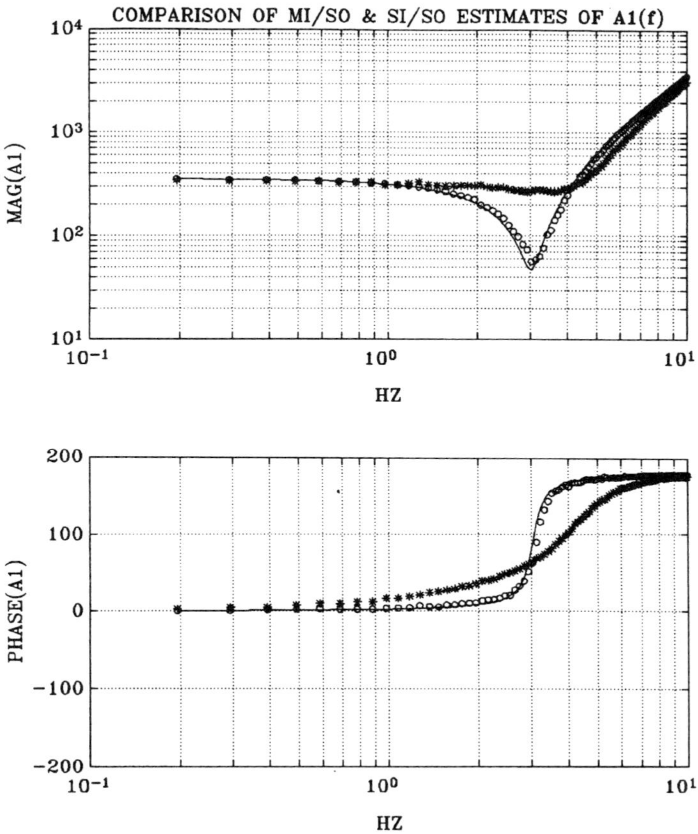

Figure 5.60.

5.5 CONCLUSIONS FROM SIMULATION WORK

Work in this chapter simulated SDOF linear and nonlinear systems where the physical parameters of the linear and nonlinear damping terms have frequency-dependent coefficients with memory of an exponential or exponential-cosine form. Other types of memory can be analyzed by following similar procedures.

Four conclusions from this simulation work are as follows:

1. For SDOF linear systems, the SI/SO linear method can correctly identify the system frequency response functions.

2. For SDOF nonlinear systems, the Reverse MI/SO Technique can correctly identify the system frequency response functions.

3. For SDOF nonlinear systems, SI/SO based optimum linear frequency response function estimates (as done in conventional modal analysis work) can give erroneous results for desired physical parameters compared to correct results obtained by the Reverse MI/SO Technique.
4. This serious mistake between SI/SO and Reverse MI/SO results will occur in practice when possible nonlinear effects are undetected because the ordinary linear coherence function is close to unity. It can also occur when possible memory effects are ignored since the wrong SI/SO results for the nonmemory case obtained here cannot be made equivalent to the correct Reverse MI/SO results.

5.6 SIX-DOF NONLINEAR MODELS FOR OCEAN ENGINEERING APPLICATIONS

This section shows how to formulate six-DOF nonlinear differential equations of motion for ocean engineering applications (references 18, 31, 40, 43, 51) that involve six physical inputs, namely, three force inputs (surge, sway, and heave in the x, y, and z directions) and three moment inputs (roll, pitch, and yaw about the x, y, and z axes), together with six physical response outputs, namely, three motion outputs (surge, sway and heave) and three angular rotation outputs (roll, pitch, and yaw). It is shown how to replace these six-DOF nonlinear system physical models, as well as two special cases, with equivalent mathematical reverse dynamics MI/MO and MI/SO linear models that can be solved by well-established linear random data analysis procedures. Similar types of nonlinear system models with equivalent direct or reverse MI/MO and MI/SO linear models apply to problems in aerospace, automotive, and other fields.

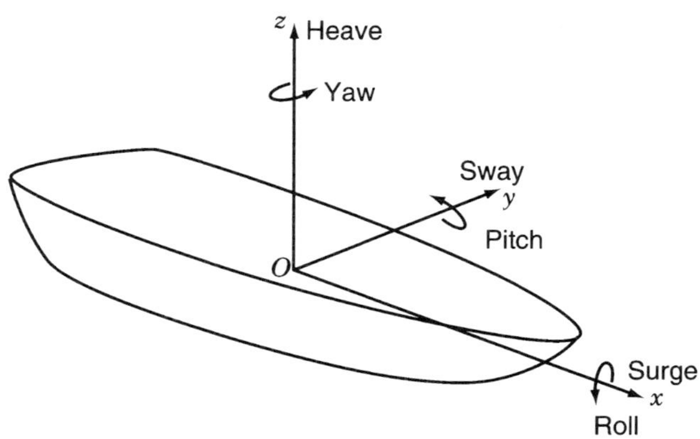

Figure 5.61. Translational and rotational axes for six-DOF ship motions.

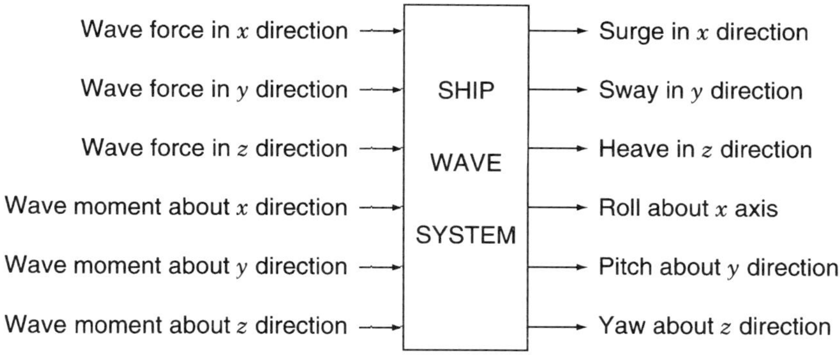

Figure 5.62. Physical input/output model for six-DOF ship motions.

The translational and rotational axes for six-DOF ship motions are shown in Figure 5.61. None of these output motions occur independently of the other motions, but instead they are coupled together as functions of the incoming ocean wave records that can arrive with random amplitudes, frequencies, and directions. The pitch and heave motions are dominant in "head" seas while the roll and sway motions are dominant in "beam" seas. For "seakeeping" applications, the emphasis is on heave, pitch, and roll motions. For "maneuvering" applications, the emphasis is on surge, sway, and yaw motions. The surge and roll motions can also be of concern in both seakeeping and maneuvering problems. The six output motions shown in Figure 5.61 are all measurable quantities.

The three wave forces in the x, y, and z directions and the three wave moments about the x, y, and z axes constitute the six input records in Figure 5.62. They are all functions of incoming ocean wave records acting on the ship that are not pictured in Figure 5.62. The hydrodynamic mechanism that converts these incoming ocean wave records into the various wave forces and moments can be very complicated. It is assumed in Figure 5.62 that appropriate time history records for the wave forces and moments are available from either theory, computation, or direct measurement. Alternative models to Figure 5.62 can be studied where the six inputs consisting of wave forces and moments are replaced by different inputs that represent measured incoming ocean wave records. This will define more basic models that can relate these measured ocean wave input records to the measured six-DOF ship motion output records.

For the six motion and angular rotation response outputs described by the vector $\mathbf{q}(t)$, the general nonlinear physical models formulated here consider nonlinear stiffness terms of form $\mathbf{q}^3(t)$ and nonlinear damping terms of form $\dot{\mathbf{q}}(t)|\dot{\mathbf{q}}(t)|$. When cross-terms in $\mathbf{q}^3(t)$ and $\dot{\mathbf{q}}(t)|\dot{\mathbf{q}}(t)|$ can be neglected, this general nonlinear problem is mathematically equivalent to a reverse dynamics 18-input/6-output linear system that can usually be broken down into two distinct

9-input/3-output linear systems. The 18 mathematical inputs are the original 6 physical outputs $\mathbf{q}(t)$ plus the 6 associated $\mathbf{q}^3(t)$ terms and the 6 associated $\dot{\mathbf{q}}(t)|\dot{\mathbf{q}}(t)|$ terms. The 6 mathematical outputs are the original 3 physical force inputs and the 3 physical moment inputs.

Two special nonlinear system physical cases are discussed in Section 5.6.2 whose equations follow from the general formulas.

Case 1. Moored ship connected at both ends.

Case 2. Unrestrained ship in head seas.

For these two cases, the nonlinear systems can be replaced by 4-input/1-output linear systems and 3-input/1-output linear systems that are practical to solve.

Relations for the six-DOF nonlinear ship model displayed in Figure 5.62 will now be developed, assuming knowledge of the six input wave forces and wave moments, and knowledge of the six output motion responses and rotation responses. The notation and subscript numbering in the following equations use the customary naval architecture convention.

5.6.1 Nonlinear Equations of Motion

Assume that the six-DOF matrix nonlinear differential equation describing the nonlinear ship motions is

$$(\mathbf{m} + \mathbf{a})\ddot{\mathbf{q}}(t) + \mathbf{c}\dot{\mathbf{q}}(t) + \mathbf{k}\mathbf{q}(t) + \mathbf{r}\mathbf{q}^3(t) + \mathbf{d}\dot{\mathbf{q}}(t)|\dot{\mathbf{q}}(t)| = \mathbf{Q}(t) \tag{5.92}$$

where $\mathbf{q}(t)$, $\mathbf{q}^3(t)$, $\dot{\mathbf{q}}(t)|\dot{\mathbf{q}}(t)|$, and $\mathbf{Q}(t)$ are 6×1 column matrices defined by

$$\begin{aligned}
\mathbf{q}(t) &= [x(t),\ y(t),\ z(t),\ \phi(t),\ \theta(t),\ \psi(t)]^{\mathrm{T}} \\
\mathbf{q}^3(t) &= [x^3(t),\ y^3(t),\ z^3(t),\ \phi^3(t),\ \theta^3(t),\ \psi^3(t)]^{\mathrm{T}} \\
\dot{\mathbf{q}}(t)|\dot{\mathbf{q}}(t)| &= [\dot{x}(t)|\dot{x}(t)|,\ \dot{y}(t)|\dot{y}(t)|,\ \dot{z}(t)|\dot{z}(t)|,\ \dot{\phi}(t)|\dot{\phi}(t)|,\ \dot{\theta}(t)|\dot{\theta}(t)|,\ \dot{\psi}(t)|\dot{\psi}(t)|]^{\mathrm{T}} \\
\mathbf{Q}(t) &= [X(t),\ Y(t),\ Z(t),\ K(t),\ M(t),\ N(t)]^{\mathrm{T}}
\end{aligned} \tag{5.93}$$

Equation (5.92) was chosen as representative of nonlinear ship motions to illustrate use of the new technique; different equations for other types of nonlinear systems can be handled similarly.

The physical terms for $\mathbf{q}(t)$ in Figures 5.61 and 5.62 are as follows:

$x(t)$ = surge (longitudinal) motion output in the x direction.
$y(t)$ = sway (lateral) motion output in the y direction.
$z(t)$ = heave (vertical) motion output in the z direction.
$\phi(t)$ = roll angular rotation output about the x axis.
$\theta(t)$ = pitch angular rotation output about the y axis.
$\psi(t)$ = yaw angular rotation output about the z axis.

The physical terms for $\mathbf{Q}(t)$ in Figures 5.61 and 5.62 are as follows:

$X(t)$ = surge (longitudinal) force input in the x direction.
$Y(t)$ = sway (lateral) force input in the y direction.
$Z(t)$ = heave (vertical) force input in the z direction.
$K(t)$ = roll angular moment input about the x axis.
$M(t)$ = pitch angular moment input about the y axis.
$N(t)$ = yaw angular moment input about the z axis.

The quantities $\mathbf{m}$, $\mathbf{a}$, $\mathbf{c}$, $\mathbf{k}$, $\mathbf{r}$, and $\mathbf{d}$ are each 6×6 matrices where

$$\begin{aligned} \mathbf{m} &= \text{mass matrix} \\ \mathbf{a} &= \text{added mass matrix} \\ (\mathbf{m} + \mathbf{a}) &= \text{virtual mass matrix} \\ \mathbf{c} &= \text{linear damping matrix} \\ \mathbf{k} &= \text{linear spring (stiffness) matrix} \\ \mathbf{r} &= \text{nonlinear spring (stiffness) matrix} \\ \mathbf{d} &= \text{nonlinear damping matrix} \end{aligned}$$

Equation (5.92) respresents a six-DOF *linear* differential equation of motion if both $\mathbf{r}$ and $\mathbf{d}$ have all zero terms. Thus, linear results are special cases of this more general nonlinear model of Eq. (5.92). The nonlinear spring term can be some different function of $\mathbf{q}(t)$ than the $\mathbf{q}^3(t)$ of Eq. (5.93), and the nonlinear damping term can be some different nonlinear function of $\dot{\mathbf{q}}(t)$ than the $\dot{\mathbf{q}}(t)|\dot{\mathbf{q}}(t)|$ of Eq. (5.93), as required to make these results fit other special problems of interest.

Assume that the origin of the axes in a ship coincides with the center of mass m of the ship. The 6×6 mass matrix $\mathbf{m}$ is then given by

$$\mathbf{m} = \begin{bmatrix} m & 0 & 0 & 0 & 0 & 0 \\ 0 & m & 0 & 0 & 0 & 0 \\ 0 & 0 & m & 0 & 0 & 0 \\ 0 & 0 & 0 & I_{xx} & 0 & -I_{xz} \\ 0 & 0 & 0 & 0 & I_{yy} & 0 \\ 0 & 0 & 0 & -I_{xz} & 0 & I_{zz} \end{bmatrix} \tag{5.94}$$

where I_{xx}, I_{yy}, and I_{zz} are the principal moments of inertia about the x, y, and z axes, respectively; and where I_{xz} is the product of inertia, which is zero if the ship has fore- and aft-symmetry.

For a laterally symmetric ship, the added mass matrix **a** has the form

$$\mathbf{a} = \begin{bmatrix} a_{11} & 0 & a_{13} & 0 & a_{15} & 0 \\ 0 & a_{22} & 0 & a_{24} & 0 & a_{26} \\ a_{31} & 0 & a_{33} & 0 & a_{35} & 0 \\ 0 & a_{42} & 0 & a_{44} & 0 & a_{46} \\ a_{51} & 0 & a_{53} & 0 & a_{55} & 0 \\ 0 & a_{62} & 0 & a_{64} & 0 & a_{66} \end{bmatrix} \tag{5.95}$$

where the a_{ij} terms have the following meanings:

a_{11}, a_{13}, a_{15} = surge added mass due to the surge, heave, and pitch accelerations, respectively.

a_{22}, a_{24}, a_{26} = sway added mass due to the sway, roll, and yaw accelerations, respectively.

a_{31}, a_{33}, a_{35} = heave added mass due to the surge, heave, and pitch accelerations, respectively.

a_{42}, a_{44}, a_{46} = roll added moment due to the sway, roll, and yaw accelerations, respectively.

a_{51}, a_{53}, a_{55} = pitch added moment due to the surge, heave, and pitch accelerations, respectively.

a_{62}, a_{64}, a_{66} = yaw added moment due to the sway, roll, and yaw accelerations, respectively.

For a laterally symmetric ship, the 6×6 *linear* damping matrix **c** has the form

$$\mathbf{c} = \begin{bmatrix} c_{11} & 0 & c_{13} & 0 & c_{15} & 0 \\ 0 & c_{22} & 0 & c_{24} & 0 & c_{26} \\ c_{31} & 0 & c_{33} & 0 & c_{35} & 0 \\ 0 & c_{42} & 0 & c_{44} & 0 & c_{46} \\ c_{51} & 0 & c_{53} & 0 & c_{55} & 0 \\ 0 & c_{62} & 0 & c_{64} & 0 & c_{66} \end{bmatrix} \tag{5.96}$$

where the c_{ij} terms have the following meanings:

c_{11}, c_{13}, c_{15} = surge damping coefficient due to the linear surge, heave, and pitch velocities, respectively.

c_{22}, c_{24}, c_{26} = sway damping coefficient due to the linear sway, roll, and yaw velocities, respectively.

c_{31}, c_{33}, c_{35} = heave damping coefficient due to the linear surge, heave, and pitch velocities, respectively.

c_{42}, c_{44}, c_{46} = roll damping coefficient due to the linear sway, roll, and yaw velocities, respectively.

c_{51}, c_{53}, c_{55} = pitch damping coefficient due to the linear surge, heave, and pitch velocities, respectively.

c_{62}, c_{64}, c_{66} = yaw damping coefficient due to the linear sway, roll, and yaw velocities, respectively.

For a laterally symmetric ship, the 6×6 *linear* spring (stiffness) matrix $\mathbf{k}$ has the general form

$$\mathbf{k} = \begin{bmatrix} k_{11} & 0 & k_{13} & 0 & k_{15} & 0 \\ 0 & k_{22} & 0 & k_{24} & 0 & k_{26} \\ k_{31} & 0 & k_{33} & 0 & k_{35} & 0 \\ 0 & k_{42} & 0 & k_{44} & 0 & k_{46} \\ k_{51} & 0 & k_{53} & 0 & k_{55} & 0 \\ 0 & k_{62} & 0 & k_{64} & 0 & k_{66} \end{bmatrix} \tag{5.97}$$

where the k_{ij} terms have the following meanings:

k_{11}, k_{13}, k_{15} = surge spring coefficient due to the linear surge, heave, and pitch motions, respectively.

k_{22}, k_{24}, k_{26} = sway spring coefficient due to the linear sway, roll, and yaw motions, respectively.

k_{31}, k_{33}, k_{35} = heave spring coefficient due to the linear surge, heave, and pitch motions, respectively.

k_{42}, k_{44}, k_{46} = roll spring coefficient due to the linear sway, roll, and yaw motions, respectively.

k_{51}, k_{53}, k_{55} = pitch spring coefficient due to the linear surge, heave, and pitch motions, respectively.

k_{62}, k_{64}, k_{66} = yaw spring coefficient due to the linear sway, roll, and yaw motions, respectively.

For a symmetric ship, neglecting terms of second-order, assume that the 6×6 *nonlinear* spring matrix $\mathbf{r}$ due to nonlinear motion terms given by the vector $\mathbf{q}^3(t)$ defined in Eq. (5.93), or by some other nonlinear function of $\mathbf{q}(t)$, has the diagonal form

$$\mathbf{r} = \begin{bmatrix} r_{11} & 0 & 0 & 0 & 0 & 0 \\ 0 & r_{22} & 0 & 0 & 0 & 0 \\ 0 & 0 & r_{33} & 0 & 0 & 0 \\ 0 & 0 & 0 & r_{44} & 0 & 0 \\ 0 & 0 & 0 & 0 & r_{55} & 0 \\ 0 & 0 & 0 & 0 & 0 & r_{66} \end{bmatrix} \tag{5.98}$$

where the terms $r_{ij} = 0$ for $i \neq j$ and where the diagonal terms r_{ii} have the following meanings:

r_{11} = surge spring coefficient due to the nonlinear surge motion.
r_{22} = sway spring coefficient due to the nonlinear sway motion.
r_{33} = heave spring coefficient due to the nonlinear heave motion.
r_{44} = roll spring coefficient due to the nonlinear roll motion.
r_{55} = pitch spring coefficient due to the nonlinear pitch motion.
r_{66} = yaw spring coefficient due to the nonlinear yaw motion.

For a symmetric ship, neglecting terms of second-order, assume that the 6×6 *nonlinear* damping matrix **d** due to nonlinear velocity terms given by the vector $\dot{\mathbf{q}}(t)|\dot{\mathbf{q}}(t)|$ defined in Eq. (5.93), or by some other nonlinear function of $\dot{\mathbf{q}}(t)$, has the diagonal form

$$\mathbf{d} = \begin{bmatrix} d_{11} & 0 & 0 & 0 & 0 & 0 \\ 0 & d_{22} & 0 & 0 & 0 & 0 \\ 0 & 0 & d_{33} & 0 & 0 & 0 \\ 0 & 0 & 0 & d_{44} & 0 & 0 \\ 0 & 0 & 0 & 0 & d_{55} & 0 \\ 0 & 0 & 0 & 0 & 0 & d_{66} \end{bmatrix} \tag{5.99}$$

where the terms $d_{ij} = 0$ for $i \neq j$ and the diagonal terms d_{ii} have the following meanings:

d_{11} = surge damping coefficient due to the nonlinear surge velocity.
d_{22} = sway damping coefficient due to the nonlinear sway velocity.
d_{33} = heave damping coefficient due to the nonlinear heave velocity.
d_{44} = roll damping coefficient due to the nonlinear roll velocity.
d_{55} = pitch damping coefficient due to the nonlinear pitch velocity.
d_{66} = yaw damping coefficient due to the nonlinear yaw velocity.

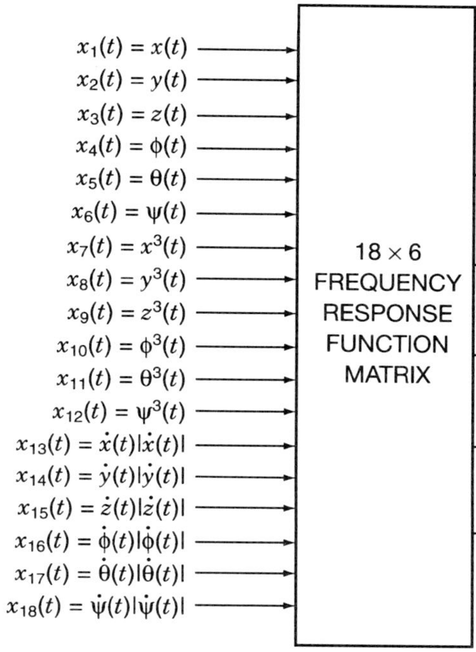

Figure 5.63. Mathematical reverse dynamics MI/MO linear model for six-DOF ship motion problems.

The most general possible equivalent mathematical reverse dynamics MI/MO linear model to Eq. (5.92) is shown in Figure 5.63, where there are eighteen inputs $x_i(t)$, $i = 1$ to 18, and six outputs $y_j(t)$, $j = 1$ to 6. This Reverse MI/MO model can require a maximum of $18 \times 6 = 108$ different frequency response functions (FRFs) to relate how each of the eighteen inputs can lead to the six outputs. However, from orthogonality and other symmetry conditions, the 18-input/6-output MI/MO linear model usually breaks down into two distinct 9-input/3-output linear models as shown in Figure 5.64, where each of these models requires a maximum of $9 \times 3 = 27$ different FRFs. In various cases of physical interest, such as the two cases discussed in the next section, only a subset of these 27 FRFs will be involved because many terms will be zero.

Recognition of this equivalence between many proposed nonlinear differential equations of motion and their mathematically equivalent reverse dynamics MI/MO linear models is a significant result. When these MI/MO linear models are valid, as can be verified by computation of appropriate ordinary coherence functions, many practical benefits follow. One can now greatly extend past any current methods, which are based on either (1) higher-order multidimensional procedures that are difficult to compute and interpret or (2) time-domain iterative procedures that can only estimate constant physical parameters. Some

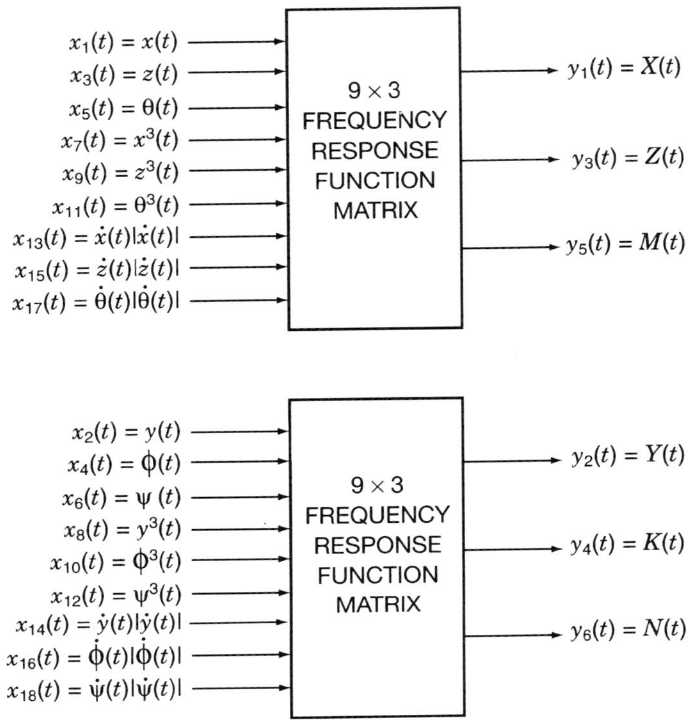

Figure 5.64. Two smaller Reverse MI/MO linear models that can replace Figure 5.63 for many six-DOF ship motion problems.

features of this new approach compared to other techniques are:

- It is not restricted to sinusoidal or other deterministic data but instead applies to measured physical input/output data with arbitrary amplitude and frequency properties.
- The nonlinear damping and stiffness coefficients that are determined from the measured data can be functions of frequency.
- Good statistical estimates can be obtained with limited amounts of data because only ordinary spectral density functions are computed.

5.6.2 Two Ocean Engineering Cases

All of the above matrices can take on special forms for ocean engineering applications of interest. Two such physical cases will now be discussed in some detail, where it will be shown how to determine from measured input/output random data the various coefficients that can yield physical parameters. These two cases represent (1) a moored ship that is connected at both ends and (2) an unrestrained ship in head seas.

Case 1. Moored Ship Connected at Both Ends. Consider a moored symmetric ship as shown in Figure 5.61 with motions only in the xz plane. For this case, with second-order effects neglected and using the vector definitions in Eq. (5.93), the matrix equation for the coupled surge, heave, and pitch motions is

$$\begin{bmatrix} m+a_{11} & 0 & a_{15} \\ 0 & m+a_{33} & 0 \\ a_{51} & 0 & I_{yy}+a_{55} \end{bmatrix} \begin{bmatrix} \ddot{x}(t) \\ \ddot{z}(t) \\ \ddot{\theta}(t) \end{bmatrix} + \begin{bmatrix} c_{11} & 0 & c_{15} \\ 0 & c_{33} & 0 \\ c_{51} & 0 & c_{55} \end{bmatrix} \begin{bmatrix} \dot{x}(t) \\ \dot{z}(t) \\ \dot{\theta}(t) \end{bmatrix}$$
$$+ \begin{bmatrix} k_{11} & 0 & k_{15} \\ 0 & k_{33} & 0 \\ k_{51} & 0 & k_{55} \end{bmatrix} \begin{bmatrix} x(t) \\ z(t) \\ \theta(t) \end{bmatrix} + \begin{bmatrix} r_{11} & 0 & 0 \\ 0 & r_{33} & 0 \\ 0 & 0 & r_{55} \end{bmatrix} \begin{bmatrix} x^3(t) \\ z^3(t) \\ \theta^3(t) \end{bmatrix}$$
$$+ \begin{bmatrix} d_{11} & 0 & 0 \\ 0 & d_{33} & 0 \\ 0 & 0 & d_{55} \end{bmatrix} \begin{bmatrix} \dot{x}(t)|\dot{x}(t)| \\ \dot{z}(t)|\dot{z}(t)| \\ \dot{\theta}(t)|\dot{\theta}(t)| \end{bmatrix} = \begin{bmatrix} X(t) \\ Z(t) \\ M(t) \end{bmatrix} \tag{5.100}$$

Certain coefficients are set equal to zero for the following reasons:

$a_{13} = c_{13} = k_{13} = 0$	Assumes that surge coefficients are independent of heave effects
$a_{31} = c_{31} = k_{31} = 0$	Assumes that heave coefficients are independent of surge effects
$a_{35} = c_{35} = k_{35} = 0$	Assumes that heave coefficients are independent of pitch effects
$a_{53} = c_{53} = k_{53} = 0$	Assumes that pitch coefficients are independent of heave effects

This matrix Eq. (5.100) for a moored ship is now the same as

$$(m+a_{11})\ddot{x}(t) + c_{11}\dot{x}(t) + k_{11}x(t) + r_{11}x^3(t) + d_{11}\dot{x}(t)|\dot{x}(t)|$$
$$+a_{15}\ddot{\theta}(t) + c_{15}\dot{\theta}(t) + k_{15}\theta(t) = X(t) \tag{5.101}$$

$$(m+a_{33})\ddot{z}(t) + c_{33}\dot{z}(t) + k_{33}z(t) + r_{33}z^3(t) + d_{33}\dot{z}(t)|\dot{z}(t)| = Z(t) \tag{5.102}$$

$$a_{51}\ddot{x}(t) + c_{51}\dot{x}(t) + k_{51}x(t) + (I_{yy} + a_{55})\ddot{\theta}(t) + c_{55}\dot{\theta}(t)$$
$$+k_{55}\theta(t) + r_{55}z^3(t) + d_{55}\dot{\theta}(t)|\dot{\theta}(t)| = M(t) \tag{5.103}$$

The equivalent mathematical reverse dynamics input/output models to Eqs. (5.101)–(5.103) are governed by the relations

$$A_{1,1}(f)X_1(f) + A_{1,5}(f)X_5(f) + A_{1,7}(f)X_7(f) + A_{1,13}(f)X_{13}(f) = Y_1(f) \tag{5.104}$$

$$A_{3,3}(f)X_3(f) + A_{3,9}(f)X_9(f) + A_{3,15}(f)X_{15}(f) = Y_3(f) \tag{5.105}$$

$$A_{5,1}(f)X_1(f) + A_{5,5}(f)X_5(f) + A_{5,11}(f)X_{11}(f) + A_{5,17}(f)X_{17}(f) = Y_5(f) \tag{5.106}$$

The nine input records in Eqs. (5.104)–(5.106) are

$x_1(t) = x(t)$ with $X_1(f)$ = Fourier transform of $x_1(t)$

$x_3(t) = z(t)$ with $X_3(f)$ = Fourier transform of $x_3(t)$

$x_5(t) = \theta(t)$ with $X_5(f)$ = Fourier transform of $x_5(t)$

$x_7(t) = x^3(t)$ with $X_7(f)$ = Fourier transform of $x_7(t)$

$x_9(t) = z^3(t)$ with $X_9(f)$ = Fourier transform of $x_9(t)$

$x_{11}(t) = \theta^3(t)$ with $X_{11}(f)$ = Fourier transform of $x_{11}(t)$

$x_{13}(t) = \dot{x}(t)|\dot{x}(t)|$ with $X_{13}(f)$ = Fourier transform of $x_{13}(t)$

$x_{15}(t) = \dot{z}(t)|\dot{z}(t)|$ with $X_{15}(f)$ = Fourier transform of $x_{15}(t)$

$x_{17}(t) = \dot{\theta}(t)|\dot{\theta}(t)|$ with $X_{17}(f)$ = Fourier transform of $x_{17}(t)$

and the three output records are

$y_1(t) = X(t)$ with $Y_1(f)$ = Fourier transform of $y_1(t)$

$y_3(t) = Z(t)$ with $Y_3(f)$ = Fourier transform of $y_3(t)$

$y_5(t) = M(t)$ with $Y_5(f)$ = Fourier transform of $y_5(t)$

The 11 FRFs in Eqs. (5.104)–(5.106) are

$$\begin{aligned} A_{1,1}(f) &= \text{FRF for output } y_1(t) \text{ from input } x_1(t) \\ &= k_{11} + j(2\pi f)c_{11} - (2\pi f)^2(m + a_{11}) \end{aligned} \tag{5.107}$$

$$\begin{aligned} A_{1,5}(f) &= \text{FRF for output } y_1(t) \text{ from input } x_5(t) \\ &= k_{15} + j(2\pi f)c_{15} - (2\pi f)^2 a_{15} \end{aligned} \tag{5.108}$$

$$A_{1,7}(f) = \text{FRF for output } y_1(t) \text{ from input } x_7(t) = r_{11} \tag{5.109}$$

$$A_{1,13}(f) = \text{FRF for output } y_1(t) \text{ from input } x_{13}(t) = d_{11} \tag{5.110}$$

$$\begin{aligned} A_{3,3}(f) &= \text{FRF for output } y_3(t) \text{ from input } x_3(t) \\ &= k_{33} + j(2\pi f)c_{33} - (2\pi f)^2(m + a_{33}) \end{aligned} \tag{5.111}$$

$$A_{3,9}(f) = \text{FRF for output } y_3(t) \text{ from input } x_9(t) = r_{33} \tag{5.112}$$

$$A_{3,15}(f) = \text{FRF for output } y_3(t) \text{ from input } x_{15}(t) = d_{33} \tag{5.113}$$

$$A_{5,1}(f) = \text{FRF for output } y_5(t) \text{ from input } x_1(t) = k_{51} + j(2\pi f)c_{51} - (2\pi f)^2 a_{51} \tag{5.114}$$

$$A_{5,5}(f) = \text{FRF for output } y_5(t) \text{ from input } x_5(t) = k_{55} + j(2\pi f)c_{55} - (2\pi f)^2(I_{yy} + a_{55}) \tag{5.115}$$

$$A_{5,11}(f) = \text{FRF for output } y_5(t) \text{ from input } x_{11}(t) = r_{55} \tag{5.116}$$

$$A_{5,17}(f) = \text{FRF for output } y_5(t) \text{ from input } x_{17}(t) = d_{55} \tag{5.117}$$

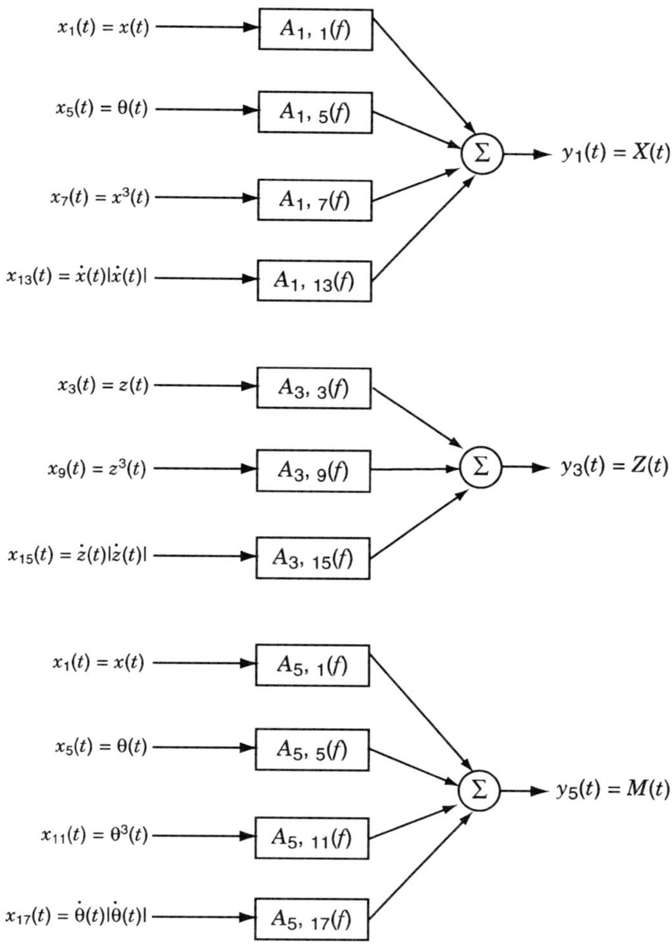

Figure 5.65. Equivalent Reverse MI/SO linear models for Case 1.

As shown in Figure 5.65, Eq. (5.104) is a 4-input/1-output linear model, Eq. (5.105) is a 3-input/1-output linear model, and Eq. (5.106) is another 4-input/1-output linear model where the 11 frequency response functions can be determined from the measured input/output data for well-defined MI/SO linear models. Computation of these 11 frequency response functions gives the coefficients from Eqs. (5.107)–(5.117) that define the physical parameters in the nonlinear differential equations of motion of the coupled surge, heave, and pitch motions for this moored ship. In particular, this practical analysis can determine whether or not the nonlinear coefficients r_{11}, r_{33}, r_{55}, d_{11}, d_{33}, and d_{55} are constants or are functions of frequency.

Similar equations and illustrations can be derived to compute the coefficients that define the physical parameters in the nonlinear differential equations of motion for the coupled sway, roll, and yaw motions of a moored ship. Reasonable assumptions should be made to simplify the work by neglecting second-order terms that can be omitted. The main limitation in applying this new approach will occur in problems where the measured physical response signals lead to mathematical input signals that are highly coherent at desired frequencies for the equivalent Reverse MI/SO linear analysis.

Case 2. Unrestrained Ship in Head Seas. Consider a laterally symmetric ship moving in head seas where the heave motion $z(t)$ and the pitch motion $\theta(t)$ are dominant. Assume that the surge motion $x(t)$ is an order of magnitude smaller than the heave and pitch motions so that it can be ignored. For this case, with second-order terms and the I_{xy} term neglected, the matrix relation for coupled heave and pitch motions is

$$\begin{bmatrix} m + a_{33} & a_{35} \\ a_{53} & I_{yy} + a_{55} \end{bmatrix} \begin{bmatrix} \ddot{z}(t) \\ \ddot{\theta}(t) \end{bmatrix} + \begin{bmatrix} c_{33} & c_{35} \\ c_{53} & c_{55} \end{bmatrix} \begin{bmatrix} \dot{z}(t) \\ \dot{\theta}(t) \end{bmatrix} + \begin{bmatrix} k_{33} & k_{35} \\ k_{51} & k_{55} \end{bmatrix} \begin{bmatrix} z(t) \\ \theta(t) \end{bmatrix}$$
$$+ \begin{bmatrix} r_{33} & 0 \\ 0 & r_{55} \end{bmatrix} \begin{bmatrix} z^3(t) \\ \theta^3(t) \end{bmatrix} + \begin{bmatrix} d_{33} & 0 \\ 0 & d_{55} \end{bmatrix} \begin{bmatrix} \dot{z}(t)|\dot{z}(t)| \\ \dot{\theta}(t)|\dot{\theta}(t)| \end{bmatrix} = \begin{bmatrix} Z(t) \\ M(t) \end{bmatrix} \tag{5.118}$$

This matrix Eq. (5.118) for a moving ship is the same as

$$(m + a_{33})\ddot{z}(t) + c_{33}\dot{z}(t) + k_{33}z(t) + r_{33}z^3(t) + d_{33}\dot{z}(t)|\dot{z}(t)|$$
$$+ a_{35}\ddot{\theta}(t) + c_{35}\dot{\theta}(t) + k_{35}\theta(t) = Z(t) \tag{5.119}$$

$$a_{53}\ddot{z}(t) + c_{53}\dot{z}(t) + k_{53}z(t) + (I_{yy} + a_{55})\ddot{\theta}(t) + c_{55}\dot{\theta}(t)$$
$$+ k_{55}\theta(t) + r_{55}\theta^3(t) + d_{55}\dot{\theta}(t)|\dot{\theta}(t)| = M(t) \tag{5.120}$$

The equivalent mathematical reverse dynamics input/output linear models to Eqs. (5.119) and (5.120) are governed by the relations

$$A_{3,3}(f)X_3(f) + A_{3,5}(f)X_5(f) + A_{3,9}(f)X_9(f) + A_{3,15}(f)X_{15}(f) = Y_3(f) \tag{5.121}$$

$$A_{5,3}(f)X_3(f) + A_{5,5}(f)X_5(f) + A_{5,11}(f)X_{11}(f) + A_{5,17}(f)X_{17}(f) = Y_5(f) \tag{5.122}$$

where the six input records are

$$\begin{aligned}
x_3(t) &= z(t) && \text{with } X_3(f) = \text{Fourier transform of } x_3(t)\\
x_5(t) &= \theta(t) && \text{with } X_5(f) = \text{Fourier transform of } x_5(t)\\
x_9(t) &= z^3(t) && \text{with } X_9(f) = \text{Fourier transform of } x_9(t)\\
x_{11}(t) &= \theta^3(t) && \text{with } X_{11}(f) = \text{Fourier transform of } x_{11}(t)\\
x_{15}(t) &= \dot{z}(t)|\dot{z}(t)| && \text{with } X_{15}(f) = \text{Fourier transform of } x_{15}(t)\\
x_{17}(t) &= \dot{\theta}(t)|\dot{\theta}(t)| && \text{with } X_{17}(f) = \text{Fourier transform of } x_{17}(t)
\end{aligned}$$

and the two output records are

$$\begin{aligned}
y_3(t) &= Z(t) && \text{with } Y_3(f) = \text{Fourier transform of } y_3(t)\\
y_5(t) &= M(t) && \text{with } Y_5(f) = \text{Fourier transform of } y_5(t)
\end{aligned}$$

The eight FRFs in Eqs. (5.121) and (5.122) are

$$\begin{aligned}
A_{3,3}(f) &= \text{FRF for output } y_3(t) \text{ from input } x_3(t)\\
&= k_{33} + j(2\pi f)c_{33} - (2\pi f)^2(m + a_{33})
\end{aligned} \tag{5.123}$$

$$\begin{aligned}
A_{3,5}(f) &= \text{FRF for output } y_3(t) \text{ from input } x_5(t)\\
&= k_{35} + j(2\pi f)c_{35} - (2\pi f)^2 a_{35}
\end{aligned} \tag{5.124}$$

$$A_{3,9}(f) = \text{FRF for output } y_3(t) \text{ from input } x_9(t) = r_{33} \tag{5.125}$$

$$A_{3,15}(f) = \text{FRF for output } y_3(t) \text{ from input } x_{15}(t) = d_{33} \tag{5.126}$$

$$\begin{aligned}
A_{5,3}(f) &= \text{FRF for output } y_5(t) \text{ from input } x_3(t)\\
&= k_{53} + j(2\pi f)c_{53} - (2\pi f)^2 a_{53}
\end{aligned} \tag{5.127}$$

$$\begin{aligned}
A_{5,5}(f) &= \text{FRF for output } y_5(t) \text{ from input } x_5(t)\\
&= k_{55} + j(2\pi f)c_{55} - (2\pi f)^2(I_{yy} + a_{55})
\end{aligned} \tag{5.128}$$

$$A_{5,11}(f) = \text{FRF for output } y_5(t) \text{ from input } x_{11}(t) = r_{55} \tag{5.129}$$

$$A_{5,17}(f) = \text{FRF for output } y_5(t) \text{ from input } x_{17}(t) = d_{55} \tag{5.130}$$

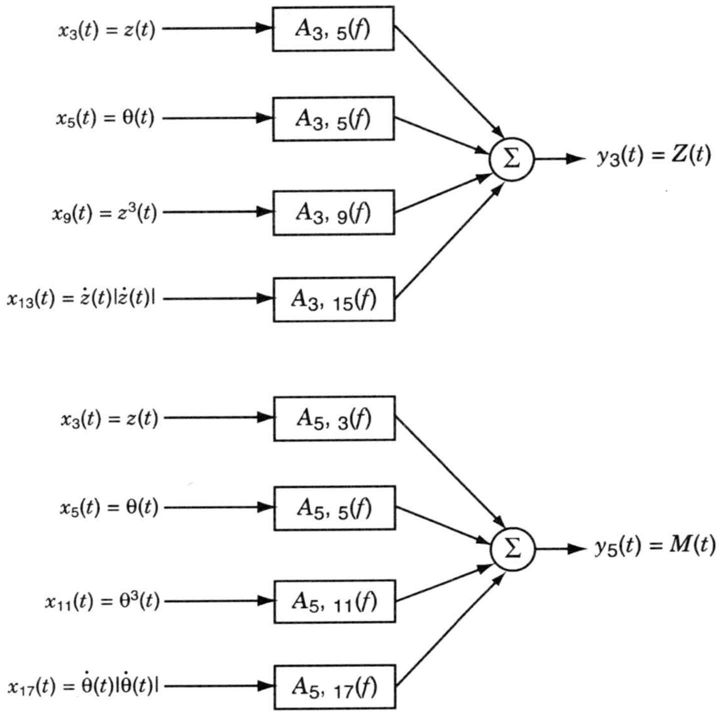

Figure 5.66 Equivalent Reverse MI/SO linear models for Case 2.

As illustrated in Figure 5.66, Eqs. (5.121) and (5.122) are each seen to be 4-input/1-output linear models where the eight frequency response functions shown can be determined from the measured input/output data for well-defined MI/SO linear models. This then gives the coefficients from Eqs. (5.123)–(5.130) that define the physical parameters in the nonlinear differential equations of motion of the coupled heave and pitch motions for this moving ship. This practical analysis can determine whether or not the nonlinear coefficients r_{33}, r_{55}, d_{33}, and d_{55} are constants or are functions of frequency.

Similar equations and illustrations can be derived to compute the coefficients that define the physical parameters in the nonlinear differential equations of motion for the coupled sway, roll, and yaw motions of a moving ship. Reasonable assumptions should be made to simplify the work by neglecting second-order terms that can be omitted. As in Case 1, the main limitation in applying this new approach will occur in problems where the measured physical response signals lead to mathematical input signals that are highly coherent at desired frequencies for the equivalent Reverse MI/SO linear analysis.

5.6.3 Summary of Results

1. This material has demonstrated that a wide class of six-DOF nonlinear differential equations of motion for ocean engineering applications can be converted into mathematically equivalent reverse dynamics MI/MO and MI/SO linear models. These linear models can be solved by standard spectral techniques to yield nonlinear system amplitude and frequency properties.
2. No restrictive assumptions are required regarding the spectral properties or the probability nature of the measured data. Practical results can be obtained without requiring large amounts of data.
3. Ordinary and cumulative coherence functions can be computed to confirm the validity of assumed nonlinear models, and to assess the relative importance of proposed linear and nonlinear terms as a function of frequency.
4. Two special cases have been discussed and illustrated for (a) a moored ship that is connected at both ends and (b) an unrestrained ship in head seas. These cases indicate how to identify desired physical parameters from measured input/output data.
5. Reverse MI/SO Techniques are practical techniques to solve nonlinear system problems in many engineering and scientific fields.

This concludes Chapter 5.

6

NONLINEAR SYSTEM RESPONSE PROPERTIES OF A NAVAL FRIGATE FROM MEASURED OCEAN ENGINEERING DATA

This chapter and Chapter 7 discuss results obtained from two experimental test programs conducted at the U.S. Naval Academy (USNA) Hydromechanics Laboratory in Annapolis, Maryland, in 1994 and 1995. Chapter 6 covers the first series of tests in June 1994, a program of one week duration involving a scale model of a naval frigate where the frigate was subjected to waves in the single direction of beam seas. Chapter 7 covers the second series of tests in July 1995, an extended program of two weeks duration involving a scale model of a naval barge where the barge was subjected to waves in the three directions of head seas, beam seas, and quartering seas. The main objective in both test programs was to demonstrate how Direct and Reverse MI/SO Techniques can provide new practical significant ways to identify nonlinear system response properties using measured ocean engineering data. These test programs are discussed here in detail as guidelines for work that should be done with real physical data for engineering and scientific investigations in other fields of concern to readers of this book.

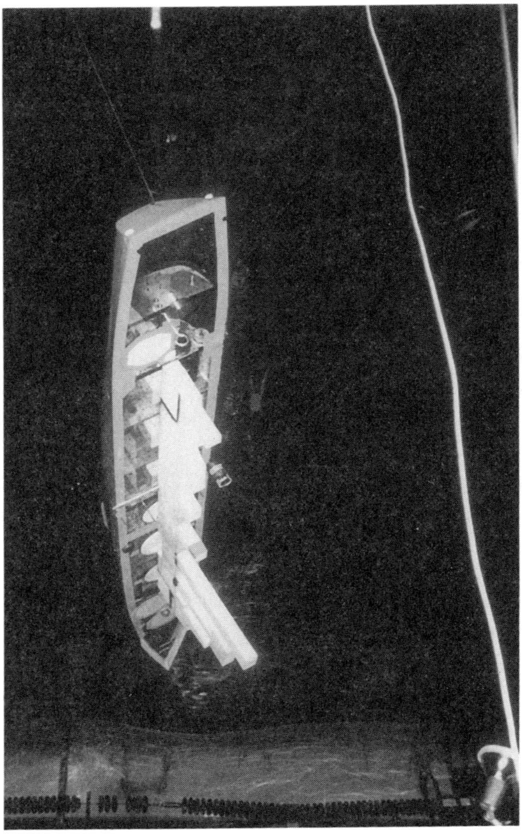

Scale Model of a Naval Frigate

6.1 TEST PROGRAM IN JUNE 1994

In previous studies for the Naval Facilities Engineering Service Center (NFESC) in Port Heuneme, California, performed by Bendat, Coppolino, and Palo (references 6 and 7), the Reverse MI/SO Technique for nonlinear system analysis and identification was shown to be applicable to a wide variety of mathematically defined nonlinear systems. These past studies, such as the work in Chapter 5, used computer-generated numerical data from simulated nonlinear systems as opposed to measured physical data from real nonlinear systems. The first report to NFESC on the application of the Reverse MI/SO Technique using measured ocean engineering data on real nonlinear systems was from experiments performed at the Hydromechanics Laboratory of the U.S. Naval Academy in Annapolis, Maryland, in June 1994. From these data, the Direct MI/SO Technique was also investigated.

Results from analysis of the test data collected in the June 1994 program demonstrate that the Reverse MI/SO Technique provides an accurate practical method to identify the dynamic properties of various desired nonlinear physical systems by converting nonlinear models into equivalent linear models. This is done by reversing the roles of physical inputs and outputs so that they become mathematical inputs and outputs in the resulting MI/SO data analysis. Chapters 4 and 5 in this book discuss and illustrate both the Reverse MI/SO Technique and the Direct MI/SO Technique. Significant features are stated in the lecture of Chapter 3.

In June 1994, a scale model of a naval frigate was placed in an inclined position such that nonlinear system response motions could occur. Measurements were taken of the physical excitation to the ship, represented by the instantaneous wave amplitude near the ship, and three physical response motions due to this excitation, represented by the sway, heave, and roll of the ship. From established ocean engineering considerations, different types of single and multiple degree of freedom (SDOF and MDOF) models defined by linear and nonlinear integrodifferential equations of motion were proposed, with unknown coefficients that could be constants or frequency-dependent. Ordinary and cumulative coherence functions were computed from the linear and nonlinear terms to assess the goodness of the fit. Application of the Direct and Reverse MI/SO Techniques using the measured ocean engineering data identified desired nonlinear system properties such as:

1. The relative contribution of each proposed linear and nonlinear term versus frequency.
2. Whether coefficients are constants or frequency-dependent.
3. How the cumulative coherence functions improve by including nonlinear terms.
4. How the cumulative coherence functions change by using the Direct MI/SO Technique versus the Reverse MI/SO Technique.

6.2 TEST OBJECTIVES

The primary objective in these June 1994 USNA tests was to demonstrate how the Reverse MI/SO Technique and the Direct MI/SO Technique can identify important nonlinear system response properties better than other existing nonlinear techniques for many desired naval applications. Previous NFESC studies showed this to be true for computer-simulated artificial data and it was now deemed appropriate to use actual ocean engineering data from a real nonlinear physical problem of concern to NFESC and USNA. Chapter 6 has analytical and graphical results from this work.

Meetings and correspondence between Dr. Paul Palo of NFESC and people associated with USNA from March to June 1994 were devoted to defining the USNA tests to be conducted and the ocean engineering data to be collected. The main goal was to examine phenomena with known nonlinearities, which are not yet fully understood but still can be described using fairly simple nonlinear integrodifferential equations of motion that contain possible frequency-dependent coefficients. The Reverse MI/SO Technique and/or the Direct MI/SO Technique would be applied to the measured excitation and response data to then correctly identify the various nonlinear terms and properties of the possible frequency-dependent coefficients. Important nonlinear system results would thus be obtained quickly and efficiently to demonstrate the power and practical benefits of the Reverse and Direct MI/SO Techniques over other currently used procedures that do not yield these results.

Two test goals were stated at the beginning of the work:

1. To collect "high-quality" ocean engineering data that could be used for future nonlinear studies as well as for the present effort.
2. To collect ocean engineering data from a scale model of a naval vessel that allows for a succession of mathematical equations of motion where various linear and nonlinear terms can be included in the proposed Reverse or Direct MI/SO data processing.

One quantitative measure of "high-quality" data is statistical confidence bounds in desired estimates from knowledge of associated coherence functions. The random error in coherence function estimates is reduced by having as much data as possible in any test. Hence, to achieve the first goal, it was deemed appropriate to conduct a small number of tests each with a large amount of data, rather than a large number of tests each with a small amount of data. It was also deemed appropriate to excite the model over as wide a range of frequencies as possible rather than over a set of overlapping narrow frequency bands. This is a critical requirement to be able to define desired frequency response functions over their full frequency range as well as to make the data more useful for later nonlinear studies on systems other than naval vessels.

To achieve the second goal, it is known that ship dynamic motions constitute a rich family for study because they include SDOF and MDOF systems, with various linear and nonlinear terms that can have constant or frequency-dependent coefficients. In particular, the lateral modes of a naval vessel can be modeled very easily starting with SDOF linear or nonlinear roll motions with constant coefficients or frequency-dependent coefficients, and progressing to coupled MDOF nonlinear sway, heave, and roll motions with constant or frequency-dependent coefficients. This allows for the formulation of a number of coupled linear and nonlinear integrodifferential equations of motion to evaluate the Reverse and Direct MI/SO Techniques in many ways.

6.3 TEST DESCRIPTION

The large model basin at the USNA Hydromechanics Laboratory is a 380-foot-long towing tank equipped with a dual-flap hydraulically driven wavemaker that can generate wave amplitudes of various desired wave frequencies and amplitudes. This tank is 26 feet wide and 16 feet deep. Reference 61 discusses this facility and the tests that were conducted. The naval vessel chosen by USNA for these tests was a 1:36 scale 11-foot model of a FFG-7 hull. This hull was placed 100 feet downstream from the wavemaker and moored at both ends. It was subjected to unidirectional seas with zero forward speed to provide lateral mode tests of the vessel.

Some appendages were removed from the FFG hull and weights added so as to incline the vessel to produce nonlinear system response effects. The placement and mooring of the ship restricted its response motions to be primarily the sway, heave, and roll of the ship. Simultaneous recordings were made of these three response motions, together with the instantaneous wave amplitude excitation at the ship. These four measurements of wave amplitude, sway, heave, and roll were the only physical ocean engineering data collected to conduct all of the later Reverse MI/SO and Direct MI/SO data processing work. The surge, pitch, and yaw motions of the ship were not measured during the tests but were observed to be small. A diagram of the wave basin model test setup (not to scale) is drawn below.

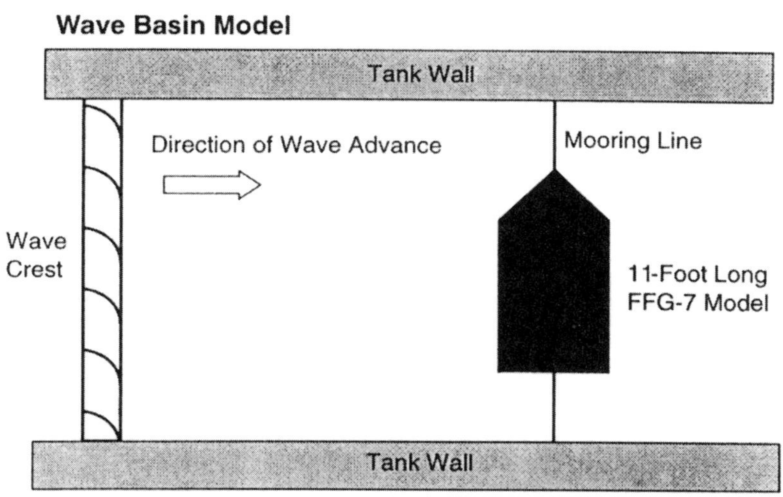

6.4 EQUATIONS OF MOTION FOR A NAVAL FRIGATE

For the experiments on the scale model of the naval frigate used here, there is an excitation proportional to the instantaneous wave amplitude $w(t)$ that causes three motion outputs, represented by the heave $z(t)$, the sway $y(t)$, and the roll angle $\phi(t)$. This test setup makes it reasonable to formulate models that can describe the following response motions:

- Independent heave, sway, and roll motions
- Coupled heave, sway, and roll motions

Nine different types of SDOF and MDOF linear and nonlinear integrodifferential equations of motion were proposed for the test data as follows:

1. Reverse SI/SO linear equation for independent heave motions.
2. Reverse SI/SO linear equation for independent sway motions.
3. Reverse SI/SO linear equation for independent roll motions.
4. Reverse MI/SO nonlinear equation for coupled heave motions.
5. Reverse MI/SO nonlinear equation for coupled sway motions.
6. Reverse MI/SO nonlinear equation for coupled roll motions.
7. Direct MI/SO nonlinear equation for coupled heave motions.
8. Direct MI/SO nonlinear equation for coupled sway motions.
9. Direct MI/SO nonlinear equation for coupled roll motions.

Each of these proposed types of equations of motion will now be discussed in some detail to show exactly how the Reverse and Direct MI/SO Techniques are applied in each situation. The notation and subscript numbering in the following equations use the customary naval architecture convention, as per Section 5.6 in this book.

6.4.1 Reverse SI/SO Linear Equation for Heave Motions

If the relative immersion of the vessel due to the wave amplitude is not considered, then a basic SDOF equation for heave motions is

$$F_{\mathrm{I}}(t) + F_{\mathrm{D}}(t) + F_{\mathrm{B}}(t) = F_{\mathrm{E}}(t) \tag{6.1}$$

where

$$\begin{aligned} F_{\mathrm{E}}(t) &= \text{external force} = w(t) \\ F_{\mathrm{I}}(t) &= \text{inertial force} = (M + a_{33})\ddot{z}(t) \\ F_{\mathrm{D}}(t) &= \text{damping force} = c_{33}\dot{z}(t) \\ F_{\mathrm{B}}(t) &= \text{buoyancy force} = k_{33}z(t) \end{aligned}$$

Thus, the proposed linear heave reverse differential equation of motion with constant damping and stiffness coefficients is

$$(M + a_{33})\ddot{z}(t) + c_{33}\dot{z}(t) + k_{33}z(t) = w(t) \tag{6.2}$$

This an be generalized to include frequency-dependent damping and stiffness coefficients using the reverse integrodifferential equation

$$(M + a_{33})\ddot{z}(t) + \int c_{33}(\tau)\dot{z}(t - \tau)\, d\tau + \int k_{33}(\tau)z(t - \tau)\, d\tau = w(t) \tag{6.3}$$

Taking the Fourier transforms of both sides yields

$$A(f)X(f) = Y(f) \tag{6.4}$$

where

$$X(f) = H(f) = \text{Fourier transform of } z(t) \tag{6.5}$$

$$Y(f) = W(f) = \text{Fourier transform of } w(t) \tag{6.6}$$

$$A(f) = K_{33}(f) - (2\pi f)^2(M + a_{33}) + j(2\pi f)C_{33}(f) \tag{6.7}$$

The functions $K_{33}(f)$ and $C_{33}(f)$ are Fourier transforms of $k_{33}(\tau)$ and $c_{33}(\tau)$, respectively. The constant coefficient case occurs when both $K_{33}(f)$ and $C_{33}(f)$ are constants, and a frequency-dependent coefficient case occurs when either of these functions varies with frequency.

Equation (6.4) is a reverse dynamic single-input/single-output (SI/SO) linear model with the mathematical input $X(f)$ of Eq. (6.5) passing through the $A(f)$ of Eq. (6.7) to give the mathematical output $Y(f)$ of Eq. (6.6).

6.4.2 Reverse SI/SO Linear Equation for Sway Motions

The proposed linear sway reverse differential equation of motion with constant damping and stiffness coefficients is

$$(M + a_{22})\ddot{y}(t) + c_{22}\dot{y}(t) + k_{22}y(t) = w(t) \tag{6.8}$$

As above, this can easily be extended to cover cases where the damping and stiffness coefficients are frequency-dependent. It follows that the linear sway reverse integrodifferential equation of motion can be expressed by the Fourier transform relation

$$A(f)X(f) = Y(f) \tag{6.9}$$

The functions $X(f)$, $Y(f)$, and $A(f)$ are now

$$X(f) = S(f) = \text{Fourier transform of } y(t) \tag{6.10}$$

$$Y(f) = W(f) = \text{Fourier transform of } w(t) \tag{6.11}$$

$$A(f) = K_{22}(f) - (2\pi f)^2(M + a_{22}) + j(2\pi f)C_{22}(f) \tag{6.12}$$

where $K_{22}(f)$ and $C_{22}(f)$ are Fourier transforms of $k_{22}(\tau)$ and $c_{22}(\tau)$, respectively. The constant coefficient case occurs when both $K_{22}(f)$ and $C_{22}(f)$ are constants, and a frequency-dependent coefficient case occurs when either of these functions varies with frequency.

Equation (6.9) is a reverse dynamic single-input/single-output linear model with the mathematical input defined by the $X(f)$ here of Eq. (6.10) passing through the linear system defined by the frequency response function $A(f)$ of Eq. (6.12) to give the mathematical output defined by the $Y(f)$ here of Eq. (6.11).

6.4.3 Reverse SI/SO Linear Equation for Roll Motions

The proposed linear roll reverse differential equation of motion with constant coefficients can be represented by

$$(I + a_{44})\ddot{\phi}(t) + c_{44}\dot{\phi}(t) + k_{44}\phi(t) = w(t) \tag{6.13}$$

where, if desired, the wave excitation $w(t)$ can be replaced here by $\dot{w}(t)$ based on the assumption that the excitation for roll motions is proportional to the derivative of the instantaneous wave slope.

This relation extends to cover cases where the coefficients are frequency-dependent when the linear roll reverse integrodifferential equation of motion is expressed by the Fourier transform relation

$$A(f)X(f) = Y(f) \tag{6.14}$$

The functions $X(f)$, $Y(f)$, and $A(f)$ are now

$$X(f) = R(f) = \text{Fourier transform of } \phi(t) \tag{6.15}$$

$$Y(f) = W(f) = \text{Fourier transform of } w(t) \tag{6.16}$$

$$A(f) = K_{44}(f) - (2\pi f)^2(I + a_{44}) + j(2\pi f)C_{44}(f) \tag{6.17}$$

where $K_{44}(f)$ and $C_{44}(f)$ are Fourier transforms of $k_{44}(\tau)$ and $c_{44}(\tau)$, respectively. The constant coefficient case occurs when both $K_{44}(f)$ and $C_{44}(f)$ are constants, and a frequency-dependent coefficient case occurs when either of these functions varies with frequency.

Equation (6.14) is a reverse dynamic single-input/single-output linear model with the mathematical input defined by the $X(f)$ here of Eq. (6.15) passing through the linear system defined by the frequency response function $A(f)$ of Eq. (6.17) to give the mathematical output defined by the $Y(f)$ here of Eq. (6.16).

6.4.4 Reverse MI/SO Nonlinear Equation for Heave Motions

The Reverse MI/SO nonlinear equation for heave motions due to coupled heave and roll assumes that a mathematical heave motion input $z(t)$ produces a mathematical wave motion output $w(t)$. From fluid dynamic fundamentals and engineering considerations, two general nonlinearities can be proposed. The first nonlinear viscous drag term consists of the relative fluid velocity multiplied by the absolute value of the relative fluid velocity, and the second nonlinear buoyancy term consists of roll displacement multiplied by the absolute value of roll displacement. Thus, the nonlinear equation due to coupled heave and roll motions includes a linear force proportional to the heave $z(t)$, a nonlinear hydrodynamic drag force proportional to $\dot{\zeta}(t)|\dot{\zeta}(t)|$, where $\dot{\zeta}(t) = \dot{z}(t) - \dot{w}(t)$, and a nonlinear roll restoring force proportional to $\phi(t)|\phi(t)|$. The proposed Reverse MI/SO nonlinear integrodifferential equation of motion for heave motions is

$$A_1(f)X_1(f) + A_2(f)X_2(f) + A_3(f)X_3(f) = Y(f) \tag{6.18}$$

where

$$X_1(f) = H(f) = \text{Fourier transform of } z(t) \tag{6.19}$$

$$X_2(f) = \text{Fourier transform of } \dot{\zeta}(t)|\dot{\zeta}(t)| \tag{6.20}$$

$$X_3(f) = \text{Fourier transform of } \phi(t)|\phi(t)| \tag{6.21}$$

$$Y(f) = W(f) = \text{Fourier transform of } w(t) \tag{6.22}$$

$$A_1(f) = \text{FRF for } z(t) \tag{6.23}$$

$$A_2(f) = \text{FRF for } \dot{\zeta}(t)|\dot{\zeta}(t)| \tag{6.24}$$

$$A_3(f) = \text{FRF for } \phi(t)|\phi(t)| \tag{6.25}$$

Equation (6.18) is a reverse dynamic three-input/single-output linear model with the mathematical inputs defined by the $X(f)$ terms in Eqs. (6.19)–(6.21) passing through the linear systems defined by the frequency response functions (FRFs) of Eqs. (6.23)–(6.25), to give the mathematical output defined by the $Y(f)$ of Eq. (6.22). The frequency response function $A_1(f)$ can be quite general here and may or may not be represented by a typical SDOF mechanical type of linear system such as Eq. (6.7). Similarly, the frequency response functions $A_2(f)$ and $A_3(f)$ are also arbitrary and will generally have frequency-dependent coefficients rather than constant coefficients.

6.4.5 Reverse MI/SO Nonlinear Equation for Sway Motions

The Reverse MI/SO nonlinear equation for sway motions due to coupled heave, sway, and roll assumes that a mathematical sway motion input $y(t)$ produces a mathematical wave motion output $w(t)$. From the same considerations as before, two general nonlinearities can be proposed. The first nonlinear term consists of the relative fluid velocity multiplied by the absolute value of the relative fluid velocity, and the second nonlinear term consists of the roll displacement multiplied by the absolute value of roll displacement. Thus, the nonlinear equation due to coupled heave, sway, and roll motions includes a linear force proportional to the sway $y(t)$, a nonlinear hydrodynamic drag force proportional to $\dot{\zeta}(t)|\dot{\zeta}(t)|$, and a nonlinear roll restoring force proportional to $\phi(t)|\phi(t)|$. The proposed Reverse MI/SO nonlinear integrodifferential equation of motion for sway motions is

$$A_1(f)X_1(f) + A_2(f)X_2(f) + A_3(f)X_3(f) = Y(f) \tag{6.26}$$

where

$$X_1(f) = S(f) = \text{Fourier transform of } y(t) \tag{6.27}$$

$$X_2(f) = \text{Fourier transform of } \dot{\zeta}(t)|\dot{\zeta}(t)| \tag{6.28}$$

$$X_3(f) = \text{Fourier transform of } \phi(t)|\phi(t)| \tag{6.29}$$

$$Y(f) = W(f) = \text{Fourier transform of } w(t) \tag{6.30}$$

$$A_1(f) = \text{FRF for } y(t) \tag{6.31}$$

$$A_2(f) = \text{FRF for } \dot{\zeta}(t)|\dot{\zeta}(t)| \tag{6.32}$$

$$A_3(f) = \text{FRF for } \phi(t)|\phi(t)| \tag{6.33}$$

Equation (6.26) is a reverse dynamic three-input/single-output linear model with the mathematical inputs defined by the $X(f)$ terms in Eqs. (6.27)–(6.29) passing through the linear systems defined by the frequency response functions of Eqs. (6.31)–(6.33), to give the mathematical output defined by the $Y(f)$ of Eq. (6.30). The frequency response function $A_1(f)$ can be quite general here and may or may not be represented by a typical SDOF mechanical type of linear system such as Eq. (6.7). Similarly, the frequency response functions $A_2(f)$ and $A_3(f)$ are also arbitrary and will generally have frequency-dependent coefficients rather than constant coefficients.

6.4.6 Reverse MI/SO Nonlinear Equation for Roll Motions

The Reverse MI/SO nonlinear equation for roll motions due to coupled heave and roll assumes that a mathematical roll motion input $\phi(t)$ produces a mathematical wave motion output $w(t)$. From fluid dynamic fundamentals and engineering considerations, two general nonlinearities can be proposed. The first nonlinear term consists of the relative fluid velocity multiplied by the absolute value of the relative fluid velocity, and the second nonlinear term consists of roll displacement multiplied by the absolute value of roll displacement. Thus, the nonlinear equation due to coupled heave and roll motions includes a linear force proportional to the roll $\phi(t)$, a nonlinear hydrodynamic drag force proportional to $\dot{\zeta}(t)|\dot{\zeta}(t)|$, and a nonlinear roll restoring force proportional to $\phi(t)|\phi(t)|$. The proposed Reverse MI/SO nonlinear integrodifferential equation of motion for roll motions is

$$A_1(f)X_1(f) + A_2(f)X_2(f) + A_3(f)X_3(f) = Y(f) \tag{6.34}$$

where

$$X_1(f) = R(f) = \text{Fourier transform of } \phi(t) \tag{6.35}$$

$$X_2(f) = \text{Fourier transform of } \dot{\zeta}(t)|\dot{\zeta}(t)| \tag{6.36}$$

$$X_3(f) = \text{Fourier transform of } \phi(t)|\phi(t)| \tag{6.37}$$

$$Y(f) = W(f) = \text{Fourier transform of } w(t) \tag{6.38}$$

$$A_1(f) = \text{FRF for } \phi(t) \tag{6.39}$$

$$A_2(f) = \text{FRF for } \dot{\zeta}(t)|\dot{\zeta}(t)| \tag{6.40}$$

$$A_3(f) = \text{FRF for } \phi(t)|\phi(t)| \tag{6.41}$$

Equation (6.34) is a reverse dynamic three-input/single-output linear model with the mathematical inputs defined by the $X(f)$ terms in Eqs. (6.35)–(6.37) passing through the linear systems defined by the frequency response functions of Eqs. (6.39)–(6.41), to give the mathematical output defined by the $Y(f)$ of Eq. (6.38). The frequency response function $A_1(f)$ can be quite general here and may or may not be represented by a typical SDOF mechanical type of linear system such as Eq. (6.7). As before, the frequency response functions $A_2(f)$ and $A_3(f)$ are also arbitrary and will generally have frequency-dependent coefficients rather than constant coefficients.

6.4.7 Direct MI/SO Nonlinear Equation for Heave Motions

The Direct MI/SO nonlinear equation for heave motions due to coupled heave and roll retains the same two nonlinear terms $X_2(f)$ and $X_3(f)$ as before in Eq. (6.18) but interchanges $X_1(f)$ and $Y(f)$ so that the physical wave motion input $w(t)$ produces the physical heave motion output $z(t)$. This gives completely different results for the three frequency response functions. For the heave motion output $z(t)$, the nonlinear equation now includes a linear force input proportional to the wave input $w(t)$, a nonlinear hydrodynamic drag force input proportional to $\dot{\zeta}(t)|\dot{\zeta}(t)|$, where $\dot{\zeta}(t) = \dot{z}(t) - \dot{w}(t)$, and a nonlinear roll restoring force input proportional to $\phi(t)|\phi(t)|$. Thus, the proposed Direct MI/SO nonlinear integrodifferential equation of motion for heave motions is now

$$A_1(f)X_1(f) + A_2(f)X_2(f) + A_3(f)X_3(f) = Y(f) \tag{6.42}$$

where

$$X_1(f) = W(f) = \text{Fourier transform of } w(t) \tag{6.43}$$

$$X_2(f) = \text{Fourier transform of } \dot{\zeta}(t)|\dot{\zeta}(t)| \tag{6.44}$$

$$X_3(f) = \text{Fourier transform of } \phi(t)|\phi(t)| \tag{6.45}$$

$$Y(f) = H(f) = \text{Fourier transform of } z(t) \tag{6.46}$$

$$A_1(f) = \text{FRF for } w(t) \tag{6.47}$$

$$A_2(f) = \text{FRF for } \dot{\zeta}(t)|\dot{\zeta}(t)| \tag{6.48}$$

$$A_3(f) = \text{FRF for } \phi(t)|\phi(t)| \tag{6.49}$$

Equation (6.42) is a direct dynamic three-input/single-output linear model with the physical inputs defined by the $X(f)$ terms in Eqs. (6.43)–(6.45) passing through the linear systems defined by the frequency response functions of Eqs. (6.47)–(6.49), to give the physical output defined by the $Y(f)$ of Eq. (6.46). The three frequency response functions here are all arbitrary and will generally have coefficients that are frequency-dependent rather than constants.

6.4.8 Direct MI/SO Nonlinear Equation for Sway Motions

The Direct MI/SO nonlinear equation for sway motions due to coupled heave, sway, and roll retains the same two nonlinear terms $X_2(f)$ and $X_3(f)$ as before in Eq. (6.26) but interchanges $X_1(f)$ and $Y(f)$ so that the physical wave motion input $w(t)$ produces the physical sway motion output $y(t)$. This gives completely different results for the three frequency response functions. For the sway motion output $y(t)$, the nonlinear equation now includes a linear force input proportional to the wave input $w(t)$, a nonlinear hydrodynamic drag force input proportional to $\dot{\zeta}(t)|\dot{\zeta}(t)|$, and a nonlinear roll restoring force input proportional to $\phi(t)|\phi(t)|$. Thus, the proposed Direct MI/SO nonlinear integrodifferential equation of motion for sway motions is now

$$A_1(f)X_1(f) + A_2(f)X_2(f) + A_3(f)X_3(f) = Y(f) \tag{6.50}$$

where

$$X_1(f) = W(f) = \text{Fourier transform of } w(t) \tag{6.51}$$

$$X_2(f) = \text{Fourier transform of } \dot{\zeta}(t)|\dot{\zeta}(t)| \tag{6.52}$$

$$X_3(f) = \text{Fourier transform of } \phi(t)|\phi(t)| \tag{6.53}$$

$$Y(f) = S(f) = \text{Fourier transform of } y(t) \tag{6.54}$$

$$A_1(f) = \text{FRF for } w(t) \tag{6.55}$$

$$A_2(f) = \text{FRF for } \dot{\zeta}(t)|\dot{\zeta}(t)| \tag{6.56}$$

$$A_3(f) = \text{FRF for } \phi(t)|\phi(t)| \tag{6.57}$$

Equation (6.50) is a direct dynamic three-input/single-output linear model with the physical inputs defined by the $X(f)$ terms in Eqs. (6.51)–(6.53) passing through the linear systems defined by the frequency response functions of Eqs. (6.55)–(6.57), to give the physical output defined by the $Y(f)$ of Eq. (6.54). The three frequency response functions here are all arbitrary and will generally have coefficients that are frequency-dependent rather than constants.

6.4.9 Direct MI/SO Nonlinear Equation for Roll Motions

The Direct MI/SO nonlinear equation for roll motions due to coupled heave and roll retains the same two nonlinear terms $X_2(f)$ and $X_3(f)$ as before in Eq. (6.34) but interchanges $X_1(f)$ and $Y(f)$ so that the physical wave motion input $w(t)$ produces the physical roll motion output $\phi(t)$. This gives completely different results for the three frequency response functions. For the roll motion output $\phi(t)$, the nonlinear equation now includes a linear force input proportional to the wave input $w(t)$, a nonlinear hydrodynamic drag force input proportional to $\dot{\zeta}(t)|\dot{\zeta}(t)|$, and a nonlinear roll restoring force input proportional to $\phi(t)|\phi(t)|$. Thus, the proposed Direct MI/SO nonlinear integrodifferential equation of motion for roll motions is

$$A_1(f)X_1(f) + A_2(f)X_2(f) + A_3(f)X_3(f) = Y(f) \tag{6.58}$$

where

$$X_1(f) = W(f) = \text{Fourier transform of } w(t) \tag{6.59}$$

$$X_2(f) = \text{Fourier transform of } \dot{\zeta}(t)|\dot{\zeta}(t)| \tag{6.60}$$

$$X_3(f) = \text{Fourier transform of } \phi(t)|\phi(t)| \tag{6.61}$$

$$Y(f) = R(f) = \text{Fourier transform of } \phi(t) \tag{6.62}$$

$$A_1(f) = \text{FRF for } w(t) \tag{6.63}$$

$$A_2(f) = \text{FRF for } \dot{\zeta}(t)|\dot{\zeta}(t)| \tag{6.64}$$

$$A_3(f) = \text{FRF for } \phi(t)|\phi(t)| \tag{6.65}$$

Equation (6.58) is a direct dynamic three-input/single-output linear model with the physical inputs defined by the $X(f)$ terms in Eqs. (6.59)–(6.61) passing through the linear systems defined by the frequency response functions of Eqs. (6.63)–(6.65), to give the physical output defined by the $Y(f)$ of Eq. (6.62). The three frequency response functions here are all arbitrary and will generally have coefficients that are frequency-dependent rather than constants.

6.5 RESULTS FROM ANALYSIS OF FRIGATE DATA

Results from analysis of the measured naval frigate test data are based on the following four recordings:

$w(t)$ = wave input excitation to the ship with Fourier transform $W(f)$

$z(t)$ = heave response motion of the ship with Fourier transform $H(f)$

$y(t)$ = sway response motion of the ship with Fourier transform $S(f)$

$\phi(t)$ = roll response motion of the ship with Fourier transform $R(f)$

TABLE 6.1 Digital Data Processing Parameters

f_s = sampling frequency = 15 Hz
$\Delta t = 1/f_s$ = time interval between samples = 0.067
$f_c = f_s/2$ = Nyquist cutoff frequency = 7.5 Hz
N = number of sample values per subrecord = 512
$T = N\,\Delta t$ = time of each subrecord = 34.13 s
n_d = number of distinct (disjoint, different) subrecords = 50
$N_{total} = n_d T$ = total number of sample values = 25,600
$T_{total} = n_d T$ = total time for all subrecords = 1707 s
$\Delta f = 1/T$ = bandwidth resolution = 0.03 Hz

TABLE 6.2 Figures from Analysis of Frigate Data

Figures 6.1–6.4	Four time history records
Figures 6.5–6.8	Four probability density functions
Figures 6.9–6.12	Four autospectral density functions
Figures 6.13–6.18	Six ordinary coherence functions
Figure 6.19	Reverse SI/SO results for Section 6.4.1
Figure 6.20	Reverse SI/SO results for Section 6.4.2
Figure 6.21	Reverse SI/SO results for Section 6.4.3
Figures 6.22–6.25	Reverse MI/SO results for Section 6.4.4
Figures 6.26–6.29	Reverse MI/SO results for Section 6.4.5
Figures 6.30–6.33	Reverse MI/SO results for Section 6.4.6
Figures 6.34–6.37	Direct MI/SO results for Section 6.4.7
Figures 6.38–6.41	Direct MI/SO results for Section 6.4.8
Figures 6.42–6.45	Direct MI/SO results for Section 6.4.9

TABLE 6.3 Symbols on Figures for Frigate Data

H = Fourier transform of heave motion $z(t)$
S = Fourier transform of sway motion $y(t)$
R = Fourier transform of roll motion $\phi(t)$
W = Fourier transform of wave motion $w(t)$
P = H − W = Fourier transform of relative heave motion $\zeta(t) = z(t) - w(t)$
VP = Fourier transform of relative heave velocity $\dot{\zeta}(t) = \dot{z}(t) - \dot{w}(t)$
|VP| = Fourier transform of absolute value of relative velocity $|\dot{\zeta}(t)|$
VP ∗ |VP| = Fourier transform of $\dot{\zeta}(t)|\dot{\zeta}(t)|$
|R| = Fourier transform of absolute value of roll motion $|\phi(t)|$
R ∗ |R| = Fourier transform of $\phi(t)|\phi(t)|$
NS = N_{total} = total number of sample values
ND = n_d = number of distinct subrecords
DF = Δf = bandwidth resolution

Table 6.1 lists the digital data processing parameters for all of the cases. A bandpass digital filter from 0.075 to 3.75 Hz with sharp cutoffs was employed to remove a low-frequency drift that was observed in the sway recording due to the mooring lines, and to reduce aliasing that occurs in numerically differentiated data. A Hanning window was applied to reduce leakage effects, together with overlap processing of 50% to recover data that was lost due to the Hanning window. MATLAP/ITAP programs were used to analyze the measured data.

Table 6.2 cross-lists the figures obtained from analysis of the test data. This table is included to aid in finding various desired different results. Table 6.3 defines the meaning of various symbol labels on the figures.

6.5.1 Preliminary Data Analysis

From the four recordings of the wave input $w(t)$, the heave response $z(t)$, the sway response $y(t)$, and the roll response $\phi(t)$, the preliminary data analysis consists of the following graphical results from each of the four records:

1. Figures 6.2–6.4 show typical filtered time history records from the first 40 seconds of data. The total amount of filtered data used for each record was approximately 1707 seconds and was digitized into 25,600 total number of sample values.

2. Figures 6.5–6.8 show probability density function (PDF) estimates obtained from the total filtered data of 25,600 samples for each record as solid curves, with their associated standardized Gaussian probability density functions as dashed curves.

The wave input PDF in Figure 6.5 and the roll response PDF in Figure 6.8 appear to be non-Gaussian, while the heave response PDF in Figure 6.6 and the sway response PDF in Figure 6.7 appear to be Gaussian. These results indicate that nonlinear operations probably occur between the wave input data to the roll response data, and that linear operations probably occur between the wave input data to the heave response data or to the sway response data.

3. Figures 6.9–6.12 show autospectral density function estimates obtained from the total filtered data of 25,600 samples for each record, where the number of distinct averages is $n_d = 50$, and the bandwidth resolution is $\Delta f = 0.03$ Hz.

The wave input autospectrum in Figure 6.9 and the heave response autospectrum in Figure 6.10 look like bandwidth-limited white noise between 0.20 and 0.65 Hz. The sway response autospectrum in Figure 6.11 and the roll response autospectrum in Figure 6.12 look like narrow-band noise between 0.45 and 0.65 Hz with a peak value at about 0.52 Hz.

4. Figures 6.13–6.18 show six different ordinary coherence function estimates obtained between any two of the four filtered records, as computed from autospectra and cross-spectra estimates where $n_d = 50$ and $\Delta f = 0.03$ Hz. Coherence values above 0.80 in Figures 6.13–6.18 indicate possible strong linear relationships. They were found to occur as follows:

The coherence function between heave and wave in Figure 6.13 is above 0.80 for frequencies between 0.17 and 0.30 Hz.

The coherence function between sway and wave in Figure 6.14 is above 0.80 for frequencies between 0.20 and 0.27 Hz.

The coherence function between roll and wave in Figure 6.15 is below 0.55 for the entire frequency range.

The coherence function between sway and heave in Figure 6.16 is above 0.80 for frequencies between 0.20 and 0.43 Hz and for frequencies between 0.58 and 1.5 Hz. A sharp dip occurs around 0.50 Hz.

The coherence function between roll and heave in Figure 6.17 is above 0.80 for frequencies between 0.58 and 1.1 Hz. A sharp dip occurs near 0.52 Hz.

The coherence function between roll and sway in Figure 6.18 is above 0.80 for frequencies between 0.50 and 1.2 Hz.

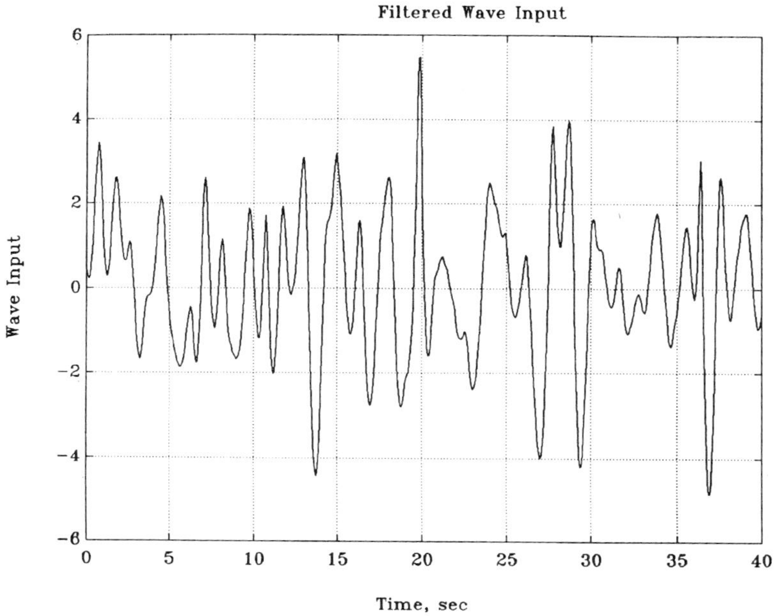

Figure 6.1.

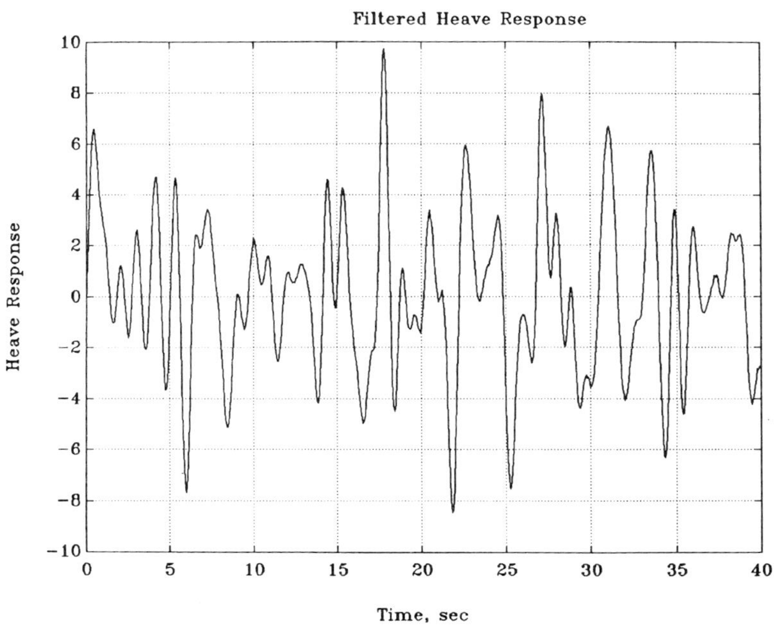

Figure 6.2.

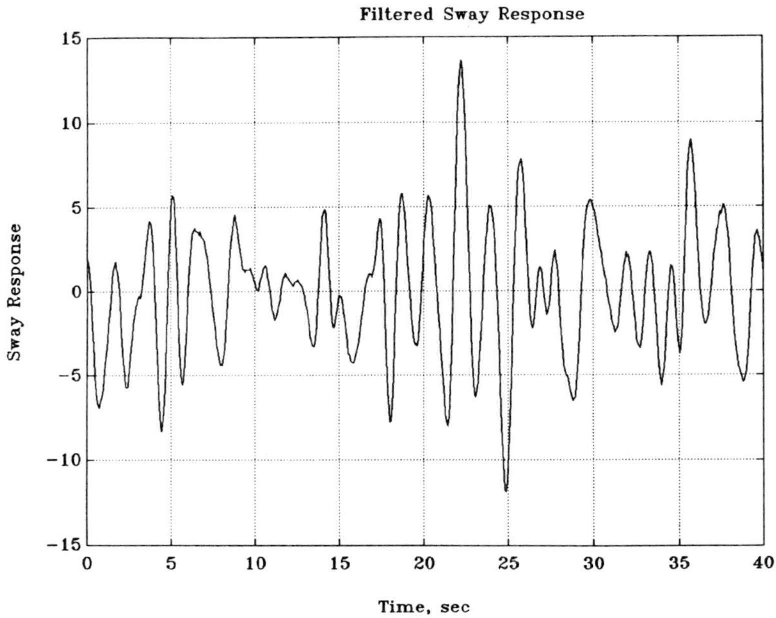

Figure 6.3.

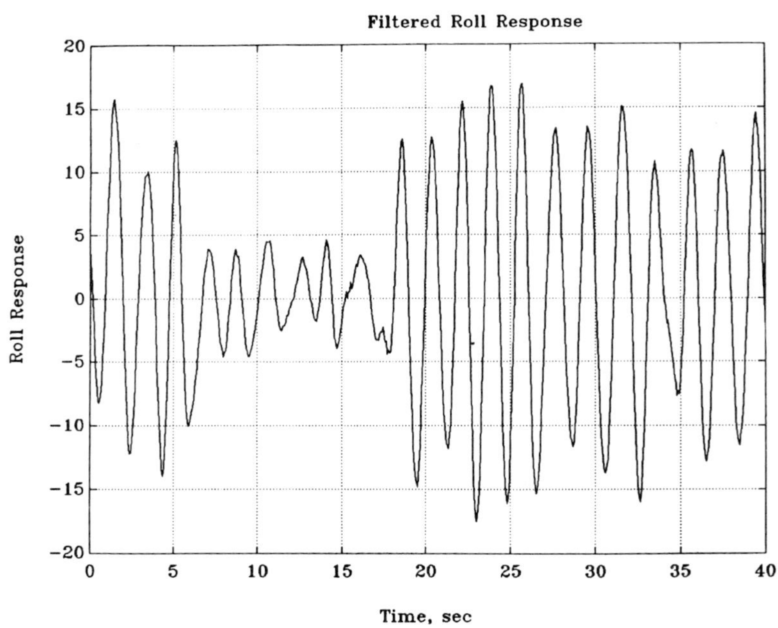

Figure 6.4.

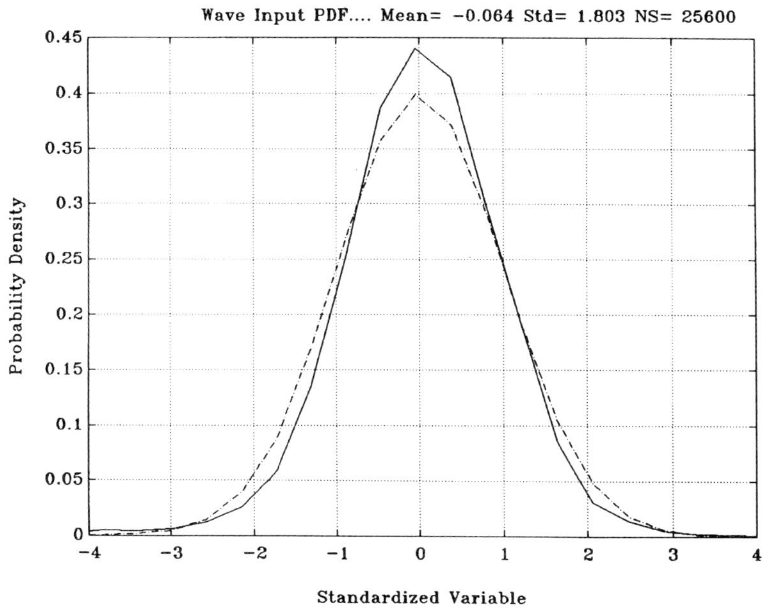

Figure 6.5.

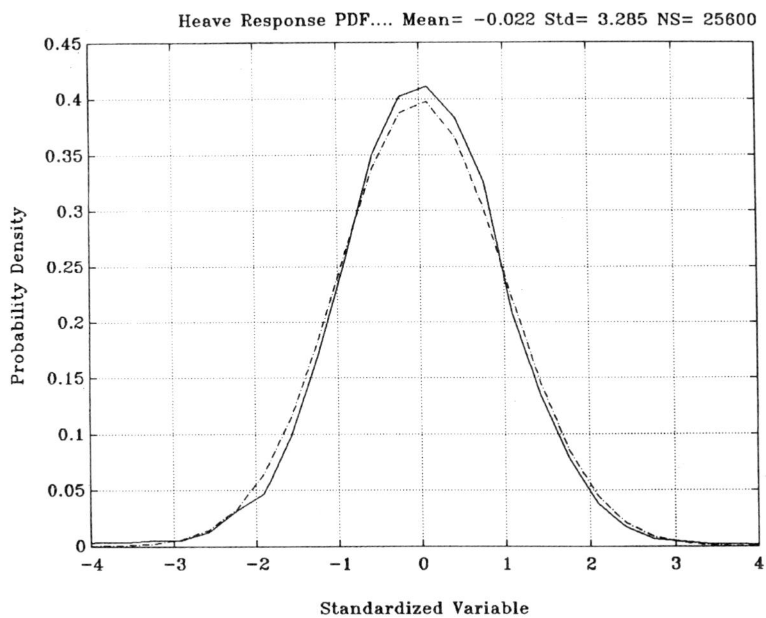

Figure 6.6.

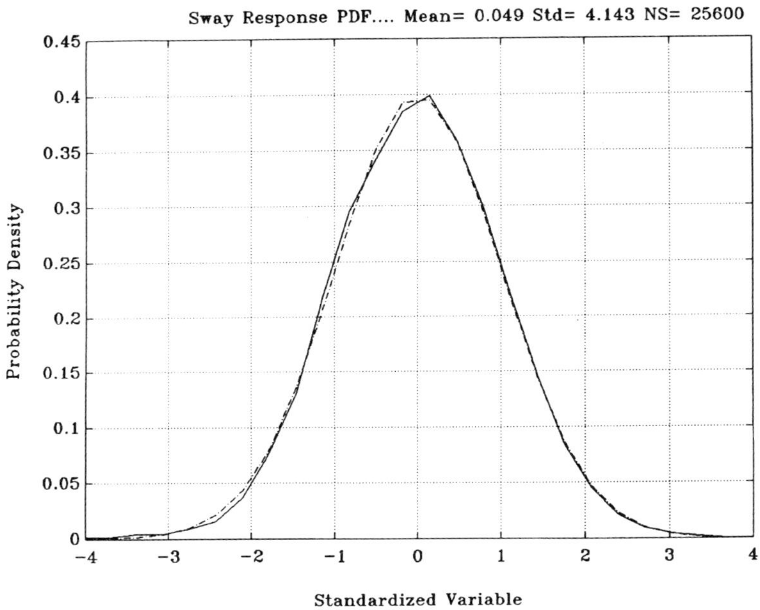

Figure 6.7.

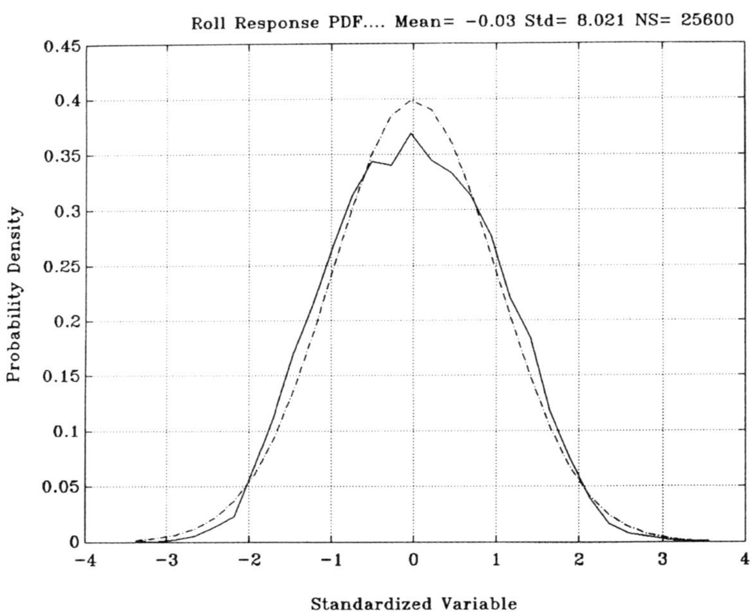

Figure 6.8.

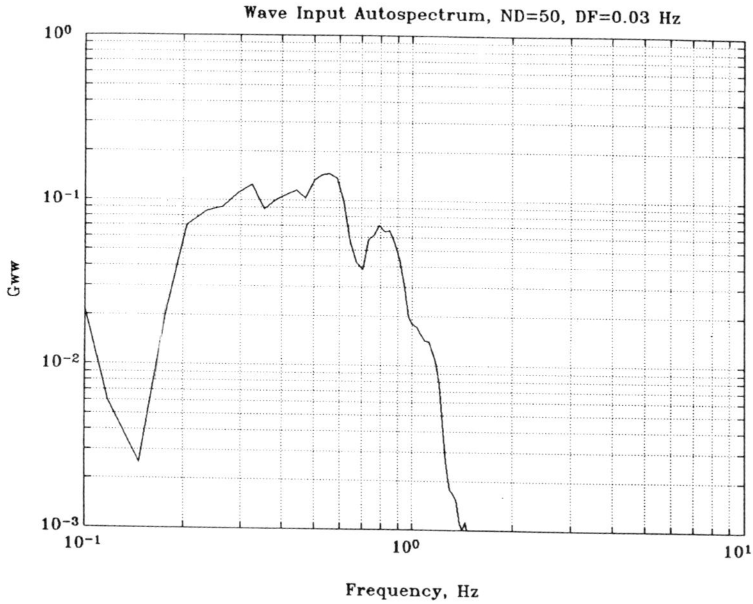

Figure 6.9.

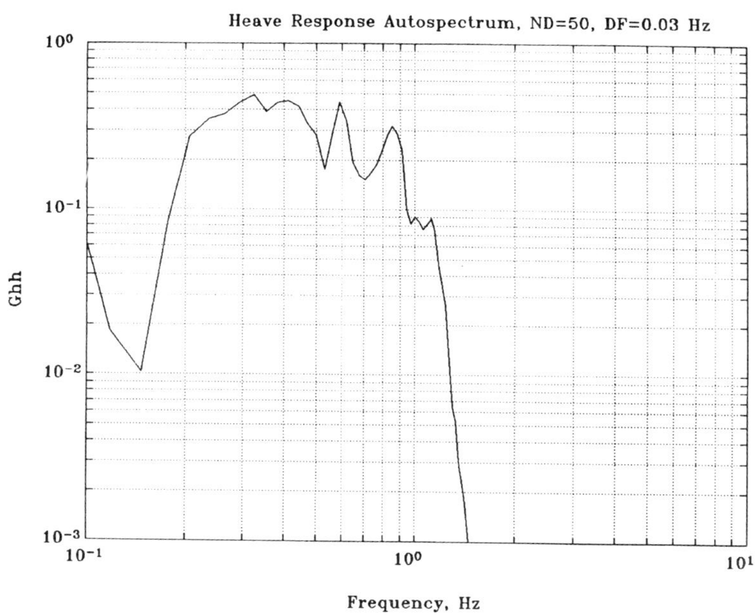

Figure 6.10.

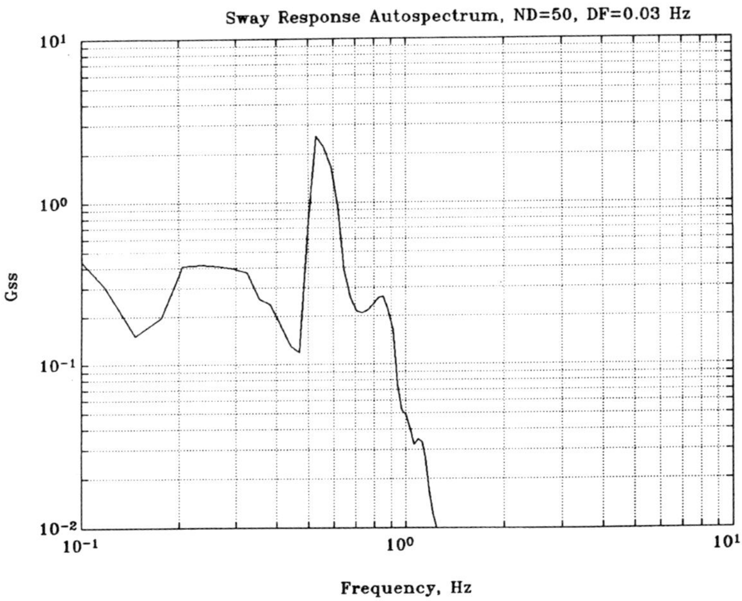

Figure 6.11.

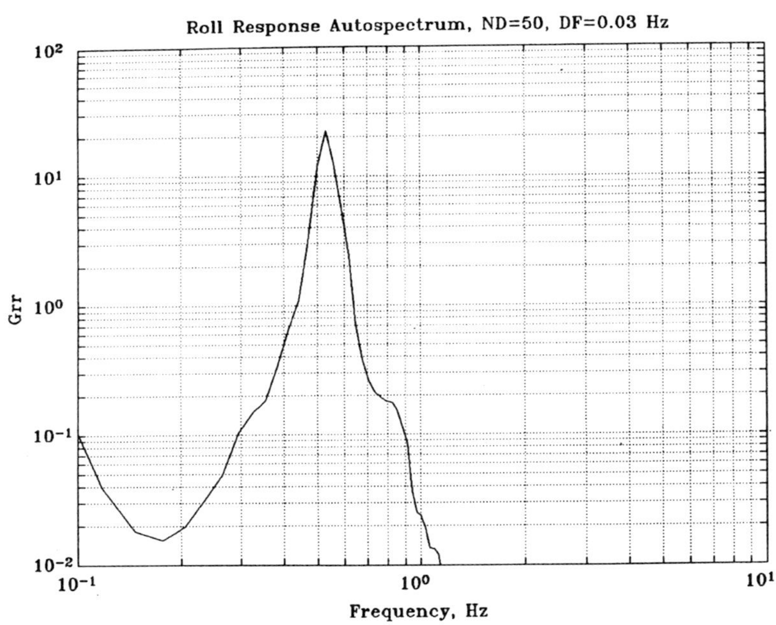

Figure 6.12.

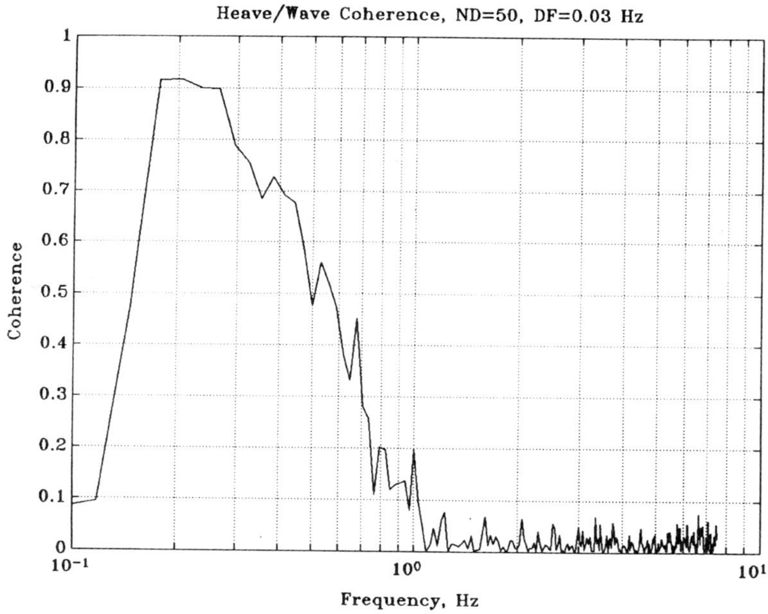

Figure 6.13.

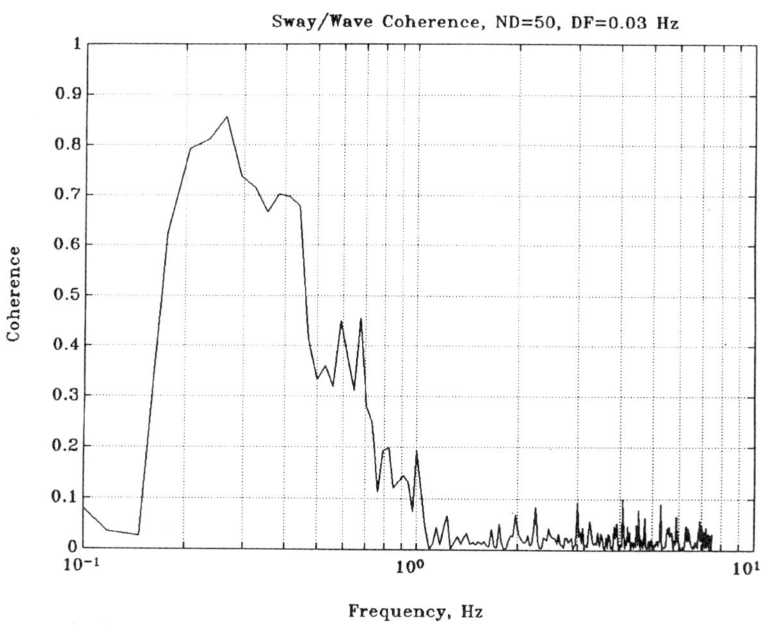

Figure 6.14.

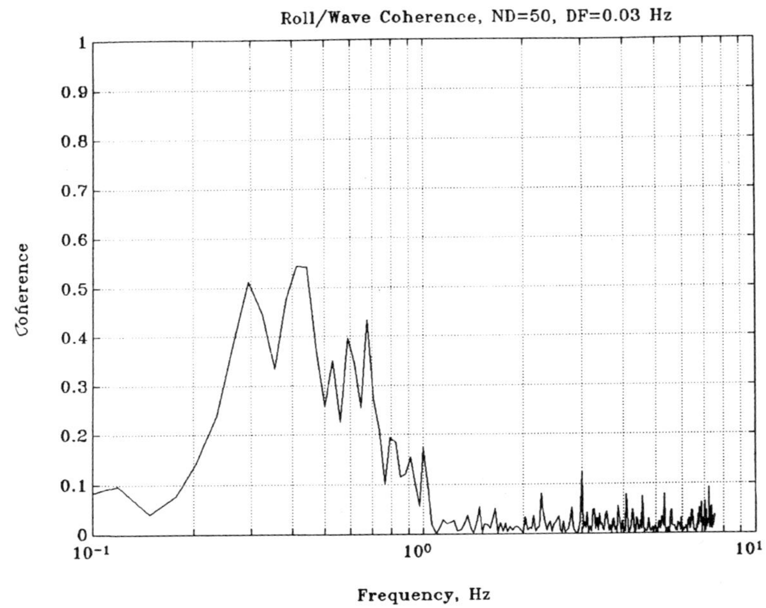

Figure 6.15.

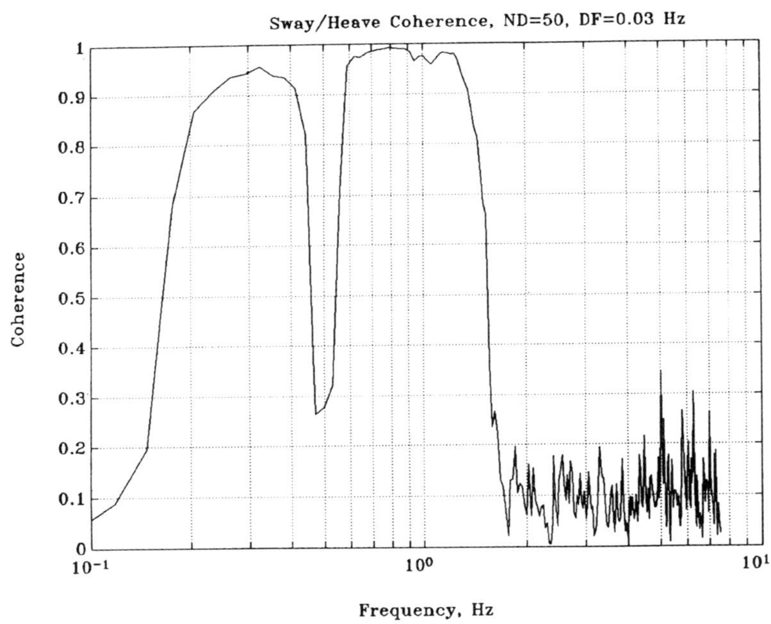

Figure 6.16.

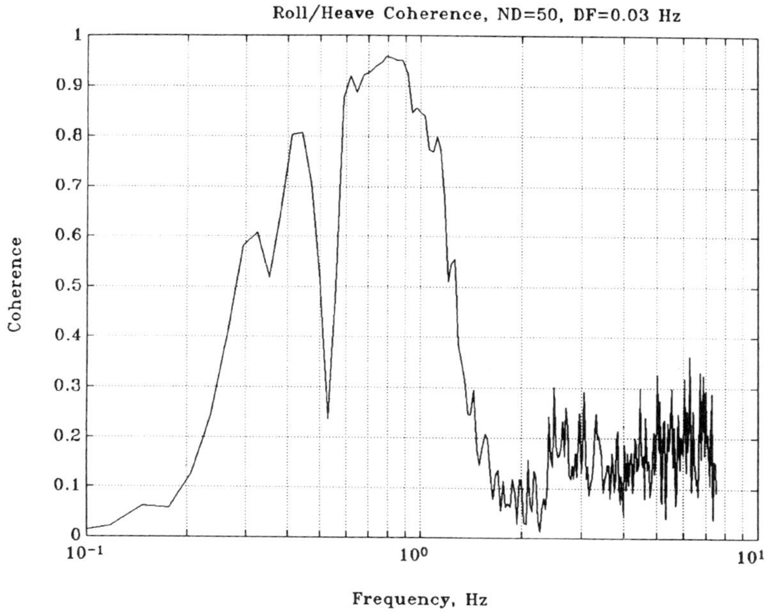

Figure 6.17.

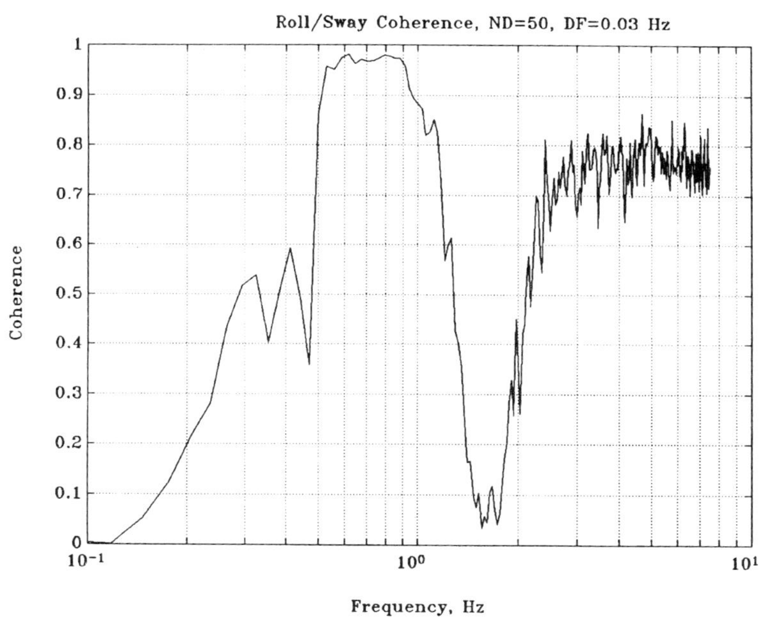

Figure 6.18.

6.5.2 Reverse SI/SO Analysis

A Reverse SI/SO model is drawn below for the case of the mathematical input $Z(f)$ to the mathematical output $W(f)$, where $Z(f)$ represents the mathematical heave input $z(t)$ and $W(f)$ represents the mathematical wave response output $w(t)$. Similar mathematical models apply when mathematical sway or roll inputs replace the mathematical heave input.

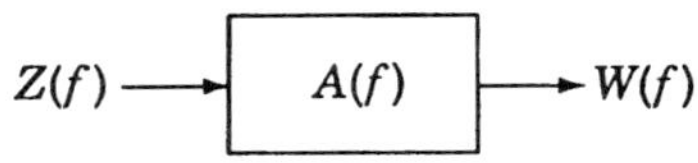

Reverse SI/SO linear analysis was conducted between selected pairs of the four filtered records as per the formulas in Sections 6.4.1–6.4.3.

From Eqs. (6.1)–(6.7) in Section 6.4.1, Figure 6.19 shows the Reverse SI/SO result for the FRF magnitude and phase estimates of $A(f)$ between the mathematical heave input $z(t)$ to the mathematical wave response $w(t)$. The ordinary coherence function estimate of Figure 6.13 indicates that these linear results might be meaningful in the frequency range between 0.15 and 0.50 Hz where the coherence function is above 0.50.

From Eqs. (6.8)–(6.12) in Section 6.4.2, Figure 6.20 shows the Reverse SI/SO result for the FRF magnitude and phase estimates of $A(f)$ between the mathematical sway input $y(t)$ to the mathematical wave response $w(t)$. The ordinary coherence function estimate of Figure 6.14 indicates that these linear results might be meangingful in the frequency range between 0.17 and 0.45 Hz where the coherence function is above 0.50.

From Eqs. (6.13)–(6.17) in Section 8.4.3, Figure 6.21 shows the Reverse SI/SO result for the FRF magnitude and phase estimates of $A(f)$ between the mathematical roll input $\phi(t)$ to the mathematical wave response $w(t)$. The low ordinary coherence function estimates in Figure 6.15 indicate that these particular linear results should not be considered meaningful for any frequency.

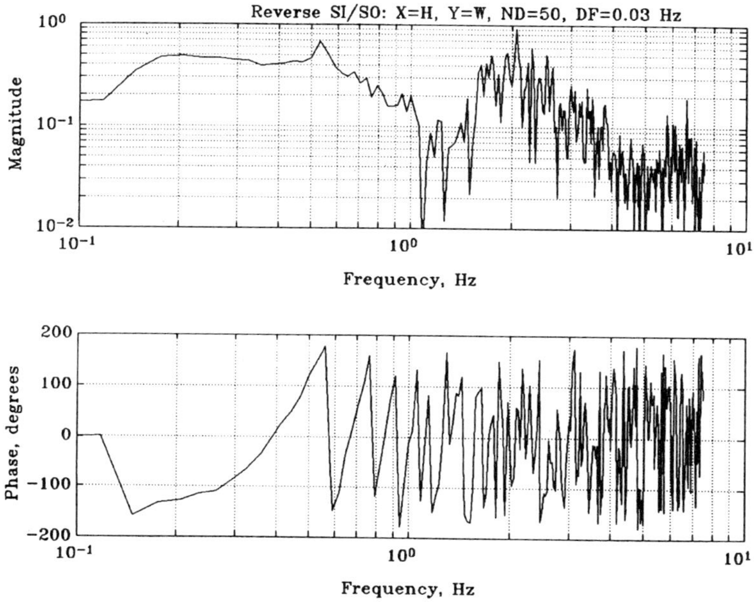

Figure 6.19.

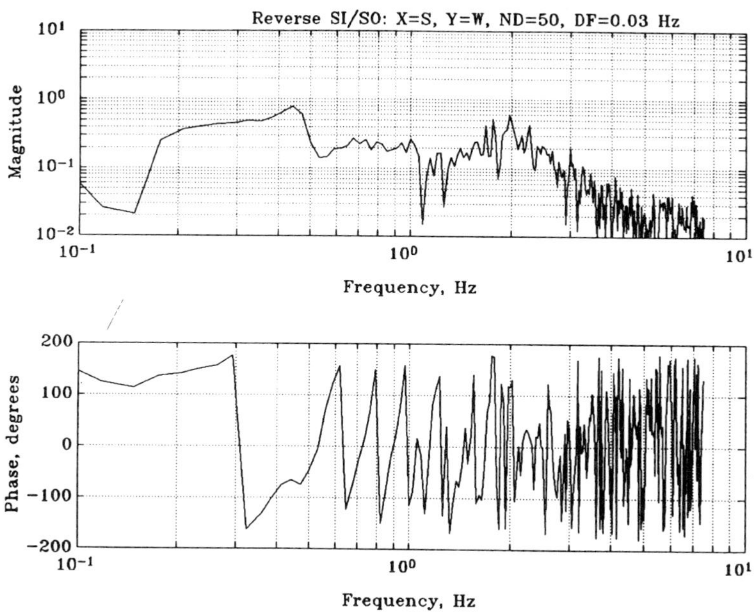

Figure 6.20.

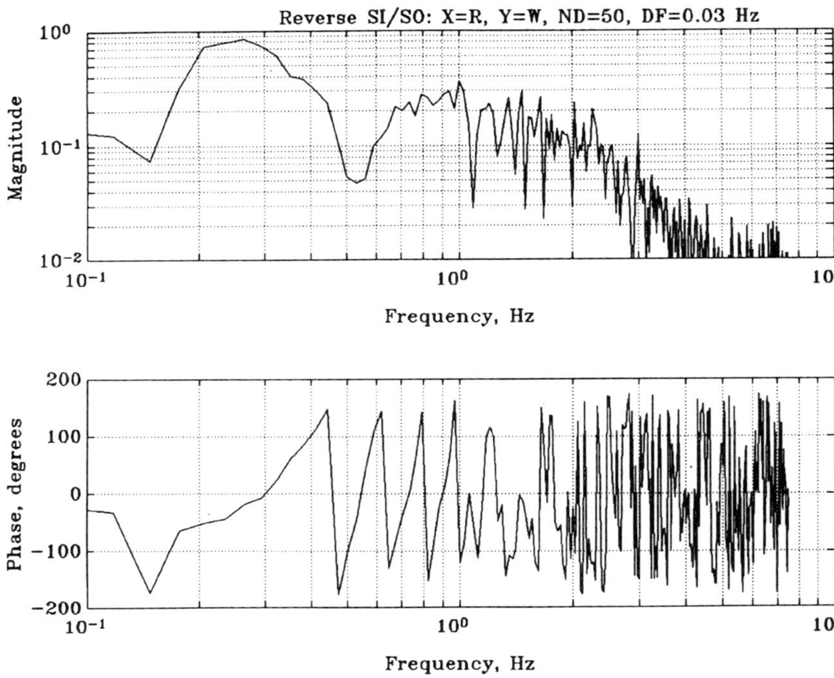

Figure 6.21.

6.5.3 Reverse MI/SO Analysis

A Reverse MI/SO three-input/single-output model is drawn below for the case of the three mathematical inputs $Z(f)$, $X_2(f)$, and $X_3(f)$ to the mathematical wave response output $W(f)$, where $Z(f)$ represents the mathematical heave input. Similar mathematical models apply to cases when the mathematical heave input is replaced by the mathematical sway or roll inputs.

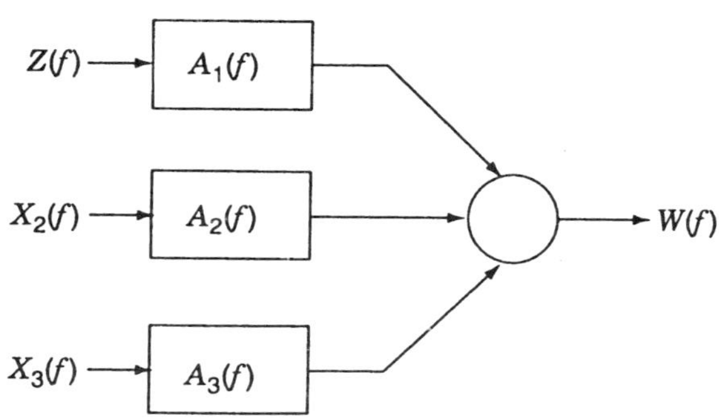

Reverse MI/SO analysis was conducted using the four filtered records as per the formulas in Sections 6.4.4–6.4.6.

Equations (6.18)–(6.25) in Section 6.4.4 represent Reverse MI/SO nonlinear relations for heave motions due to coupled heave and roll. For this case, Figures 6.22–6.24 show results for the FRF magnitude and phase estimates to the mathematical wave response $w(t)$:

Of $A_1(f)$ due to the first mathematical heave input $z(t)$

Of $A_2(f)$ due to the second mathematical input $\dot{\zeta}(t)|\dot{\zeta}(t)|$

Of $A_3(f)$ due to the third mathematical input $\phi(t)|\phi(t)|$

where $\dot{\zeta}(t) = \dot{z}(t) - \dot{w}(t)$. The ordinary and cumulative coherence function estimates in Figure 6.25 indicate that these results are meaningful in the frequency range between 0.15 and 1.2 Hz. Note that the cumulative coherence function is above 0.80 in the frequency range from 0.17 to 0.60 Hz. Comparison of Figure 6.25 with Figure 6.13 shows how the cumulative coherence function improves by adding on the second nonlinear term and also the particular frequencies where this nonlinear term is important.

Equations (6.26)–(6.33) in Section 6.4.5 represent Reverse MI/SO nonlinear relations for sway motions due to coupled heave, sway, and roll. For this case, Figures 6.26–6.28 show results for the FRF magnitude and phase estimates to the mathematical wave response $w(t)$:

Of $A_1(f)$ due to the first mathematical sway input $y(t)$

Of $A_2(f)$ due to the second mathematical input $\dot{\zeta}(t)|\dot{\zeta}(t)|$

Of $A_3(f)$ due to the third mathematical input $\phi(t)|\phi(t)|$

The ordinary and cumulative coherence function estimates in Figure 6.29 indicate that these results are meaningful for all three inputs in the frequency range between 0.17 and 1.2 Hz. Thus, the results for $A_1(f)$, $A_2(f)$, and $A_3(f)$ should be used here. Note that the cumulative coherence function is above 0.80 in the frequency range from 0.20 to 0.45 Hz. Comparison of Figure 6.29 with Figure 6.14 shows how the cumulative coherence function improves by adding on the second nonlinear term and the particular frequencies where this nonlinear term is important.

Equations (6.34)–(6.41) in Section 6.4.6 represent Reverse MI/SO nonlinear relations for roll motions due to coupled heave and roll. For this case, Figures 6.30–6.32 show results for the FRF magnitude and phase estimates to the mathematical wave response $w(t)$:

Of $A_1(f)$ due to the first mathematical roll input $\phi(t)$

Of $A_2(f)$ due to the second mathematical input $\dot{\zeta}(t)|\dot{\zeta}(t)|$

Of $A_3(f)$ due to the third mathematical input $\phi(t)|\phi(t)|$

The ordinary and cumulative coherence function estimates in Figure 6.33 indicate that these results are meaningful for the first two inputs only in the frequency range between 0.20 and 1.0 Hz. Thus, the result for the third input should not be used here. Note that the cumulative coherence function is above 0.80 in the narrow frequency range from 0.38 to 0.47 Hz. Comparison of Figure 6.33 with Figure 6.15 shows how the cumulative coherence function improves by adding on the second nonlinear input term and the particular frequencies where this nonlinear input term is important.

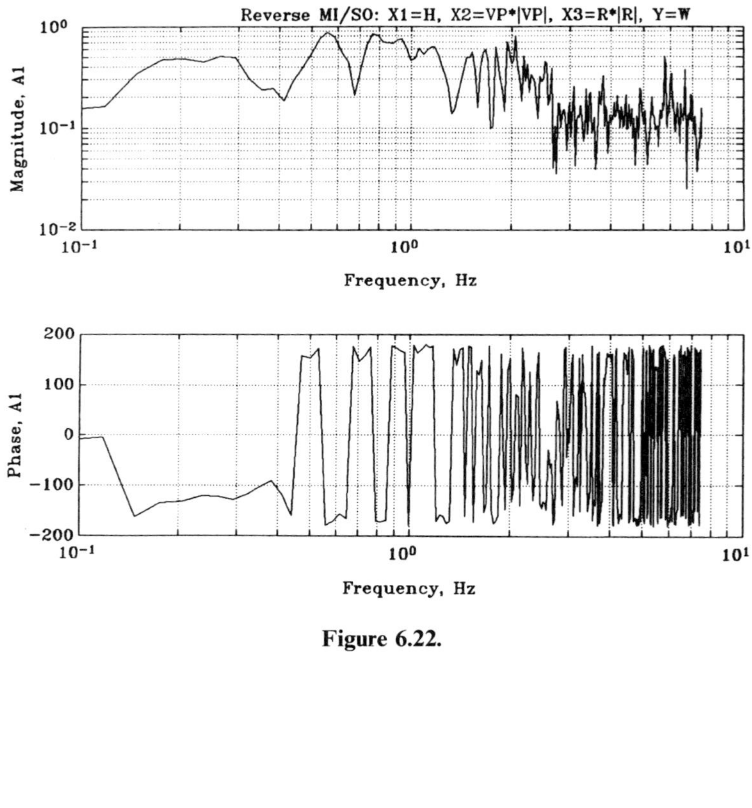

Figure 6.22.

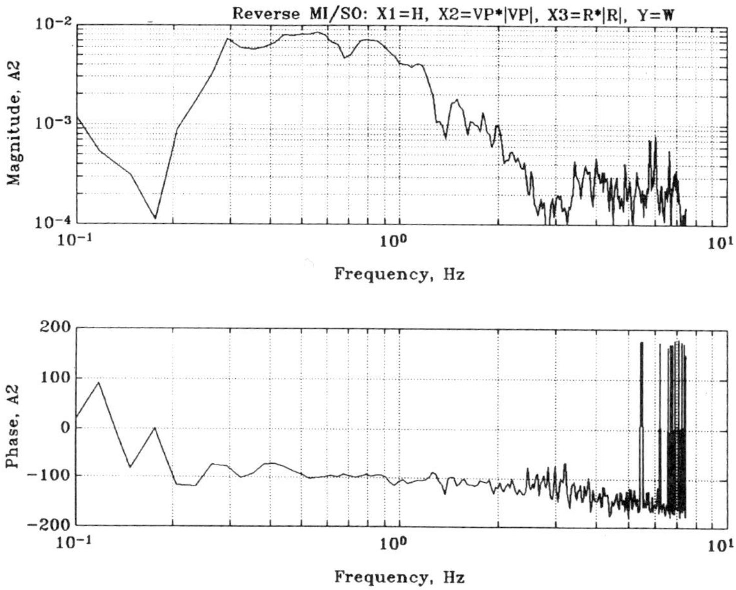

Figure 6.23.

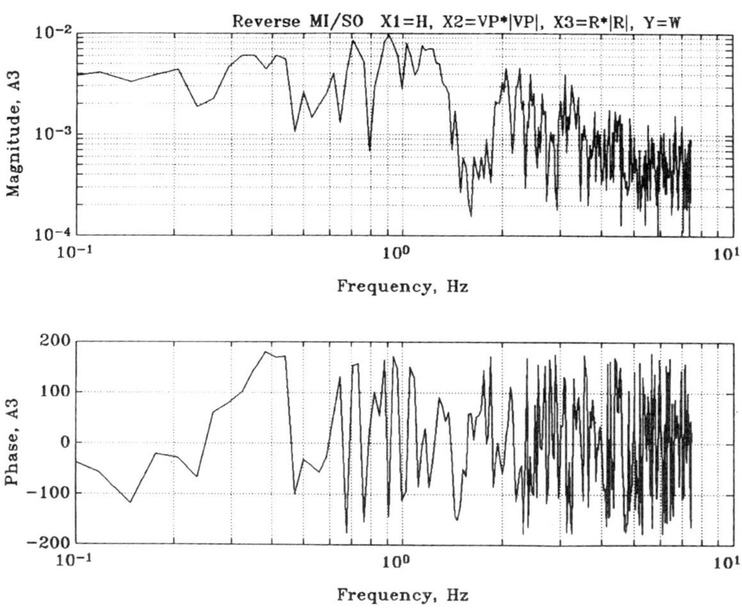

Figure 6.24.

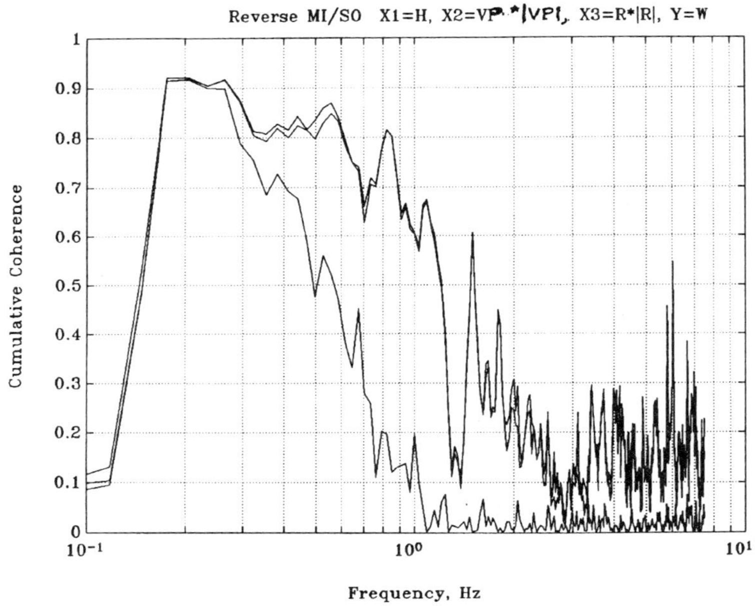

Figure 6.25.

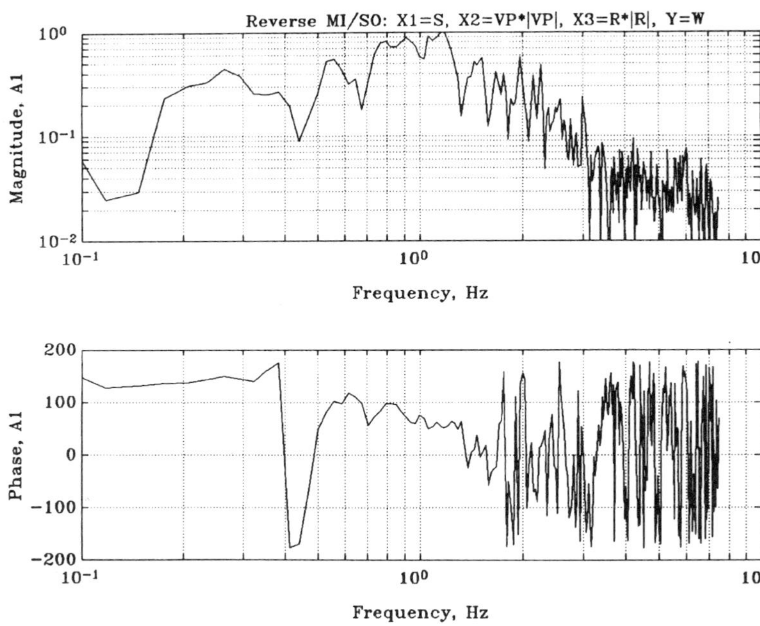

Figure 6.26.

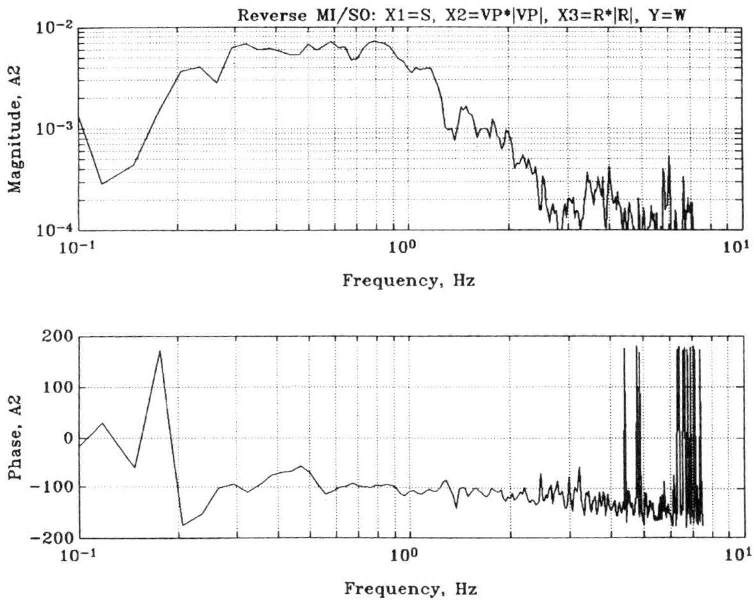

Figure 6.27.

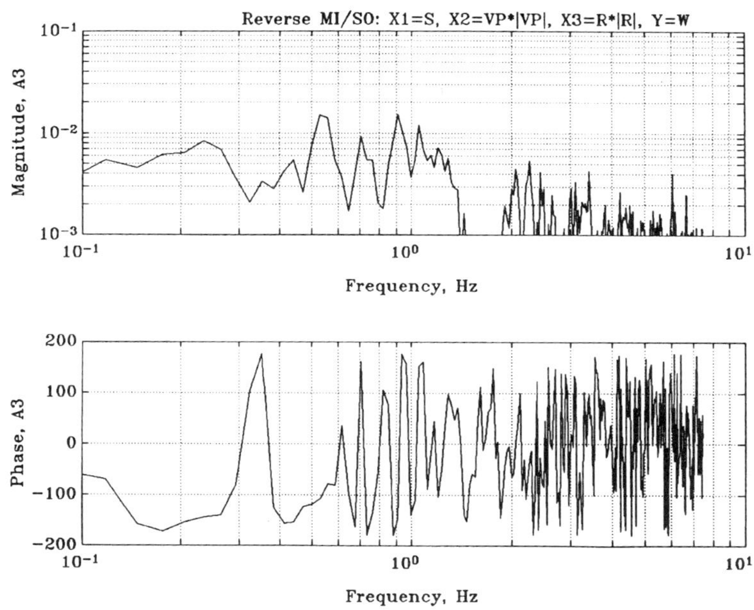

Figure 6.28.

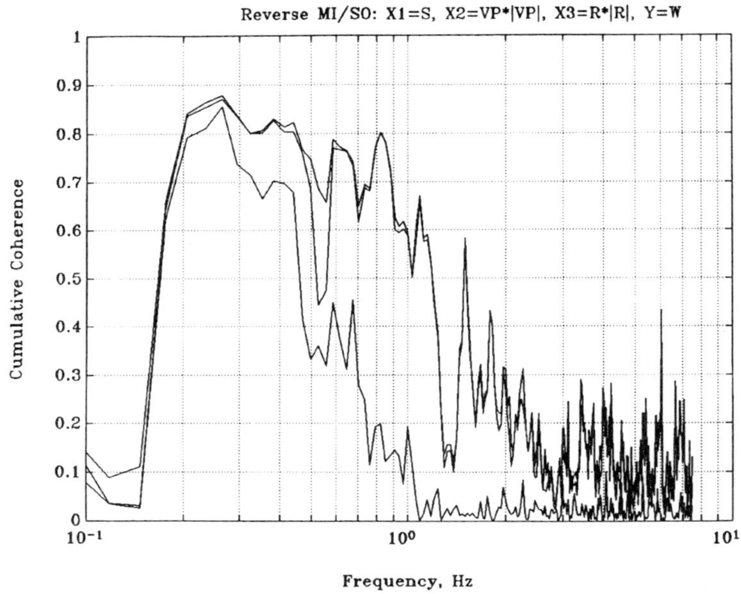

Figure 6.29.

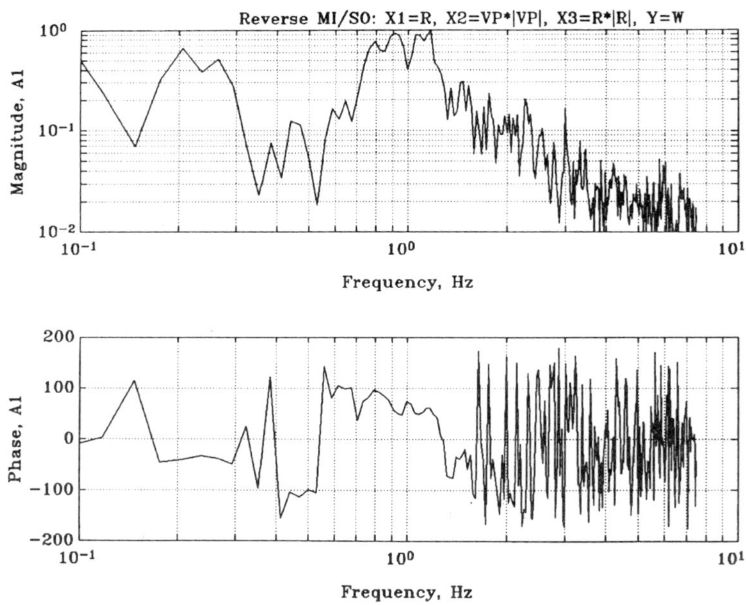

Figure 6.30.

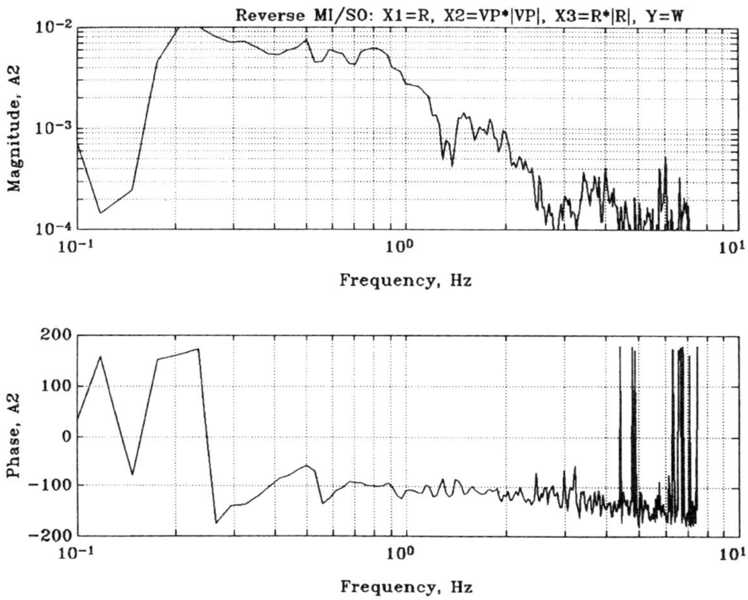

Figure 6.31.

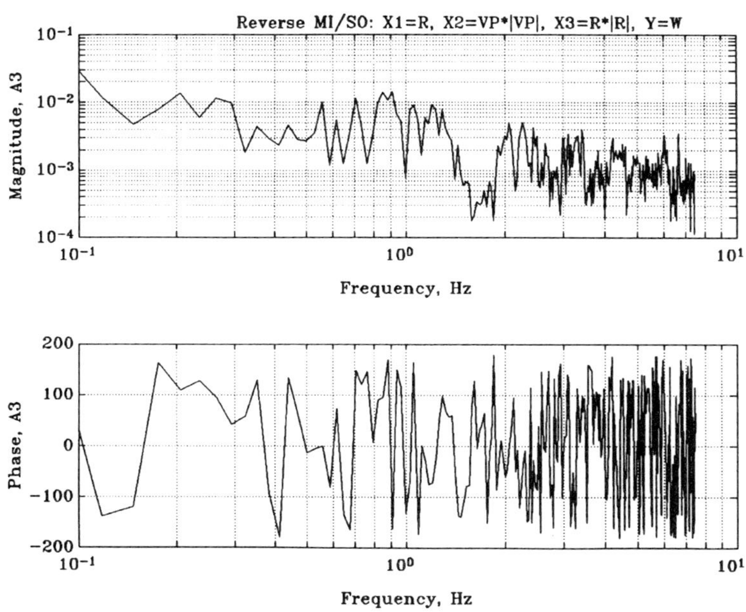

Figure 6.32.

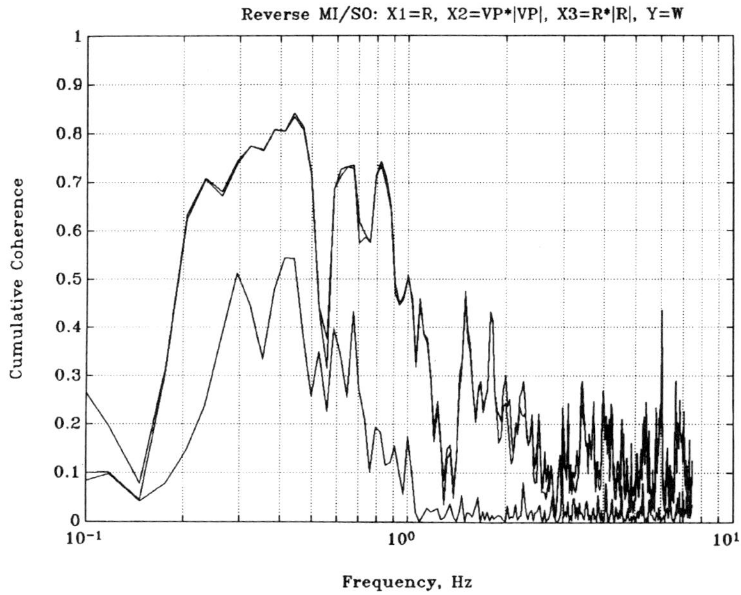

Figure 6.33.

6.5.4 Direct MI/SO Analysis

A Direct MI/SO three-input/single-output model is drawn below for the case of the three physical inputs $W(f)$, $X_2(f)$, and $X_3(f)$ to the physical heave response output $Z(f)$. Similar physical models apply to cases when the physical heave response output is replaced by the physical sway or roll response outputs.

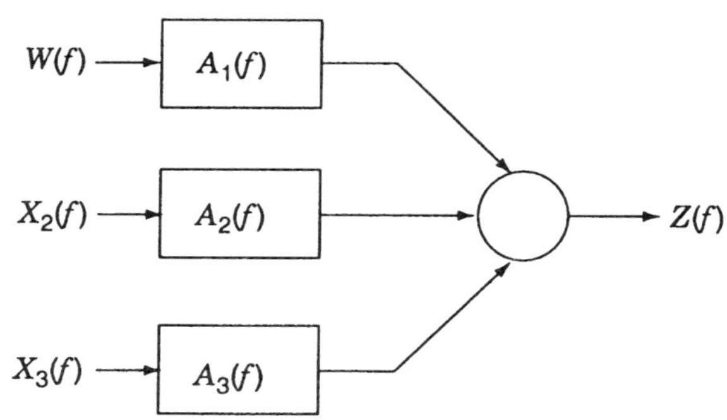

Direct MI/SO analysis was conducted using the four filtered records as per the formulas in Sections 6.4.7–6.4.9.

Equations (6.42)–(6.49) in Section 6.4.7 represent Direct MI/SO nonlinear relations for physical heave motions due to coupled heave and roll. For this case, Figures 6.34–6.36 show results for the FRF magnitude and phase estimates to the physical heave response $z(t)$:

Of $A_1(f)$ due to the first physical wave input $w(t)$

Of $A_2(f)$ due to the second physical input $\dot{\zeta}(t)|\dot{\zeta}(t)|$

Of $A_3(f)$ due to the third physical input $\phi(t)|\phi(t)|$

where $\dot{\zeta}(t) = \dot{z}(t) - \dot{w}(t)$. The ordinary and cumulative coherence function estimates in Figure 6.37 indicate that these results are meaningful mainly for the first two inputs in the frequency range between 0.17 and 1.3 Hz where the cumulative coherence function is above 0.80. Comparison of Figure 6.37 with Figure 6.25 shows that this Direct MI/SO analysis gives higher values for the cumulative coherence function over a wider frequency range than the corresponding Reverse MI/SO analysis.

Equations (6.50)–(6.57) in Section 6.4.8 represent Direct MI/SO nonlinear relations for physical sway motions due to coupled heave, sway, and roll. For this case, Figures 6.38–6.40 show results for the FRF magnitude and phase estimates to the physical sway response $y(t)$:

Of $A_1(f)$ due to the first physical wave input $w(t)$

Of $A_2(f)$ due to the second physical input $\dot{\zeta}(t)|\dot{\zeta}(t)|$

Of $A_3(f)$ due to the third physical input $\phi(t)|\phi(t)|$

The ordinary and cumulative coherence function estimates in Figure 6.41 indicate that these results are meaningful for all three inputs in the frequency range between 0.20 and 1.3 Hz where the cumulative coherence function is above 0.80. Comparison of Figure 6.41 with Figure 6.29 shows that this Direct MI/SO analysis gives higher values for the cumulative coherence function over a wider frequency range than the corresponding Reverse MI/SO analysis.

Equations (6.58)–(6.65) in Section 6.4.9 represent Direct MI/SO nonlinear relations for physical roll motions due to coupled heave and roll. For this case, Figures 6.42–6.44 show results for the FRF magnitude and phase estimates to the physical roll response $\phi(t)$:

Of $A_1(f)$ due to the first physical wave input $w(t)$

Of $A_2(f)$ due to the second physical input $\dot{\zeta}(t)|\dot{\zeta}(t)|$

Of $A_3(f)$ due to the third physical input $\phi(t)|\phi(t)|$

The ordinary and cumulative coherence function estimates in Figure 6.45 indicate that these results are meaningful for all three inputs in the frequency range between 0.04 and 1.3 Hz. Thus, all of the results for $A_1(f)$, $A_2(f)$, and $A_3(f)$ should be used here. Note that the cumulative coherence function is above 0.80 in the frequency range from 0.04 to 0.10 Hz, and again from 0.30 to 1.0 Hz. Comparison of Figure 6.45 with Figure 6.33 shows that this Direct MI/SO analysis gives higher values for the cumulative coherence function over a wider frequency range than the corresponding Reverse MI/SO analysis.

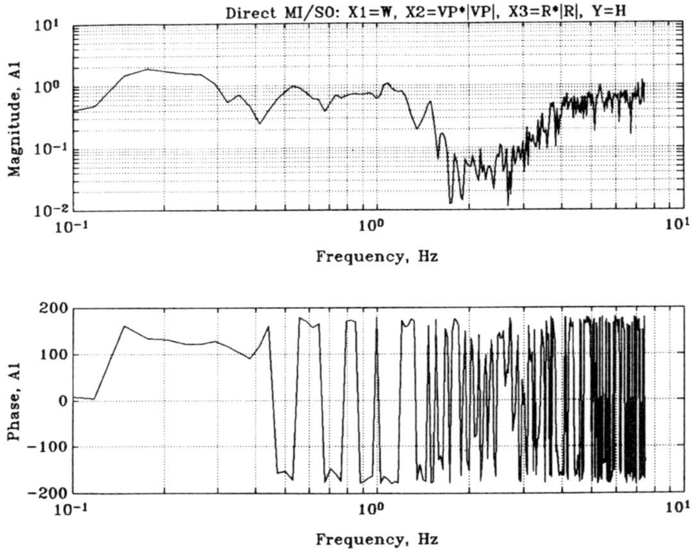

Figure 6.34.

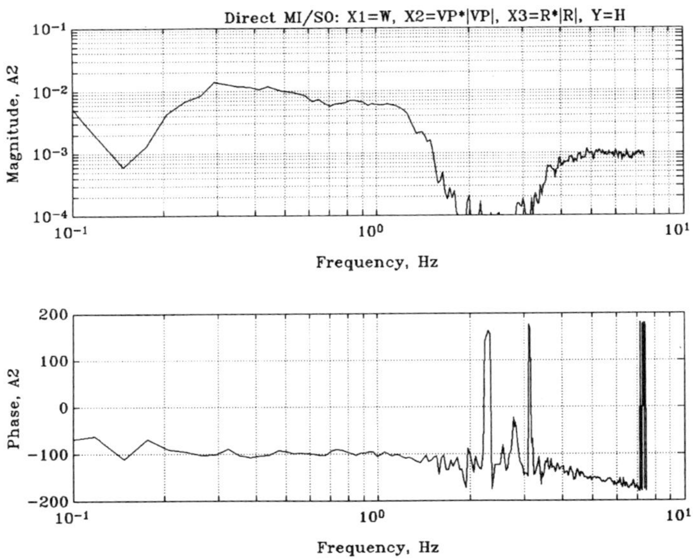

Figure 6.35.

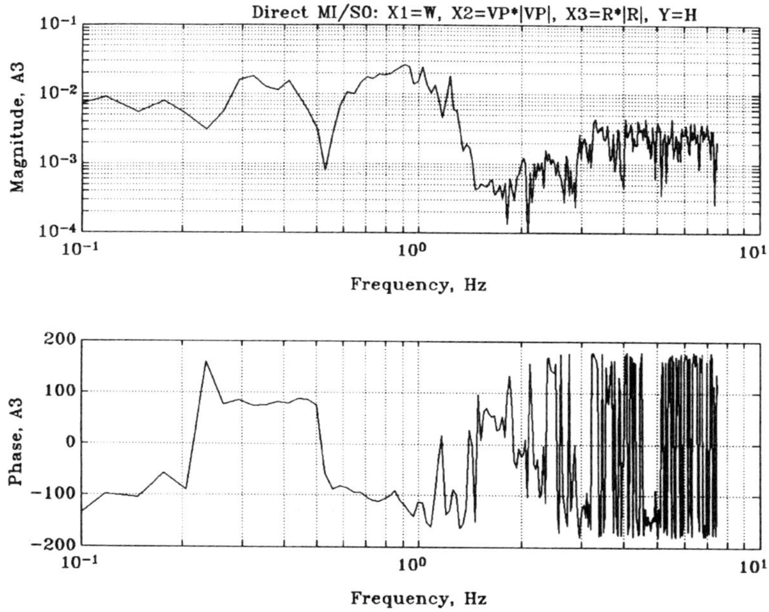

Figure 6.36.

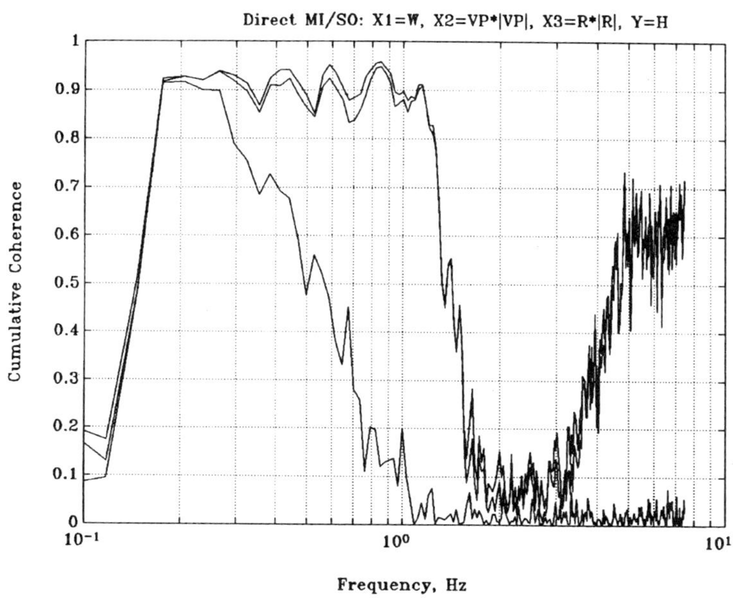

Figure 6.37.

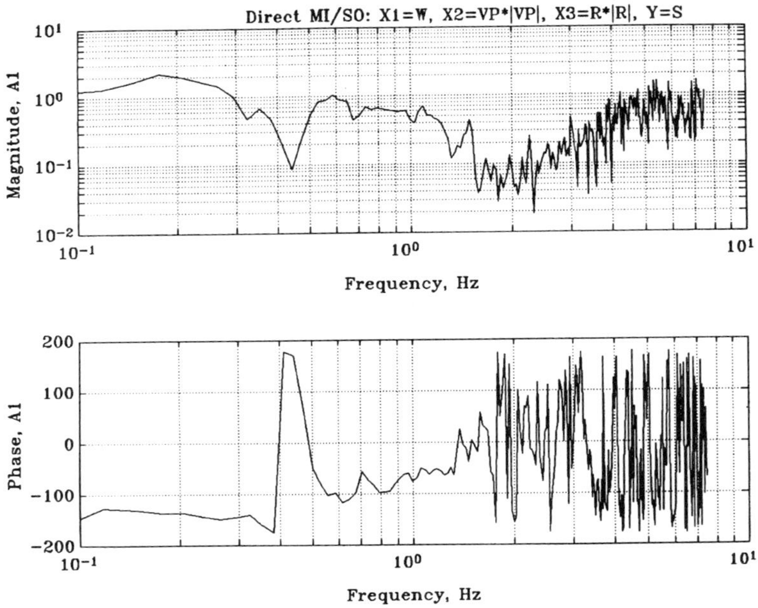

Figure 6.38.

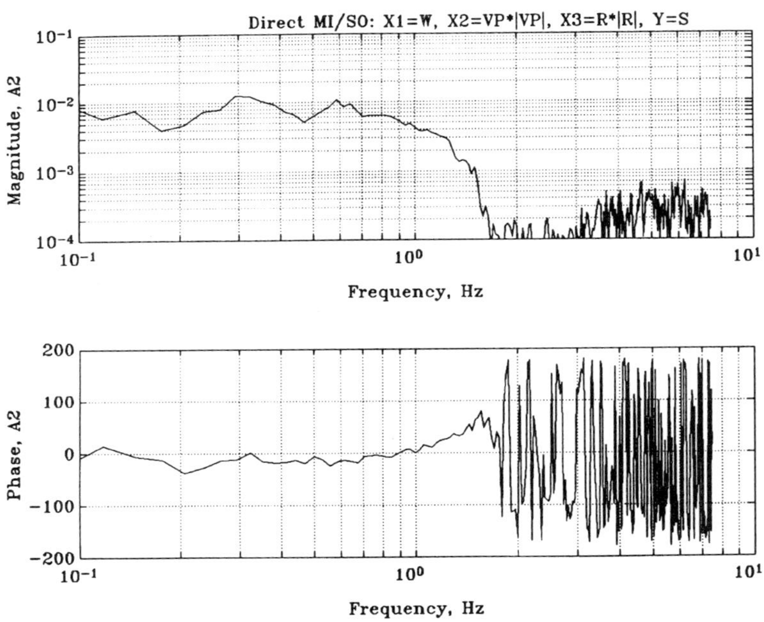

Figure 6.39.

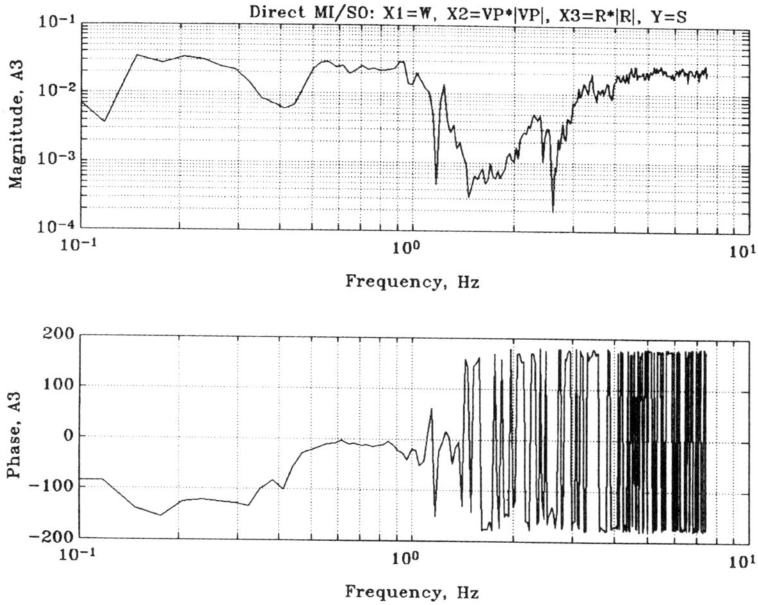

Figure 6.40.

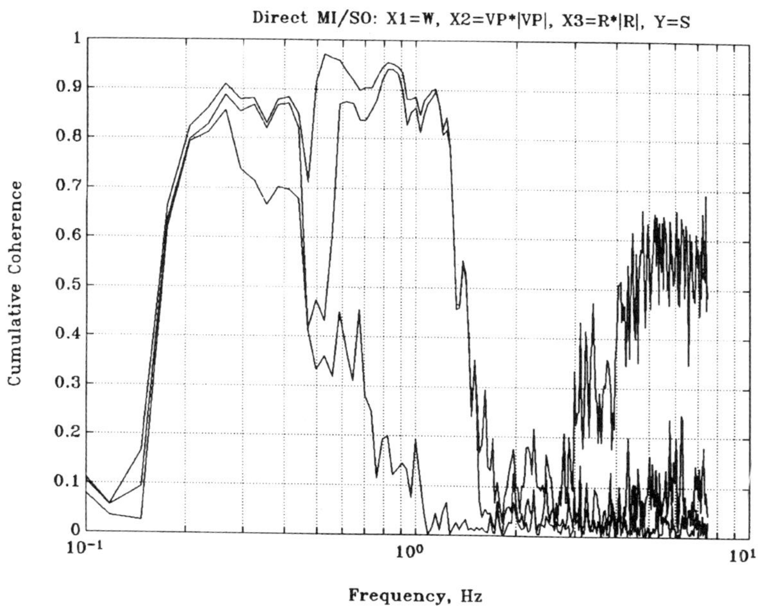

Figure 6.41.

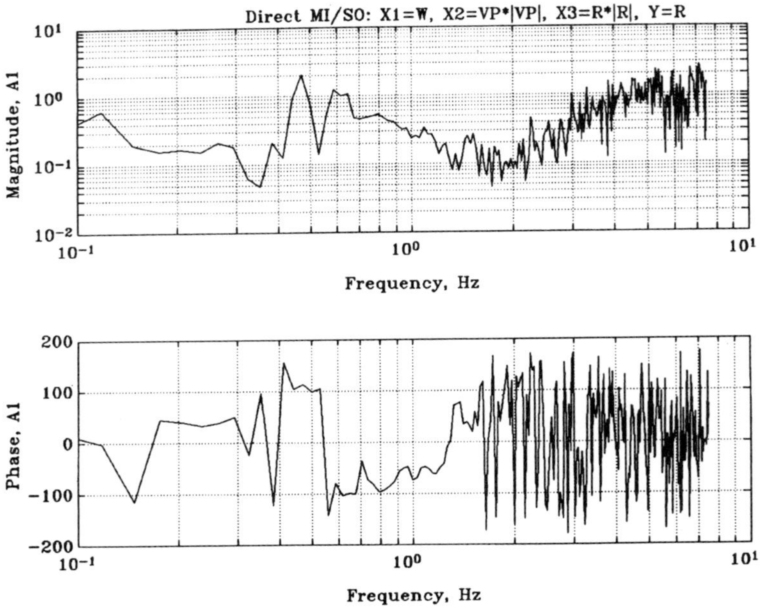

Figure 6.42.

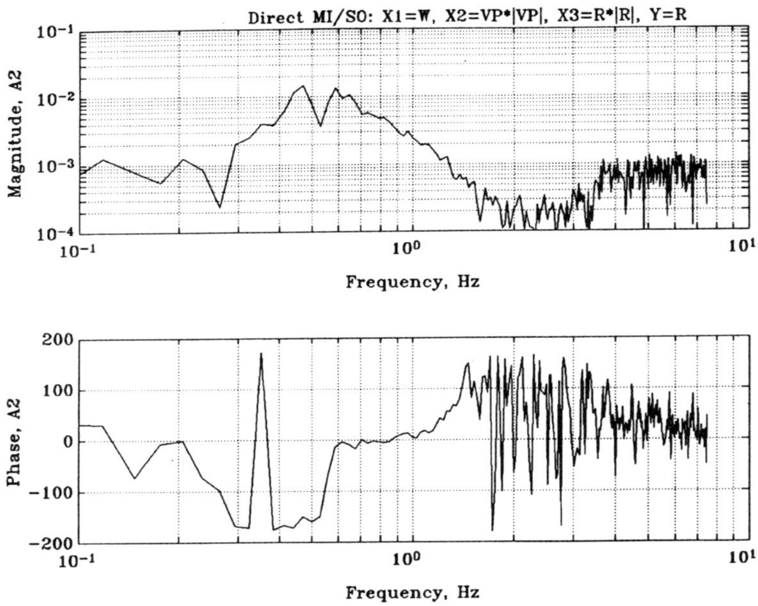

Figure 6.43.

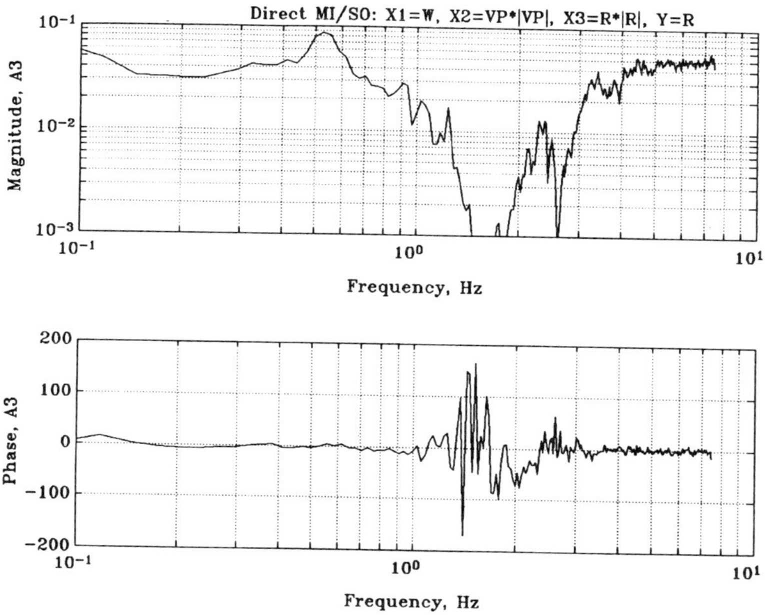

Figure 6.44.

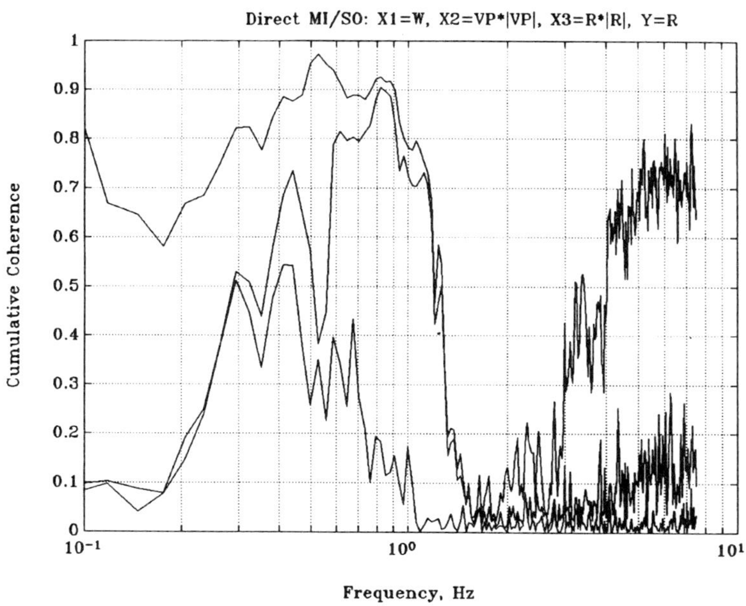

Figure 6.45.

6.6 CONCLUSIONS FROM FRIGATE TESTS

The analytical material and numerical results using the measured ocean engineering data from the NFESC/USNA naval frigate test program in June 1994 demonstrate and prove the following conclusions:

1. Actual measured data as opposed to computer-simulated (artificial) data can identify the relative contributions of each linear and nonlinear term in various proposed nonlinear systems by using Direct and Reverse MI/SO Techniques.
2. Computed coefficients for the linear and nonlinear terms in proposed nonlinear integrodifferential equations of motion are nearly always frequency-dependent rather than constants.
3. With either the Direct or Reverse MI/SO Technique, computed cumulative coherence functions give improved results over wider frequency ranges when nonlinear terms are included and they show the particular frequencies where the nonlinear terms are important.
4. For nonlinear terms involving a naval frigate, the data obtained here produced higher cumulative coherence values with the Direct MI/SO Technique than with the corresponding Reverse MI/SO Technique.
5. For nonlinear terms involving a naval barge, to be discussed in Chapter 7, the Reverse MI/SO Technique produced higher cumulative coherence values than the corresponding Direct MI/SO Technique.

6.7 MAC/RAN SYSTEM PROGRAMS

A small part of the naval frigate test data was analyzed using the MAC/RAN System, a PC-based set of computer programs for the analysis of random data, developed by Paul Agbabian from University Software Systems in Los Angeles, California. This work was done to check some of the MATLAB/ITAP results in Section 6.5 and to demonstrate certain capabilities of the MAC/RAN System.

Table 6.4 lists the digital data processing parameters for this MAC/RAN analysis. Linear trend removal was employed with highpass filtering from

TABLE 6.4 Parameters for MAC/RAN Analysis

f_s = sampling frequency = 15 Hz
N_{total} = total number of sample values = 12,000
T_{total} = total record length = 800 s
n_d = number of frequency averages = 25
$\Delta f = n_d/T_{\text{total}}$ = bandwidth resolution = 0.0312 Hz

TABLE 6.5 Figures from MAC/RAN Analysis

Figures 6.46–6.49	Four probability density functions
Figures 6.50–6.53	Four autospectral density functions
Figures 6.54–6.56	Three ordinary coherence functions
Figure 6.57	Reverse MI/SO model
Figures 6.58–6.61	Reverse MI/SO results
Figure 6.62	Direct MI/SO model
Figures 6.63–6.66	Direct MI/SO results

TABLE 6.6 Labels on MAC/RAN Figures

$w(t)$ = wave motion
$z(t)$ = heave motion
$y(t)$ = sway motion
$r = r(t)$ = roll motion
$r\|r\|$ = roll restoring force
$p = p(t)$ = relative heave motion = $z(t) - w(t)$
$vp = vp(t)$ = relative heave velocity = $\dot{z}(t) - \dot{w}(t)$
$vp\|vp\|$ = hydrodynamic drag force

0.075 Hz. Spectral estimates were obtained by frequency-wise averaging of single-segment Fourier transforms of the total record lengths of input and output records. No Hanning or other window was used or required.

Table 6.5 lists figures obtained from the MAC/RAN analysis of the naval frigate test data. Table 6.6 defines the meaning of various labels on the figures.

6.7.1 Preliminary Data Analysis

Preliminary data analysis of the naval frigate data consists of results for probability density functions in Figures 6.46–6.49, results for autospectral density functions in Figures 6.50–6.53, and results for ordinary coherence functions in Figures 6.54–6.56. These plots from the MAC/RAN analysis should be compared with related plots obtained from the previous MATLAB/ITAP analysis in Figures 6.5–6.15. Differences found here are due to the fact that the same data set and the same digital data processing parameters were not used.

The probability density function estimates were plotted from 3000 sample values of the measured data from each record, together with their associated Gaussian (unstandarized) probability density functions. A chi-square goodness-of-fit test for normality was applied to each record at the $\alpha = 0.05$ level of significance, reference 1. The hypothesis of normality was accepted for the wave input data but rejected for the three response data.

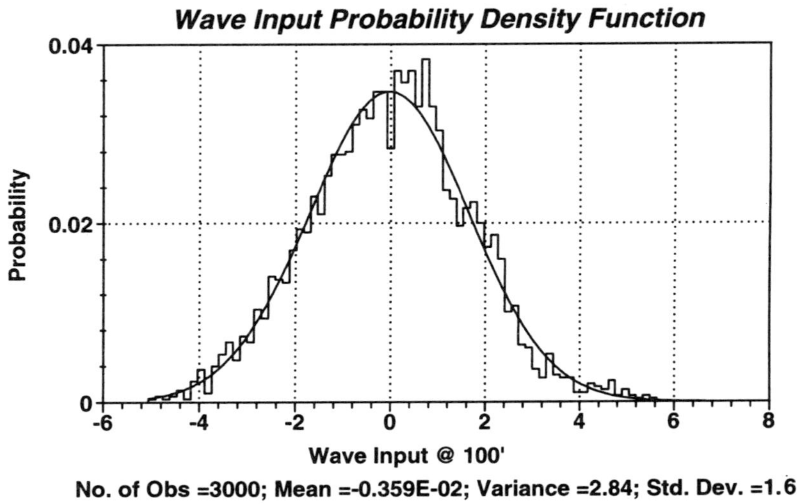

Figure 6.46.

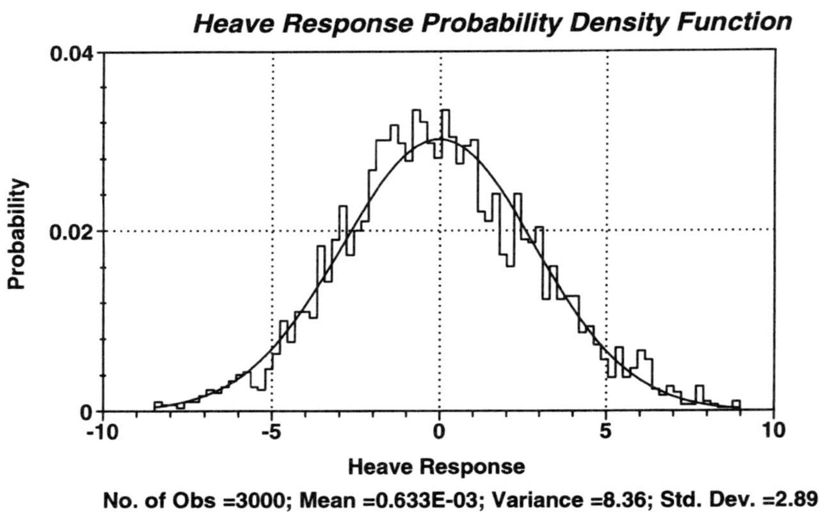

Figure 6.47.

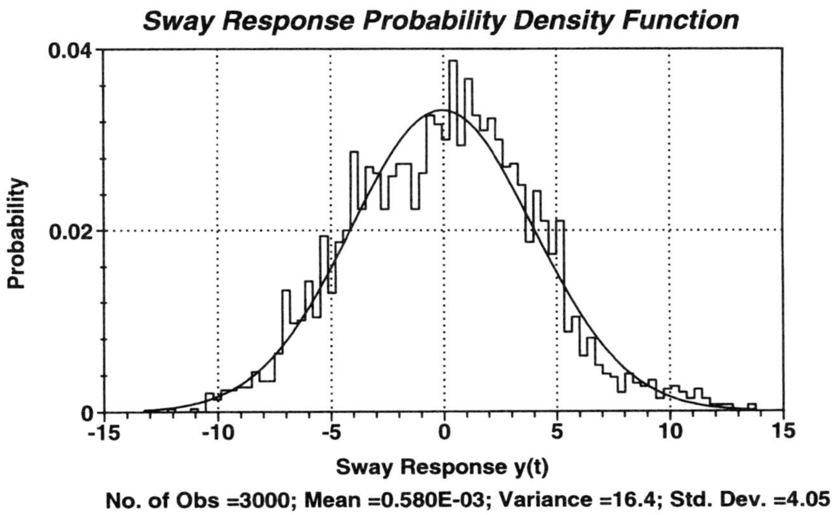

Figure 6.48.

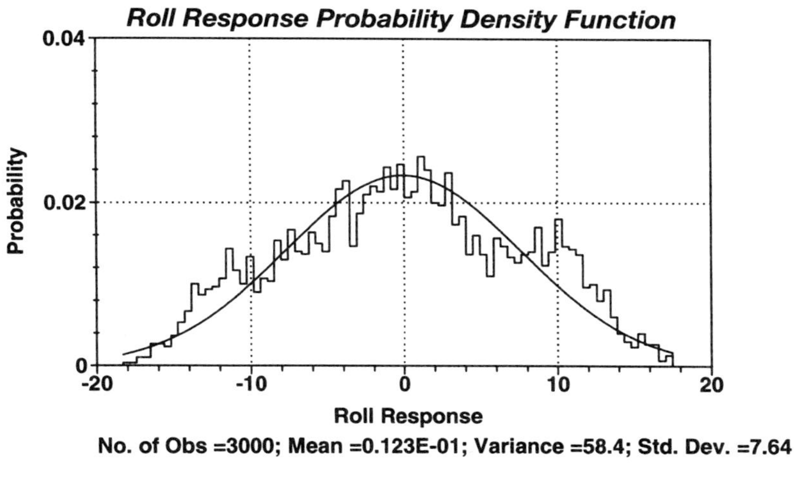

Figure 6.49.

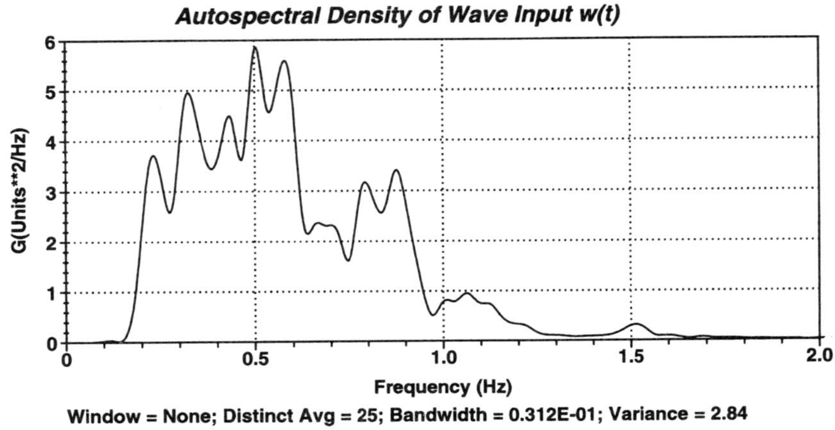

Figure 6.50.

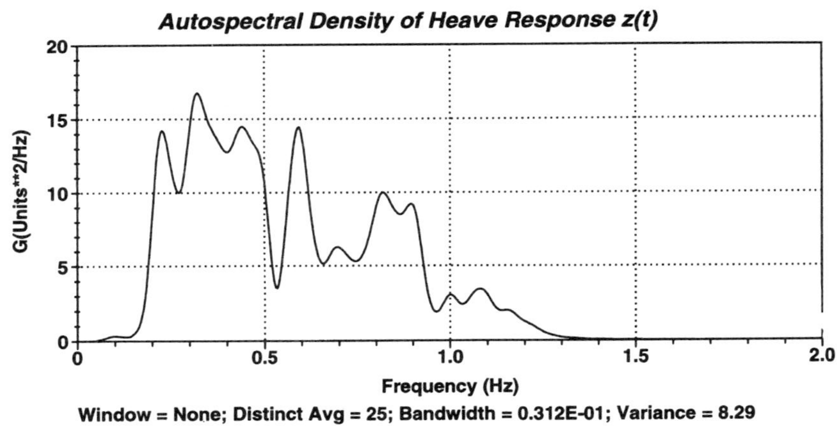

Figure 6.51.

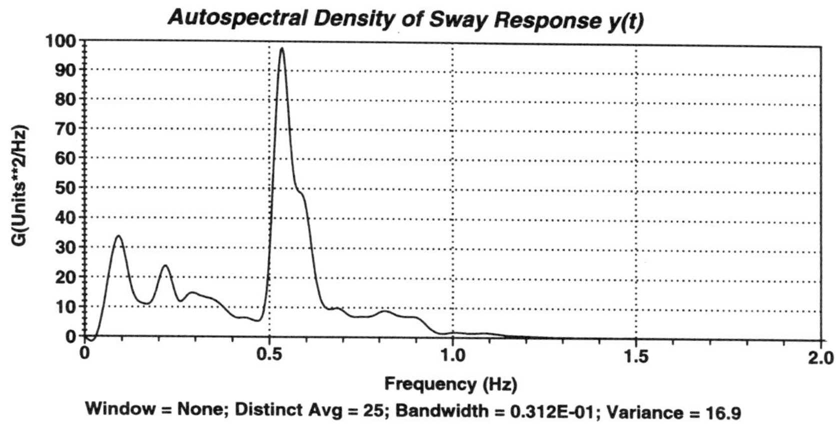

Figure 6.52.

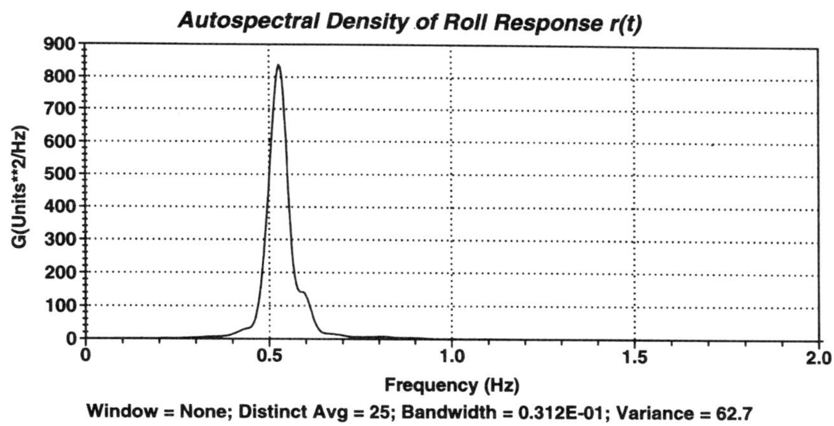

Figure 6.53.

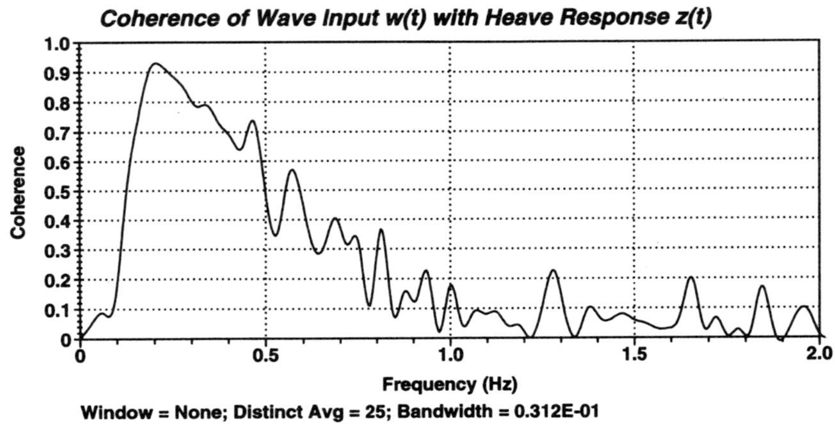

Figure 6.54.

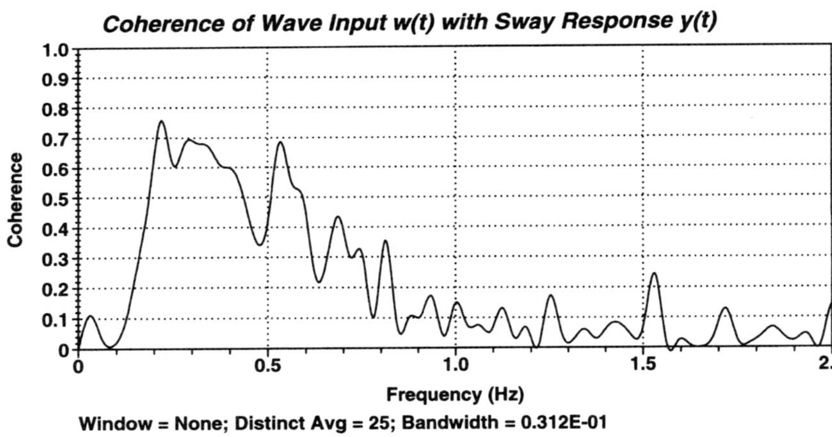

Figure 6.55.

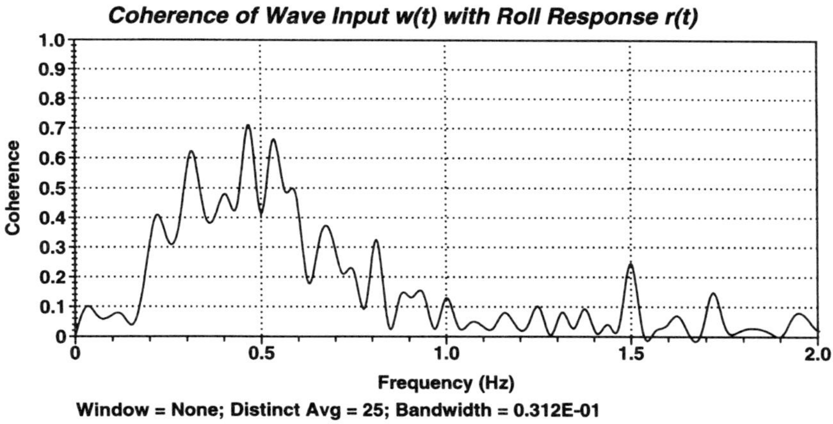

Figure 6.56.

6.7.2 Reverse MI/SO Analysis

The Reverse MI/SO Mathematical Model is shown in Figure 6.57 for a mathematical wave output from a mathematical sway input due to coupled heave, sway, and roll. The three-input/single-output Fourier transform equation for this mathematical model is

$$A_1(f)X_1(f) + A_2(f)X_2(f) + A_3(f)X_3(f) = Y(f) \tag{6.66}$$

where

$$X_1(f) = S(f) = \text{Fourier transform of sway } y(t) \tag{6.67}$$

$$X_2(f) = \text{Fourier transform of } vp|vp| \tag{6.68}$$

$$X_3(f) = \text{Fourier transform of } r|r| \tag{6.69}$$

$$Y(f) = W(f) = \text{Fourier transform of wave } w(t) \tag{6.70}$$

$$A_1(f) = \text{FRF from } X_1(f) \text{ to } Y(f) \tag{6.71}$$

$$A_2(f) = \text{FRF from } X_2(f) \text{ to } Y(f) \tag{6.72}$$

$$A_3(f) = \text{FRF from } X_3(f) \text{ to } Y(f) \tag{6.73}$$

Results from this Reverse MI/SO analysis are shown in Figures 6.58–6.61. These plots from the MAC/RAN analysis should be compared with related plots obtained from the previous analysis in Figures 6.26–6.29. The strong similarity for the cumulative coherence functions in Figures 6.29 and 6.61 indicates that these particular MATLAB/ITAP results and MAC/RAN results are in agreement.

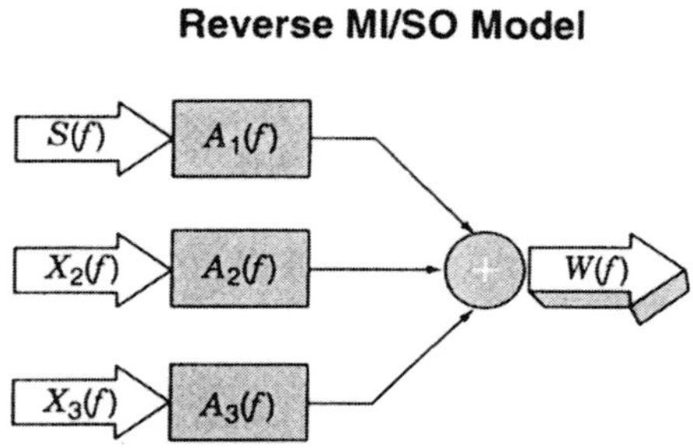

Figure 6.57.

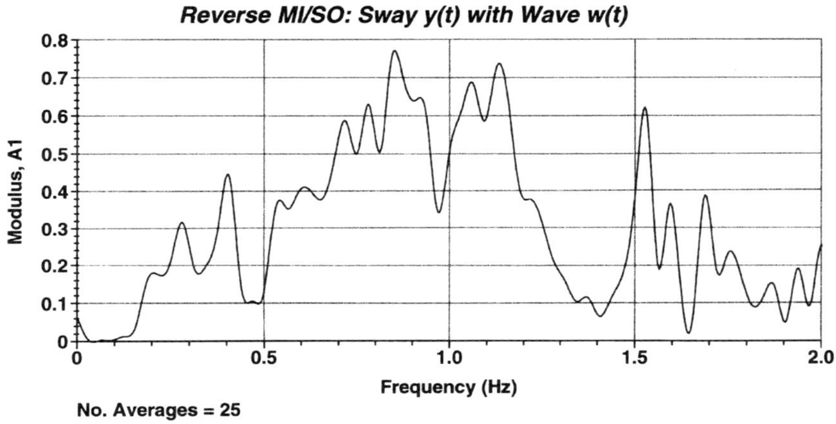

Figure 6.58.

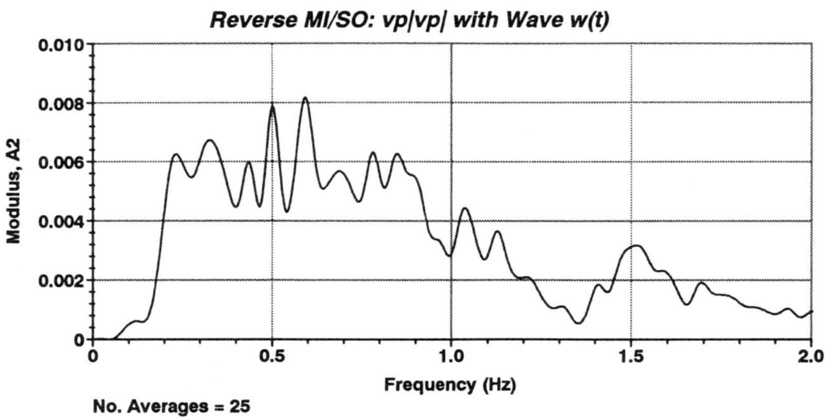

Figure 6.59.

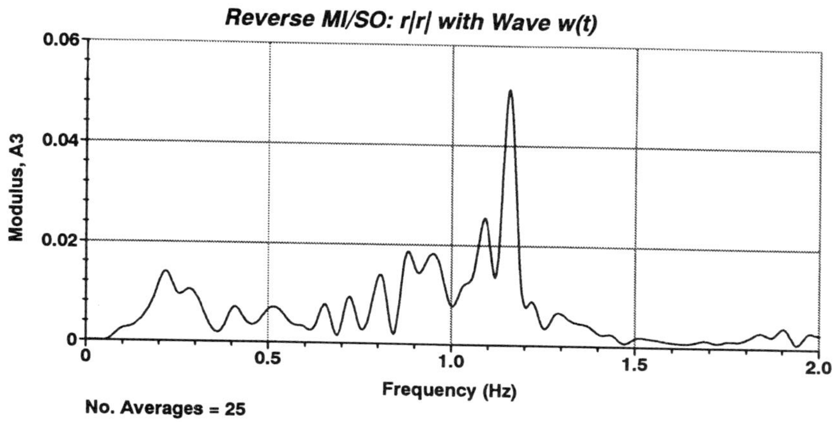

Figure 6.60.

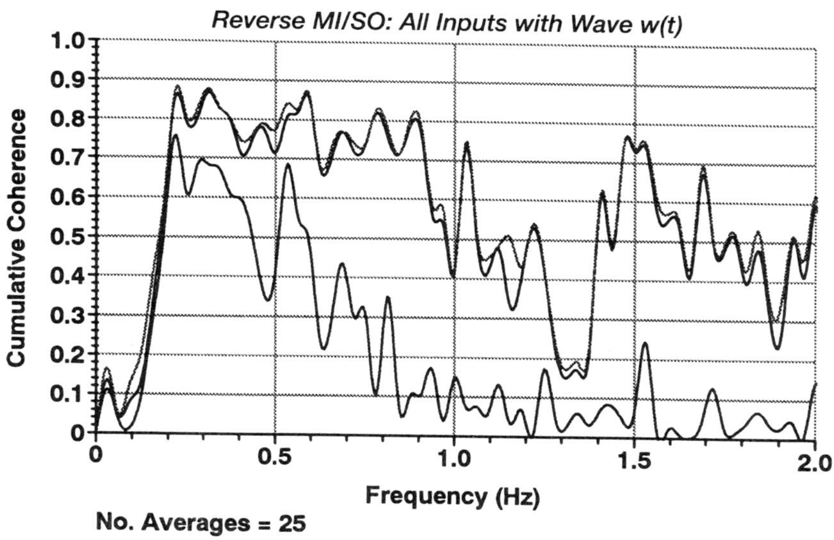

Figure 6.61.

6.7.3 Direct MI/SO Analysis

The Direct MI/SO Physical Model is shown in Figure 6.62 for a physical roll response output from a physical wave motion input due to coupled heave and roll. The three-input/single-output Fourier transform equation for this physical model is

$$A_1(f)X_1(f) + A_2(f)X_2(f) + A_3(f)X_3(f) = Y(f) \tag{6.74}$$

where

$$X_1(f) = W(f) = \text{Fourier transform of wave } w(t) \tag{6.75}$$

$$X_2(f) = \text{Fourier transform of } vp|vp| \tag{6.76}$$

$$X_3(f) = \text{Fourier transform of } r|r| \tag{6.77}$$

$$Y(f) = R(f) = \text{Fourier transform of roll } r(t) \tag{6.78}$$

$$A_1(f) = \text{FRF from } X_1(f) \text{ to } Y(f) \tag{6.79}$$

$$A_2(f) = \text{FRF from } X_2(f) \text{ to } Y(f) \tag{6.80}$$

$$A_3(f) = \text{FRF from } X_3(f) \text{ to } Y(f) \tag{6.81}$$

Results from this Direct MI/SO analysis are shown in Figures 6.63–6.66. These plots from the MAC/RAN analysis should be compared with related plots obtained from the previous analysis in Figures 6.42–6.45. The strong similarity for the cumulative coherence functions in Figures 6.45 and 6.66 indicates that these particular MATLAB/ITAP results and MAC/RAN results are in agreement.

This concludes Chapter 6.

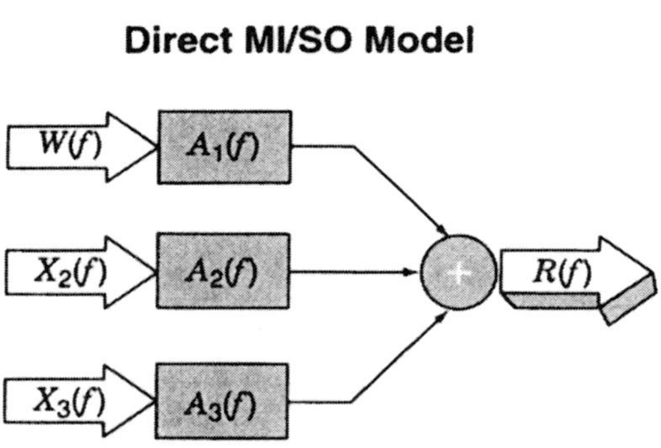

Figure 6.62.

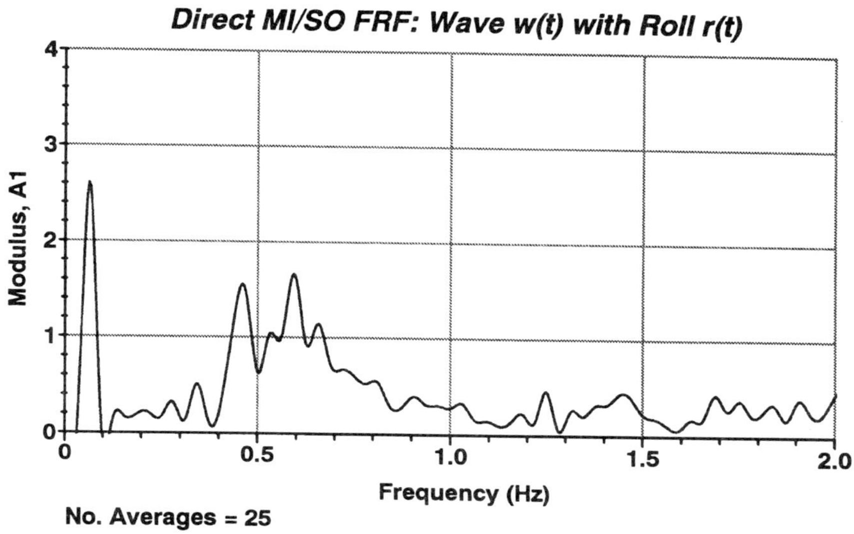

Figure 6.63.

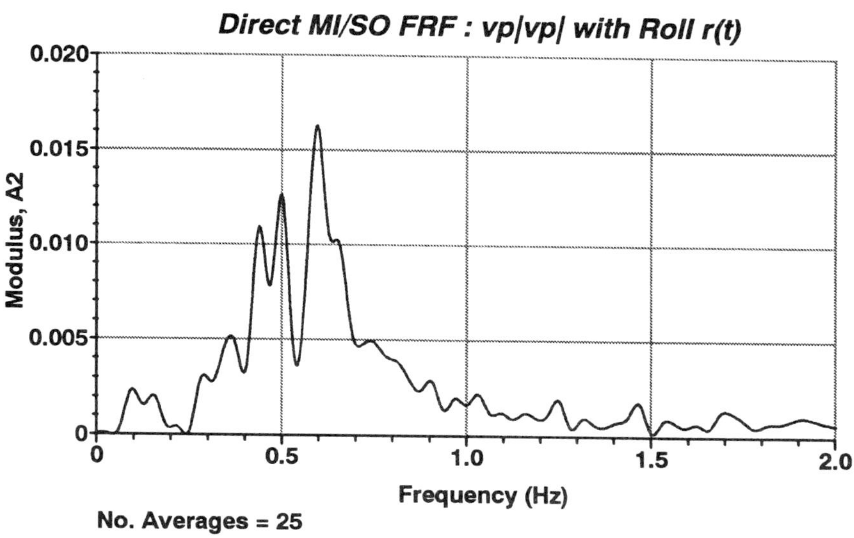

Figure 6.64.

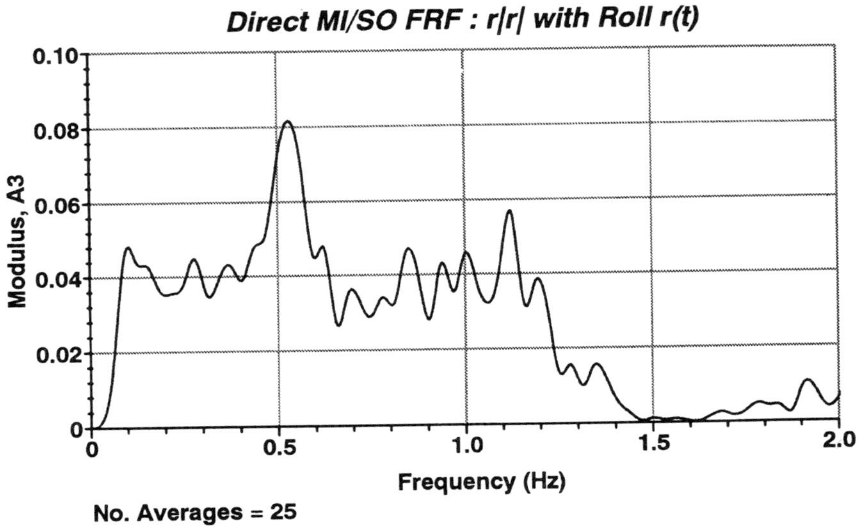

Figure 6.65.

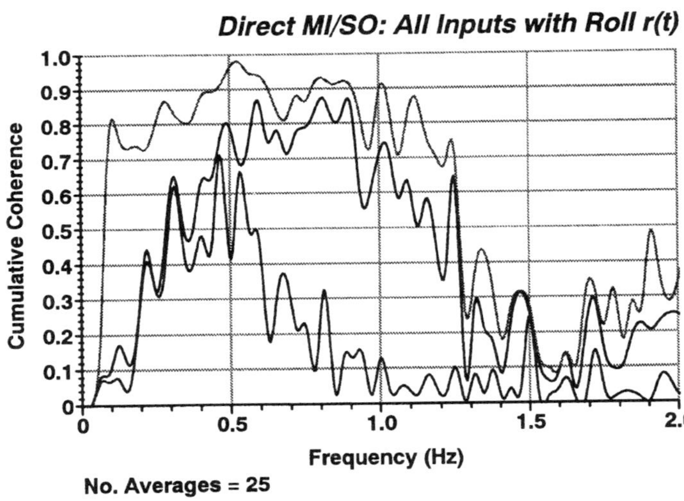

Figure 6.66.

7

NONLINEAR SYSTEM RESPONSE PROPERTIES OF A NAVAL BARGE FROM MEASURED OCEAN ENGINEERING DATA

This chapter contains results from the second experimental test program conducted at the Hydromechanics Laboratory of the U.S. Naval Academy in July 1995. Tests were designed using a scale model of a free rectangular barge to extend the work done in the first program described in Chapter 6. The barge was subjected to small amplitude waves and to large amplitude waves in the three different directions of head, beam, and quartering (oblique) seas to produce six different sets of experimental data. Detailed discussions are given of the various linear and nonlinear system models and response results that were obtained from analyzing these experimental data. The measured physical data from this free barge model showed that the heave responses in head, beam, and quartering seas are predominantly linear. Nonlinearity was found to be significant in roll and pitch responses in beam and quartering seas. In this 1995 test program, Reverse MI/SO Techniques gave better nonlinear results than Direct MI/SO Techniques by producing higher computed cumulative coherence functions over wider frequency ranges. This was opposite to that obtained in the previous 1994 test program where Direct MI/SO results were superior. The outcomes in this chapter and the previous chapter demonstrate the capability of these new practical MI/SO techniques to solve nonlinear system problems in many engineering and scientific fields.

Scale Model of a Naval Barge

7.1 TEST PROGRAM IN JULY 1995

The first joint NFESC/USNA project conducted in June 1994, discussed in Chapter 6, showed that Direct and Reverse Multiple-Input/Single-Output (MI/SO) Techniques for nonlinear system analysis and identification are applicable to many mathematically defined nonlinear systems. In the first test program, the nonlinear response motions of a scale model of a naval frigate were studied with a unidirectional random sea to provide lateral responses of the vessel. Work in this second test program, conducted in July 1995, involved the analysis of data collected from a scale model of a rectangular barge. Two different sets of small and large wave conditions were created for these tests where the barge was subjected to three different unidirectional random seas represented by head seas, beam seas, and quartering seas.

The small waves were waves where the amplitudes were below the freeboard of the barge. Tests with these small waves are expected to yield linear results only. The large waves were waves where a high percentage of the amplitudes exceeded the freeboard of the barge when the barge was fixed. Tests with large waves are expected to yield both linear and nonlinear results. The small waves and the large waves were each measured with the barge model either inside the test tank or lifted from the test tank to remove any possible interference effects. Basic data analyses were performed to establish the amplitude and frequency properties of these distrubed and undisturbed waves.

This chapter discusses the analysis of the data collected in July 1995 to conduct response tests on a free barge model. Futher data from a third test program was collected in July 1996 to conduct excitation tests on a fixed barge model. The results from analyzing this excitation data are not discussed here because research on these matters is still underway.

7.2 TEST OBJECTIVES

The main objective in these tests was to identify appropriate linear and nonlinear integrodifferential equations of motion that can describe the barge motions in large waves. Direct and Reverse MI/SO Techniques from Chapters 3–6 (references 3–8) were used to help identify the following forces as can be determined from various computed frequency response functions:

Linear inertial forces

Linear damping forces

Linear restoring forces

Nonlinear damping forces

Nonlinear restoring forces

Different linear and nonlinear systems were proposed to accomplish this main objective. Ordinary and cumulative coherence functions were computed to determine the relative contributions from each proposed linear and nonlinear term. Associated frequency response functions for each linear and nonlinear term were also computed.

Identification of the desired integrodifferential equations of motion requires a number of steps. For the small wave tests, standard linear analyses were performed on SI/SO linear systems as described in Section 7.4.1. This was conducted to determine how the Response Tests with Small Waves on the Free Barge Model (Test A) agree with established linear theory results. Good agreement here helps to validate the data acquisition operations. It also establishes an important linear baseline for the next step of analyzing the large wave data where further linear and nonlinear results were computed that could give new results that need to be interpreted.

For the large wave tests, Direct MI/SO Techniques and Reverse MI/SO Techniques were performed on nonlinear systems as described in Section 7.4.3 using two-input/single-output models and three-input/single-output models. This was conducted so as to try to obtain new desired linear and nonlinear results from measured data for Response Tests with Large Waves on the Free Barge Model (Test B).

7.3 TEST DESCRIPTION

The dimensions for the rectangular barge used in this test program are listed in Table 7.1. The coordinate system for the barge is shown in Figure 7.1.

All of these tests were conducted in a towing tank that is 120 feet long, 8 feet wide, and 5 feet deep. Irregular waves were generated by a dual-flap wavemaker at one end of the tank. During each test, digital records were recorded by USNA of the time series of all forces and/or motions and provided to NFESC on computer ASCII data file format. The center of the barge was located 50 feet downstream from the wavemaker. The wave height was measured with a wave probe also located 50 feet downstream from the wavemaker but placed 8 inches from the tank wall, Reference 59 describes the instrumentation used to collect the data for these tests.

The barge was subjected to three different unidirectional random seas as pictured in Figure 7.2 to provide the following results:

Head seas of 0° to provide longitudinal results

Beam seas of 90° to provide lateral results

Oblique angle of 30° to provide quartering results

Six time history records were recorded for the Response Tests on the Free Barge Model as shown in Table 7.2.

TABLE 7.1 Barge Dimensions

Length $L = 63.00$ inches
Beam $B = 19.25$ inches
Depth $D = 5.375$ inches
Draft $T = 2.70$ inches
Freeboard $F = D - T = 2.675$ inches

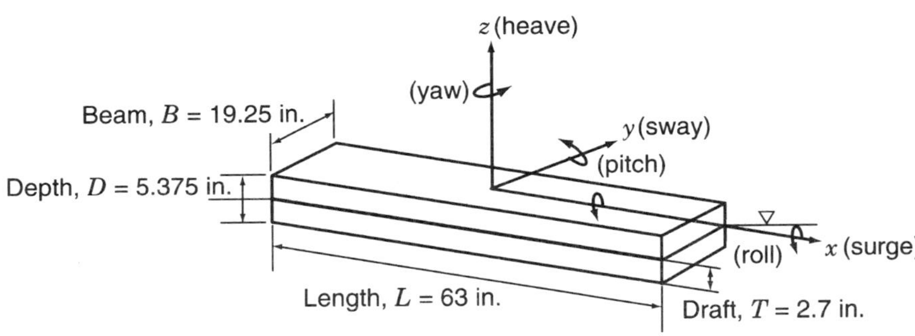

Figure 7.1. Schematic of model showing coordinate system.

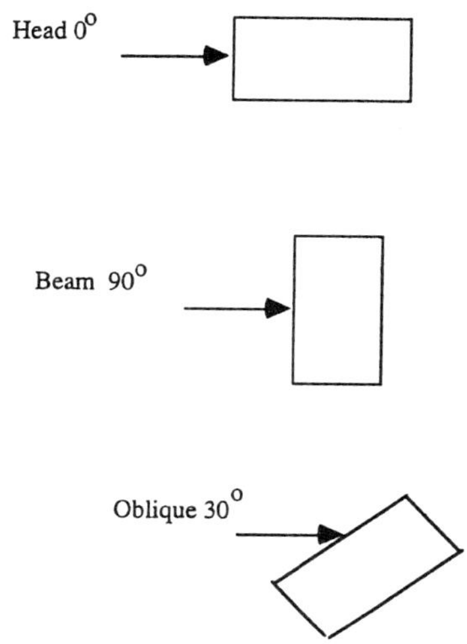

Figure 7.2. Three unidirectional random seas.

TABLE 7.2 Free Barge Recordings

X	x Force gauge	lb
Y	y Force gauge	lb
Pitch	Pitch angle (+ means bow up)	deg
Heave	Heave motion	ft
Roll	Roll angle (+ means port up, starboard down)	deg
Wave	Wave gauge	ft

In these tests, the barge was constrained in surge, sway, and yaw and permitted to move only in heave, roll, and pitch. The forces in the x and y directions were measured, together with the heave linear motion, the roll angular motion, and the pitch angular motion. The wave wire was placed in line with the center of the barge. This gave the following sets of data:

Incident wave amplitude input	$w(t)$
Surge (longitudinal) force in the x direction	$X(t)$
Sway (lateral) force in the y direction	$Y(t)$
Heave (vertical) motion output in the z direction	$z(t)$
Roll angular rotation output about the x axis	$\phi(t)$
Pitch angular rotation output about the y axis	$\theta(t)$

The surge and sway force recordings were not used in the later analysis of this free barge data. The only data of concern were given by the heave, roll, and pitch recordings, together with the recordings of the disturbed and undisturbed waves.

Discussions between the participants in this research work indicated that the test setup here made it reasonable to formulate various uncoupled and coupled Response Tests for the Small Waves (Test A) and Response Tests for the Large Waves (Test B) as described below.

Test A For small waves, identify the optimum linear system frequency response functions (FRFs) from the physical input small waves to each of the following three physical output response records:

1. Heave motion output, $z(t)$.
2. Roll angle output, $\phi(t)$.
3. Pitch angle output, $\theta(t)$.

No models are needed for the surge motion output, the sway motion output, or the yaw angle output because these three outputs were constrained to be negligible.

For the head-on waves, the analysis should be conducted only for (1) the heave motion output and (3) the pitch angle output, since the roll angle output term will be negligible.

For the beam-on waves, the analysis should be conducted only for (1) the heave motion output and (2) the roll angle output, since the pitch angle output term will be negligible.

All three available outputs should be analyzed for the case of oblique (quartering) waves.

Test B For large waves, identify the best-fit linear and nonlinear system terms with their associated frequency response functions (FRFs) for the same cases studied in Test A where various nonlinear terms should be considered.

7.4 EQUATIONS OF MOTION FOR A FREE BARGE

Three model types were formulated for the Response Tests on a Free Barge where results can be obtained for each of the wave directions (head-on, beam-on, and oblique).

- Linear SI/SO models for small waves
- Linear SI/SO models for large waves
- Nonlinear MI/SO models for large waves

Functions and equations will be stated for each of these models using ideas and notation that follow the customary naval architecture convention, as per Section 5.6 in this book.

7.4.1 Linear SI/SO Models for Small Waves

Three different single-input/single-output linear models can be proposed from the small wave physical inputs to each of the separate physical outputs represented by the heave motion, the roll angular rotation, and the pitch angular rotation. For the small wave tests, it is reasonable to assume here that a linear relationship exists between the wave amplitudes and each of the resulting motions and rotations of the free barge.

1. *Heave Motion Linear Response* $z(t)$ The proposed heave linear differential equation of motion with constant damping and stiffness coefficients is

$$(m + a_{33})\ddot{z}(t) + c_{33}\dot{z}(t) + k_{33}z(t) = w(t) \tag{7.1}$$

where $z(t)$ is the physical heave motion output, and the heave force input $Z(t)$ has been replaced by a linearly equivalent physical wave amplitude input $w(t)$. This is valid for small waves by making some linear changes that can occur in the coefficients. Equation (7.1) can be extended to cover frequency-dependent damping and stiffness coefficients by using the heave linear integrodifferential equation of motion

$$(m + a_{33})\ddot{z}(t) + \int c_{33}(\tau)\dot{z}(t - \tau)\,d\tau + \int k_{33}(\tau)z(t - \tau)\,d\tau = w(t) \tag{7.2}$$

where $c_{33}(\tau)$ and $k_{33}(\tau)$ are weighting functions that include the constants c_{33} and k_{33} as special cases.

Taking the Fourier transforms of both sides of Eq. (7.2) yields the simple mathematical formula

$$A(f)X(f) = Y(f) \tag{7.3}$$

where

$$\begin{aligned} X(f) &= \text{Fourier transform of mathematical input } z(t) \\ Y(f) &= \text{Fourier transform of mathematical output } w(t) \\ A(f) &= \text{Frequency response function from } z(t) \text{ to } w(t) \\ &= K_{33}(f) - (2\pi f)^2(m + a_{33}) + j(2\pi f)C_{33}(f) \end{aligned} \tag{7.4}$$

The functions $K_{33}(f)$ and $C_{33}(f)$ are Fourier transforms of $k_{33}(\tau)$ and $c_{33}(\tau)$, respectively. A frequency-dependent coefficient situation occurs when either

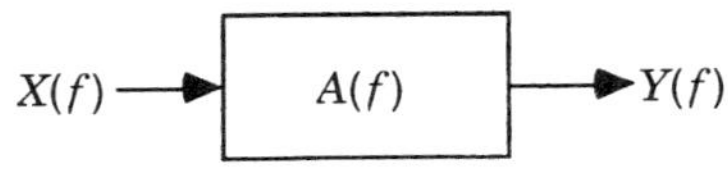

Figure 7.3. Reverse dynamic single-input/single-output system.

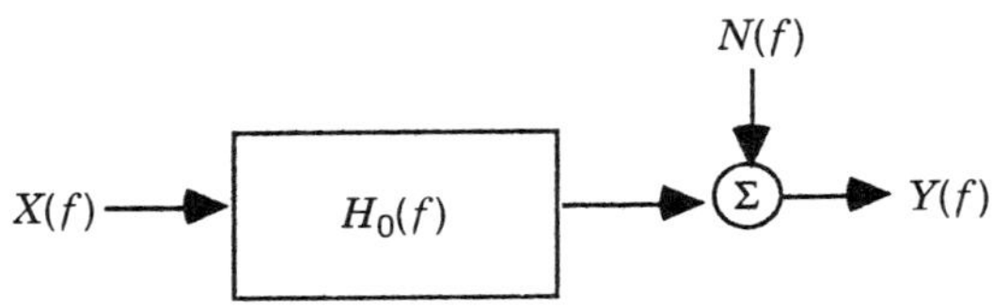

Figure 7.4. Direct dynamic single-input/single-output linear model.

$K_{33}(f)$ or $C_{33}(f)$ varies with frequency. Also, the form of $A(f)$ can be more general than described by Eq. (7.4).

Equation (7.3) is a reverse dynamic SI/SO linear model with the single mathematical input $z(t)$ defined by $X(f)$ passing through the linear system defined by $A(f)$ to give the mathematical output $w(t)$ defined by $Y(f)$. Note that $X(f)$ actually represents the physical heave motion response $z(t)$, and $Y(f)$ actually represents the physical wave excitation $w(t)$. The model for Eq. (7.3) is shown in Figure 7.3. The notation $A(f)$ is used instead of the familiar $H(f)$ to emphasize the fact that this is a reverse dynamic model.

When $X(f)$ represents the Fourier transform of the physical input $x(t) = w(t)$, and when $Y(f)$ represents the Fourier transform of the physical output $y(t) = z(t)$, one obtains the direct dynamic model of Figure 7.4. The noise term $N(f)$ represents deviations from the ideal model. When $N(f) = 0$, the optimum linear frequency function $H_o(f)$ is the reciprocal of $A(f)$, namely,

$$H_o(f) = [A(f)]^{-1} \tag{7.5}$$

Similar relations like Eqs. (7.1)–(7.5) with reverse and direct dynamic models like Figures 7.3 and 7.4 apply to the next two cases of roll rotation and pitch rotation.

2. *Roll Rotation Linear Response* $\phi(t)$ The proposed roll rotation differential equation of motion with constant damping and stiffness coefficients is

$$(m + a_{44})\ddot{\phi}(t) + c_{44}\dot{\phi}(t) + k_{44}\phi(t) = w(t) \tag{7.6}$$

where the roll angular moment $K(t)$ has been replaced by an appropriate linearly equivalent $w(t)$.

3. *Pitch Rotation Linear Response* $\theta(t)$ The proposed pitch rotation linear differential equation of motion with constant damping and stiffness

coefficients is

$$(m + a_{55})\ddot{\theta}(t) + c_{55}\dot{\theta}(t) + k_{55}\theta(t) = w(t) \tag{7.7}$$

where the pitch angular moment $M(t)$ has been replaced here by its appropriate linearly equivalent $w(t)$.

As done above in Eq. (7.1)–(7.4) for the heave motion case, one can extend Eqs. (7.6) and (7.7) to cover cases of frequency-dependent coefficients by using terms computed in models like Figure 7.3 or 7.4.

7.4.2 Linear SI/SO Models for Large Waves

The same three specified linear SI/SO models can be proposed as in Section 7.4.1 except that the wave input is now the large waves instead of the small waves. The same functions and equations can be computed as before. Results for the various frequency response functions and coherence functions should be different because the large-amplitude waves can produce nonlinear effects not considered previously.

7.4.3 Nonlinear MI/SO Models for Large Waves

Many different types of nonlinear models can be proposed. Special models of major interest should be studied due to heave, roll, and pitch motions that can occur in head seas, beam seas, and quartering seas. These models should have one linear input term and one or two parallel nonlinear input terms, where the first nonlinear term represents a nonlinear relative hydrodynamic damping force and the second nonlinear term represents a nonlinear restoring force. Direct MI/SO and reverse MI/SO techniques should be used to analyze the data. Models will be formulated for the following seven cases:

1. Heave motions in head seas.
2. Pitch motions in head seas.
3. Heave motions in beam seas.
4. Roll motions in beam seas.
5. Heave motions in quartering seas.
6. Roll motions in quartering seas.
7. Pitch motions in quartering seas.

Direct MI/SO Nonlinear Model for Two Inputs The *direct* MI/SO two-input/single-output nonlinear integrodifferential equation of motion for any of these seven cases can be represented by

$$H_1(f)X_1(f) + H_2(f)X_2(f) = Y(f) \tag{7.8}$$

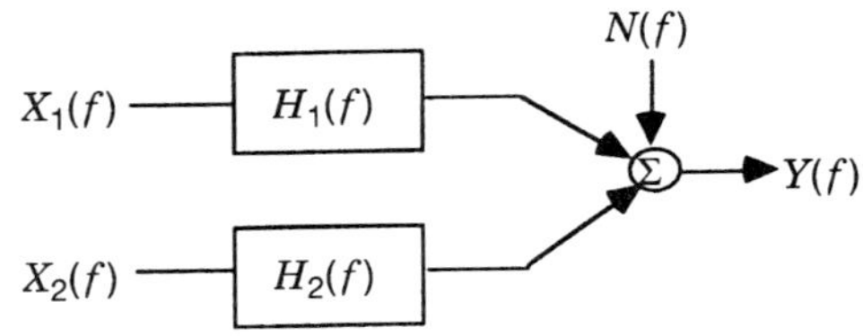

Figure 7.5. Direct dynamic two-input/single-output linear model.

where

$X_1(f)$ = Fourier transform of first physical input

$X_2(f)$ = Fourier transform of second physical (nonlinear) input

$Y(f)$ = Fourier transform of physical output

$H_1(f)$ = Frequency response function (FRF) from $X_1(f)$ to $Y(f)$

$H_2(f)$ = Frequency response function (FRF) from $X_2(f)$ to $Y(f)$

Equation (7.8) is a *direct dynamic* two-input/single-output linear model with the physical inputs defined by the two $X(f)$ terms passing through the linear systems defined by the two $H(f)$ terms, respectively, to produce the physical output defined by the $Y(f)$ term. The $H(f)$ terms can contain frequency-dependent coefficients. This model is pictured in Figure 7.5 with a noise term $N(f)$ representing deviations from the ideal model.

Direct MI/SO Nonlinear Models for Three Inputs. The *direct* MI/SO three-input/single-output nonlinear integrodifferential equation of motion for any of the seven cases listed earlier is given by

$$H_1(f)X_1(f) + H_2(f)X_2(f) + H_3(f)X_3(f) = Y(f) \tag{7.9}$$

where

$X_1(f)$ = Fourier transform of first physical input

$X_2(f)$ = Fourier transform of second physical (nonlinear) input

$X_3(f)$ = Fourier transform of third physical (nonlinear) input

$Y(f)$ = Fourier transform of physical output

$H_1(f)$ = FRF from $X_1(f)$ to $Y(f)$

$H_2(f)$ = FRF from $X_2(f)$ to $Y(f)$

$H_3(f)$ = FRF from $X_3(f)$ to $Y(f)$

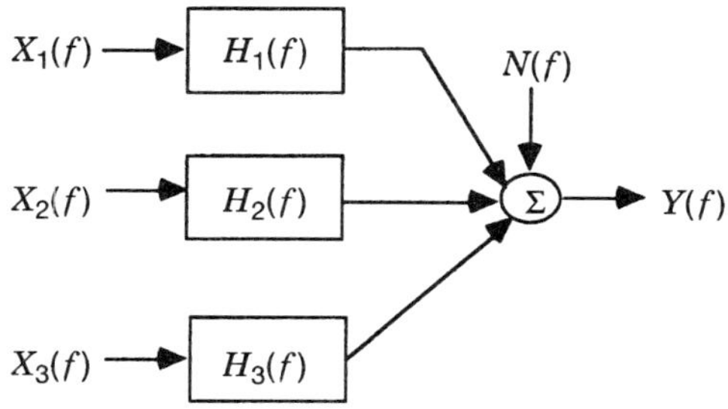

Figure 7.6. Direct dynamic three-input/single-output linear model.

Equation (7.9) is a *direct dynamic* three-input/single-output linear model with the physical inputs defined by the three $X(f)$ terms passing through the linear systems defined by the three $H(f)$ terms, respectively, to produce the physical output defined by the $Y(f)$ term. The $H(f)$ terms can contain frequency-dependent coefficients. This model is pictured in Figure 7.6 with a noise term $N(f)$ representing deviations from the ideal model.

Reverse MI/SO Nonlinear Model for Two Inputs. The *reverse* MI/SO two-input/single-output nonlinear integrodifferential equation of motion for any of the seven cases listed earlier can be represented by

$$A_1(f)X_1(f) + A_2(f)X_2(f) = Y(f) \tag{7.10}$$

where

$X_1(f)$ = Fourier transform of first mathematical input

$X_2(f)$ = Fourier transform of second mathematical (nonlinear) input

$Y(f)$ = Fourier transform of mathematical output

$A_1(f)$ = FRF from $X_1(f)$ to $Y(f)$

$A_2(f)$ = FRF from $X_2(f)$ to $Y(f)$

Equation (7.10) is a *reverse dynamic* two-input/single-output linear model with the mathematical inputs defined by the two $X(f)$ terms passing through the linear systems defined by the $A(f)$ terms, respectively, to produce the mathematical output defined by the $Y(f)$ term. The two $A(f)$ terms can contain frequency-dependent coefficients. This model is pictured in Figure 7.7 with a noise term $N(f)$ representing deviations from the ideal model. The notation $H(f)$ versus $A(f)$ represents the FRFs in Figures 7.5 and 7.7, respectively. This notation helps to distinguish direct models from reverse models.

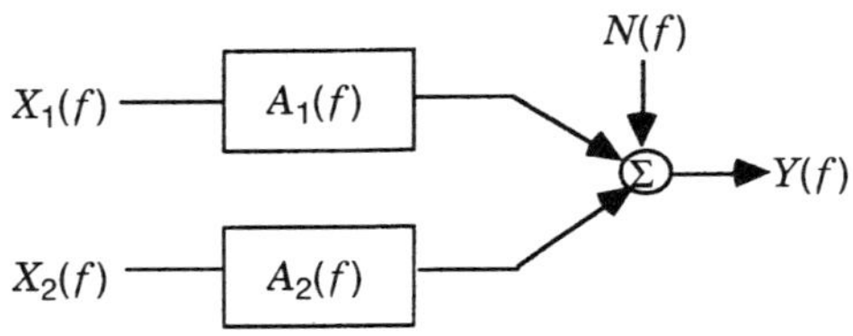

Figure 7.7. Reverse dynamic two-input/single-output linear model.

Reverse MI/SO Nonlinear Model for Three Inputs. The *reverse* MI/SO three-input/single-output nonlinear integrodifferential equation of motion for any of the seven cases listed earlier is given by

$$A_1(f)X_1(f) + A_2(f)X_2(f) + A_3(f)X_3(f) = Y(f) \tag{7.11}$$

where

$X_1(f)$ = Fourier transform of first mathematical input

$X_2(f)$ = Fourier transform of second mathematical (nonlinear) input

$X_3(f)$ = Fourier transform of third mathematical (nonlinear) input

$Y(f)$ = Fourier transform of mathematical output

$A_1(f)$ = FRF from $X_1(f)$ to $Y(f)$

$A_2(f)$ = FRF from $X_2(f)$ to $Y(f)$

$A_3(f)$ = FRF from $X_3(f)$ to $Y(f)$

Equation (7.11) is a *reverse dynamic* three-input/single-output linear model with the mathematical inputs defined by the three $X(f)$ terms passing through the linear systems defined by the three $A(f)$ terms, respectively, to produce the

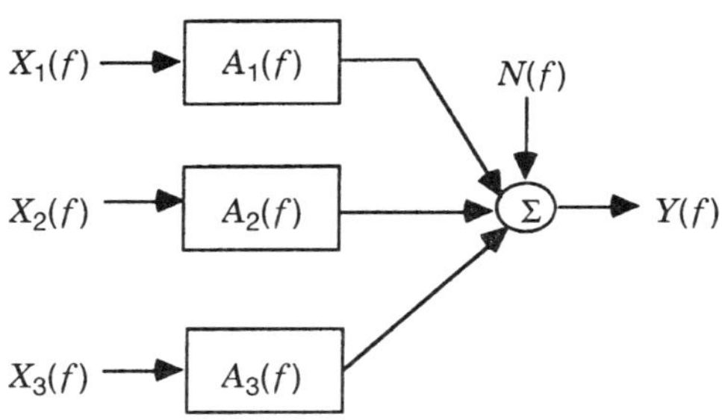

Figure 7.8. Reverse dynamic three-input/single-output linear model.

mathematical output defined by the $Y(f)$ term. The $A(f)$ terms can contain frequency-dependent coefficients. This model is pictured in Figure 7.8 with $N(f)$ representing deviations from the ideal model.

The particular forms for the input and output terms will now be stated for all seven cases of interest. Appropriate input and output records will be stated first for the Direct MI/SO models of Figures 7.5 and 7.6, and then for the Reverse MI/SO models of Figures 7.7 and 7.8. The two-input linear models in Figures 7.5 and 7.7 can be considered to be special cases of the three-input linear models in Figures 7.6 and 7.8 where the X_3 terms, the H_3 terms, and the A_3 terms are all zero.

For the heave and pitch motions in head seas and for the heave and roll motions in beam seas, both the direct and reverse two-input models as well as the direct and reverse three-input models were analyzed. In quartering seas, results showed that it was sufficient to analyze only the direct and reverse two-input models for the desired heave, roll, and pitch motions.

In the following material, a simplified notation will be used that appears on later graphical results in this chapter:

$$W = \text{Fourier transform of wave motion } w(t)$$

$$\dot{W} = \text{Fourier transform of wave motion derivative } \dot{w}(t)$$

$$Z = \text{Fourier transform of heave motion } z(t)$$

$$\dot{Z} = \text{Fourier transform of heave motion derivative } \dot{z}(t)$$

$$P = \text{Fourier transform of pitch angle } \theta(t)$$

$$R = \text{Fourier transform of roll angle } \phi(t)$$

Direct MI/SO Nonlinear Models. For the direct two-input linear models and three-input linear models, the first input term $X_1(f)$ is a physical input from the wave $w(t)$ given by

$$X_1 = W$$

The second input term $X_2(f)$ is a physical input for a nonlinear relative hydrodynamic damping force, which *for notational purposes only* will be represented by

$$X_2 = (\dot{Z} - \dot{W}) * |\dot{Z} - \dot{W}| = \text{Fourier transform of } \{\dot{z}(t) - \dot{w}(t)\}|\dot{z}(t) - \dot{w}(t)|$$

The third input term $X_3(f)$ is a physical input for a nonlinear restoring force that varies with each case of interest. This term is zero for the two-input models. The output $Y(f)$ is a physical output that varies with each case.

For given input and output records, one should solve for the cumulative coherence function and the associated values of the frequency response functions: $H_1(f)$ and $H_2(f)$ in the two-input models and $H_1(f)$, $H_2(f)$, and $H_3(f)$ in the three-input models.

1. *Heave Motions in Head Seas Using Large Wave Data*

	Direct Two-Input Model	Direct Three-Input Model
X_1	W	W
X_2	$(\dot{Z} - \dot{W}) * \lvert\dot{Z} - \dot{W}\rvert$	$(\dot{Z} - \dot{W}) * \lvert\dot{Z} - \dot{W}\rvert$
X_3	0	$P * \lvert P\rvert$
Y	Z	Z

2. *Pitch Motions in Head Seas Using Large Wave Data*

	Direct Two-Input Model	Direct Three-Input Model
X_1	W	W
X_2	$(\dot{Z} - \dot{W}) * \lvert\dot{Z} - \dot{W}\rvert$	$(\dot{Z} - \dot{W}) * \lvert\dot{Z} - \dot{W}\rvert$
X_3	0	$P * \lvert P\rvert$
Y	P	P

3. *Heave Motions in Beam Seas Using Large Wave Data*

	Direct Two-Input Model	Direct Three-Input Model
X_1	W	W
X_2	$(\dot{Z} - \dot{W}) * \lvert\dot{Z} - \dot{W}\rvert$	$(\dot{Z} - \dot{W}) * \lvert\dot{Z} - \dot{W}\rvert$
X_3	0	$R * \lvert R\rvert$
Y	Z	Z

4. *Roll Motions in Beam Seas Using Large Wave Data*

	Direct Two-Input Model	Direct Three-Input Model
X_1	W	W
X_2	$(\dot{Z} - \dot{W}) * \lvert\dot{Z} - \dot{W}\rvert$	$(\dot{Z} - \dot{W}) * \lvert\dot{Z} - \dot{W}\rvert$
X_3	0	$R * \lvert R\rvert$
Y	R	R

5. *Heave Motions in Quartering Seas Using Large Wave Data*

Direct Two-Input Model	
X_1	W
X_2	$(\dot{Z} - \dot{W}) * \lvert\dot{Z} - \dot{W}\rvert$
Y	Z

6. *Roll Motions in Quartering Seas Using Large Wave Data*

Direct Two-Input Model	
X_1	W
X_2	$(\dot{Z} - \dot{W}) * \lvert\dot{Z} - \dot{W}\rvert$
Y	R

7. *Pitch Motions in Quartering Seas Using Large Wave Data*

Direct Two-Input Model	
X_1	W
X_2	$(\dot{Z} - \dot{W}) * \lvert\dot{Z} - \dot{W}\rvert$
Y	P

Reverse MI/SO Nonlinear Models. For the reverse two-input linear models and three-input linear models, the first input term $X_1(f)$ is a mathematical input that varies with each case of interest. The second input term $X_2(f)$ is a mathematical input for a nonlinear relative hydrodynamic damping force, which *for notational purposes only* will be represented by

$$X_2 = (\dot{Z} - \dot{W}) * |\dot{Z} - \dot{W}| = \text{Fourier transform of } \{\dot{z}(t) - \dot{w}(t)\}|\dot{z}(t) - \dot{w}(t)|$$

The third input term $X_3(f)$ is a mathematical input for a nonlinear restoring force that varies with each case. This term is zero for the two-input models. The output $Y(f)$ is a mathematical output from the wave $w(t)$ given by

$$Y = W$$

For given input and output records, one should solve for the cumulative coherence function and the associated values of the frequency response functions: $A_1(f)$ and $A_2(f)$ in the two-input models and $A_1(f)$, $A_2(f)$, and $A_3(f)$ in the three-input models.

1. *Heave Motions in Head Seas Using Large Wave Data*

	Reverse Two-Input Model	Reverse Three-Input Model
X_1	Z	Z
X_2	$(\dot{Z} - \dot{W}) * \lvert\dot{Z} - \dot{W}\rvert$	$(\dot{Z} - \dot{W}) * \lvert\dot{Z} - \dot{W}\rvert$
X_3	0	$P * \lvert P\rvert$
Y	W	W

2. *Pitch Motions in Head Seas Using Large Wave Data*

	Reverse Two-Input Model	Reverse Three-Input Model
X_1	P	P
X_2	$(\dot{Z} - \dot{W}) * \lvert\dot{Z} - \dot{W}\rvert$	$(\dot{Z} - \dot{W}) * \lvert\dot{Z} - \dot{W}\rvert$
X_3	0	$P * \lvert P\rvert$
Y	W	W

3. *Heave Motions in Beam Seas Using Large Wave Data*

	Reverse Two-Input Model	Reverse Three-Input Model
X_1	Z	Z
X_2	$(\dot{Z} - \dot{W}) * \lvert\dot{Z} - \dot{W}\rvert$	$(\dot{Z} - \dot{W}) * \lvert\dot{Z} - \dot{W}\rvert$
X_3	0	$R * \lvert R\rvert$
Y	W	W

4. *Roll Motions in Beam Seas Using Large Wave Data*

	Reverse Two-Input Model	Reverse Three-Input Model
X_1	R	R
X_2	$(\dot{Z} - \dot{W}) * \lvert\dot{Z} - \dot{W}\rvert$	$(\dot{Z} - \dot{W}) * \lvert\dot{Z} - \dot{W}\rvert$
X_3	0	$R * \lvert R\rvert$
Y	W	W

5. *Heave Motions in Quartering Seas Using Large Wave Data*

Reverse Two-Input Model	
X_1	Z
X_2	$(\dot{Z} - \dot{W}) * \lvert\dot{Z} - \dot{W}\rvert$
Y	W

6. *Roll Motions in Quartering Seas Using Large Wave Data*

Reverse Two-Input Model	
X_1	R
X_2	$(\dot{Z} - \dot{W}) * \lvert\dot{Z} - \dot{W}\rvert$
Y	W

7. *Pitch Motions in Quartering Seas Using Large Wave Data*

Reverse Two-Input Model	
X_1	P
X_2	$(\dot{Z} - \dot{W}) * \lvert\dot{Z} - \dot{W}\rvert$
Y	W

7.5 RESULTS FROM ANALYSIS OF BARGE DATA

Response tests conducted on the free barge model as per Sections 7.3 and 7.4 gave recordings of the following data:

$$w(t) = \text{wave amplitude (small and large)}$$

$$z(t) = \text{heave motion}$$

$$\phi(t) = \text{roll angle}$$

$$\theta(t) = \text{pitch angle}$$

Six different sets of data were collected in the free barge tests as described in reference 59 with the notation shown in Table 7.3.

The data associated with each test condition in Table 7.3 consist of a sequence of separate 200 second tests, each with 2048 sample values. In order to avoid the processing of discontinuous data, spectral analyses were per-

TABLE 7.3 Notation for Free Barge Data Sets

Data Set	Direction	Waves
H1	Head	Small
B1	Beam	Small
Q1	Quartering	Small
H5	Head	Large
B5	Beam	Large
Q5	Quartering	Large

TABLE 7.4 Digital Data Processing Parameters

f_s = sampling frequency	= 10.24 Hz
Δt = time interval between samples = $1/f_s$	= 0.0977 s
f_c = Nyquist cutoff frequency = $f_s/2$	= 5.12 Hz
N = number of sample values per subrecord	= 256
T = time of each subrecord = $N\,\Delta t$	= 25 s
Δf = bandwidth resolution = $1/T$	= 0.04 Hz

TABLE 7.5 Total Samples and Number of Averages

Data Set	Total Samples	Number of Averages
H1	14,336	56
B1	12,544	49
Q1	17,020	70
H5	21,504	84
B5	23,298	91
Q5	19,456	76

formed without any overlap on 25 second subrecords using 256 sample values each. Table 7.4 lists the digital data processing parameters for the cases studied. A Hanning window was applied to reduce leakage effects. Table 7.5 states the total number of sample values, $N_{\text{total}} = n_d N$, and the total number of different subrecords n_d employed in the analysis of each of the six different data sets. MATLAB/ITAP programs were used to analyze the measured data.

Table 7.6 cross-lists the figures in this chapter from the analysis of the free barge data. This table is included to aid in finding various desired different results.

TABLE 7.6 Figures from Analysis of Free Barge Data

Figures 7.9–7.15	Linear results for small waves H1
Figures 7.16–7.22	Linear results for small waves B1
Figures 7.23–7.31	Linear results for small waves Q1
Figures 7.32–7.38	Linear results for large waves H5
Figures 7.39–7.45	Linear results for large waves B5
Figures 7.46–7.54	Linear results for large waves Q5
Figures 7.55–7.61	Linear FRFs for small versus large waves
Figures 7.62–7.75	Direct MI/SO nonlinear results for H5
Figures 7.76–7.89	Reverse MI/SO nonlinear results for H5
Figures 7.90–7.103	Direct MI/SO nonlinear results for B5
Figures 7.104–7.117	Reverse MI/SO nonlinear results for B5
Figures 7.118–7.126	Direct MI/SO nonlinear results for Q5
Figures 7.127–7.135	Reverse MI/SO nonlinear results for Q5

Figures 7.9–7.61 show the linear results. The circles are at frequencies where the coherence function is greater than or equal to 0.75. At these frequencies, the random errors in estimating associated frequency response functions are less than or equal to 10%.

Figures 7.62–7.135 show the nonlinear results. Direct MI/SO Techniques for the three wave directions are shown in Figures 7.62–7.75, 7.90–7.103, and 7.118–7.126. Reverse MI/SO Techniques for the three wave directions are shown in Figures 7.76–7.89, 7.104–7.117, and 7.127–7.135.

7.5.1 Linear SI/SO Analysis of Small Wave Data

Preliminary analysis of the small wave free barge data consists of time history "strip" plots, probability density function plots, and autospectral density function plots of individual sensors. For selected pairs of sensors, SI/SO spectral analysis is performed with results presented in terms of the ordinary coherence function and frequency response function amplitude and phase. The circles in these figures are at frequencies where the coherence function is greater than or equal to 0.75.

Head Sea Small Wave Data H1. Undisturbed and disturbed wave motion sensor data, shown in Figures 7.9 and 7.10, respectively, indicate that the two signals have similar Gaussian probability densities with similar autospectra and standard deviations. The frequency response function associated with undisturbed wave motion input and disturbed wave motion output, shown in Figure 7.11, has high coherence above 0.75 from 0.0–1.2 Hz. Consistent with the similarity of undisturbed and disturbed wave autospectra and standard deviations, the frequency response function in the "coherent" frequency band has roughly unit magnitude, nearly zero phase angle below 0.80 Hz, and decreasing phase at frequencies greater than 0.80 Hz.

Heave motion sensor data, shown in Figure 7.12, illustrate that this signal has a Gaussian probability density and a more narrow-band autospectrum with steeper magnitude roll-off than that associated with wave motion. The heave motion standard deviation, 0.0445, is lower than that of the undisturbed wave motion, 0.0579. The frequency response function associated with undisturbed wave motion input and heave motion output, shown in Figure 7.13, has coherence above 0.75 from 0.0–0.90 Hz. The frequency response function in the "coherent" frequency band has roughtly unit magnitude and nearly zero phase angle below 0.40 Hz. Above 0.40 Hz, the frequency response magnitude decreases with increasing frequency without any significant deviation of the phase angle from zero value.

Pitch motion sensor data, shown in Figure 7.14, illustrate that this signal has a Gaussian probability density and an autospectrum with significant magnitude in the 0.20–1.20 Hz frequency band and a 1.127 standard deviation. The frequency response function associated with undisturbed wave motion input and pitch motion output, shown in Figure 7.15, has a coherence value above 0.75 from 0.25–0.95 Hz. From 0.10–0.80 Hz, this frequency response function has a magnitude that increases in a linear manner with a 90° phase angle. Near 0.90 Hz, the frequency response reaches a peak value and the phase angle descends toward zero.

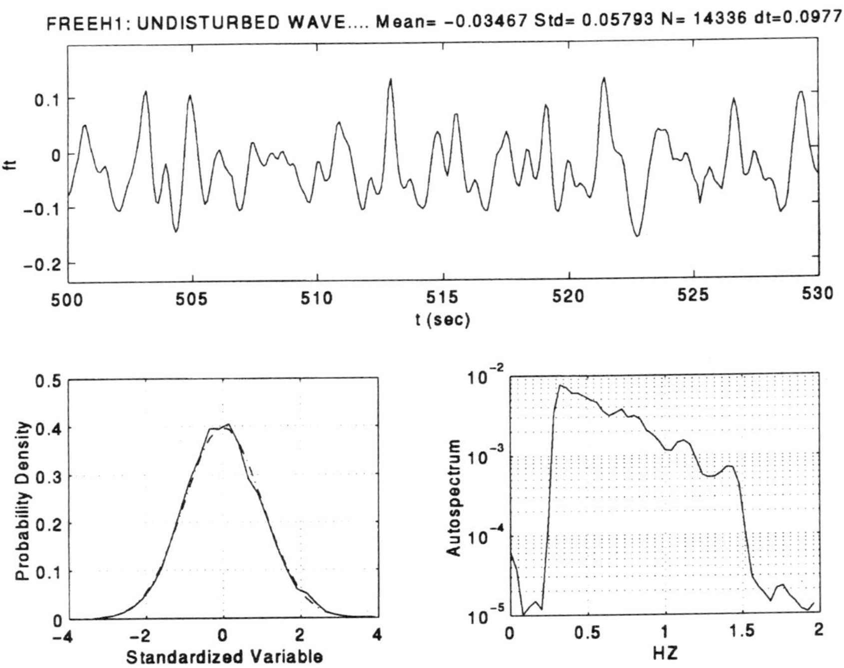

Figure 7.9.

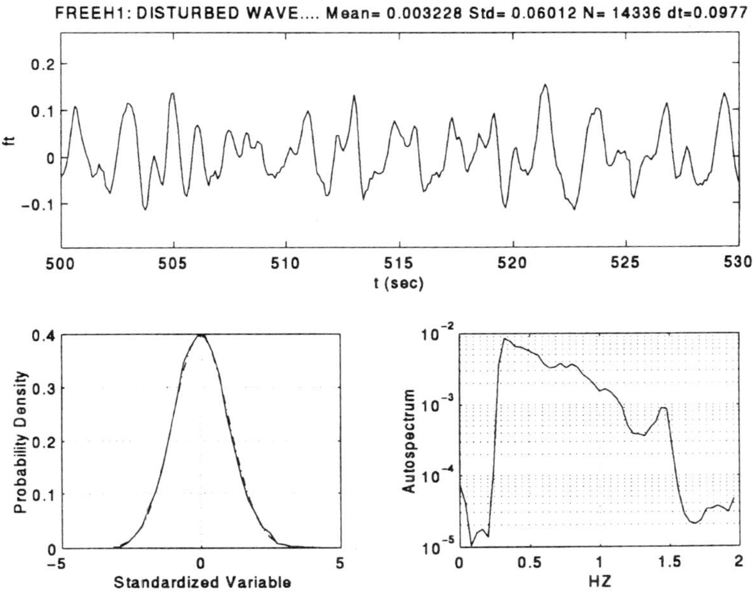

Figure 7.10.

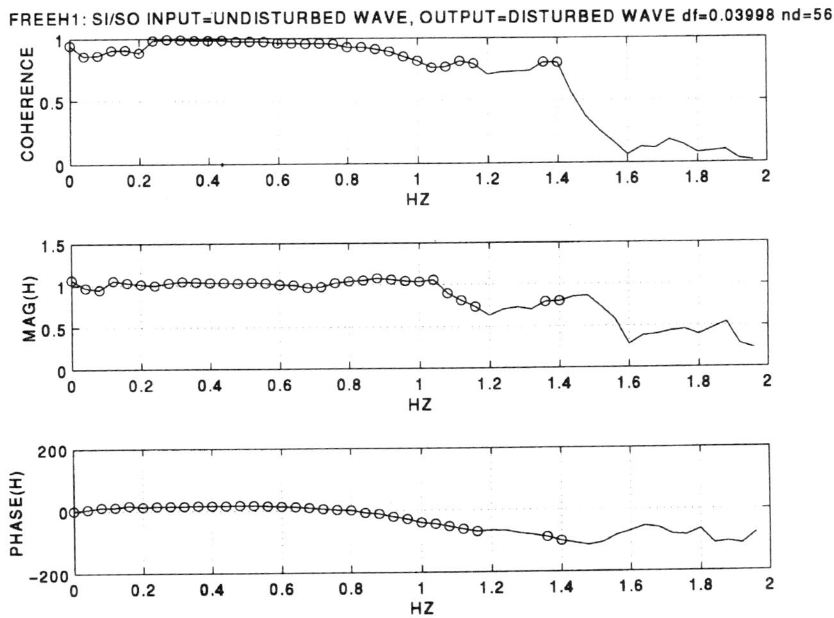

Figure 7.11.

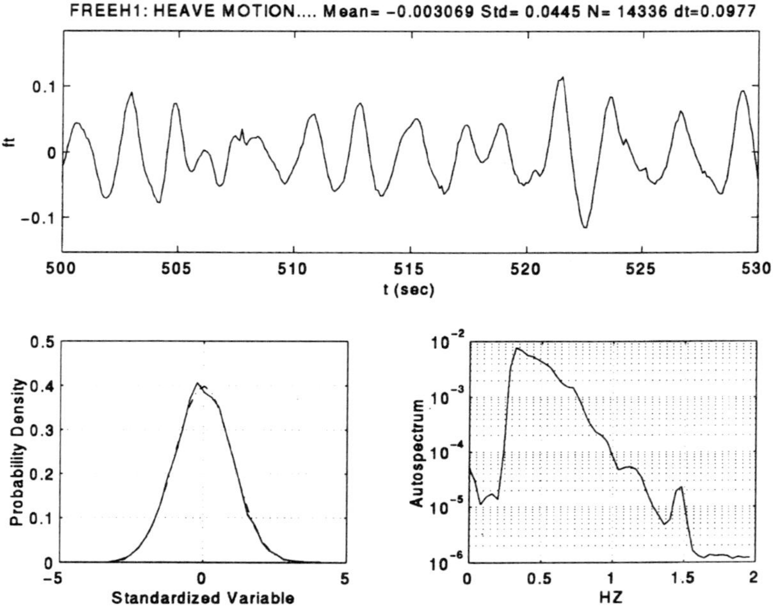

Figure 7.12.

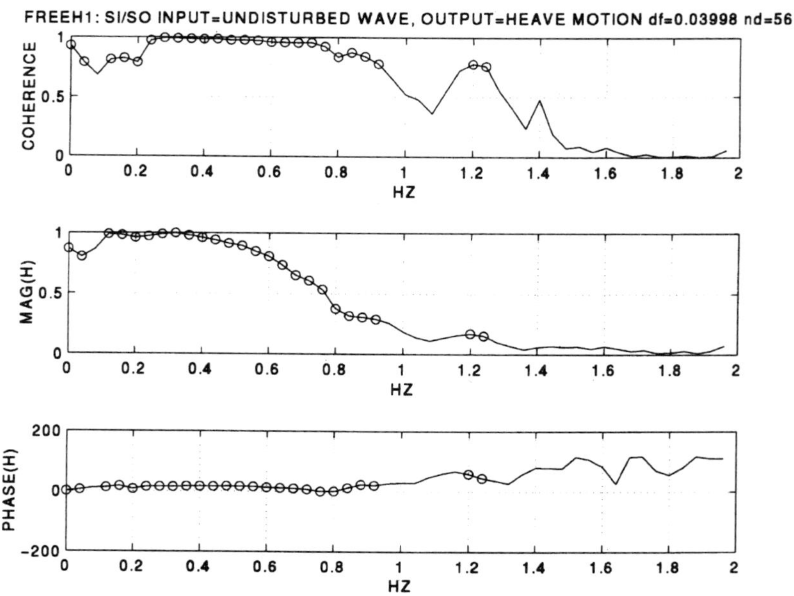

Figure 7.13.

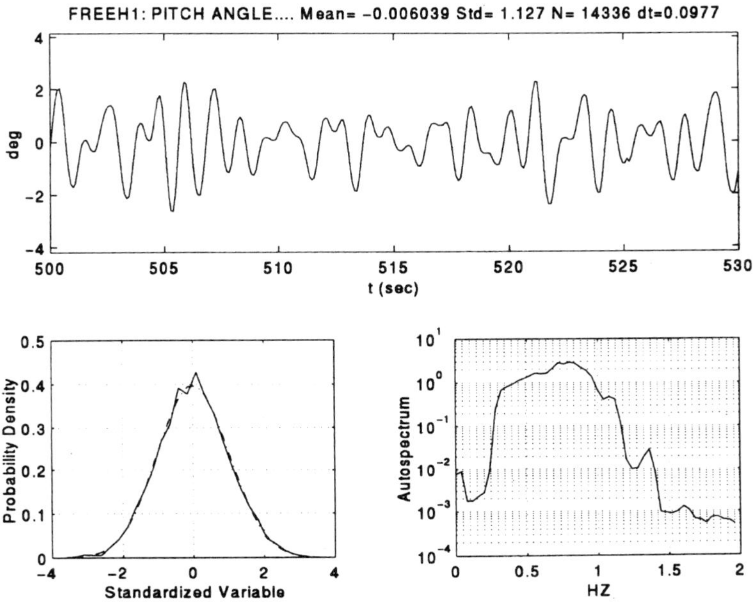

Figure 7.14.

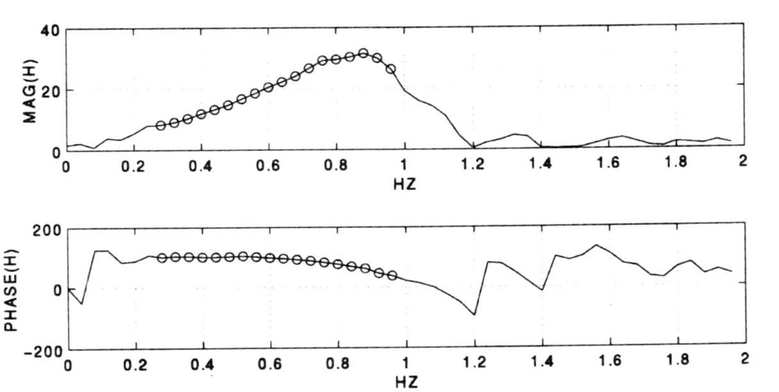

Figure 7.15.

Beam Sea Small Wave Data B1. Undisturbed and disturbed wave motion sensor data, in Figures 7.16 and 7.17, respectively, indicate that the two signals have Gaussian probability densities with similar autospectra and standard deviations. The frequency response function associated with undisturbed wave motion input and disturbed wave motion output, shown in Figure 7.18, has coherence above 0.75 from 0.0–1.0 Hz. The frequency response function in the "coherent" frequency band has roughly unit magnitude, nearly zero phase angle below 0.80 Hz, and decreasing phase at frequency greater than 0.80 Hz. This trend is similar to that noted in the head sea case (Figures 7.9–7.11).

Heave motion sensor data, shown in Figure 7.19, illustrate that this signal has a Gaussian probability density and an autospectrum with magnitude roll-off similar to that associated with wave motions. The heave motion standard deviation, 0.05858, is nearly identical to that of the undisturbed wave motion, 0.05837. The frequency response function associated with undisturbed wave motion input and heave motion output, in Figure 7.20, has coherence above 0.75 from 0.22–1.10 Hz. The frequency response function in the "coherent" frequency band has roughly unit magnitude and nearly zero phase angle below 0.80 Hz. This is a "wave-follower" behavior. Above 0.90 Hz, the frequency response magnitude decreases with increasing frequency and the phase angle decreases toward $-180°$.

Roll motion sensor data, shown in Figure 7.21, illustrate that this signal has a Gaussian probability density and an autospectrum with a peak magnitude at roughly 0.70 Hz and a 2.005 standard deviation. The frequency response function associated with undisturbed wave motion input and roll motion output, shown in Figure 7.22, has coherence above 0.75 from 0.27–1.20 Hz. The frequency response function has a magnitude that clearly peaks at about 0.70 Hz with an associated phase angle near 180°. The phase angle decreases over the 0.27–1.20 Hz frequency band. Smaller peak values of the coherence function in Figure 7.22 than in Figure 7.20 indicate that more nonlinearity is present in roll responses than in heave responses.

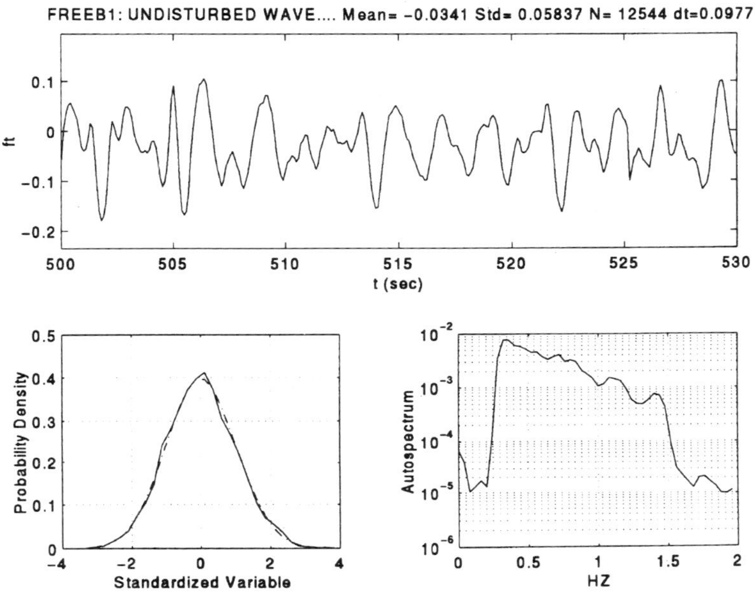

Figure 7.16.

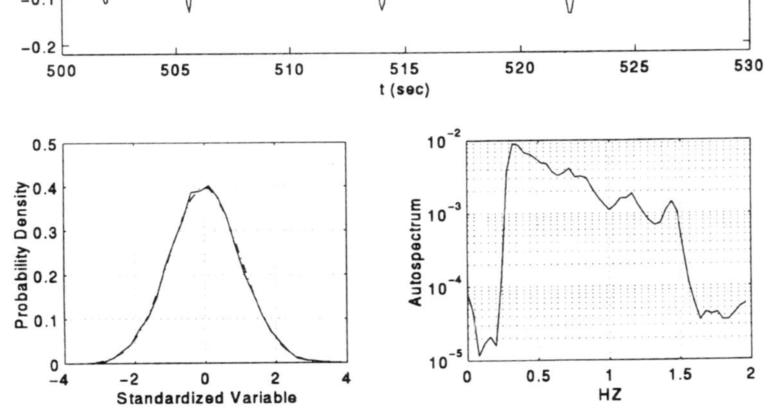

Figure 7.17.

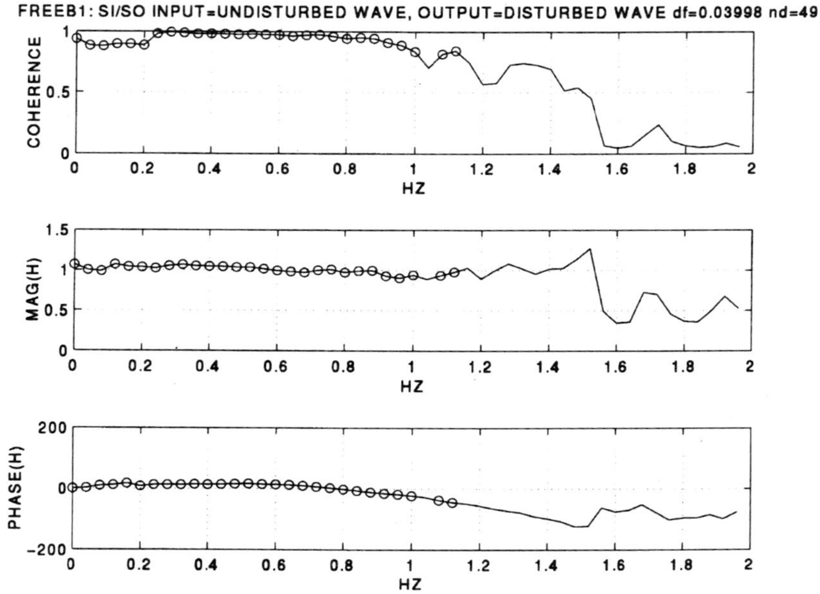

Figure 7.18.

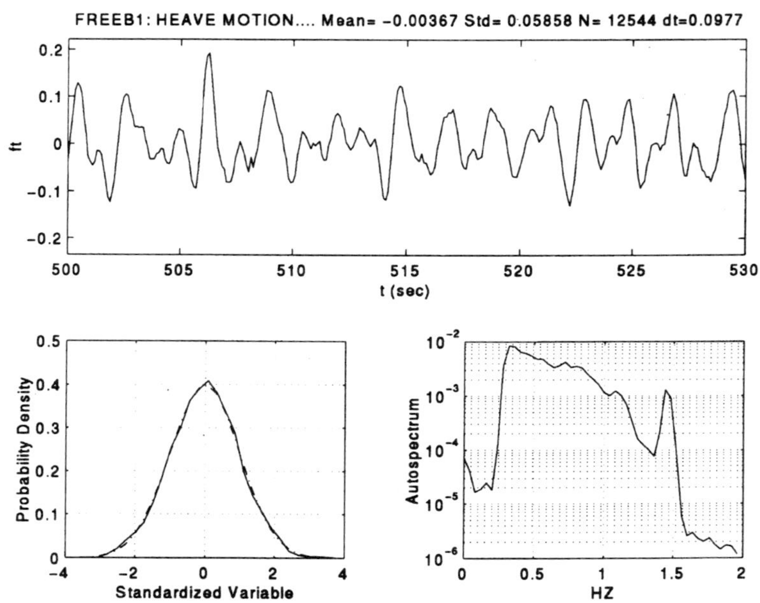

Figure 7.19.

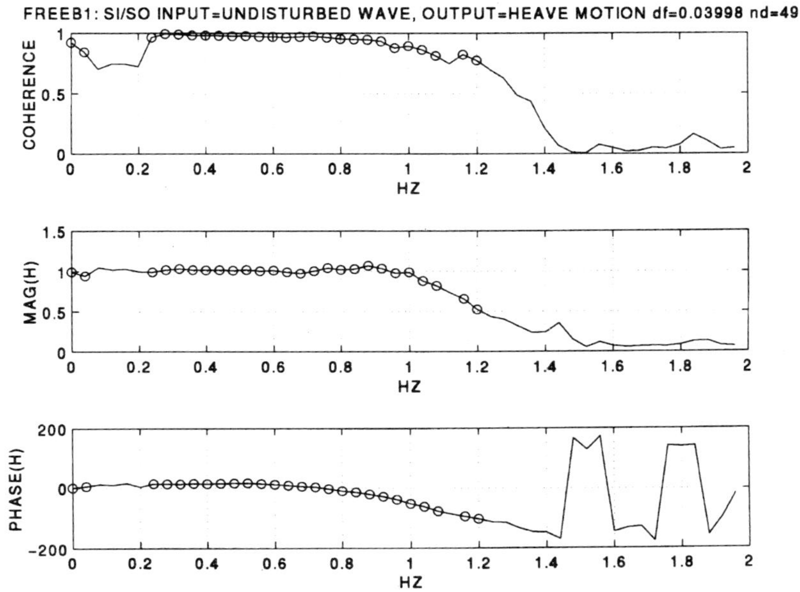

Figure 7.20.

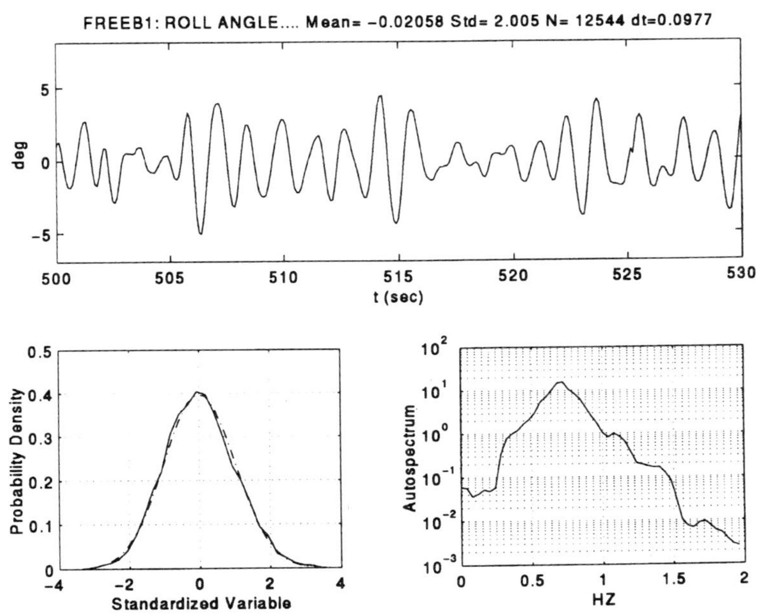

Figure 7.21.

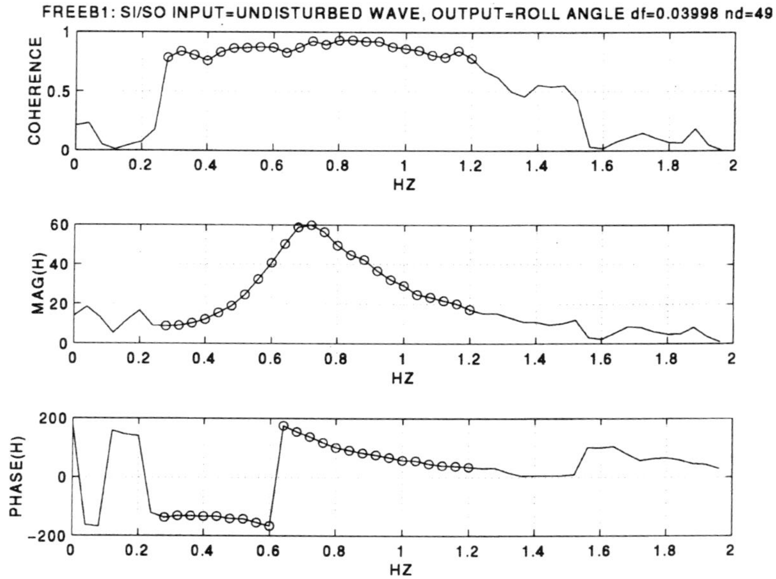

Figure 7.22.

Quartering Sea Small Wave Data Q1. Undisturbed and disturbed wave motion sensor data, shown in Figures 7.23 and 7.24, respectively, indicate that the two signals have Gaussian probability densities with similar autospectra and standard deviations. The frequency response function associated with undisturbed wave motion input and disturbed wave motion output, shown in Figure 7.25, has coherence above 0.75 from 0.0–1.0 Hz. The frequency response function in the "coherent" frequency band has roughly unit magnitude (with somewhat greater excursions from unit value than those noted for the head and beam sea cases), nearly zero phase angle below 0.80 Hz, and decreasing phase at frequencies greater than 0.80 Hz.

Heave motion sensor data, shown in Figure 7.26, illustrate that this signal has a Gaussian probability density and an autospectrum with steeper magnitude roll-off than that associated with wave motion. The heave motion standard deviation, 0.04597, is lower than that of the undisturbed wave motion, 0.05754. The frequency response function associated with undisturbed wave motion input and heave motion output, shown in Figure 7.27, has coherence above 0.75 from 0.1–0.95 Hz. The frequency response function in the "coherent" frequency band has roughly unit magnitude and nearly zero phase angle below 0.40 Hz. Above 0.40 Hz, the frequency response magnitude decreases with increasing frequency without significant deviation of the phase angle from zero value. This trend is a "wave-follower" behavior similar to that which occurs for head seas in small waves (see Figure 7.13).

Roll motion sensor data, shown in Figure 7.28, illustrate that this signal has a Gaussian probability density and an autospectrum with a peak magnitude slightly below 0.70 Hz and 0.7956 standard deviation. The frequency response function associated with undisturbed wave motion input and roll motion output, shown in Figure 7.29, has coherence above 0.75 from 0.32–0.90 Hz. This frequency response function has a magnitude that clearly peaks at about 0.680 Hz with an associated phase angle near 180°. The phase angle decreases over the 0.32–0.90 Hz frequency band. This frequency response function is similar to that noted for beam seas but with roughly one-half the beam sea peak magnitude (see Figure 7.22); this is attributed to the fact that the quartering sea encounters the barge at about 30° off the bow. Smaller peak values of the coherence function in Figure 7.29 than in Figure 7.27 indicate that more nonlinearity is present in roll responses than in heave responses.

Pitch motion sensor data, shown in Figure 7.30, illustrate that this signal has a Gaussian probability density and an autospectrum with significant magnitude in the 0.20–1.20 Hz frequency band and a 1.086 standard deviation. The frequency response function associated with undisturbed wave motion input and pitch motion output, shown in Figure 7.31, has coherence above 0.75 from 0.27–1.10 Hz. This frequency response function has a magnitude that increases in a roughly linear manner with a 90° phase angle below 0.80 Hz. At about 0.90 Hz, the frequency response function magnitude reaches a peak value with the phase angle decreasing toward zero above 0.80 Hz. This frequency response function is similar to that noted for head seas (see Figure 7.15).

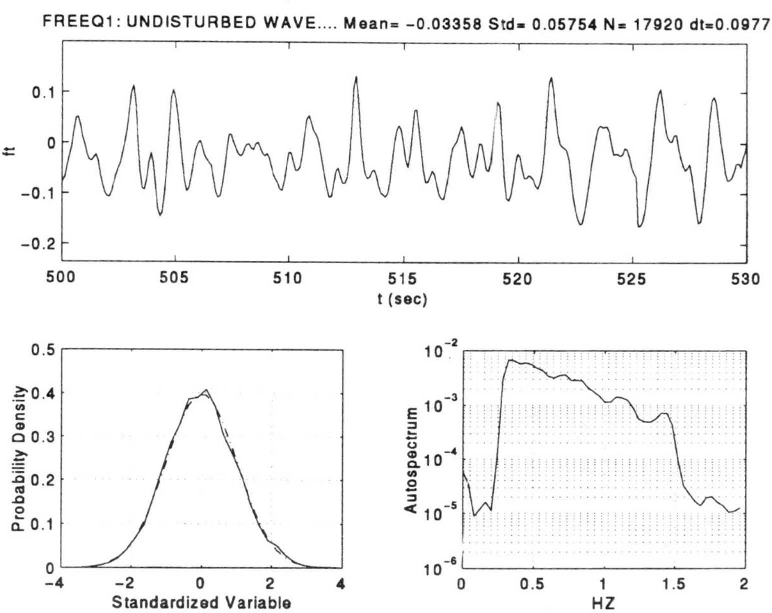

Figure 7.23.

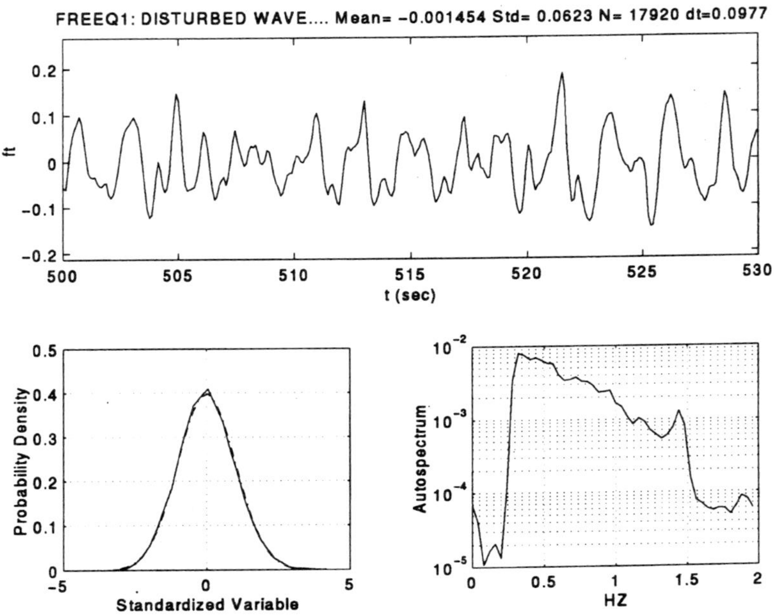

Figure 7.24.

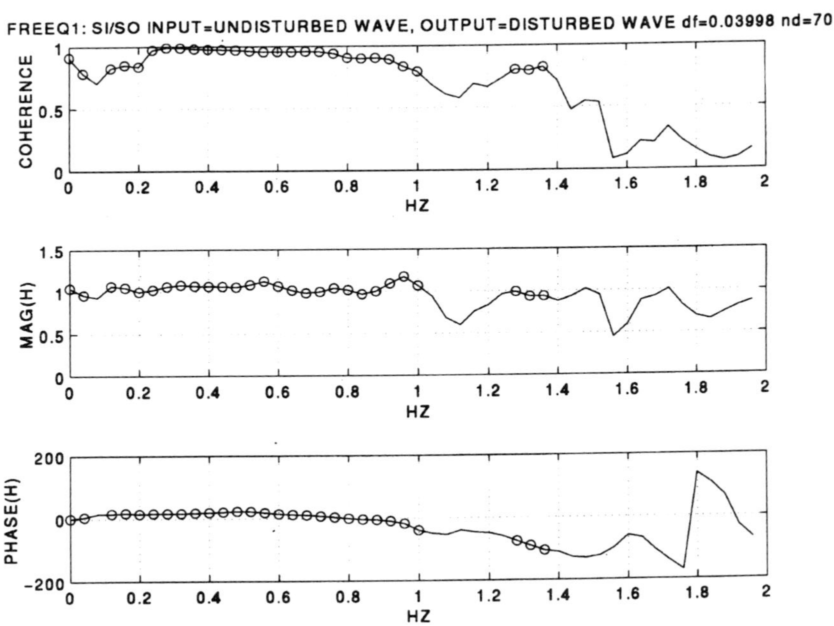

Figure 7.25.

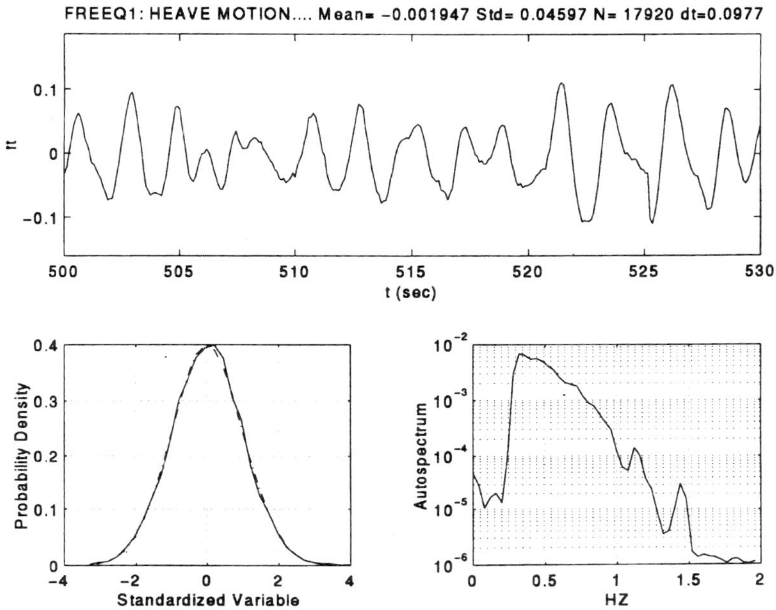

Figure 7.26.

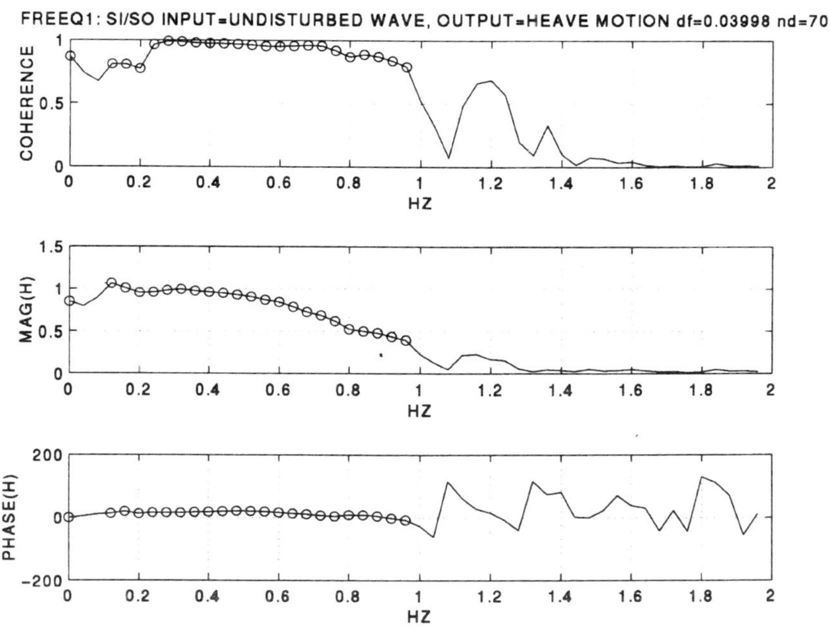

Figure 7.27.

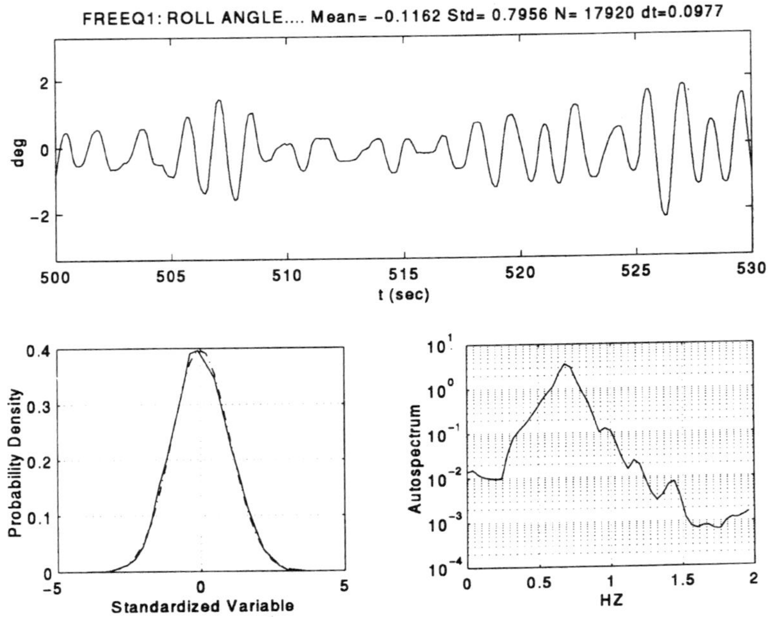

Figure 7.28.

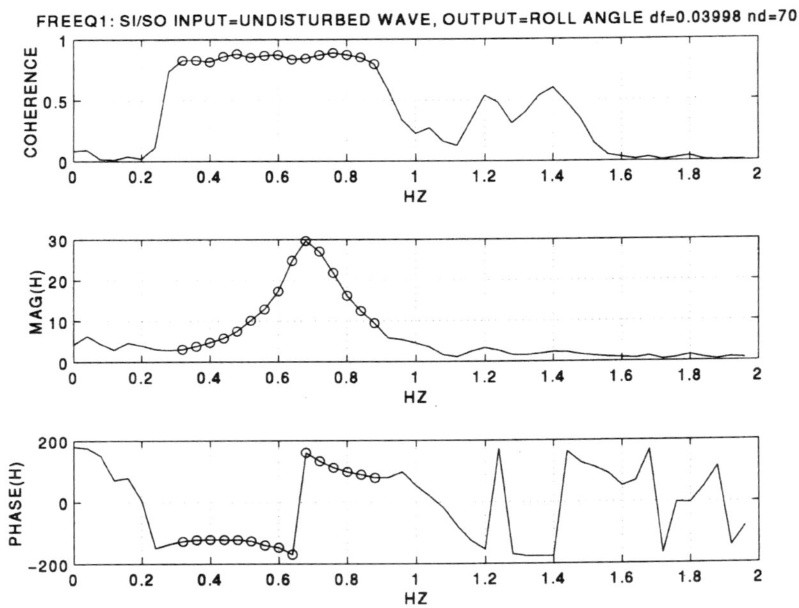

Figure 7.29.

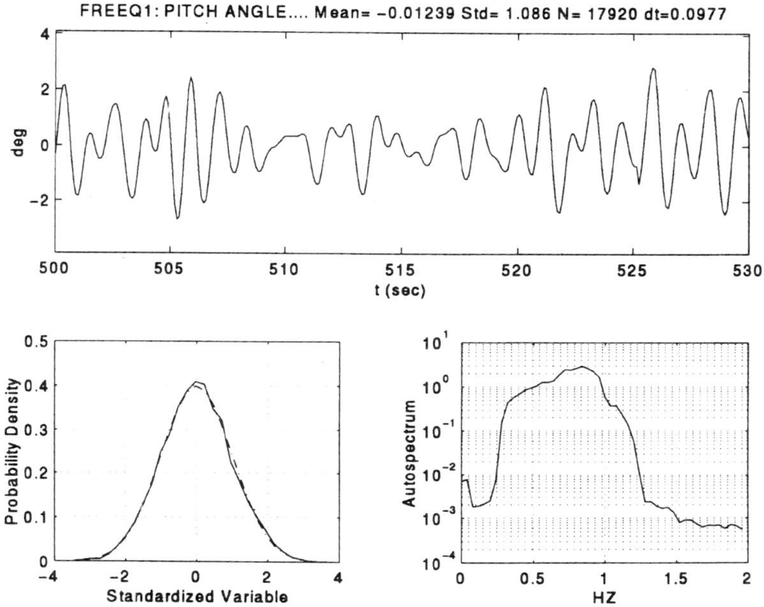

Figure 7.30.

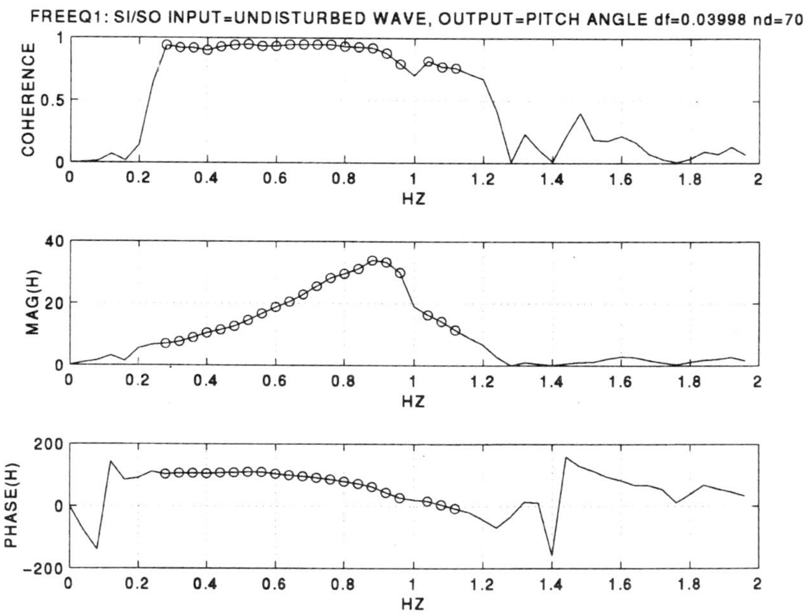

Figure 7.31.

7.5.2 Linear SI/SO Analysis of Large Wave Data

As in the case of small wave free barge data, preliminary analysis of the large wave data consists of time history "strip" plots, probability density function plots, and autospectral density function plots of individual sensors. For selected pairs of sensors, SI/SO spectral analysis is performed with results given in terms of the ordinary coherence function and frequency response function amplitude and phase. A discussion of these results is provided below.

Head Sea Large Wave Data H5. Undisturbed and disturbed wave motion sensor data, shown in Figures 7.32 and 7.33, respectively, indicate that the two signals have Gaussian probability densities with similar autospectra and standard deviations. The frequency response function associated with undisturbed wave motion input and disturbed wave motion output, shown in Figure 7.34, has coherence above 0.75 from 0.25–1.05 Hz. Consistent with the similarity of undisturbed and disturbed wave properties, the frequency response function in the "coherent" frequency band has roughly unit magnitude, nearly zero phase angle below 0.80 Hz, and negative phase angles at frequencies greater than 0.80 Hz. These results, which are similar to those noted for small head sea wave data (see Figure 7.11), indicate that the large wave input motions do not exhibit noticeable nonlinearity.

Heave motion sensor data, shown in Figure 7.35, illustrate that this signal has a Gaussian probability density and an autospectrum with steeper magnitude roll-off than that associated with wave motion. The heave motion standard deviation, 0.0731, is lower than that of the undisturbed wave motion, 0.1276. The frequency response function associated with undisturbed wave motion input and heave motion output, in Figure 7.36, has coherence above 0.75 (actually near unity) from 0.25–0.95 Hz. Consistent with the trends in undisturbed wave and heave motion autospectra and standard deviations, the frequency response function in the "coherent" frequency band has roughly unit magnitude and nearly zero phase angle below 0.40 Hz. Above 0.40 Hz, the frequency response magnitude decreases with increasing frequency without significant deviation of phase angle from zero value. These results, which are similar to those noted for small head sea wave data (see Figure 7.13), indicate that the large wave input motions do not induce noticeable heave response nonlinearity.

Pitch motion sensor data, shown in Figure 7.37, illustrate that this signal has a nearly Gaussian probability density and an autospectrum with significant magnitude in the 0.20–1.20 Hz frequency band and 2.716 standard deviation. The frequency response function associated with undisturbed wave motion input and pitch motion output, shown in Figure 7.38, has coherence above 0.75 from 0.30–1.05 Hz. This frequency response function has a magnitude that increases in a roughly linear manner and a 90° phase angle below 0.80 Hz. At about 0.90 Hz, the frequency response reaches a peak value and the phase angle descends toward zero. These results, which are similar to those noted for small head sea wave data, indicate that the large wave input motions do not induce noticeable pitch response nonlinearity (see Figure 7.15).

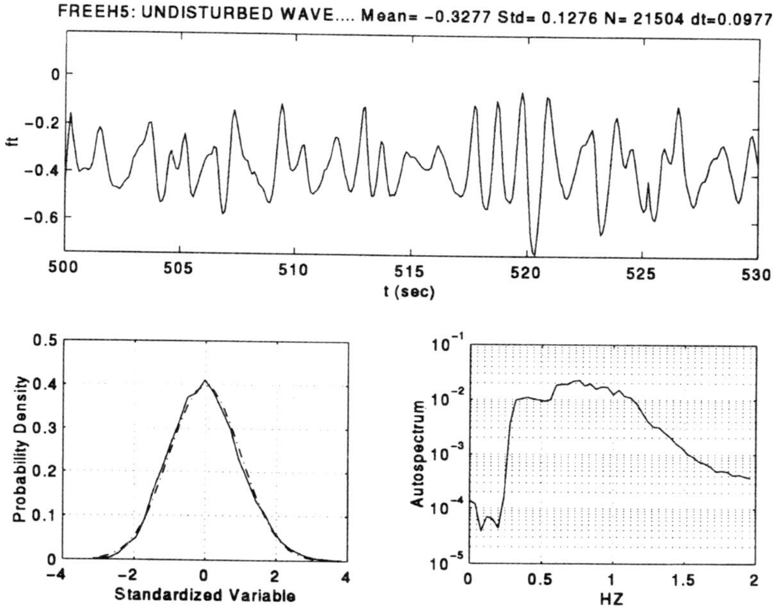

Figure 7.32.

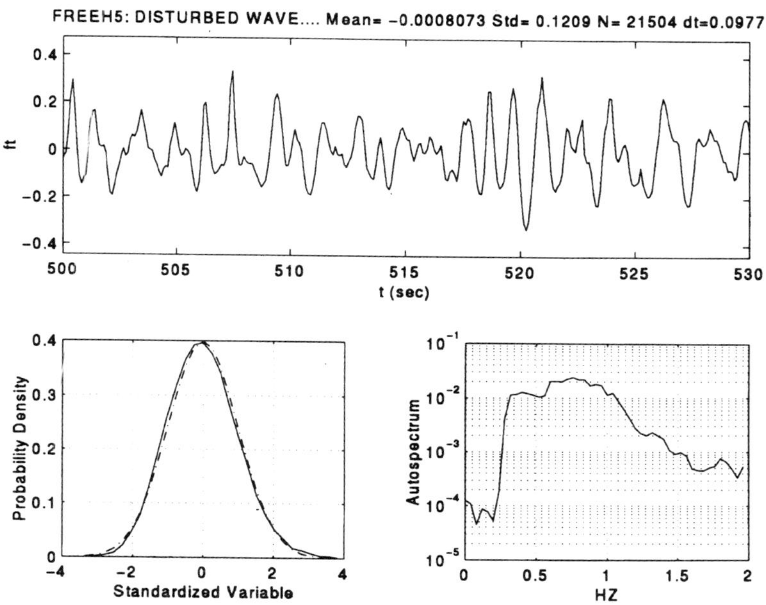

Figure 7.33.

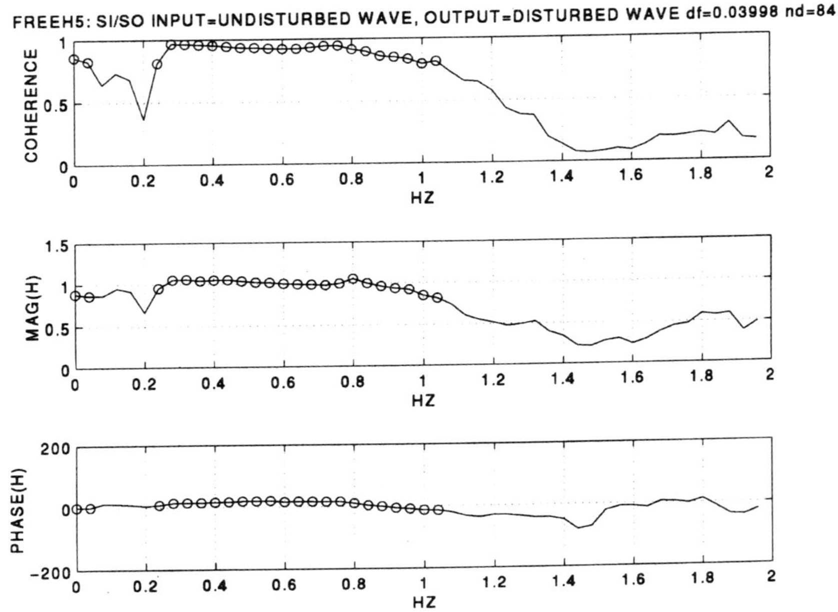

Figure 7.34.

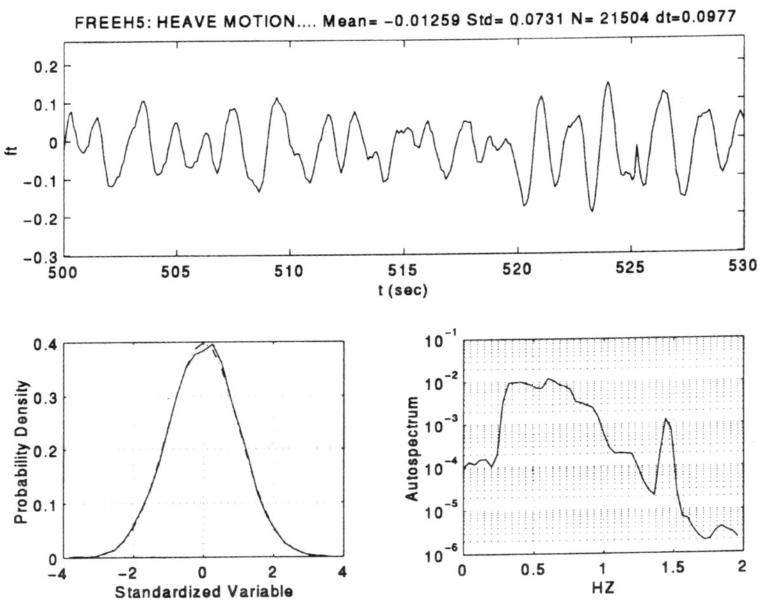

Figure 7.35.

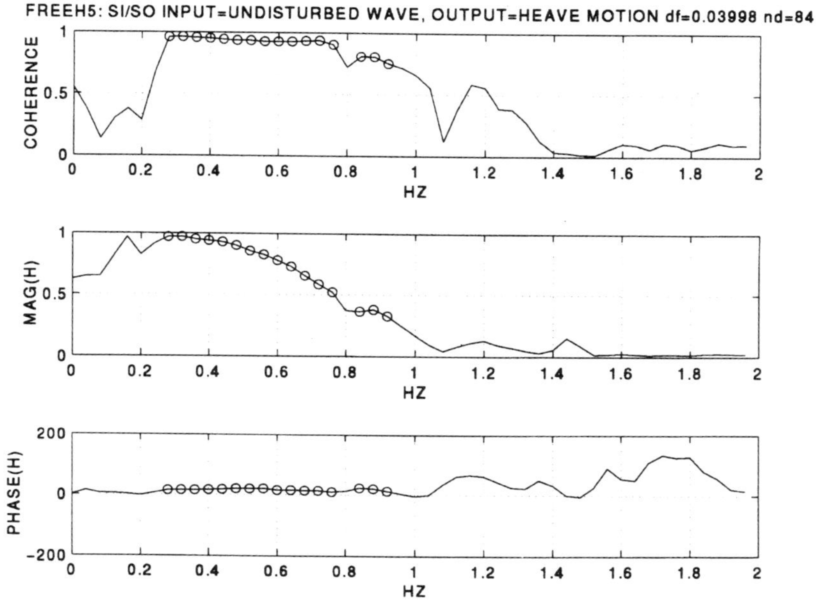

Figure 7.36.

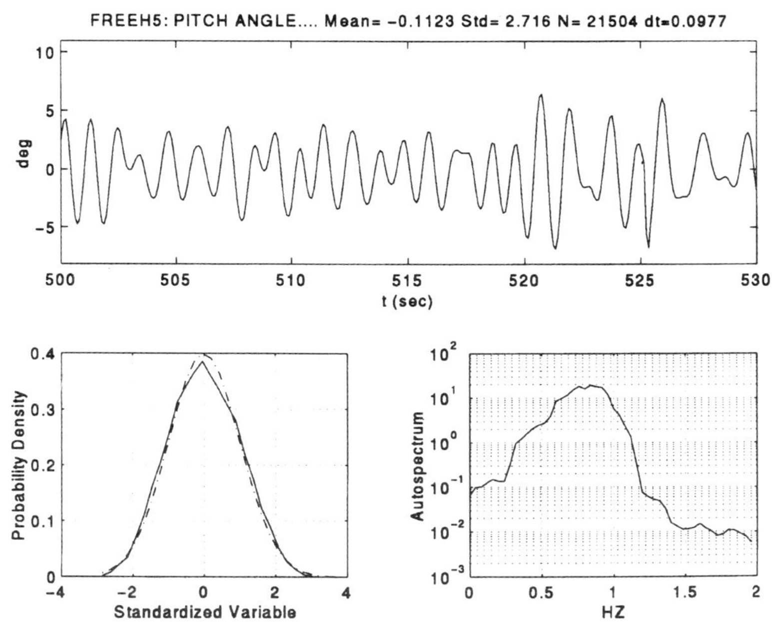

Figure 7.37.

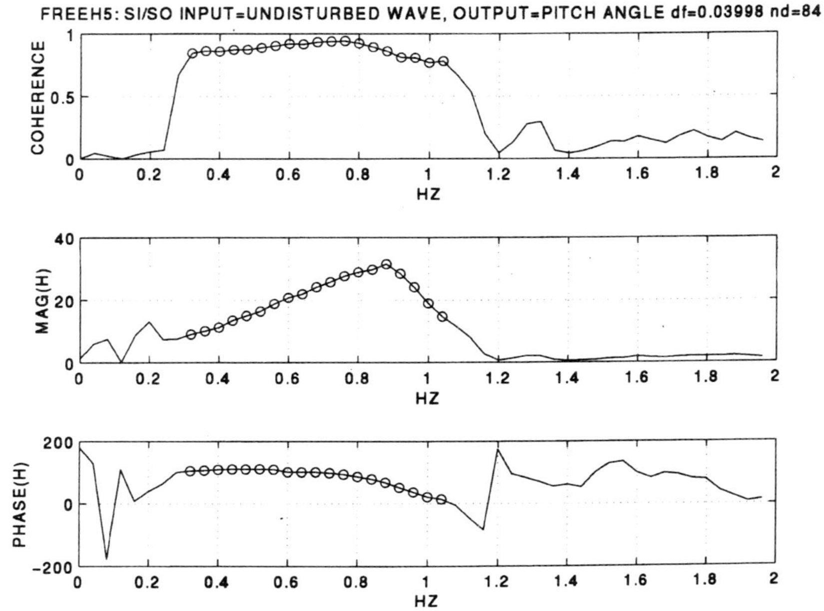

Figure 7.38.

Beam Sea Large Wave Data B5. Undisturbed and disturbed wave motion sensor data, shown in Figures 7.39 and 7.40, respectively, indicate that the two signals have Gaussian probability densities with similar autospectra and standard deviations. The frequency response function associated with undisturbed wave motion input and disturbed wave motion output, shown in Figure 7.41, has coherence above 0.75 (generally close to unity) from 0.25–0.95 Hz. Consistent with the similarity of undisturbed and disturbed wave autospectra and standard deviations, the frequency response function in the "coherent" frequency band has roughly unit magnitude and nearly zero phase angles below 0.95 Hz. This trend is similar to that noted for the small wave beam sea case (see Figure 7.18).

Heave motion sensor data, shown in Figure 7.42, illustrate that this signal has a Gaussian probability density and an autospectrum with significant magnitude in the 0.30–1.1 Hz frequency band. The heave motion standard deviation, 0.1161, is slightly lower than that of the undisturbed wave motion, 0.1269. The frequency response function associated with undisturbed wave motion input and heave motion output, in Figure 7.43, has coherence above 0.75 from 0.27–1.15 Hz. Consistent with the trends in undisturbed wave and heave motion autospectra and standard deviations, the frequency response function in the "coherent" frequency band has roughly unit magnitude and nearly zero phase angle below 0.80 Hz. Above 0.95 Hz, the frequency response magnitude decreases with increasing frequency and the phase angle decreases toward −180°. This trend is similar to that noted for the small wave beam sea case (see Figure 7.20).

Roll motion sensor data, shown in Figure 7.44, illustrate that this signal has a Gaussian probability density and an autospectrum with a peak magnitude at roughly 0.70 Hz and a 5.05 standard deviation. The frequency response function associated with undisturbed wave motion input and roll motion output, shown in Figure 7.45, has coherence above 0.75 from 0.6–1.15 Hz. This frequency response function has a magnitude that peaks at about 0.70 Hz with an associated phase angle near 180°. The roll motion frequency response function obtained here differs from the corresponding small wave beam sea case, indicating the possibility of nonlinearity (see Figure 7.22).

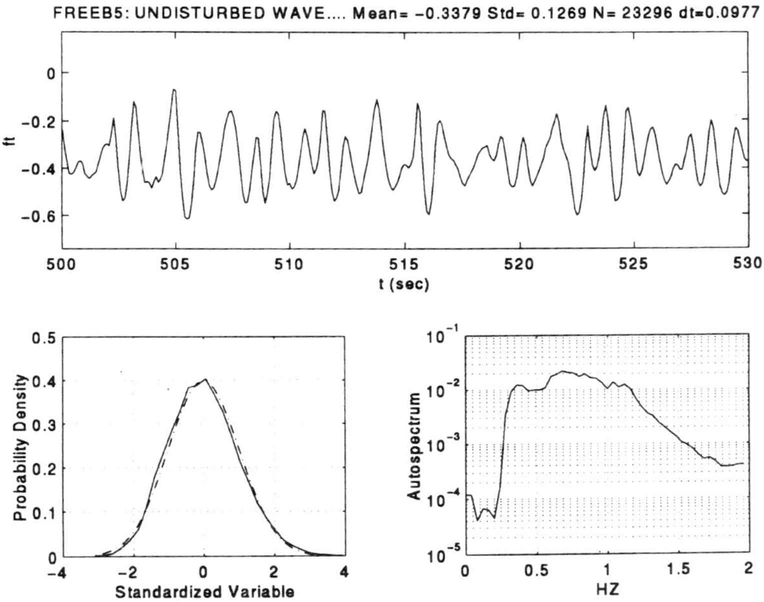

Figure 7.39.

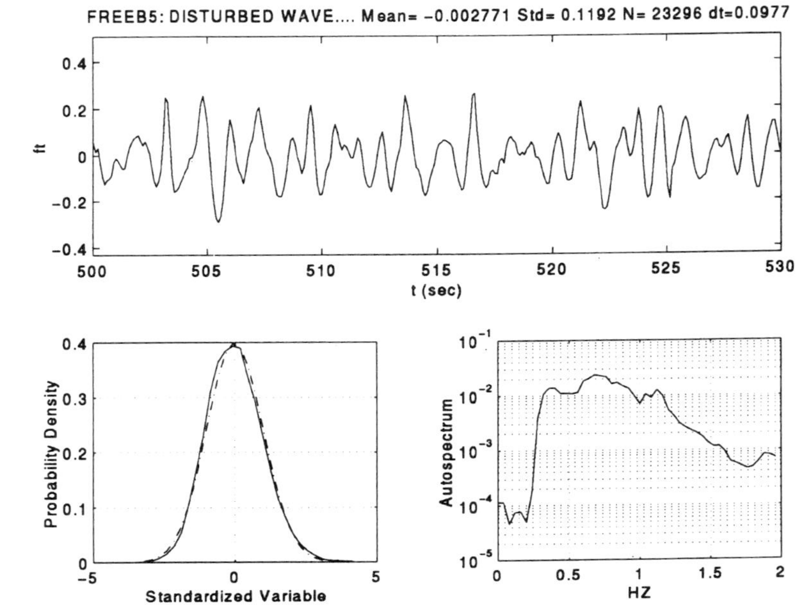

Figure 7.40.

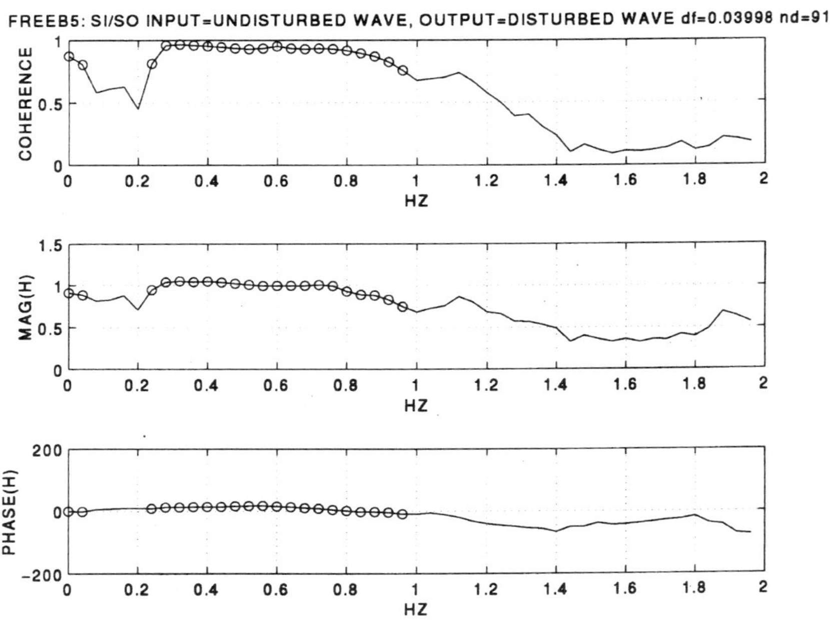

Figure 7.41.

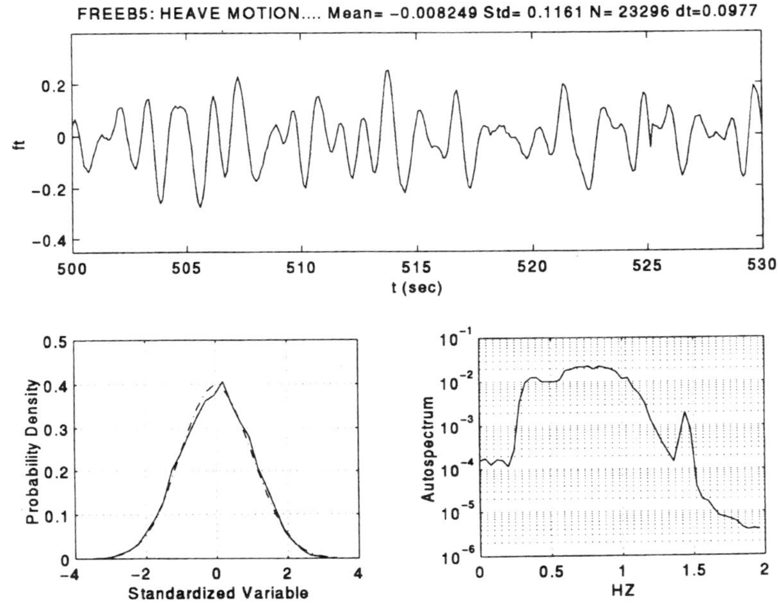

Figure 7.42.

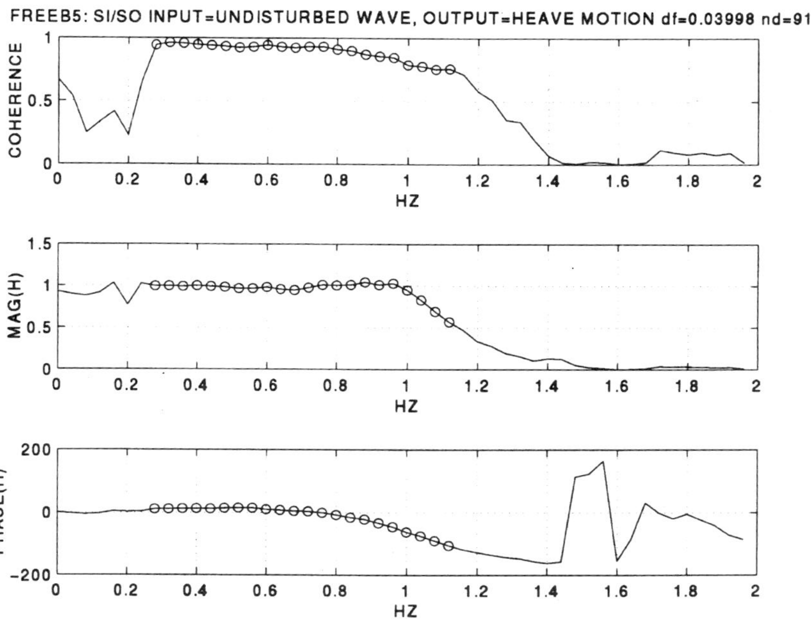

Figure 7.43.

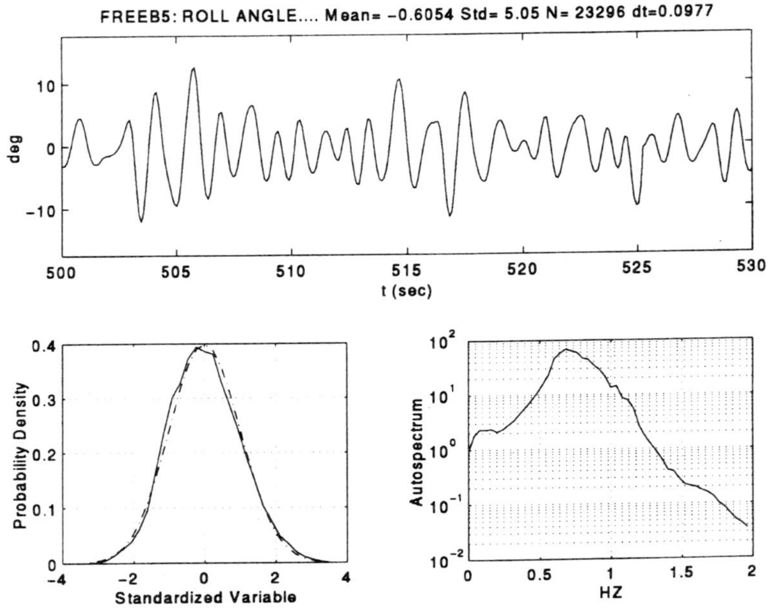

Figure 7.44.

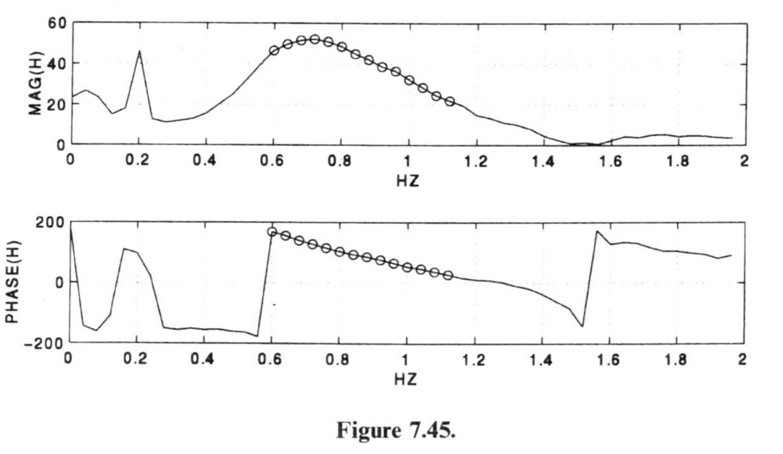

Figure 7.45.

Quartering Sea Large Wave Data Q5. Undisturbed and disturbed wave motion sensor data, in Figures 7.46 and 7.47, respectively, indicate that the two signals have Gaussian probability densities with similar autospectra and standard deviations. The frequency response function associated with undisturbed wave motion input and disturbed wave motion output, shown in Figure 7.48, has coherence above 0.75 from 0.25–1.1 Hz. The frequency response function in the "coherent" frequency band has roughly unit magnitude (with a peaking in magnitude just below 1.0 Hz) and has nearly zero phase angle for frequencies below 1.1 Hz.

Heave motion sensor data, in Figure 7.49, show that this signal has a Gaussian probability density and an autospectrum with steeper magnitude roll-off than that associated with wave motion. The heave motion standard deviation, 0.07813, is lower than that of the undisturbed wave motion, 0.1290. The frequency response function associated with undisturbed wave motion input and heave motion output, shown in Figure 7.50, has coherence above 0.75 from 0.25–0.95 Hz. Consistent with properties in undisturbed wave and heave motion data, the frequency response function in the "coherent" frequency band has roughly unit magnitude with nearly zero phase angle below 0.40 Hz. Above 0.40 Hz, the frequency response magnitude decreases with increasing frequency without significant deviation of phase angle from zero value. This trend is similar to that noted for quartering sea small wave data (see Figure 7.27).

Roll motion sensor data, shown in Figure 7.51, illustrate that this signal has a Gaussian probability density and an autospectrum with a peak magnitude slightly below 0.70 Hz and a 2.046 standard deviation. The frequency response function associated with undisturbed wave motion input and roll motion output, shown in Figure 7.52, has ordinary coherence above 0.75 over a relatively narrow frequency band from 0.55–0.90 Hz. This frequency response function has a magnitude that peaks at about 0.65 Hz with an associated phase angle near 180°. A stronger nonlinearity is indicated than in Figure 7.29.

Pitch motion sensor data, shown in Figure 7.53, illustrate that this signal has a Gaussian probability density and an autospectrum with significant magnitude in the 0.20–1.20 Hz frequency band and a 2.499 standard deviation. The frequency response function associated with undisturbed wave motion input and pitch motion output, shown in Figure 7.54, has coherence above 0.75 from 0.27–1.12 Hz. The frequency response function has a magnitude that increases in a roughly linear manner and a 90° phase angle below 0.80 Hz. At about 0.85 Hz, the frequency response magnitude reaches a peak value. The associated phase angle descends toward zero above 0.80 Hz. These results are similar to Figure 7.31.

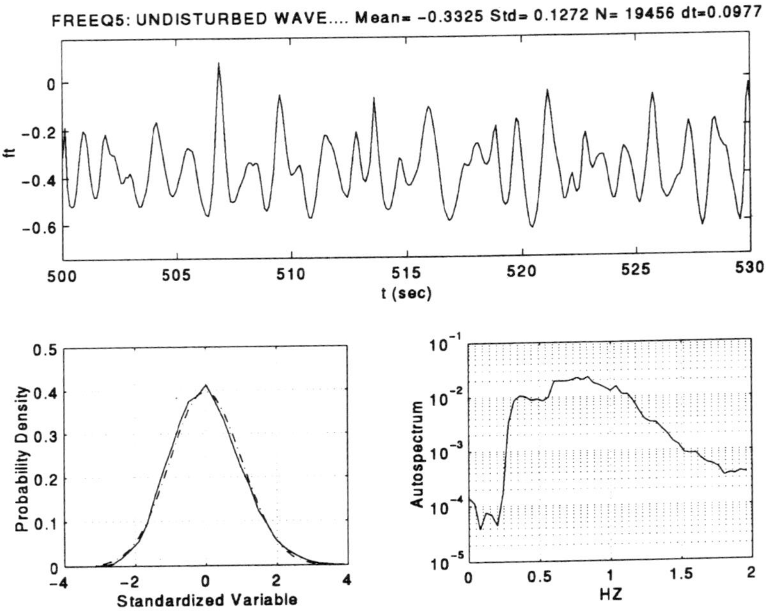

Figure 7.46.

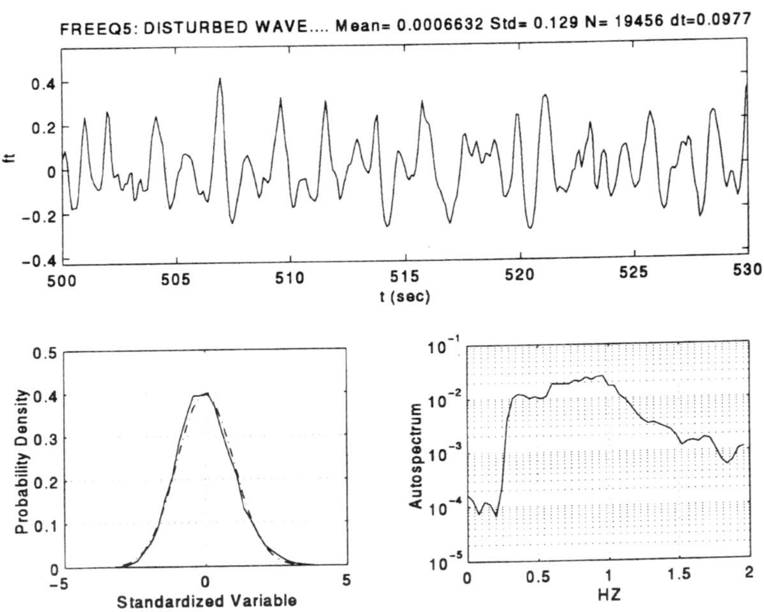

Figure 7.47.

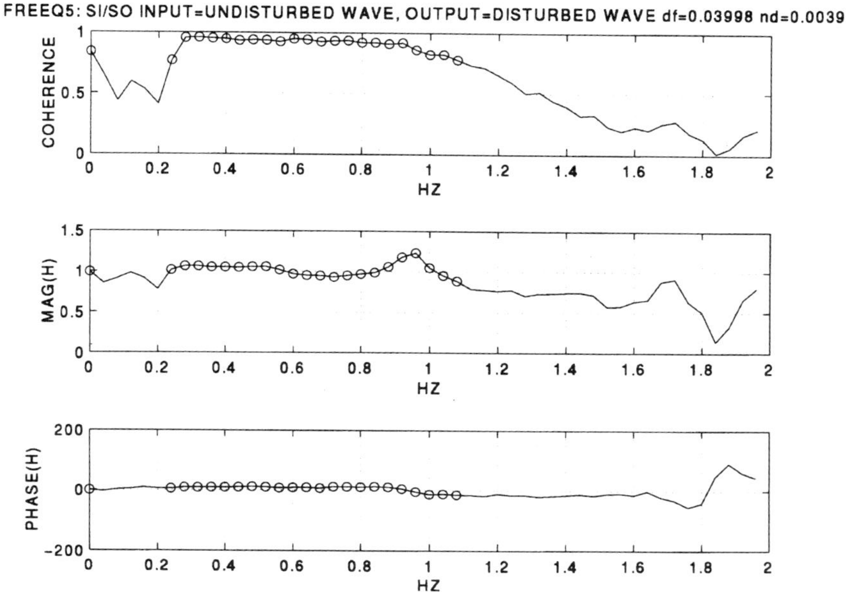

Figure 7.48.

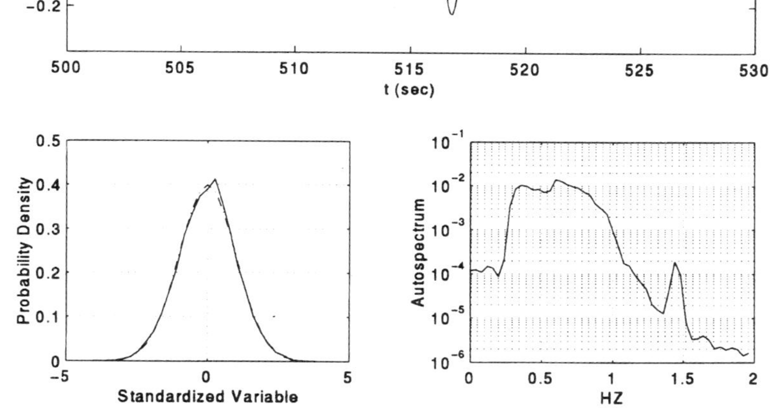

Figure 7.49.

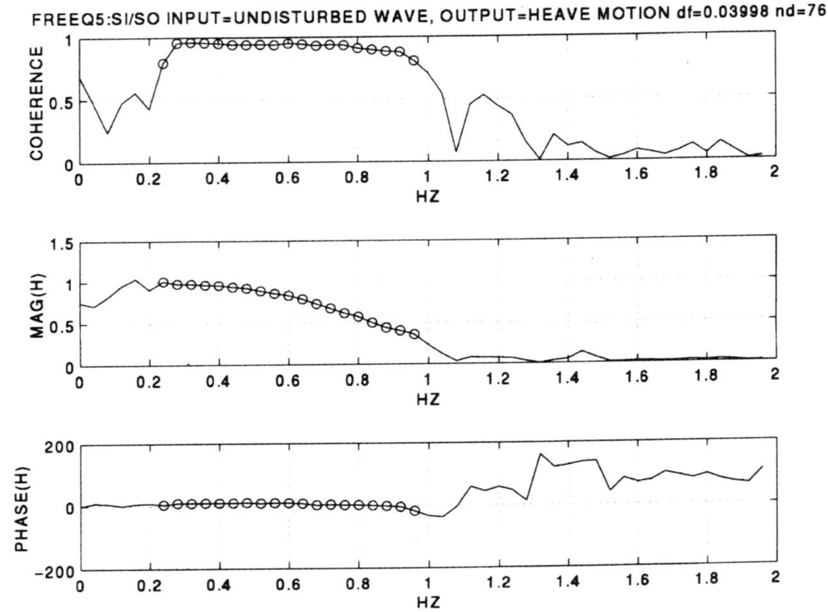

Figure 7.50.

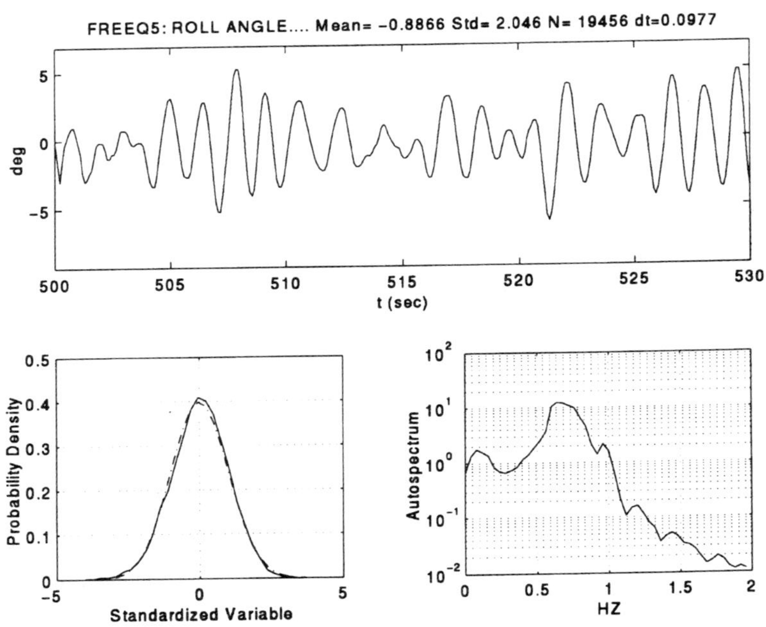

Figure 7.51.

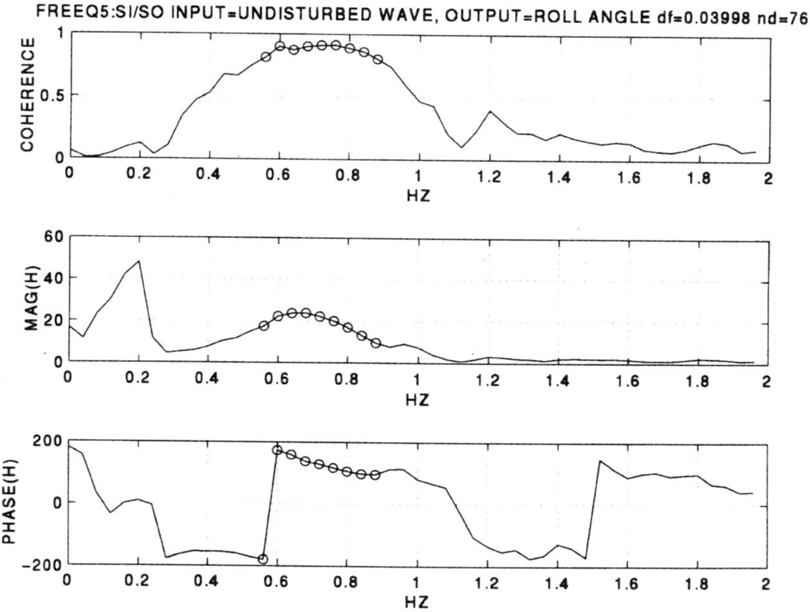

Figure 7.52.

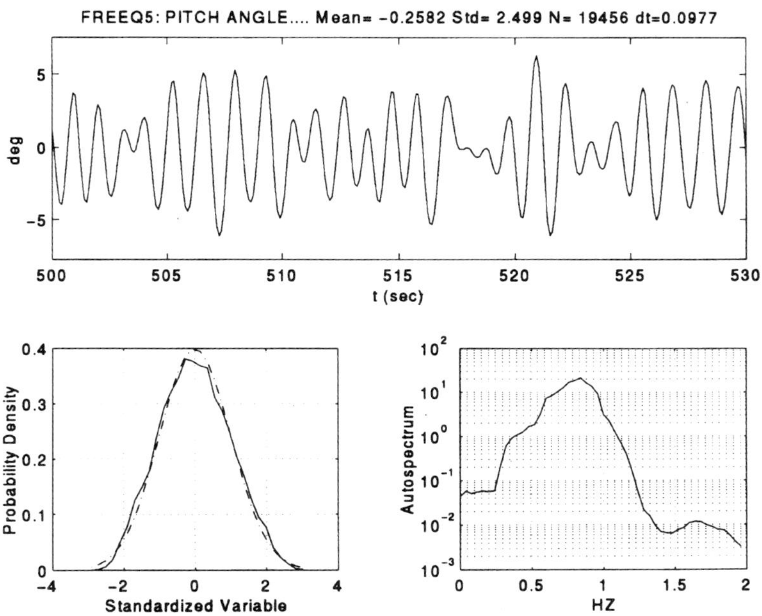

Figure 7.53.

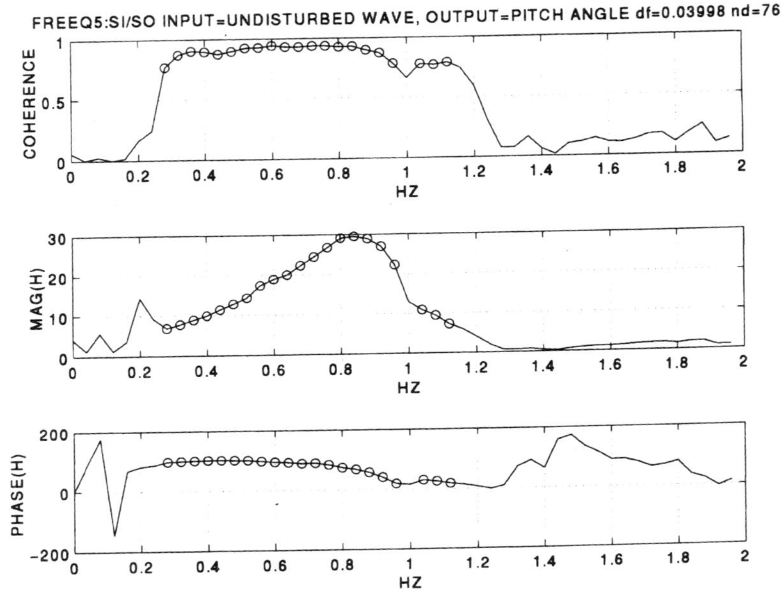

Figure 7.54.

7.5.3 Linear SI/SO Frequency Response Functions

Before proceeding with nonlinear system analysis of free barge large wave data, it is of interest to compare SI/SO based linear frequency response function estimates from small waves versus large waves. Such comparisons provide insight into the relative effectiveness of linear system results in high seas and indicate, in a preliminary manner, which responses might be affected by nonlinear behavior. In the discussion that follows, frequency response function estimates of magnitude and phase from small waves and large waves are compared with coherence functions greater than or equal to 0.75.

Free Barge Head Sea Data. Heave motion frequency response estimates, compared in Figure 7.55, indicate nearly identical estimates for low and high wave data. This suggests that nonlinearity is a negligible contributor to heave response dynamics in high-level head seas.

Pitch motion frequency response estimates, shown in Figure 7.56, as in the case of heave motion, indicate nearly identical estimates for low and high wave data. This suggests that nonlinearity is a negligible contributor to pitch response dynamics in high-level head seas.

Free Barge Beam Sea Data. Heave motion frequency response estimates, compared in Figure 7.57, indicate nearly identical estimates for low and high wave data in the frequency band below 1.05 Hz. The estimates indicate a slight deviation above 1.05 Hz. Since this trend is consistent over this frequency band, nonlinearity may be a contributor to heave response dynamics in high-level beam seas.

Roll motion frequency response estimates, shown in Figure 7.58, indicate a difference in magnitude in the frequency band associated with highest level frequency response. This suggests that nonlinearity is a contributor to roll response dynamics in high-level beam seas.

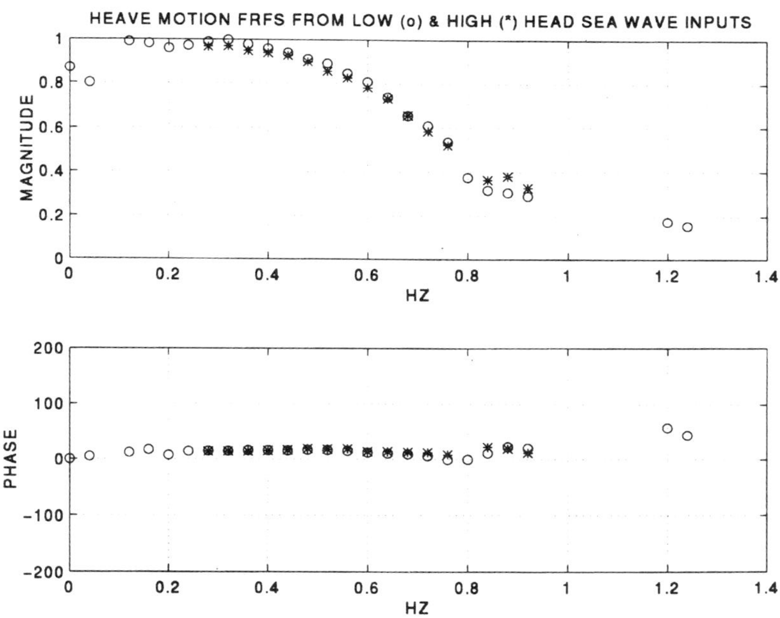

Figure 7.55.

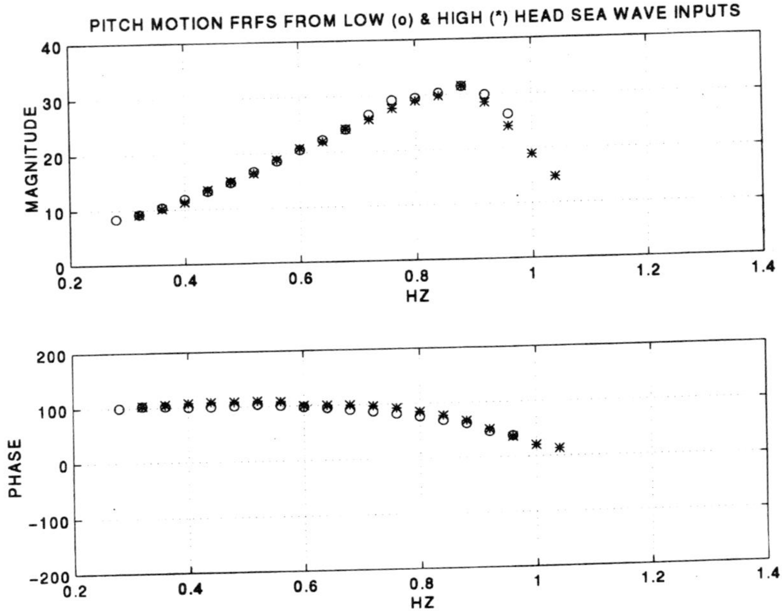

Figure 7.56.

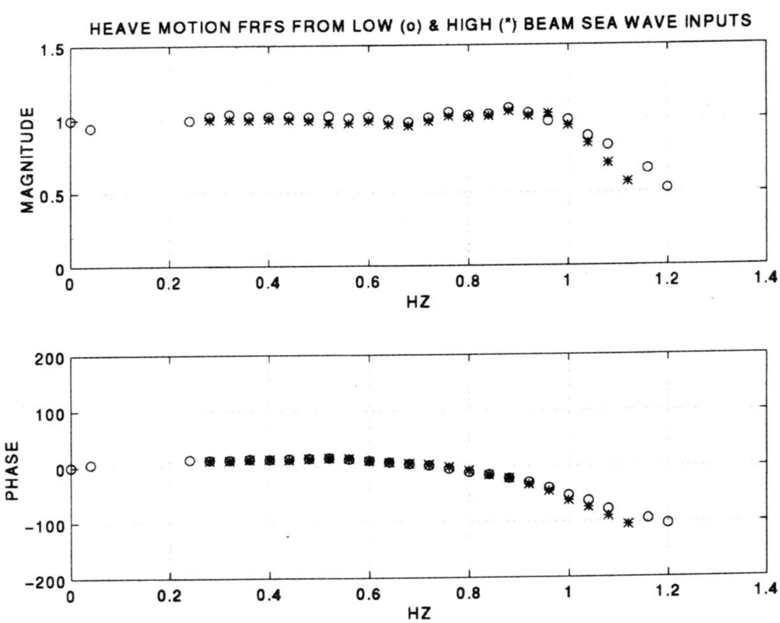

Figure 7.57.

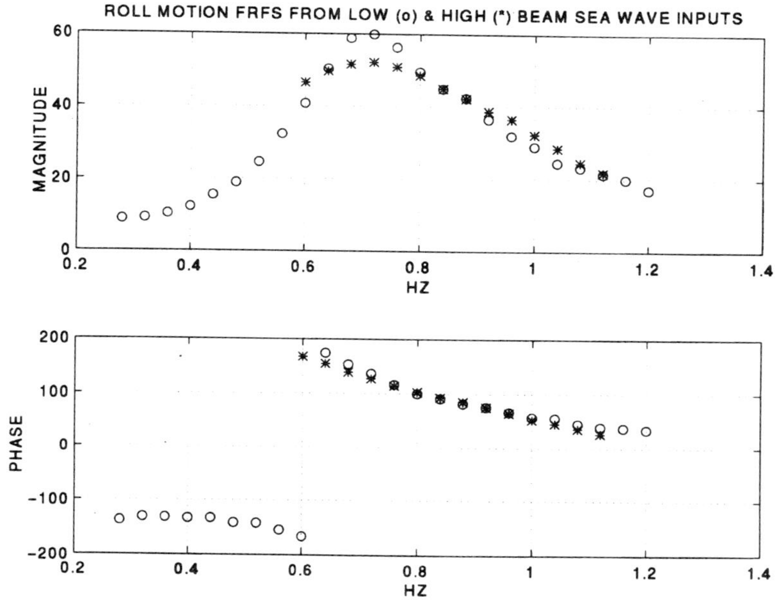

Figure 7.58.

Free Barge Quartering Sea Data. Heave motion frequency response estimates, compared in Figure 7.59, indicate nearly identical estimates for low and high wave data. As in the case of head seas, this suggests that nonlinearity is a negligible contributor to heave response dynamics, in high-level quartering seas.

Roll motion frequency response estimates, shown in Figure 7.60, indicate a difference in magnitude in the frequency band associated with the highest level frequency response. As in the case of beam seas, this suggests that nonlinearity is a contributor to roll response dynamics in high-level quartering seas.

Pitch motion frequency response estimates, shown in Figure 7.61, indicate a difference between low and high wave data in the frequency band above 0.85 Hz. This suggests that nonlinearity contributes to pitch response dynamic behavior in high-level quartering seas.

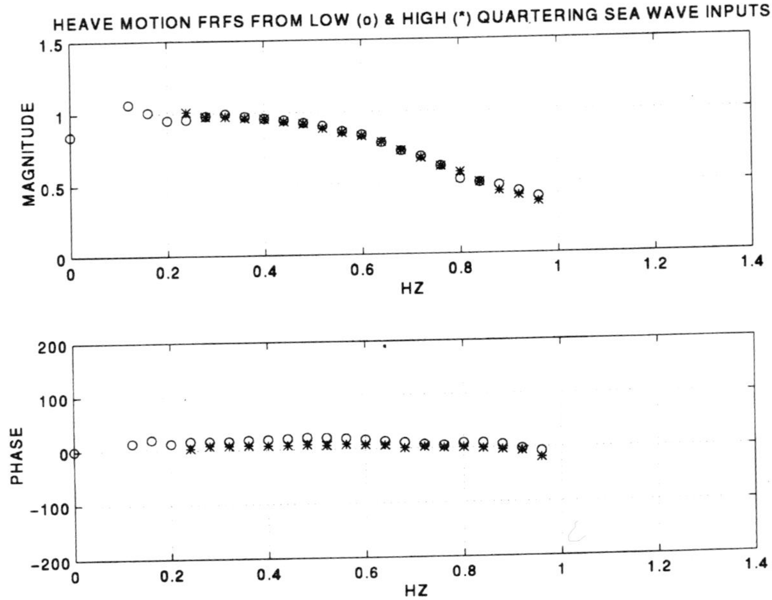

Figure 7.59.

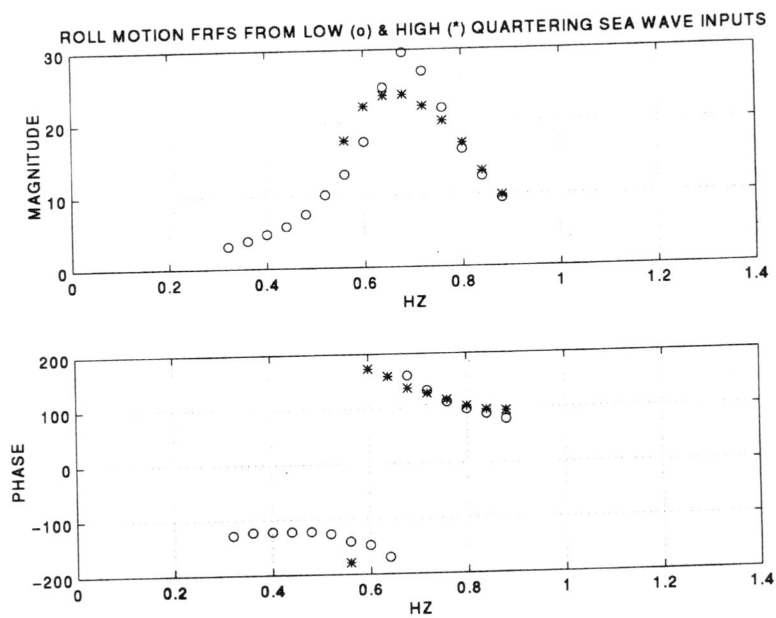

Figure 7.60.

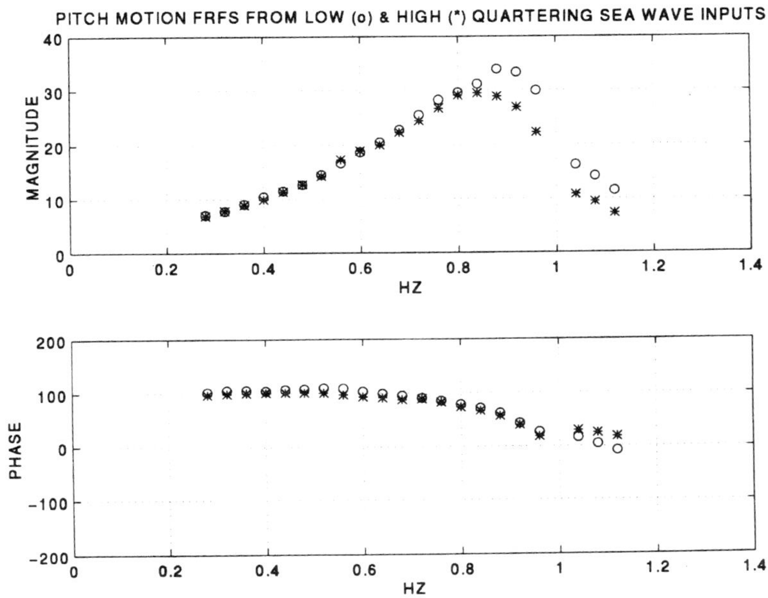

Figure 7.61.

7.5.4 Nonlinear MI/SO Analysis of Large Wave Head Sea Data

MI/SO spectral analyses have been performed on large wave response data from the viewpoint of both direct and reverse dynamic system models. A decision to focus on two specific nonlinear terms for the head sea case was made after numerical experimentation with many alternative nonlinear expressions. The two nonlinear terms are (1) a quadratic form of relative heave velocity (to describe hydrodynamic nonlinear damping) and (2) a quadratic form of pitch motion as proposed in Section 7.4.3 (to describe nonlinear stiffness). Results of the large wave, head sea data analyses are presented below as obtained from the free barge H5 data.

Direct Dynamic System Models, Two Inputs, H5 Data. Analysis for this case considers two inputs, namely, (1) the undisturbed wave motion, $w(t)$, and (2) the quadratic relative velocity. Using heave motion, $z(t)$, to express the output for the heave nonlinear differential equation, a Direct MI/SO analysis was performed. The cumulative coherence function is shown in Figure 7.62, and the linear and nonlinear frequency response functions are given in Figures 7.63 and 7.64, respectively.

Some improvement in coherence is exhibited from incorporation of the single nonlinearity over the 0.60–1.4 Hz frequency band. This might be expected since the ordinary coherence associated with the linear SI/SO case fell in this range (see Figure 7.36). The linear system frequency response, $H_1(f)$, is similar in magnitude and phase to the linear SI/SO result below 1.4 Hz, and the nonlinear system frequency response, $H_2(f)$, is of lower level in the 0.60–1.0 Hz frequency band with a $-90°$ phase angle. It is concluded here that the linear SI/SO model may be adequate to describe heave response dynamics in head seas for frequencies below 1.4 Hz. However, at about 1.45 Hz, a strong nonlinearity is clearly present in all three plots due to the nonlinear relative velocity term. Note that a narrow-band increase in the heave autospectrum of Figure 7.35 is also present at this frequency.

Using pitch motion, $\theta(t)$, to express the output for the pitch nonlinear differential equation, a Direct MI/SO analysis was conducted in a manner analogous to that described for heave motion. The cumulative coherence for that analysis is shown in Figure 7.65, and the linear and nonlinear frequency responses are given in Figures 7.66 and 7.67, respectively. Minor improvement in coherence is exhibited from incorporation of the single nonlinearity over the 0.25–1.0 Hz frequency band. The linear system frequency response, $H_1(f)$, is similar in form to the linear SI/SO result (see Figure 7.38), and the nonlinear system frequency response, $H_2(f)$, is of lower level with a peak at about 0.90 Hz.

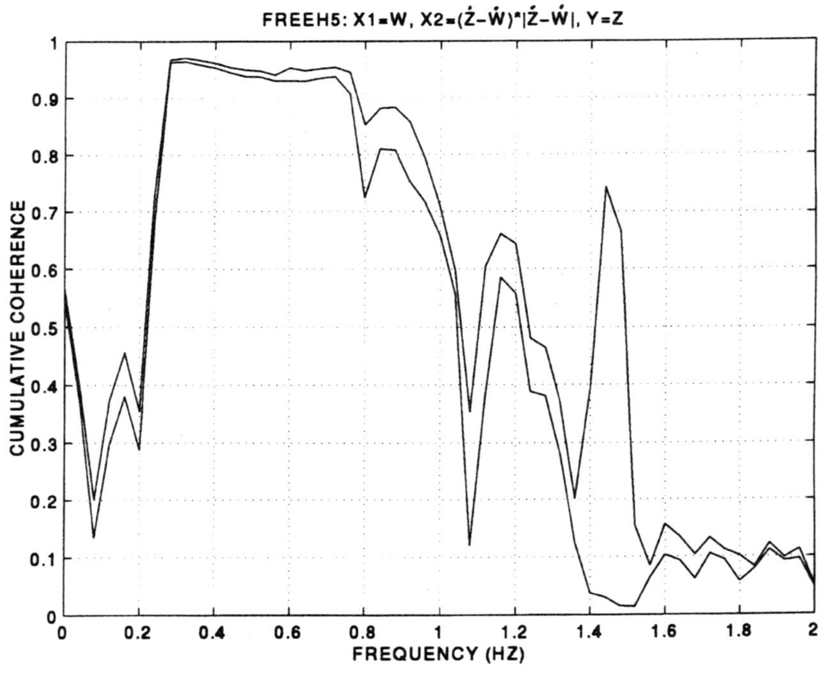

Figure 7.62.

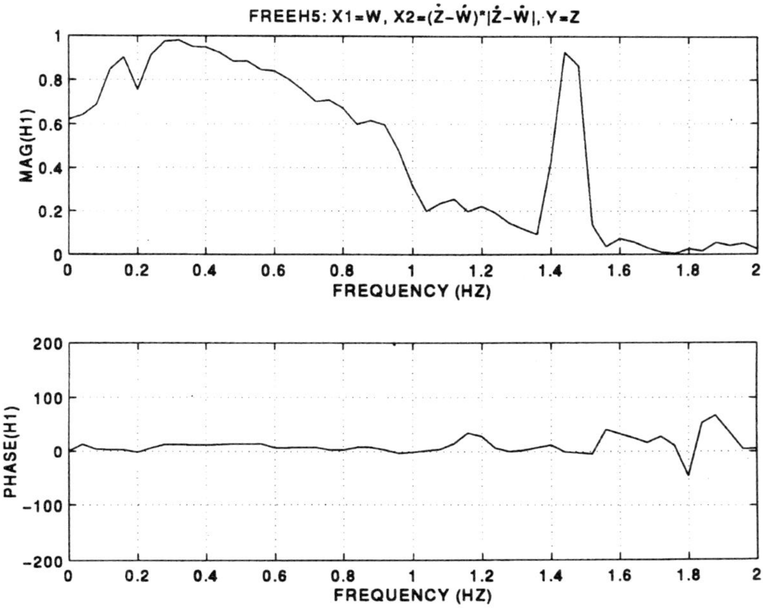

Figure 7.63.

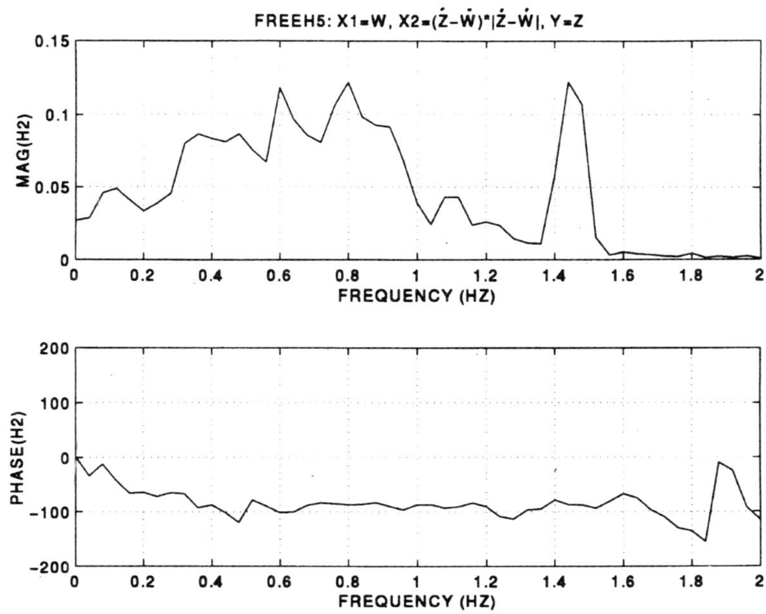

Figure 7.64.

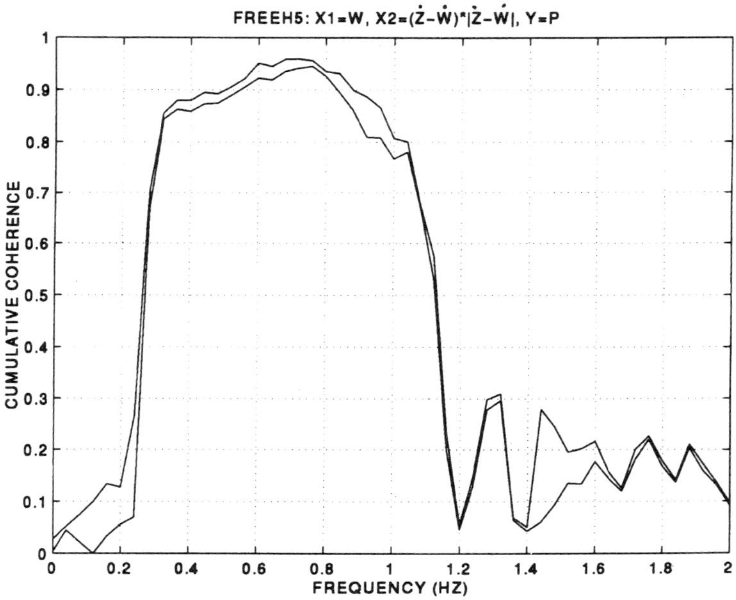

Figure 7.65.

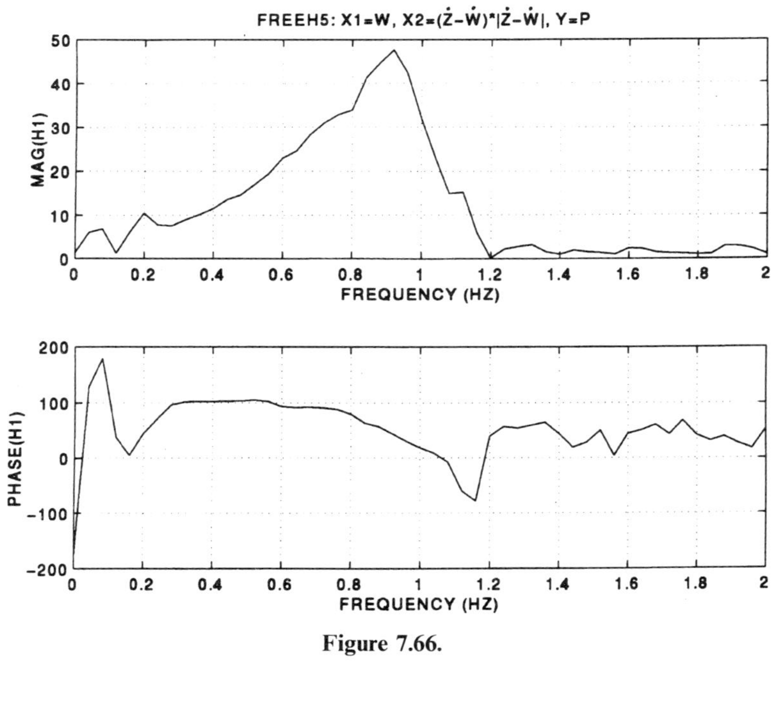

Figure 7.66.

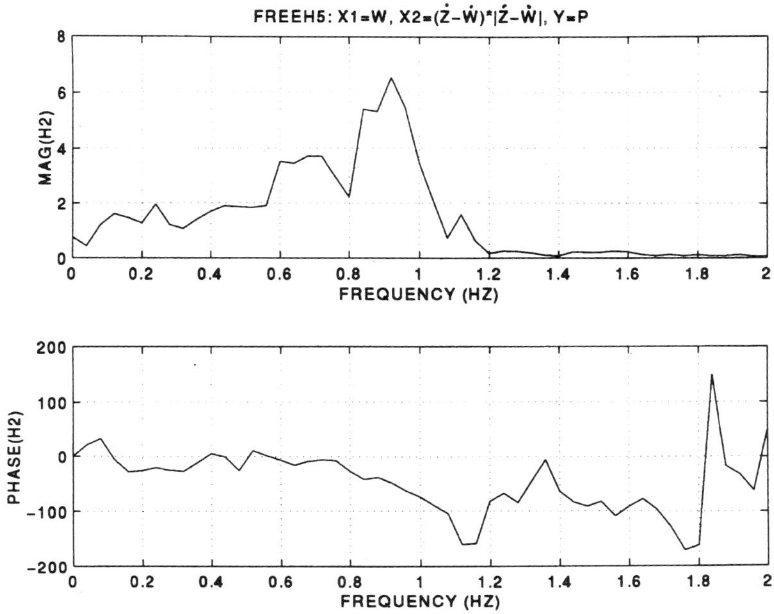

Figure 7.67.

Direct Dynamic System Models, Three Inputs, H5 Data. The present analysis considers three inputs, namely, (1) the undisturbed wave motion, $w(t)$, (2) the quadratic relative velocity, and (3) the quadratic pitch motion $\theta(t)|\theta(t)|$. Using heave motion, $z(t)$, to express the output for the heave nonlinear differential equation, a Direct MI/SO analysis was performed. The cumulative coherence function for that analysis is shown in Figure 7.68, and the linear and nonlinear frequency response functions are given in Figures 7.69, 7.70, and 7.71, respectively.

Significant improvement in coherence is exhibited from incorporation of the two nonlinearities over the 0.80–1.4 Hz frequency band. This suggests that nonlinearity, most pronounced in that particular frequency band, may be described in terms of the two hypothesized nonlinear terms. Below 1.4 Hz, the linear system frequency response, $H_1(f)$, is similar in magnitude and phase to the linear SI/SO result in Figure 7.36. All three frequency response functions show a strong peak at about 1.45 Hz that indicates a nonlinearity due to the relative velocity term.

Using pitch motion, $\theta(t)$, to express the output for the pitch nonlinear differential equation, a Direct MI/SO analysis was conducted in a manner analogous to that described for the three inputs and heave motion output. The cumulative coherence for that analysis is shown in Figure 7.72, and the linear and nonlinear frequency responses are given in Figures 7.73, 7.74, and 7.75, respectively. A definite improvement in coherence is exhibited from incorporation of the two nonlinear terms over the 0.25–1.4 Hz frequency band. The linear system frequency response, $H_1(f)$, is roughly similar to the linear SI/SO result (see Figure 7.38), and the nonlinear system frequency response functions, $H_2(f)$ and $H_3(f)$, are of lower levels with several peaks in the 0.60–1.1 Hz frequency band.

For the linear and nonlinear forces associated with the frequency response functions $H_1(f)$, $H_2(f)$ and $H_3(f)$, phase angles of zero indicate forces without damping, while phase angles of 90° or −90° indicate forces that are all damping.

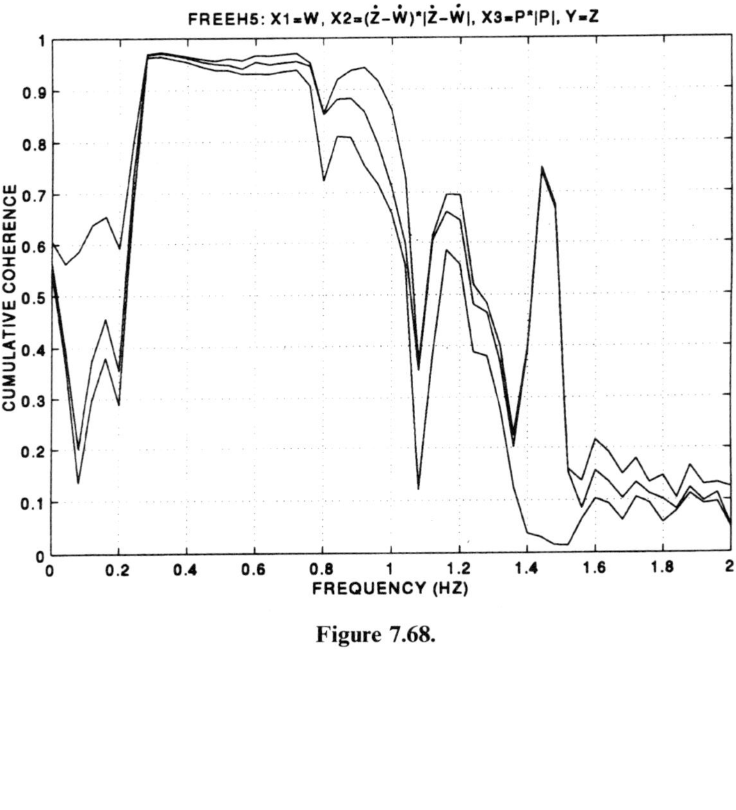

Figure 7.68.

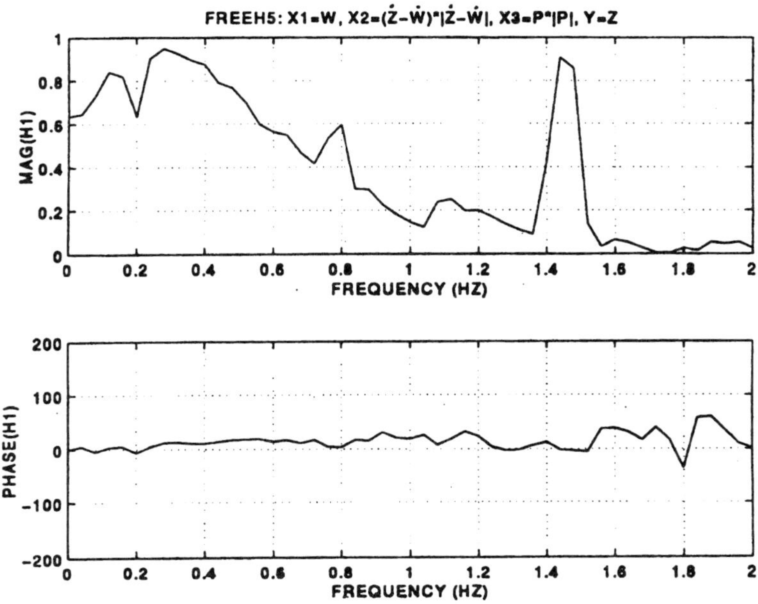

Figure 7.69.

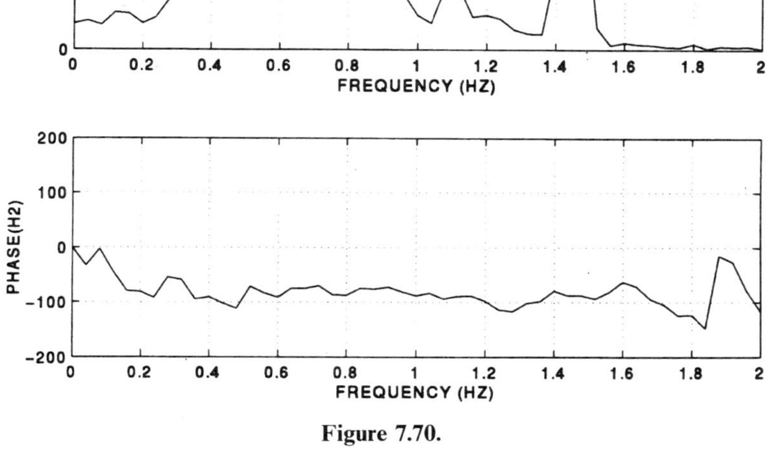

Figure 7.70.

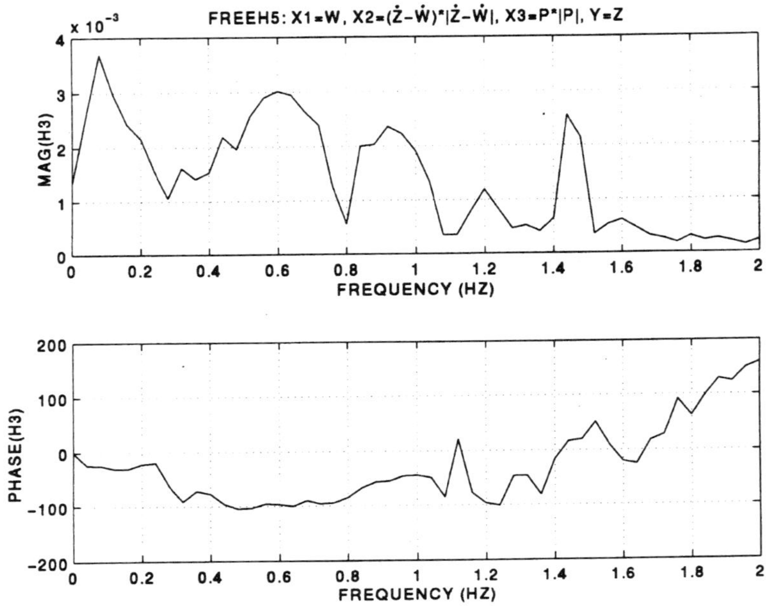

Figure 7.71.

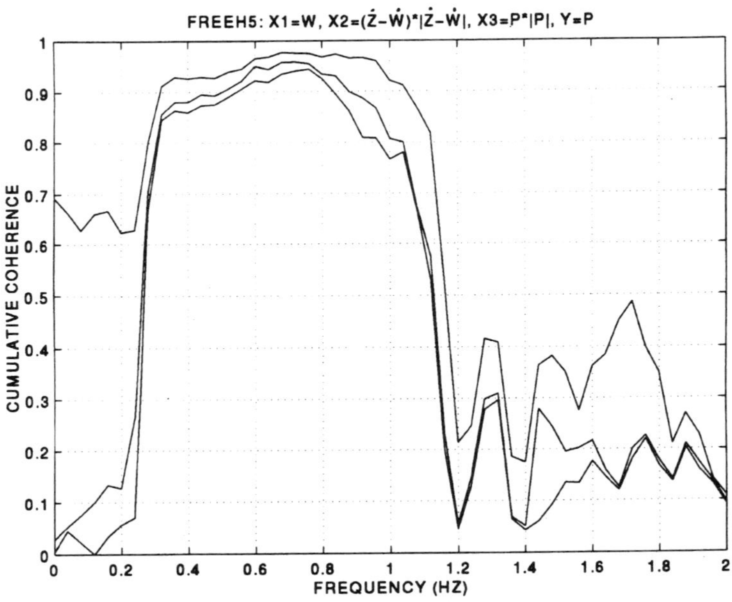

Figure 7.72.

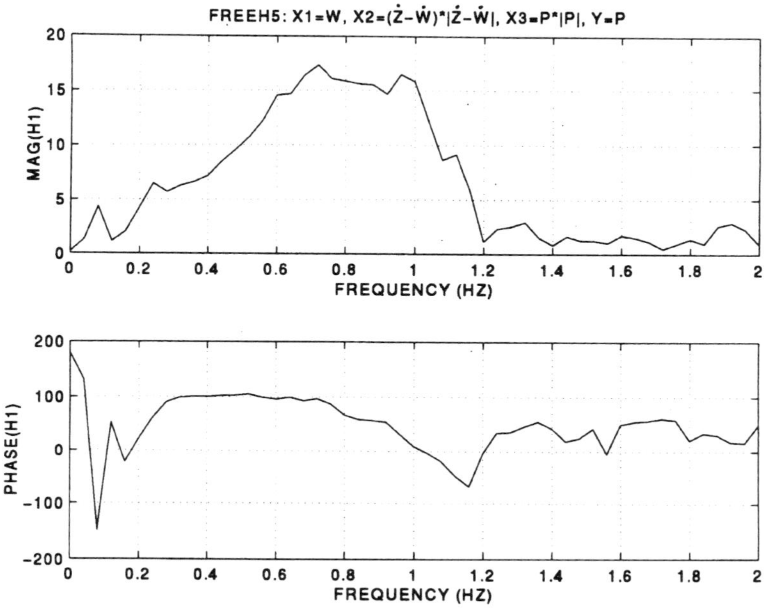

Figure 7.73.

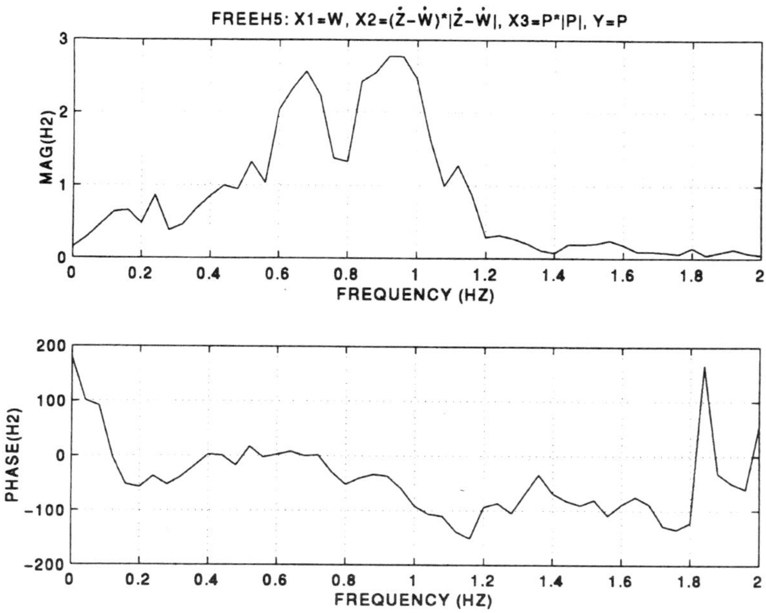

Figure 7.74.

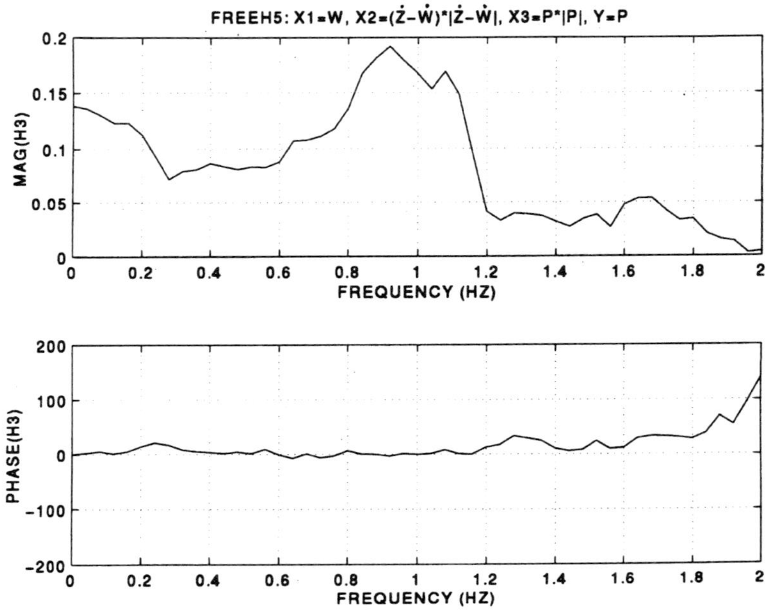

Figure 7.75.

Reverse Dynamic System Models, Two Inputs, H5 Data. The reverse dynamic system analysis considers here two inputs for the "heave" equation, namely, (1) heave motion, $z(t)$, and quadratic relative velocity. Using undisturbed wave motion, $w(t)$, to express the output for the heave nonlinear differential equation, a Reverse MI/SO analysis was performed. The cumulative coherence function for that analysis is shown in Figure 7.76, and the linear and nonlinear frequency response functions are given in Figures 7.77 and 7.78, respectively. Strong improvement in coherence is exhibited from incorporation of the single nonlinearity in the frequency band above 0.80 Hz. The linear system frequency response, $A_1(f)$, has an increasing magnitude for frequencies below 1.2 Hz with a 0° phase angle. The nonlinear frequency response, $A_2(f)$, is of lower level with a 90° phase angle, and a peak magnitude at about 0.80 Hz. For frequencies below 0.80 Hz, it appears reasonable to conclude that the linear SI/SO model adequately describes heave response dynamics. Above 0.80 Hz, the proposed nonlinear term should be included.

The reverse dynamic system analysis considers two inputs for the "pitch" equation, namely, (1) pitch motion, $\theta(t)$, and (2) quadratic relative velocity. Using undisturbed wave motion, $w(t)$, to express the output for the pitch nonlinear differential equation, a Reverse MI/SO analysis was conducted in a manner analogous to that described for the reverse heave motion equation. The cumulative coherence for that analysis is shown in Figure 7.79, and the linear and nonlinear frequency responses are given in Figures 7.80 and 7.81, respectively. Marked improvement in coherence is exhibited from incorporation of the single nonlinearity over the 0.8–1.4 Hz frequency band. The linear system frequency response, $A_1(f)$, is similar in form to the inverse of the linear SI/SO result below 1.0 Hz (see Figure 7.38). The nonlinear system frequency response, $A_2(f)$, has a 90° phase angle for frequencies above 0.30 Hz. This indicates a nonlinear damping force. A strong nonlinearity appears near 1.45 Hz.

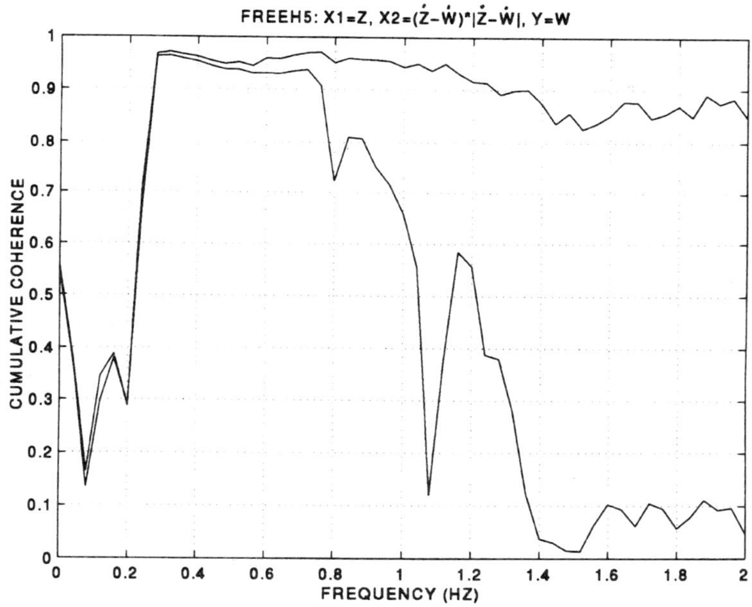

Figure 7.76.

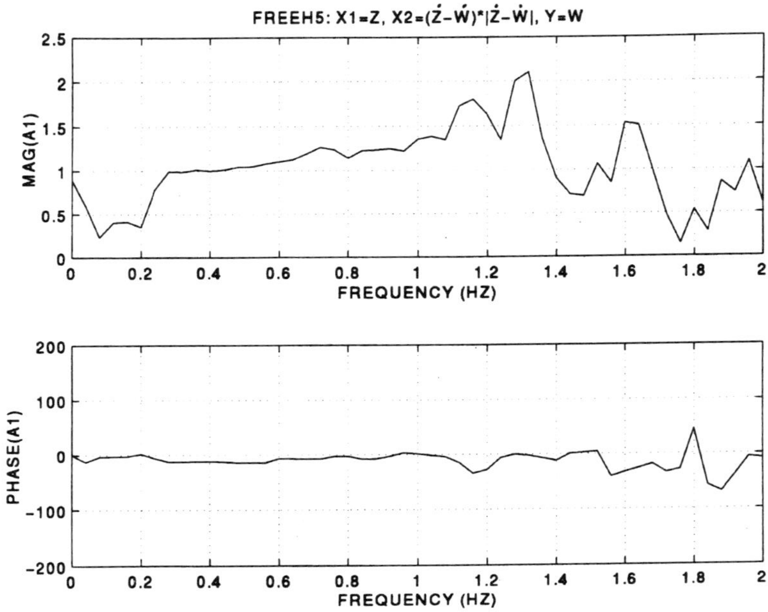

Figure 7.77.

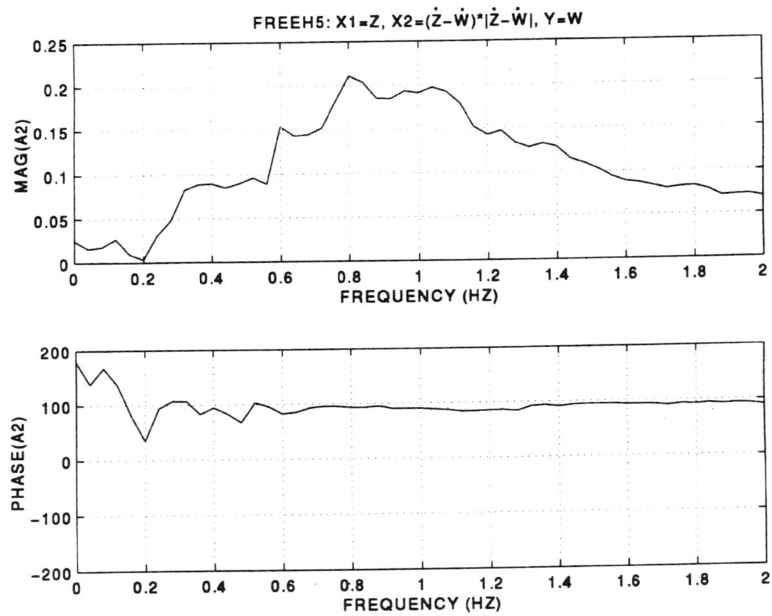

Figure 7.78.

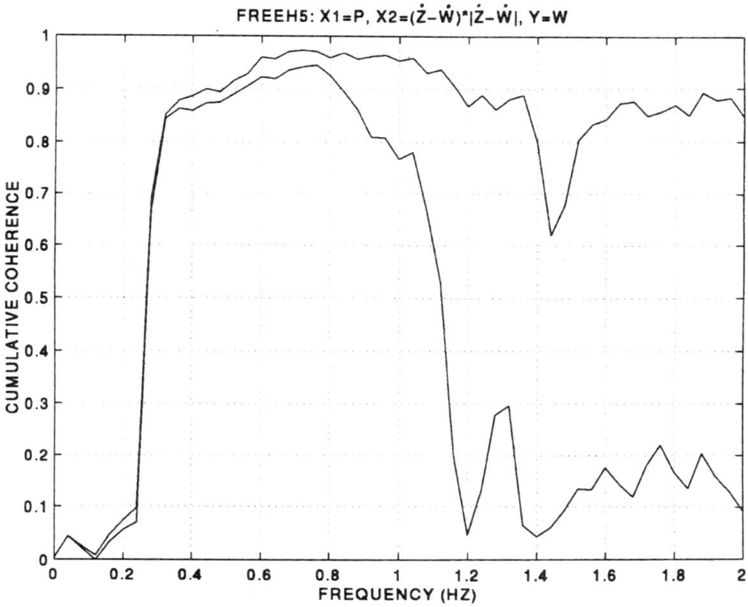

Figure 7.79.

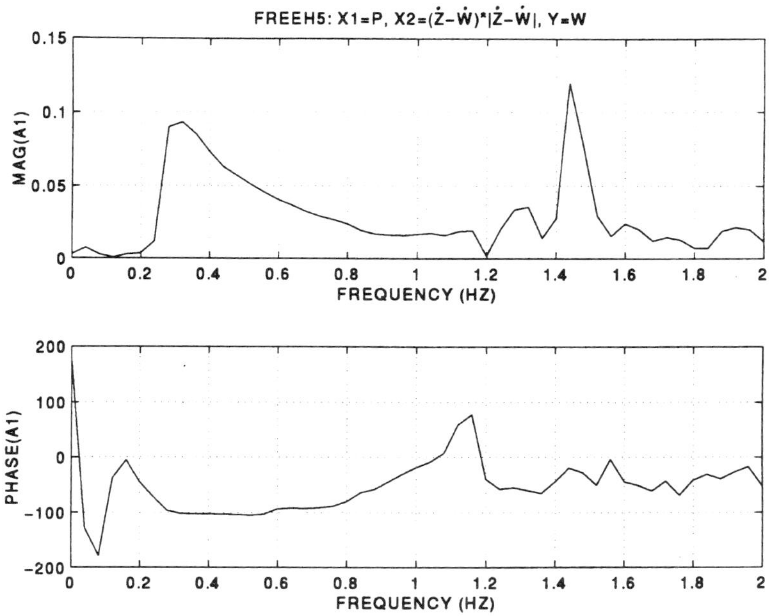

Figure 7.80.

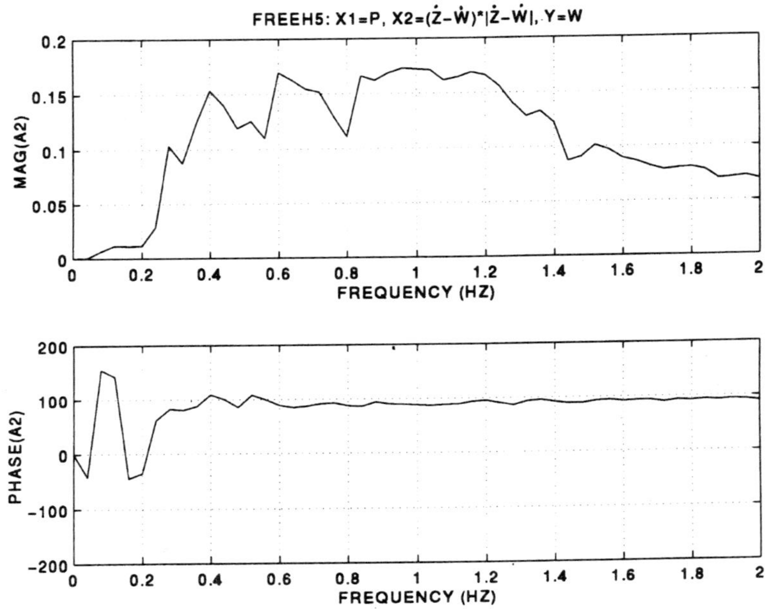

Figure 7.81.

Reverse Dynamic System Models, Three Inputs, H5 Data. The reverse dynamic model for the "heave" equation employs three inputs, namely, (1) heave motion, $z(t)$, (2) quadratic relative velocity, and (3) quadratic pitch motion. Using the undisturbed wave motion, $w(t)$, as the output for this model, a Reverse MI/SO analysis was performed. The cumulative coherence function for this analysis is shown in Figure 7.82, and the linear and nonlinear frequency response functions are given in Figures 7.83, 7.84, and 7.85, respectively. No improvement in coherence is exhibited from incorporation of the second nonlinearity over the 0.25–2.0 Hz frequency band. Therefore, the previous two-input reverse model is judged adequate. There is no need for further consideration of this three-input/single-output "heave" nonlinear model.

The reverse dynamic model for the "pitch" equation employs three inputs, namely, (1) pitch motion, $\phi(t)$, (2) quadratic relative velocity, and (3) quadratic pitch motion. Using the undisturbed wave motion, $w(t)$, as the output for this model, a Reverse MI/SO analysis was performed. The cumulative coherence for this analysis is shown in Figure 7.86, and the linear and nonlinear frequency responses are given in Figures 7.87, 7.88, and 7.89, respectively. No improvement in coherence is exhibited from incorporation of the second nonlinearity over the 0.25–2.0 Hz frequency band. Therefore, the previous two-input reverse model is judged adequate. There is no need for further consideration of this three-input reverse nonlinear model.

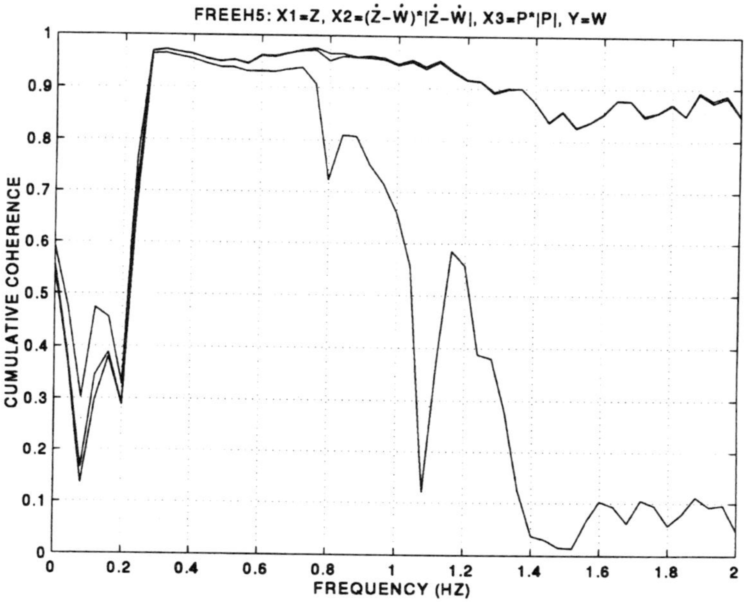

Figure 7.82.

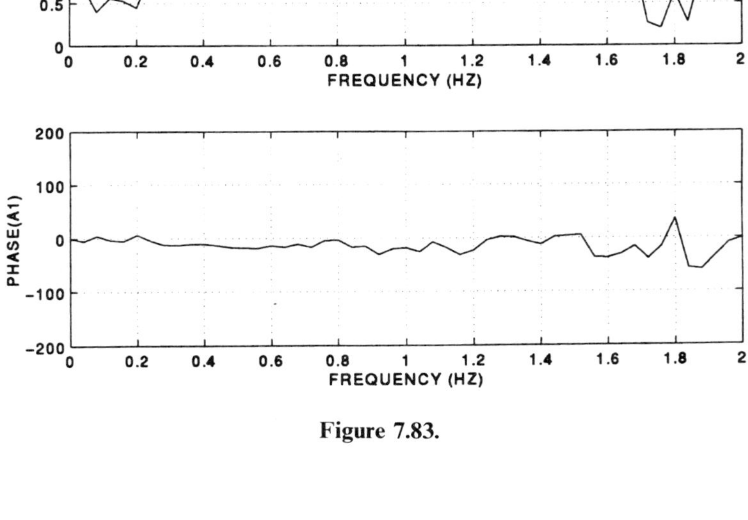

Figure 7.83.

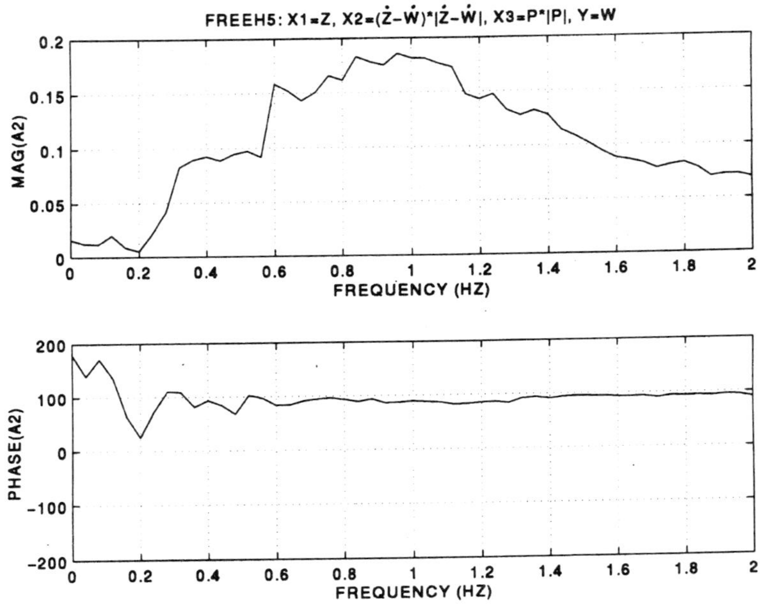

Figure 7.84.

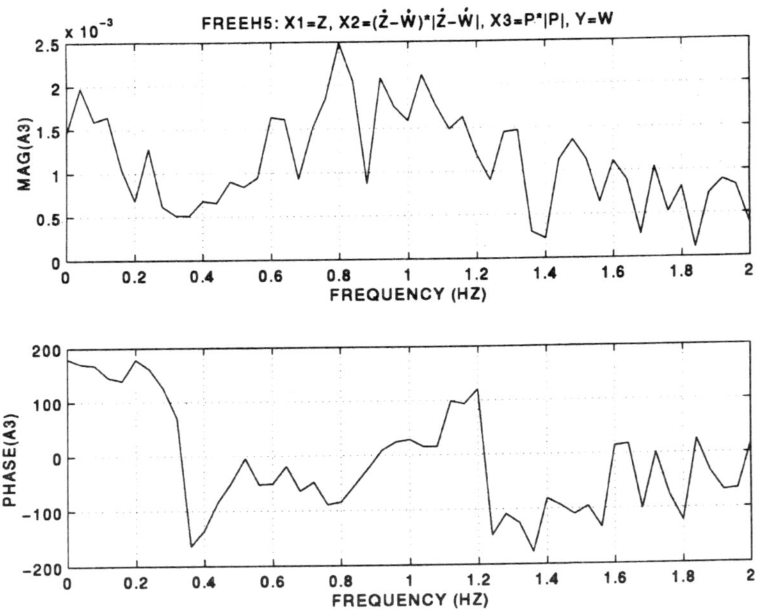

Figure 7.85.

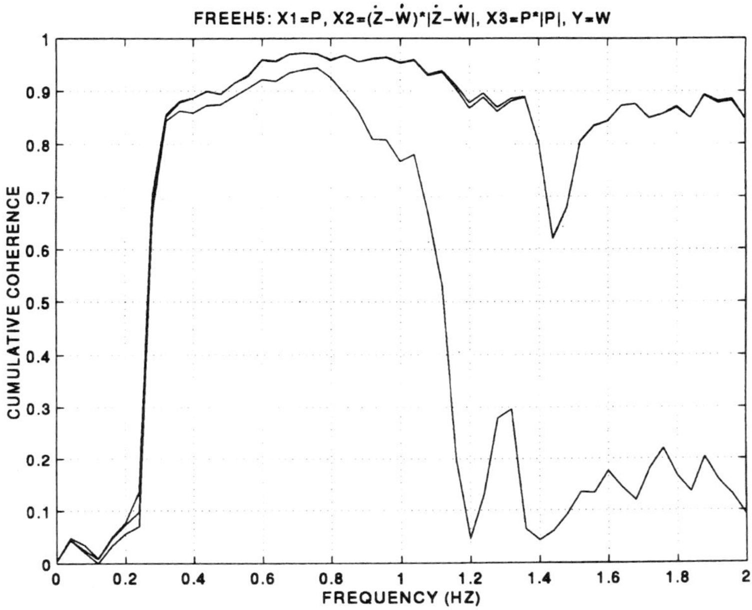

Figure 7.86.

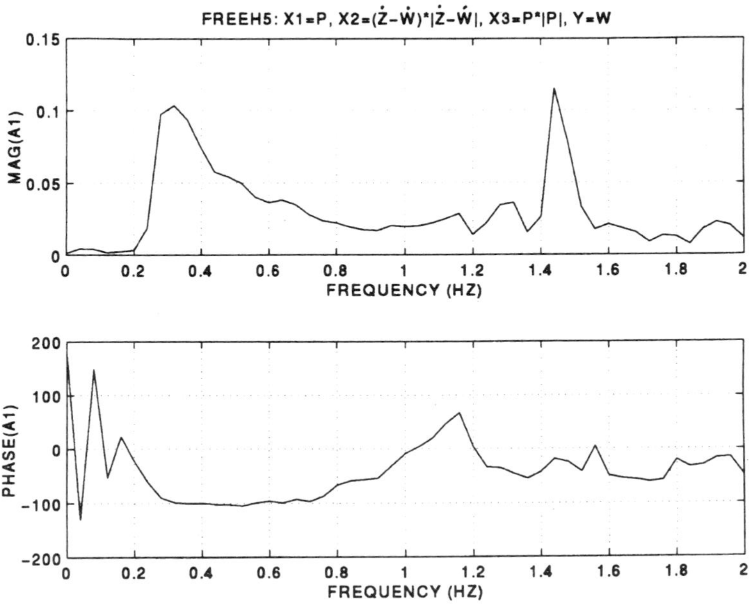

Figure 7.87.

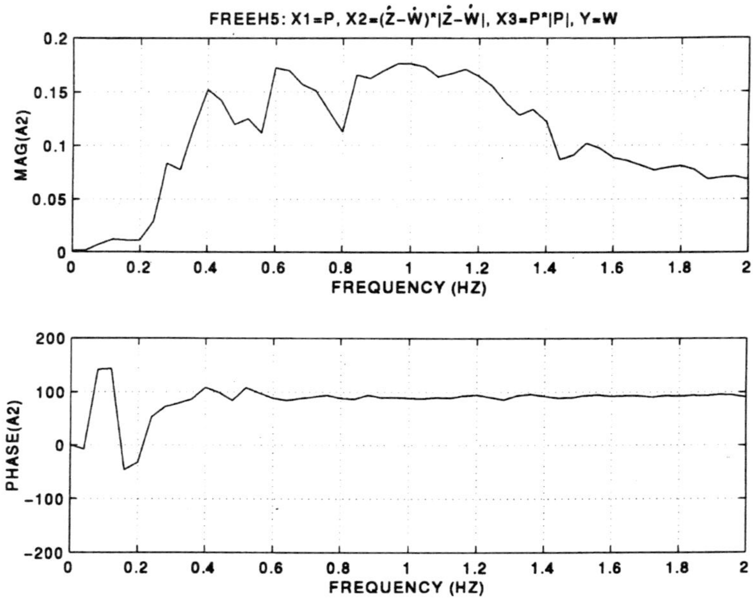

Figure 7.88.

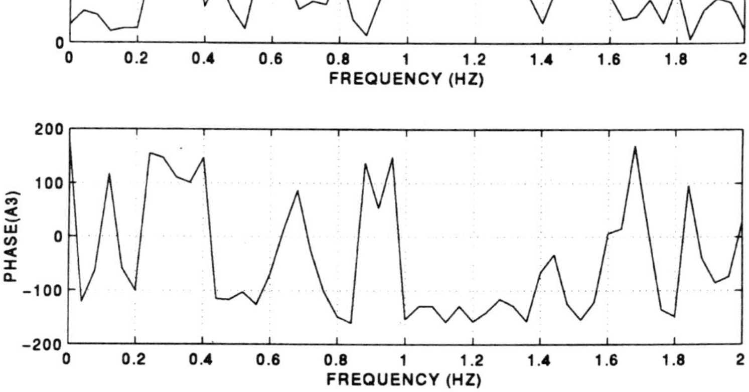

Figure 7.89.

7.5.5 Nonlinear MI/SO Analysis of Large Wave Beam Sea Data

In a manner similar to that conducted for head seas, MI/SO spectral analyses were carried out on large wave, beam sea response data from the viewpoint of both direct and reverse dynamic system models. Attention was focused on two specific nonlinear terms (after numerical experimentation with many alternative nonlinear expressions), namely, (1) a quadratic form of relative heave velocity and (2) a quadratic form of roll motion as proposed in Section 7.4.3. Results of the large wave, beam sea data analyses are presented below as obtained from the free barge B5 data.

Direct Dynamic System Models, Two Inputs, B5 Data. Direct model analysis for the heave equation considers two inputs, namely, (1) the undisturbed wave motion, $w(t)$, and (2) the quadratic relative velocity. Using heave motion, $z(t)$, to express the output, a Direct MI/SO analysis gives the cumulative coherence function shown in Figure 7.90, with the linear and nonlinear frequency response functions shown in Figures 7.91 and 7.92, respectively.

Marked improvement in coherence is exhibited due to incorporation of the single nonlinearity over the 0.30–1.2 Hz frequency band. This suggests that the quadratic relative velocity term is a significant contributor to heave dynamics in a beam sea. The linear system frequency response, $H_1(f)$, is similar in magnitude and phase to the linear SI/SO result (see Figure 7.43). The nonlinear system frequency response, $H_2(f)$, has a broad peak centered around 0.80 Hz and has about a -90° phase angle. A nonlinear effect is indicated at 1.45 Hz.

Using roll motion, $\phi(t)$, to express the output for the roll nonlinear differential equation, a Direct MI/SO analysis was conducted in a manner analogous to that described for heave motion. The cumulative coherence for that analysis is shown in Figure 7.93, and the linear and nonlinear frequency responses are given in Figures 7.94 and 7.95, respectively. Improvement in coherence is exhibited from incorporation of the single nonlinearity over the 0.6–1.2 Hz frequency band, indicating the significance of the relative velocity nonlinearity. The linear system frequency response, $H_1(f)$, is similar in form to the linear SI/SO result (see Figure 7.45), and the nonlinear system frequency response, $H_2(f)$, is of lower level with the phase angle shifting with increasing frequency from 90° at 0.60 Hz toward zero at 1.0 Hz.

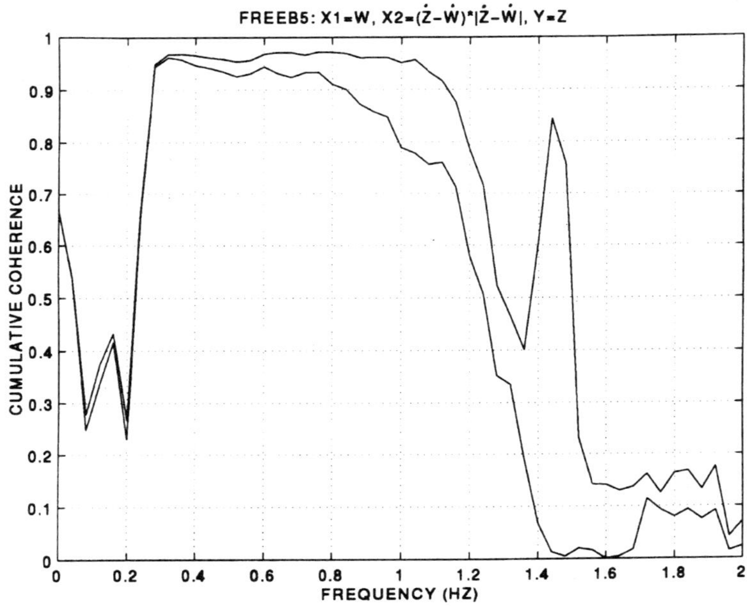

Figure 7.90.

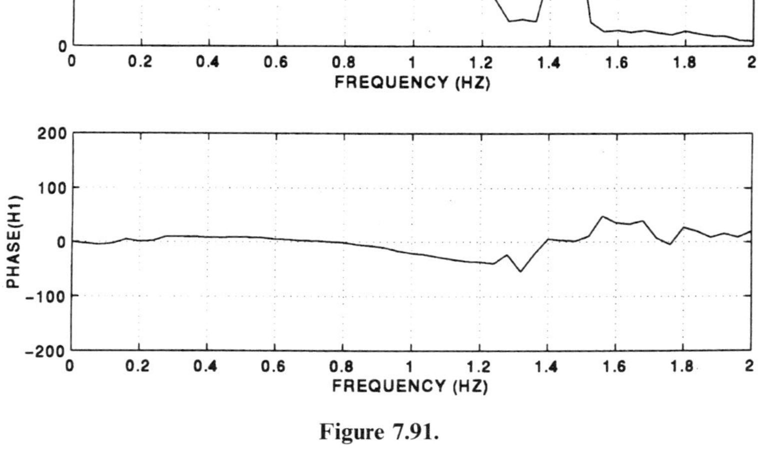

Figure 7.91.

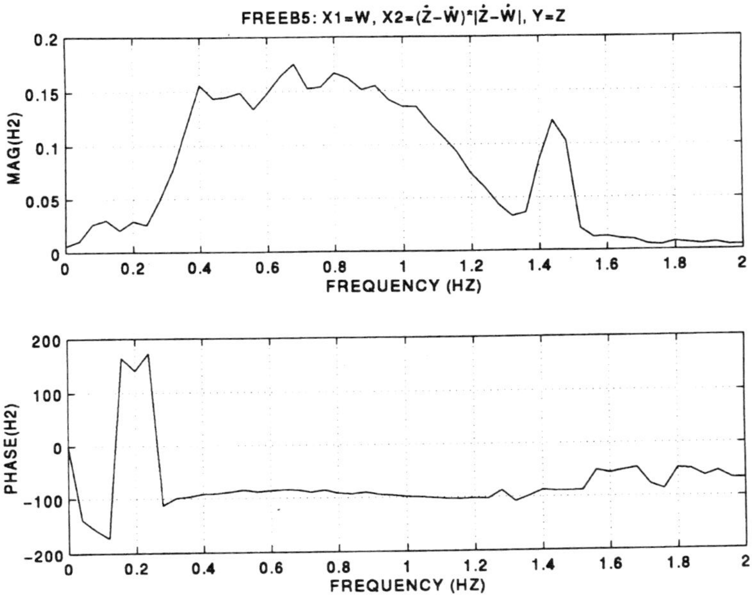

Figure 7.92.

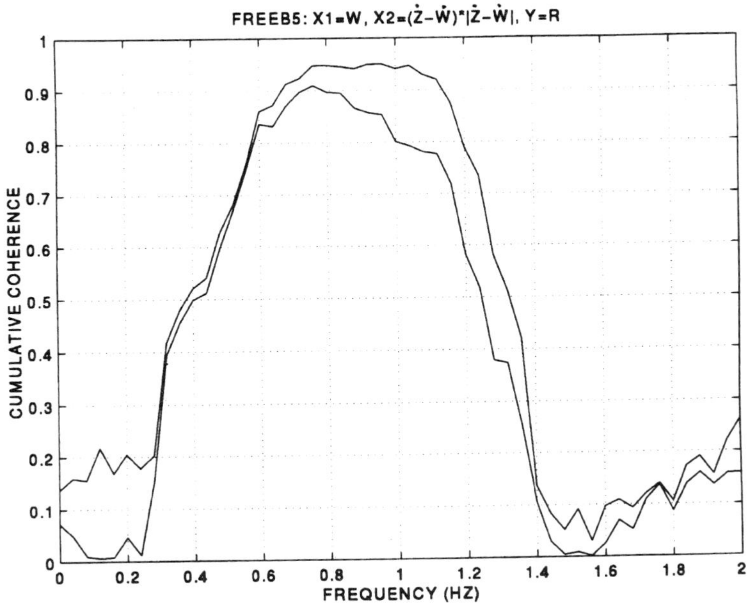

Figure 7.93.

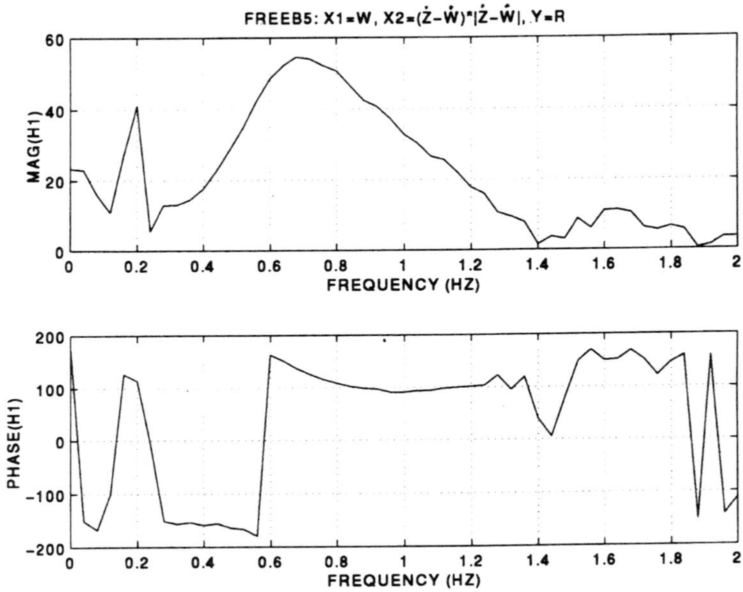

Figure 7.94.

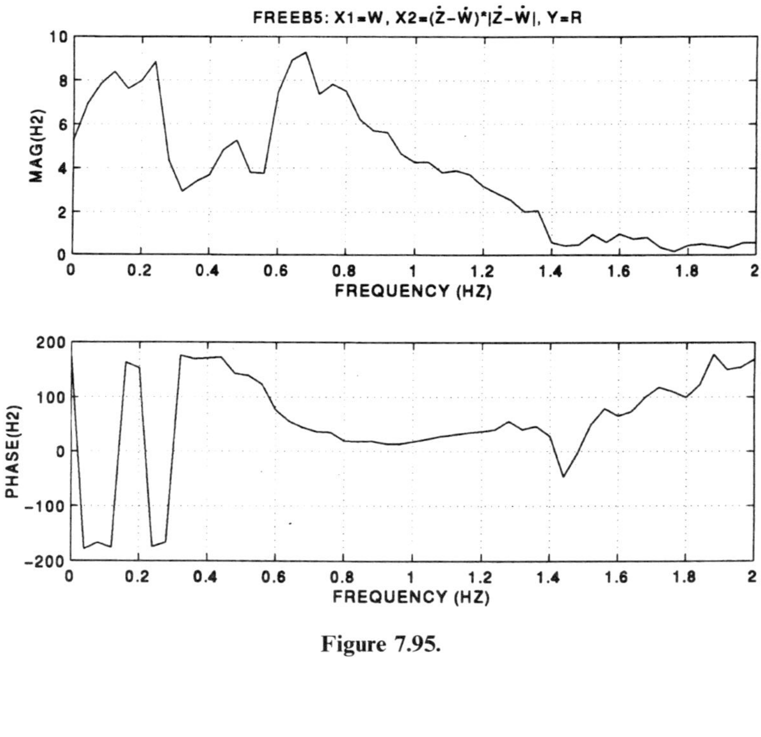

Figure 7.95.

Direct Dynamic System Models, Three Inputs, B5 Data. Direct analysis of the heave dynamics in a beam sea considers three inputs, namely, (1) undisturbed wave motion, $w(t)$, (2) quadratic relative velocity, and (3) quadratic roll motion, $\phi(t)|\phi(t)|$. Using heave motion, $z(t)$, to express the output for the heave nonlinear differential equation, a Direct MI/SO analysis gives the cumulative coherence function shown in Figure 7.96, with the linear and nonlinear frequency response functions in Figures 7.97, 7.98, and 7.99, respectively. Some improvement in coherence is exhibited from incorporation of the roll nonlinearity below 0.30 Hz and above 1.2 Hz. As before, a nonlinear effect is indicated at about 1.45 Hz.

Using roll motion, $\phi(t)$, to express the output in a beam sea for the roll nonlinear equation, a Direct MI/SO analysis was conducted in a manner analogous to that described for the three inputs and heave motion output. The cumulative coherence for that analysis is shown in Figure 7.100, with the linear and nonlinear frequency responses in Figures 7.101, 7.102, and 7.103, respectively. General improvement in the cumulative coherence function is exhibited from incorporation of the two nonlinear terms over the 0.0–1.8 Hz frequency band.

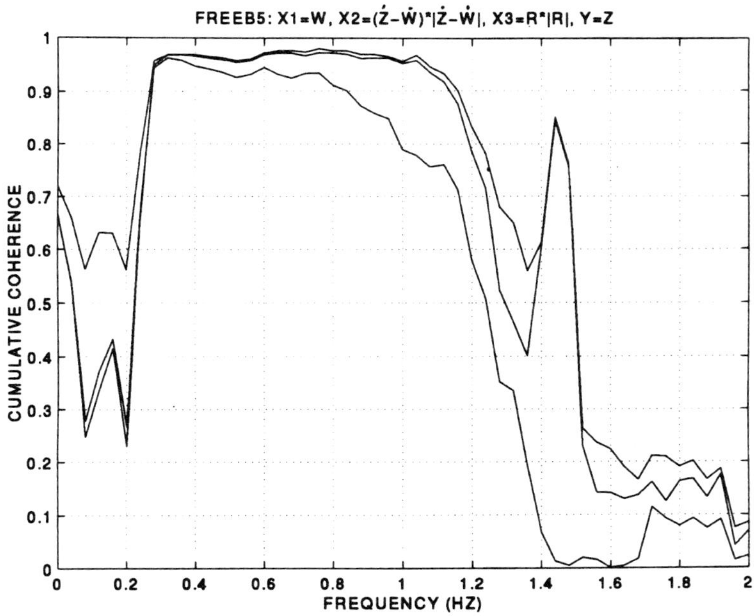

Figure 7.96.

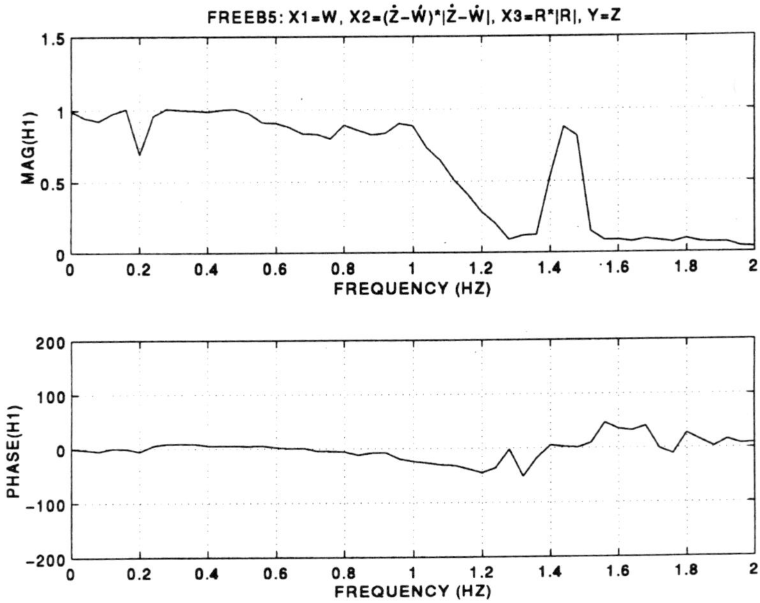

Figure 7.97.

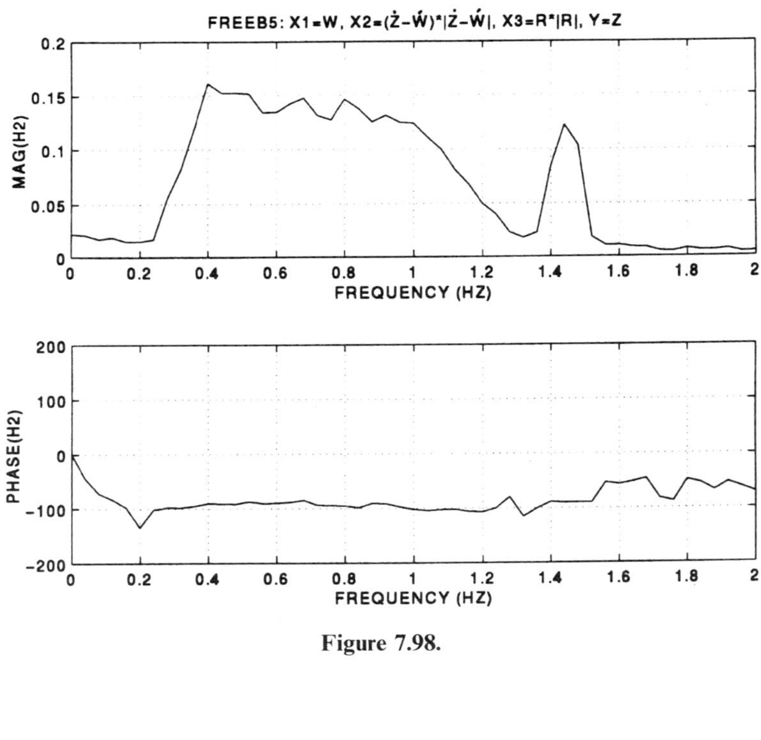

Figure 7.98.

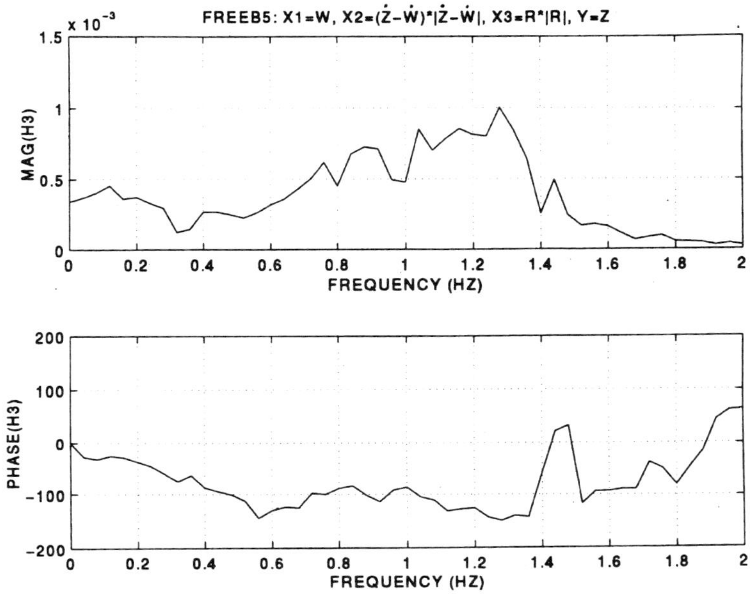

Figure 7.99.

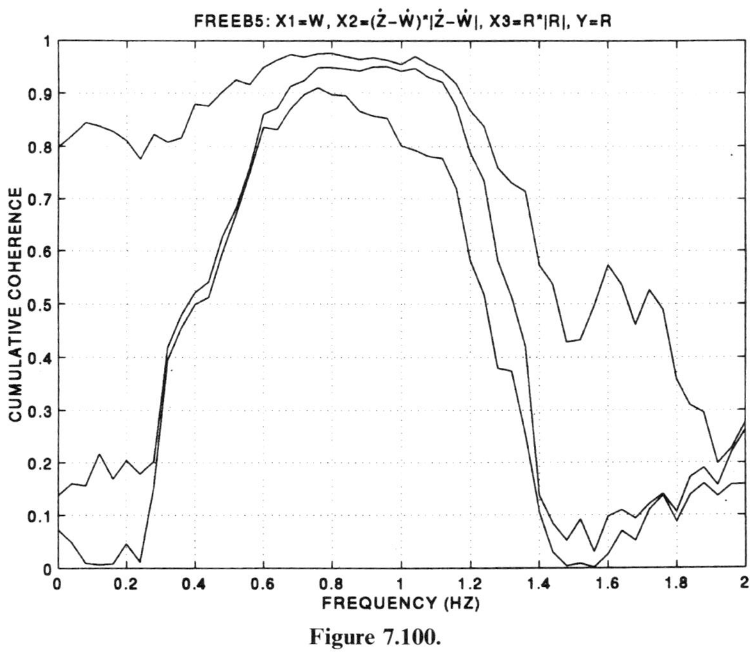

Figure 7.100.

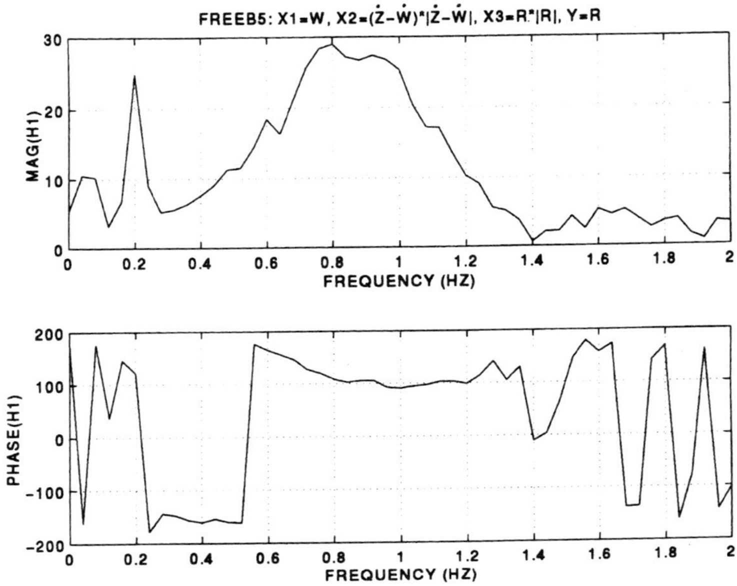

Figure 7.101.

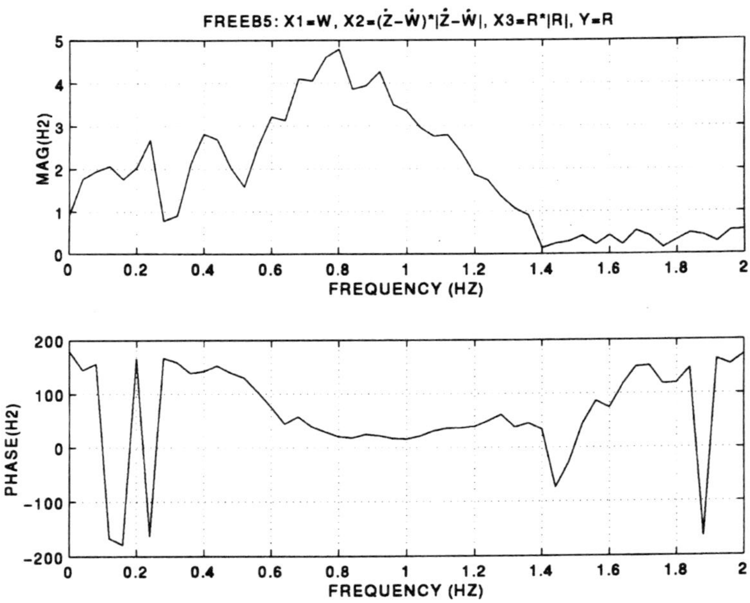

Figure 7.102.

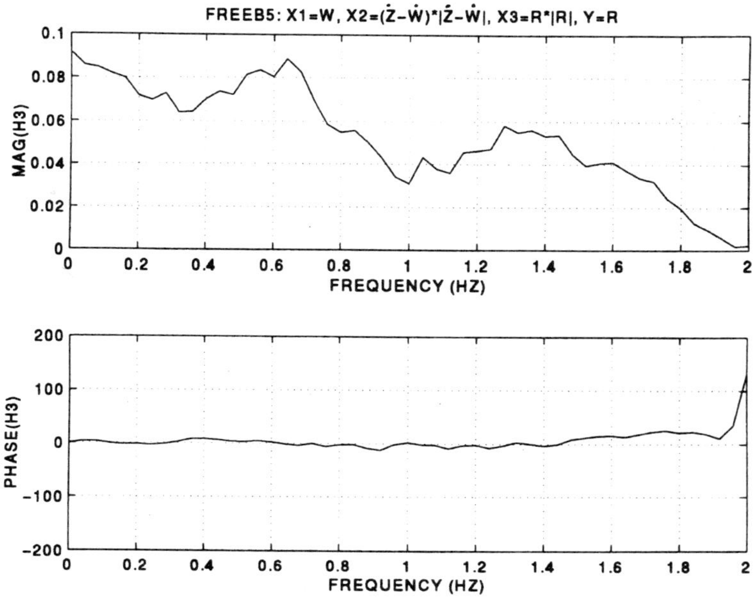

Figure 7.103.

Reverse Dynamic System Models, Two Inputs, B5 Data. The reverse dynamic system analysis considers two inputs for the "heave" equation, namely, (1) the heave motion, $z(t)$, and (2) the quadratic relative velocity. Using undisturbed wave motion, $w(t)$, to express the output for the heave nonlinear differential equation, a Reverse MI/SO analysis gives the cumulative coherence function shown in Figure 7.104, with the linear and nonlinear frequency response functions in Figures 7.105 and 7.106, respectively. Clear improvement in coherence is exhibited from incorporation of the single nonlinearity over the 0.30–2.0 Hz frequency band. The linear system frequency response, $A_1(f)$, is somewhat similar to the inverse magnitude and phase of the linear SI/SO frequency response function (see Figure 7.43). The nonlinear system frequency response, $A_2(f)$, has a lower uniform level with a 90° phase angle from 0.30 to 2.0 Hz.

The reverse dynamic system analysis considers two inputs for the "roll" equation, namely, (1) roll motion, $\phi(t)$, and (2) quadratic relative velocity. Using undisturbed wave motion, $w(t)$, to express the output for the roll nonlinear differential equation, a Reverse MI/SO analysis was conducted in a manner analogous to that described for the reverse heave motion equation. The cumulative coherence for that analysis is shown in Figure 7.107, and the linear and nonlinear frequency responses are given in Figures 7.108 and 7.109, respectively. Marked improvement in coherence is exhibited from incorporation of the single nonlinearity over the 0.60–1.4 Hz frequency band. For frequencies below 1.0 Hz, the linear system frequency response $A_1(f)$ is similar in form to the inverse of the linear SI/SO result (see Figure 7.45). The nonlinear system frequency response, $A_2(f)$, is roughly uniform with a 90° phase angle for frequencies from 0.60–1.40 Hz. Note that a strong nonlinearity shows at about 1.45 Hz.

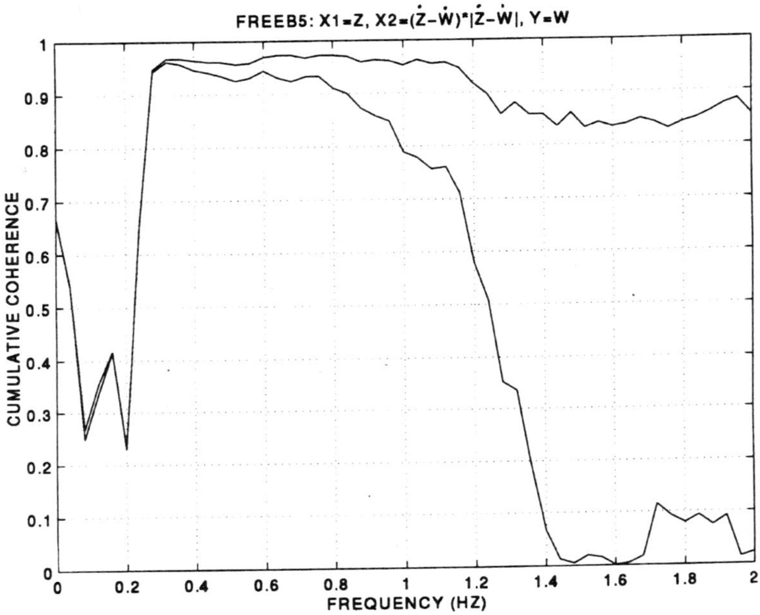

Figure 7.104.

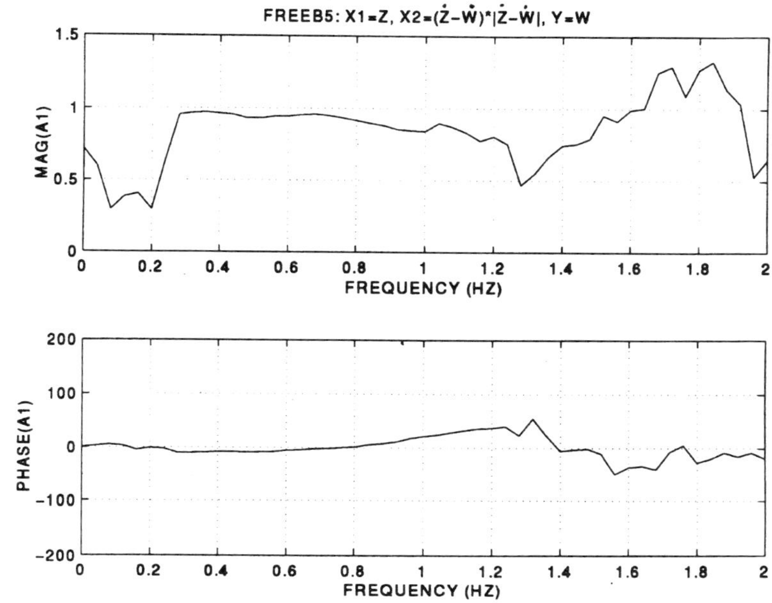

Figure 7.105.

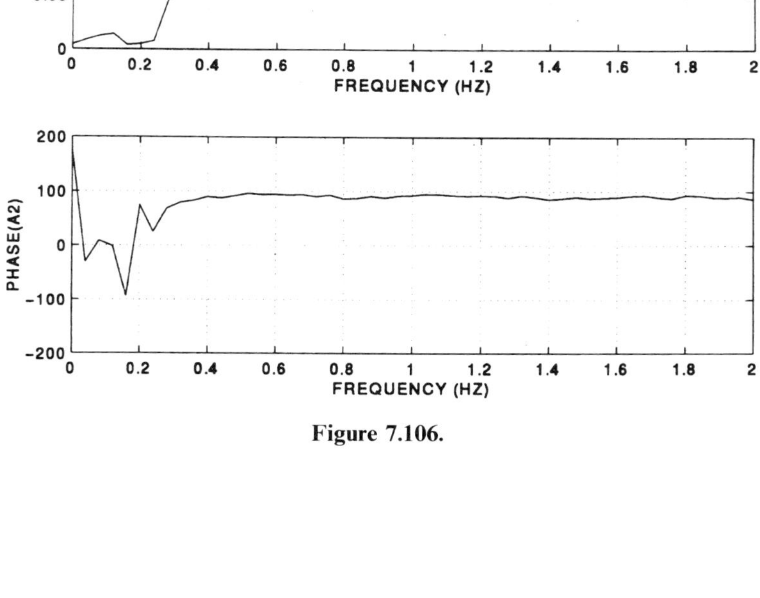

Figure 7.106.

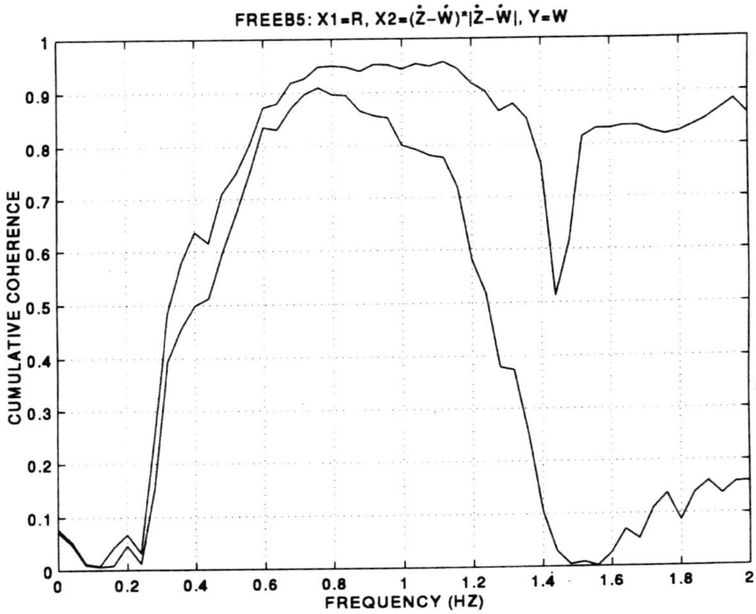

Figure 7.107.

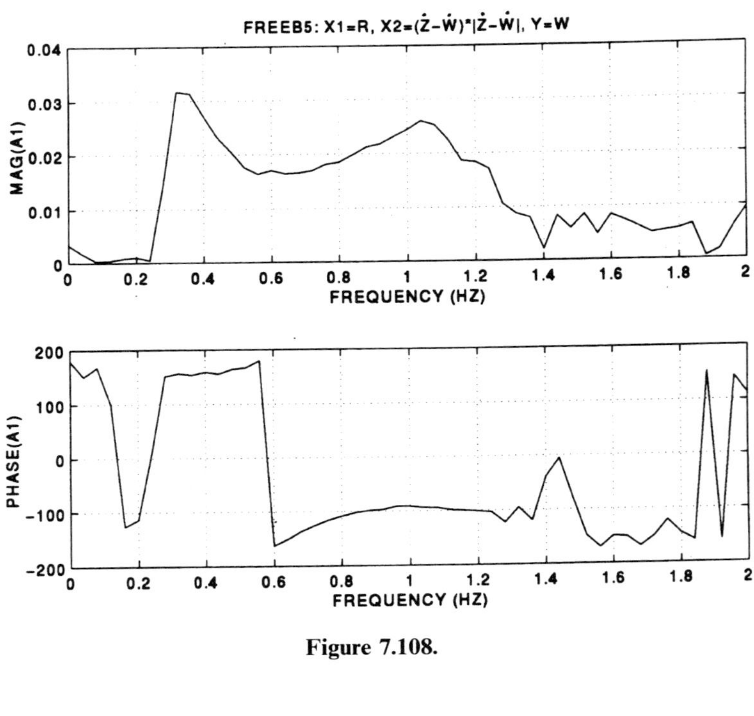

Figure 7.108.

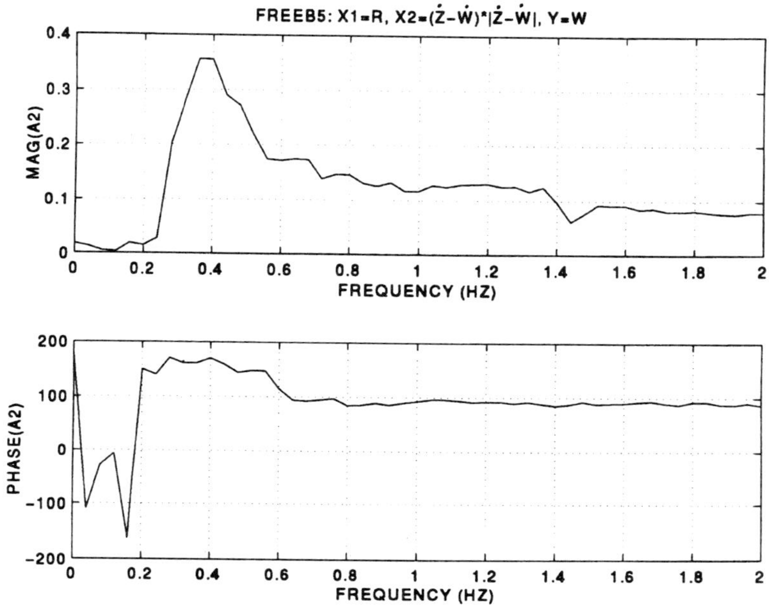

Figure 7.109.

Reverse Dynamic System Models, Three Inputs, B5 Data. The reverse dynamic model for the "heave" equation employs three inputs, namely, (1) heave motion, $z(t)$, (2) quadratic relative velocity, and (3) quadratic roll motion, $\phi(t)|\phi(t)|$. Using the undisturbed wave motion, $w(t)$, as the output for the reverse dynamic model, a Reverse MI/SO analysis was performed. The cumulative coherence function for this analysis is shown in Figure 7.110, and the linear and nonlinear frequency response functions are given in Figures 7.111, 7.112, and 7.113, respectively. No improvement in coherence is exhibited from incorporation of the second nonlinearity over the 0.30–2.0 Hz frequency band. Therefore, the previous two-input reverse dynamic model is judged adequate. There is no need for further consideration of this three-input/single-output "heave" nonlinear model.

The reverse dynamic model for the "roll" equation employs three inputs, namely, (1) roll motion, $\phi(t)$, (2) quadratic relative velocity, and (3) quadratic roll motion, $\phi(t)|\phi(t)|$. Using the undisturbed wave motion, $w(t)$, as the output for the reverse dynamic model, a Reverse MI/SO analysis was performed. The cumulative coherence function for this analysis is shown in Figure 7.114, and the linear and nonlinear frequency responses are given in Figures 7.115, 7.116, and 7.117, respectively. No improvement in coherence is exhibited from incorporation of the second nonlinearity over the 0.0–2.0 Hz frequency band. Therefore, the previous two-input reverse model for roll motions is adequate.

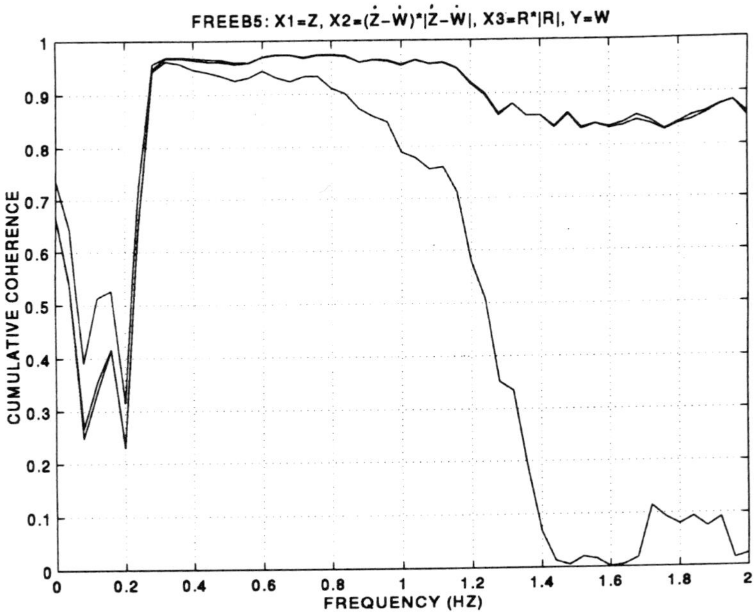

Figure 7.110.

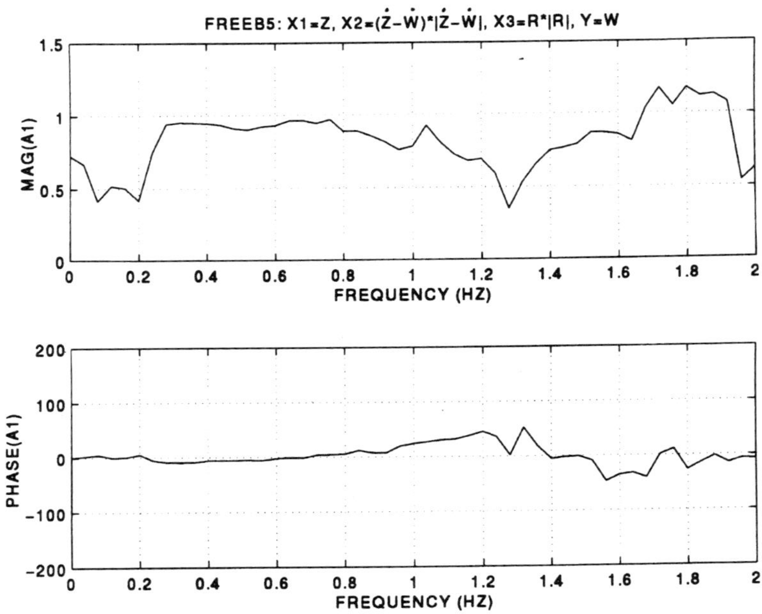

Figure 7.111.

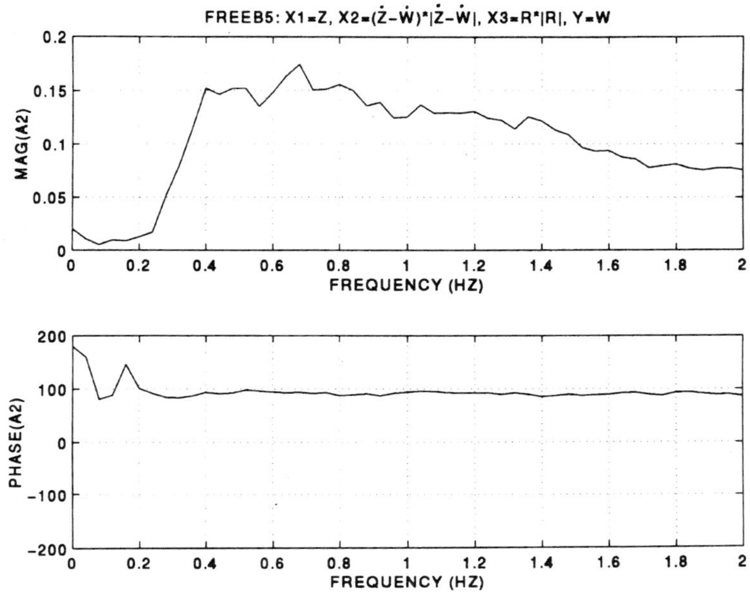

Figure 7.112.

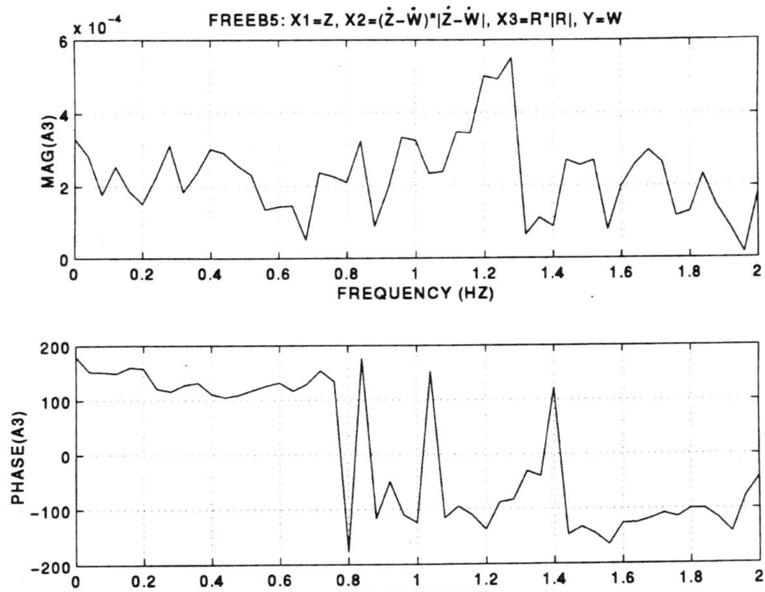

Figure 7.113.

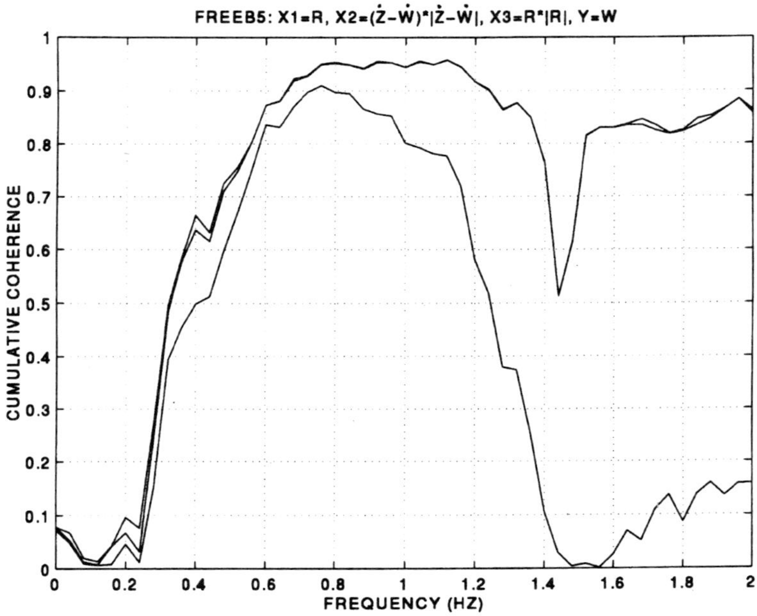

Figure 7.114.

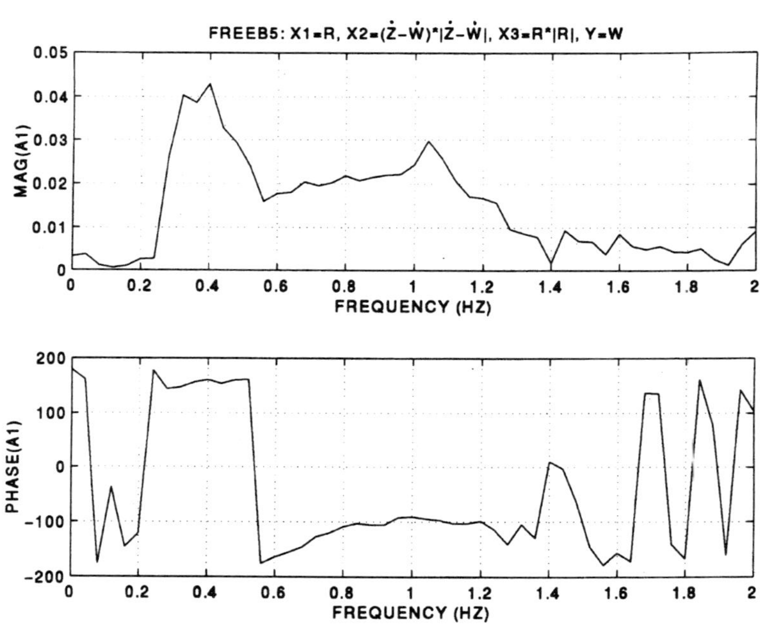

Figure 7.115.

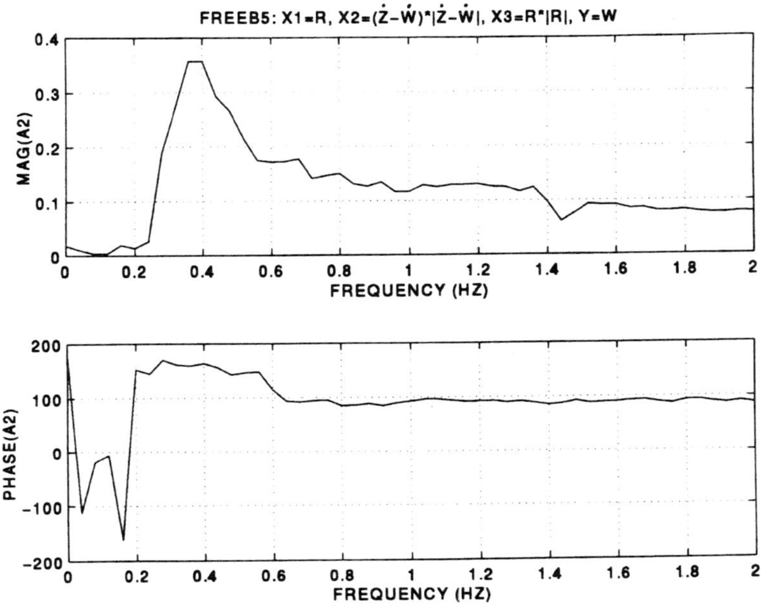

Figure 7.116.

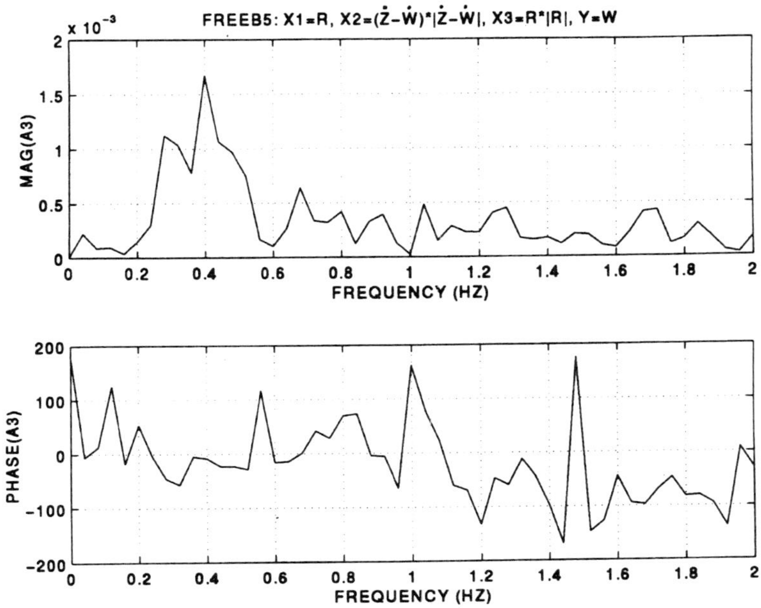

Figure 7.117.

7.5.6 Nonlinear MI/SO Analysis of Large Wave Quartering Sea Data

The majority of MI/SO analyses performed for the large wave head sea data and the large wave beam sea data indicate that only one nonlinear term is of significance, namely, the quadratic form of relative heave velocity. Hence, the quartering sea response analyses have been limited to cases of direct and reverse dynamic system models using two inputs. Results of these large wave, quartering sea data analyses are presented below as obtained from the free barge Q5 data.

Direct Dynamic System Models, Two Inputs, Q5 Data. Direct model analysis for the heave equation considers two inputs, namely, (1) undisturbed wave motion, $w(t)$, and (2) quadratic relative velocity. Using heave motion, $z(t)$, to express the output, a Direct MI/SO analysis was performed. The cumulative coherence function for that analysis is shown in Figure 7.118, and the linear and nonlinear frequency response functions are given in Figures 7.119 and 7.120, respectively. Negligible improvement in coherence is exhibited here due to incorporation of the single nonlinearity. This direct analysis would conclude that this particular nonlinear term is not significant in a quartering sea.

Using roll motion, $\phi(t)$, to express the output for the roll nonlinear differential equation, a Direct MI/SO analysis was conducted in a manner analogous to that described for heave motion. The cumulative coherence for that analysis is shown in Figure 7.121, and the linear and nonlinear frequency responses are shown in Figures 7.122 and 7.123, respectively. Negligible improvement in coherence is exhibited from incorporation of the single nonlinearity. Hence, this direct MI/SO analysis is not able to describe nonlinear roll dynamics.

Using pitch motion, $\theta(t)$, to express the output for the pitch nonlinear differential equation, a Direct MI/SO analysis was conducted as before. The cumulative coherence function is shown in Figure 7.124, with the linear and nonlinear frequency responses in Figures 7.125 and 7.126, respectively. Negligible improvement in coherence here means that this quadratic nonlinear term is not able to describe nonlinear pitch dynamics.

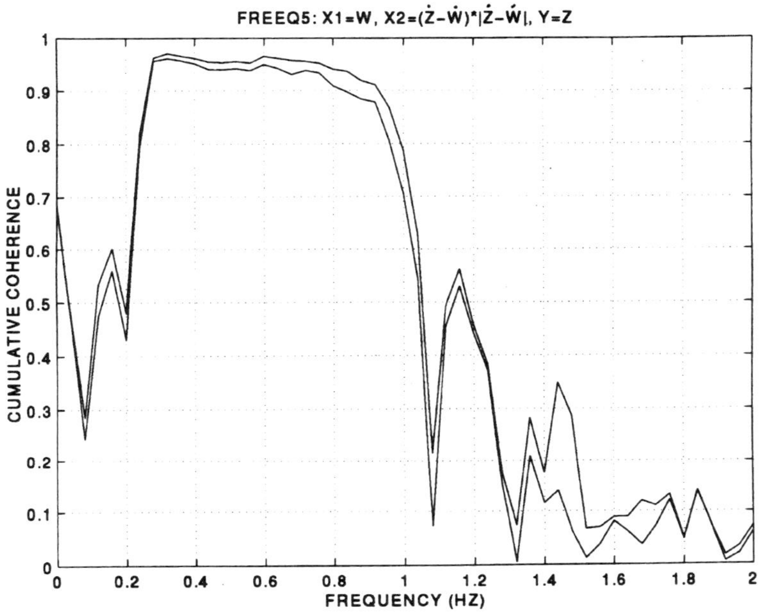

Figure 7.118.

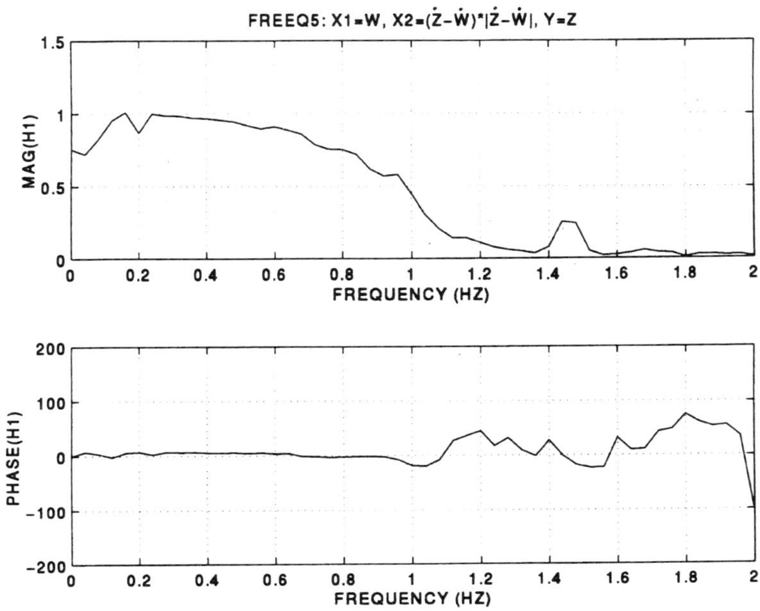

Figure 7.119.

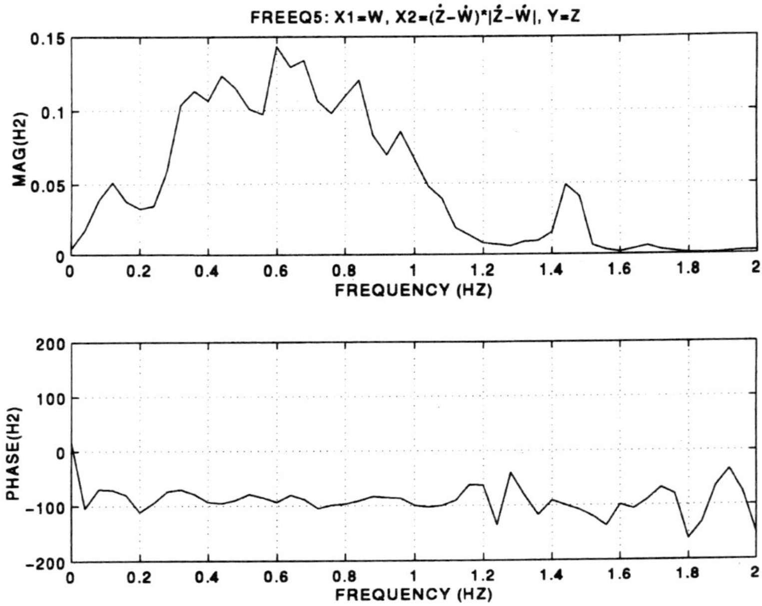

Figure 7.120.

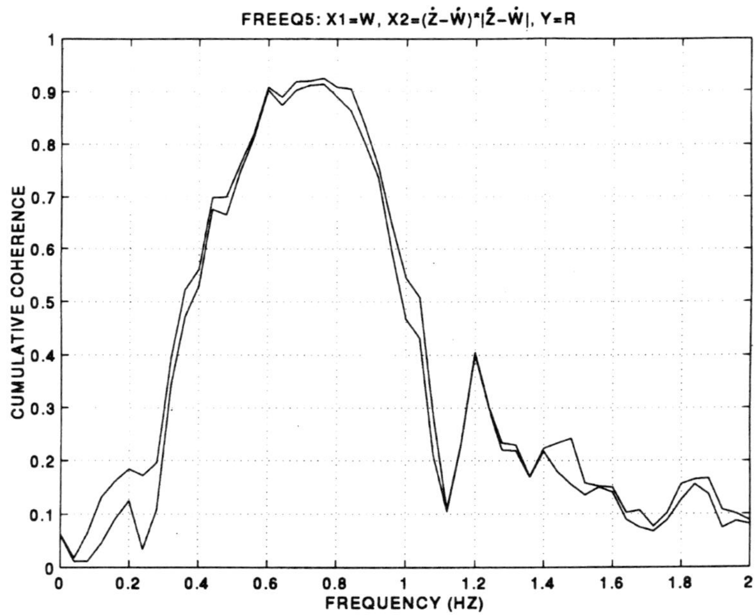

Figure 7.121.

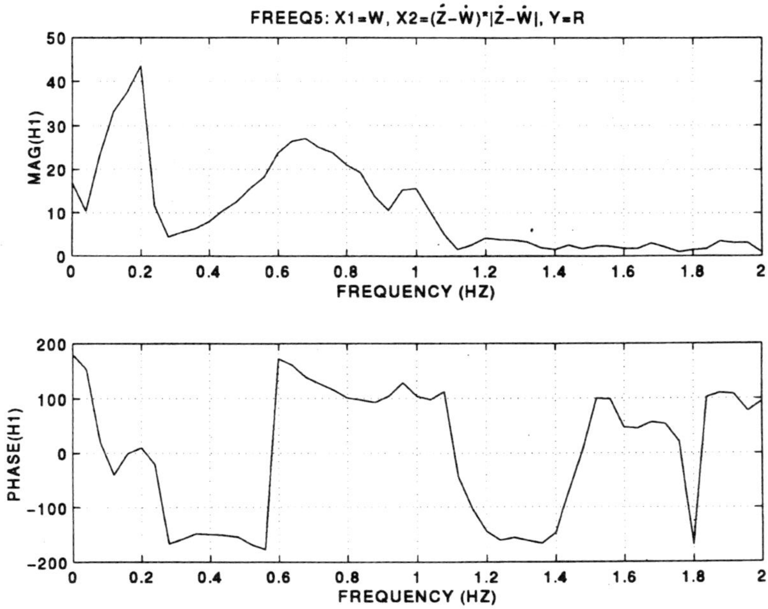

Figure 7.122.

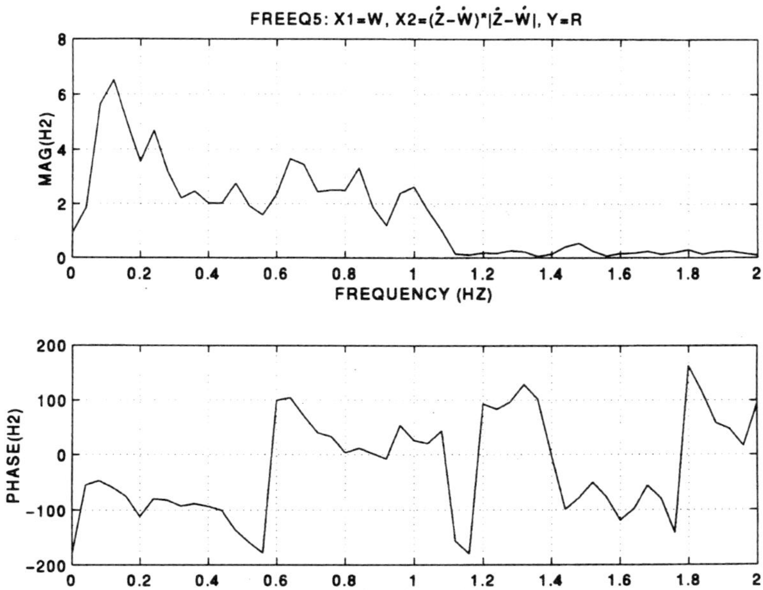

Figure 7.123.

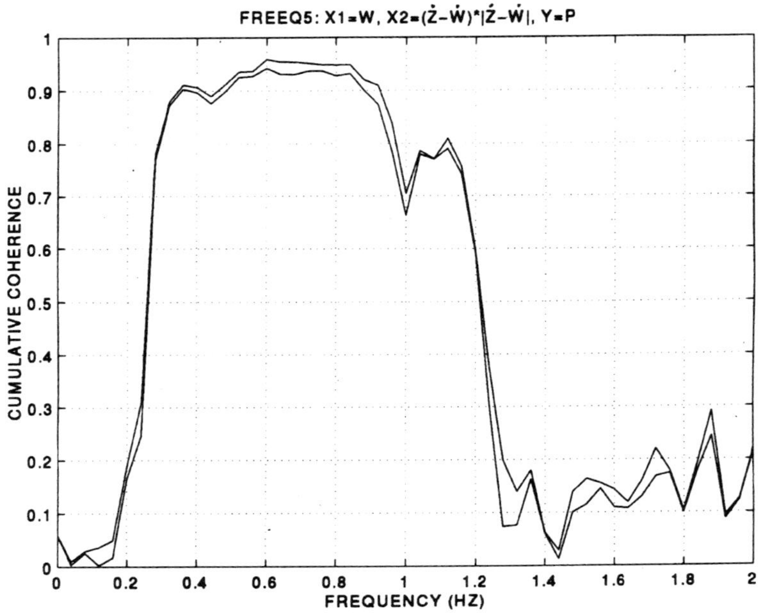

Figure 7.124.

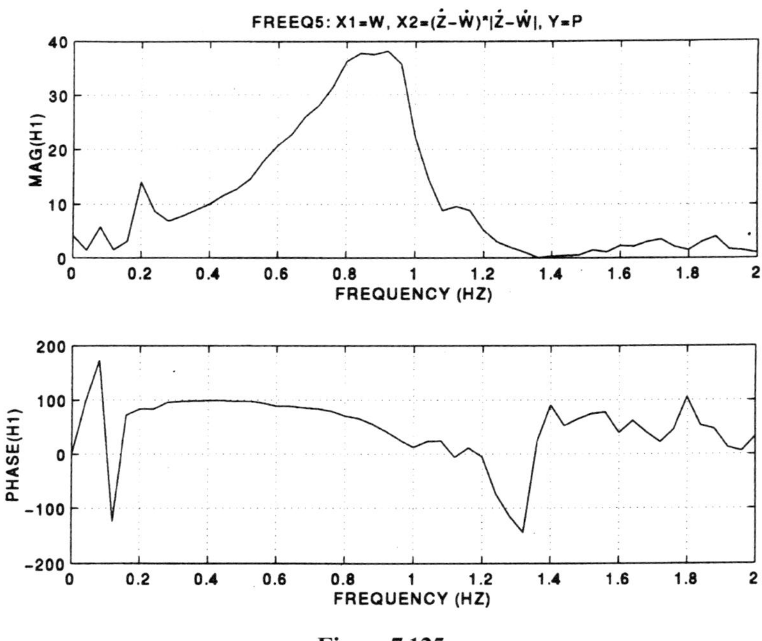

Figure 7.125.

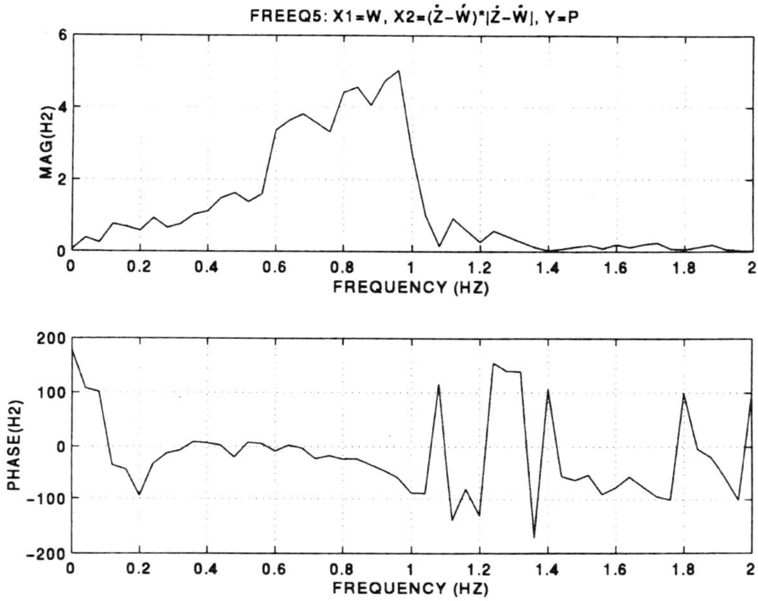

Figure 7.126.

Reverse Dynamic System Models, Two Inputs, Q5 Data. The reverse dynamic system analysis considers two inputs for the "heave" equation, namely, (1) the heave motion, $z(t)$, and (2) the quadratic relative velocity. Using undisturbed wave motion, $w(t)$, to express the output for the heave nonlinear differential equation, a Reverse MI/SO analysis was performed giving the cumulative coherence function shown in Figure 7.127 and the linear and nonlinear frequency response functions in Figures 7.128 and 7.129, respectively. Improvement in coherence is very significant from incorporation of the single nonlinearity over the 0.30–2.0 Hz frequency band. The linear system frequency response, $A_1(f)$, is similar to the inverse magnitude and phase of the linear SI/SO frequency response function (see Figure 7.50). The nonlinear system frequency response, $A_2(f)$, if of lower magnitude, with a broad peak and with a fairly constant 90° phase angle.

The reverse dynamic system analysis considers two inputs for the "roll" equation, namely, (1) roll motion, $\phi(t)$, and (2) quadratic relative velocity. Using undisturbed wave motion, $w(t)$, to express the output for the roll nonlinear differential equation, a Reverse MI/SO analysis was conducted as before. The cumulative coherence is shown in Figure 7.130, with linear and nonlinear frequency responses in Figures 7.131 and 7.132, respectively. Strong improvement in coherence is exhibited from the nonlinearity over the 0.60–2.0 Hz frequency band. The linear system frequency response, $A_1(f)$, is similar to the inverse of the linear SI/SO result (see Figure 7.52). The nonlinear system frequency response, $A_2(f)$, is fairly smooth with a 180° phase below 0.60 Hz and a 90° phase above 0.60 Hz.

The reverse dynamic system analysis considers two inputs for the "pitch" equation, namely, (1) pitch motion, $\theta(t)$, and (2) quadratic relative velocity. Using undisturbed wave motion, $w(t)$, to express the output, a Reverse MI/SO analysis was conducted as before. The cumulative coherence is shown in Figure 7.133, with linear and nonlinear frequency responses in Figures 7.134 and 7.135, respectively. Strong improvement in coherence is exhibited from the nonlinearity over the 0.6–2.0 Hz frequency band. The linear system frequency response, $A_1(f)$, is similar to the inverse of the linear SI/SO result (see Figure 7.54). The nonlinear system frequency response, $A_2(f)$, shows a broad peak with a 90° phase above 0.25 Hz.

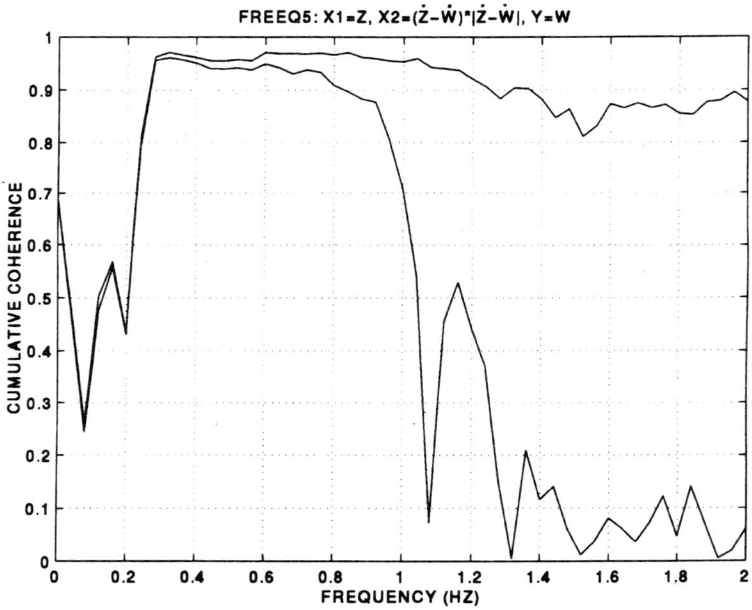

Figure 7.127.

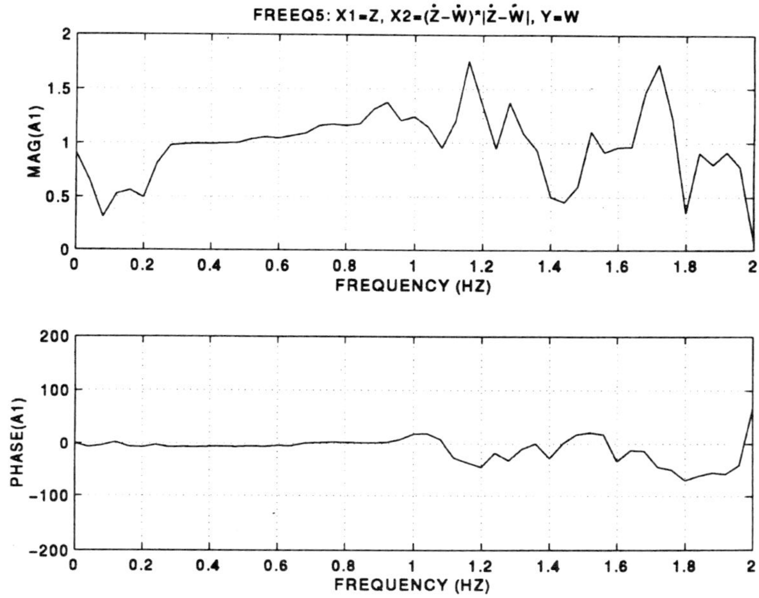

Figure 7.128.

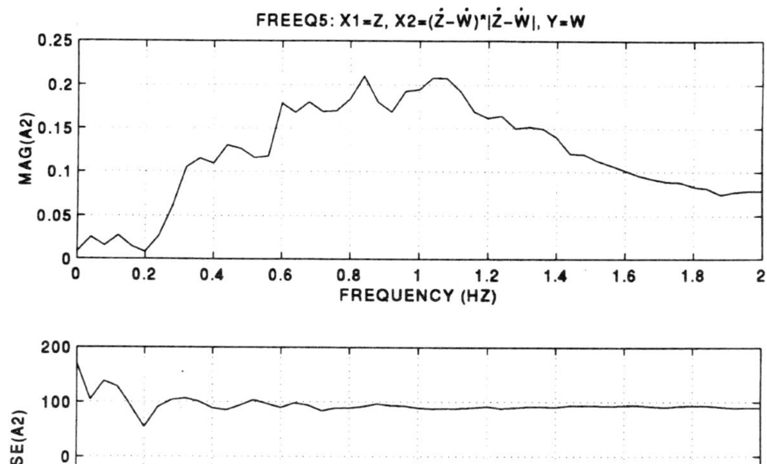

Figure 7.129.

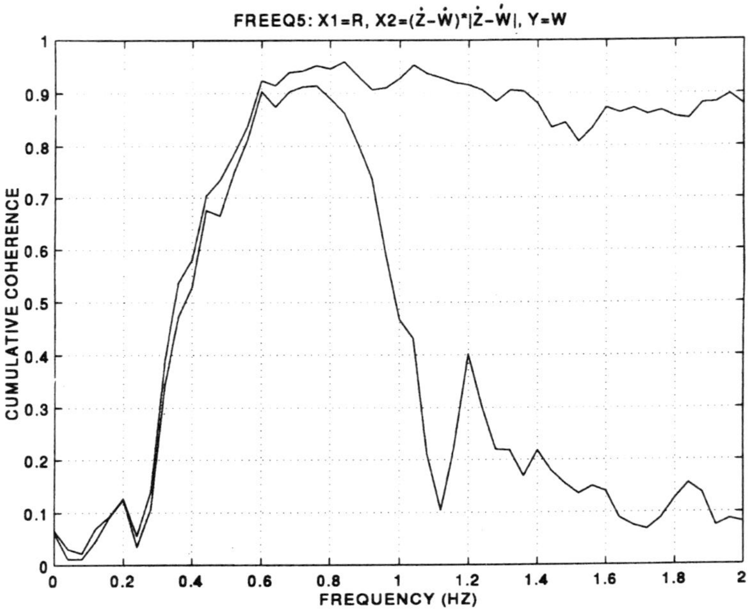

Figure 7.130.

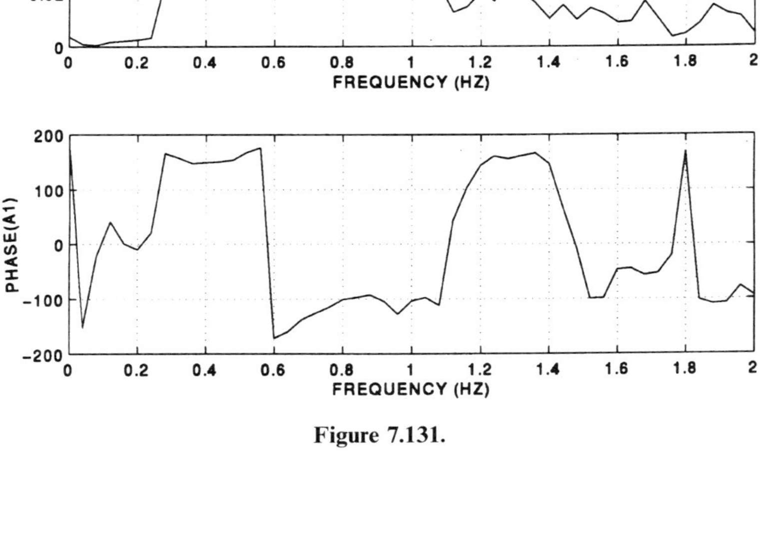

Figure 7.131.

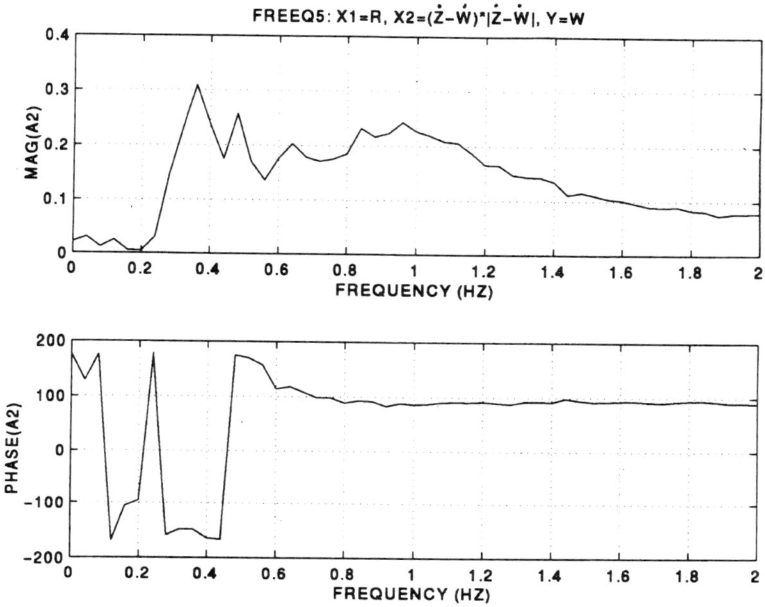

Figure 7.132.

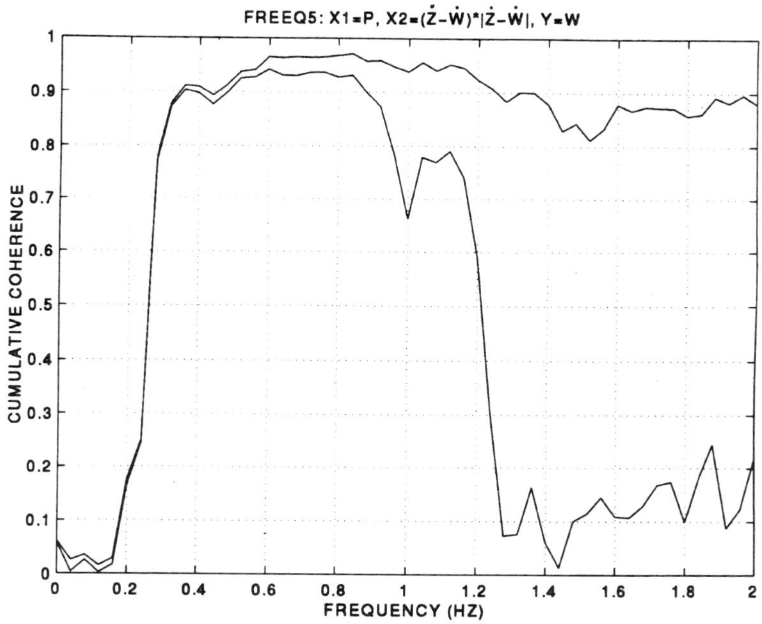

Figure 7.133.

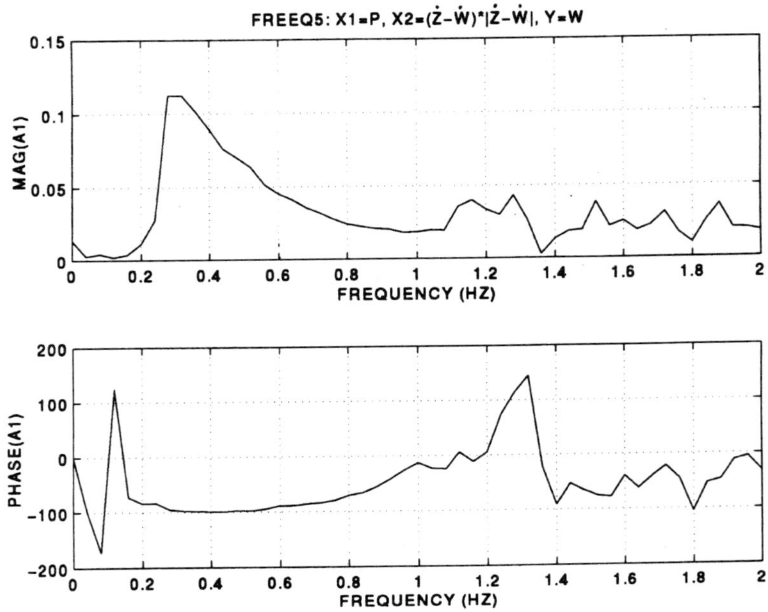

Figure 7.134.

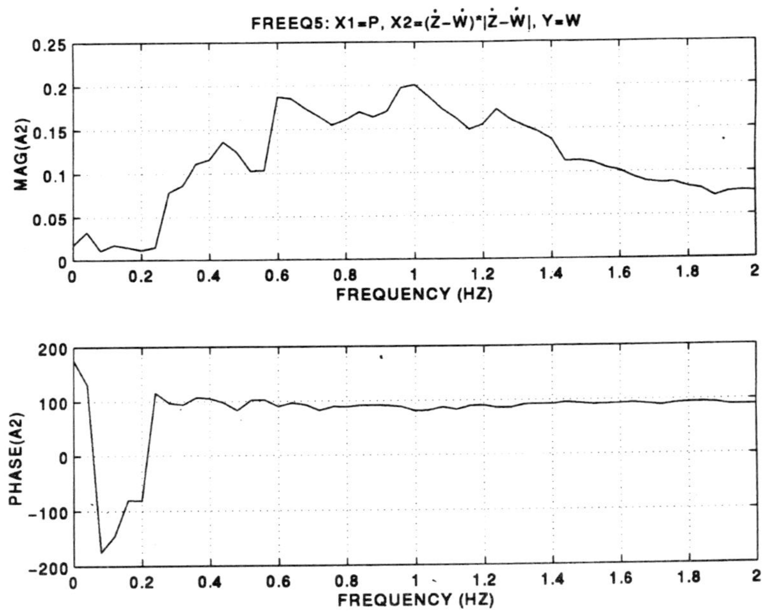

Figure 7.135.

7.6 CONCLUSIONS FROM BARGE TESTS

An extensive set of data was recorded from the NFESC/USNA naval barge test program in July 1995 to obtain heave, roll, and pitch results in head, beam, and quartering seas. Data associated with small and large wave motions were recorded for all three wave directions. Computed results from this extensive test program demonstrate and prove the following conclusions:

1. Linear SI/SO analyses of the free barge small and large wave induced data suggest that the heave response in head, beam, and quartering seas is predominantly linear. The nonlinear MI/SO analyses conducted here indicate that nonlinearity is important in the roll and pitch responses in beam and quartering seas.
2. One particular nonlinear term plays a key role. That term is associated with hydrodynamic damping described in terms of the quadratic fluid heave velocity relative to the barge.
3. In this free barge study, Reverse MI/SO Techniques gave better nonlinear results than Direct MI/SO Techniques by producing higher cumulative coherence values over wider frequency ranges. This was opposite to results found from the frigate test data in Chapter 6.
4. The work in Chapters 6 and 7 shows how the Direct and Reverse MI/SO Techniques can be applied to analyze and identify nonlinear system properties from measured data in real systems.
5. These results and other material in this book make the Direct and Reverse MI/SO Techniques the most practical general methods available today to help solve desired nonlinear system problems in many fields.

This concludes Chapter 7.

8

BILINEAR AND TRILINEAR SYSTEMS

This chapter concerns the theoretical properties of general nonlinear systems known as bilinear (second-order) and trilinear (third-order) systems, representing functional Volterra extensions of well-known properties for linear (first-order) systems. This chapter shows how to determine the multidimensional second- and third-order weighting and frequency response functions that describe these systems. Examples illustrating these results include finite-memory nonlinear systems represented by zero-memory square law systems and cubic systems that are preceded or followed by linear systems. The chapter contains material on how to synthesize a time-varying linear system by a constant-parameter bilinear system. Also discussed are multidimensional higher-order cross-correlation and cross-spectral density functions that will be used in Chapter 9 to analyze and identify nonlinear system input/output properties of Gaussian stationary random data passing through bilinear and trilinear systems. The material in Chapters 8 and 9 is needed to understand when and how these complicated Volterra techniques should be applied, and to appreciate the benefits when they can be replaced by the new, simpler, practical MI/SO techniques in Chapters 3–7.

8.1 FUNCTIONAL REPRESENTATION OF NONLINEAR SYSTEMS

The output of a nonlinear system can often be expressed as a "power series with memory" by using a Volterra functional representation as discussed in references 23, 34, 47, 50, and 52. By this representation, the output $y(t)$ due to an input $x(t)$ is given by a sum of Volterra functionals as follows:

$$y(t) = y_0 + y_1(t) + y_2(t) + y_3(t) + \cdots \tag{8.1}$$

where

$$\begin{aligned} y_0 &= \text{constant} \\ y_1(t) &= \text{linear output} \\ y_2(t) &= \text{bilinear output} \\ y_3(t) &= \text{trilinear output} \end{aligned}$$

and so on. The mean value $\bar{y} = E[y(t)]$ satisfies

$$\bar{y} = y_0 + \bar{y}_1 + \bar{y}_2 + \bar{y}_3 + \cdots \tag{8.2}$$

Now

$$y(t) - \bar{y} = [y_1(t) - \bar{y}_1] + [y_2(t) - \bar{y}_2] + [y_3(t) - \bar{y}_3] + \cdots \tag{8.3}$$

In the frequency domain, by taking Fourier transforms of both sides of Eq. (8.3), one obtains

$$Y(f) = Y_1(f) + Y_2(f) + Y_3(f) + \cdots \tag{8.4}$$

where

$$Y(f) = \mathscr{F}[y(t) - \bar{y}] = \int_{-\infty}^{\infty} [y(t) - \bar{y}]e^{-j2\pi ft}\,dt \tag{8.5}$$

with

$$\begin{aligned} Y_1(f) &= \mathscr{F}[y_1(t) - \bar{y}_1] \\ Y_2(f) &= \mathscr{F}[y_2(t) - \bar{y}_2] \end{aligned} \tag{8.6}$$

and so on.

General ways will now be developed to describe the separate quantities $y_1(t)$, $y_2(t)$, and $y_3(t)$ in the time domain, and their corresponding $Y_1(f)$, $Y_2(f)$, and $Y_3(f)$ in the frequency domain. This will be followed in Section 8.2 by examples deemed to be of physical importance. Without loss of generality, it will be assumed henceforth that $y_0 = 0$. Otherwise, results obtained here apply to $[y(t) - y_0]$ rather than to $y(t)$ by itself. Figure 8.1 illustrates the situation for third-order Volterra nonlinear input/output models where $y(t)$ does not go beyond $y_3(t)$. This is a single-input/single-output nonlinear model between the input $x(t)$ and the output $y(t)$.

In further analysis, to obtain closed-form answers, assume that $x(t)$ is a stationary (ergodic) random process such that any n random variables $x_i =$

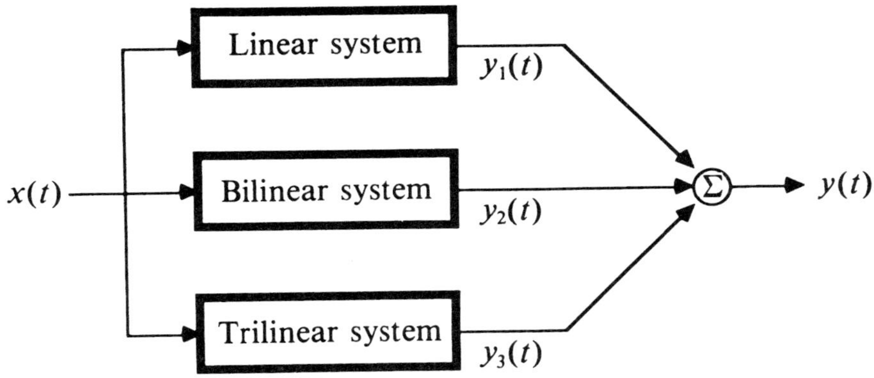

Figure 8.1. Third-order Volterra nonlinear input/output model.

$x(t_i)$, $i = 1, 2, \ldots, n$, follow an n-dimensional Gaussian (normal) distribution with zero mean values. Even-order moments of such random variables then reduce to products of second-order moments, and all odd-order moments are zero. It follows that the mean values

$$\bar{y}_1 = E[y_1(t)] = 0, \quad \bar{y}_3 = E[y_3(t)] = 0, \quad \bar{y}_2 = E[y_2(t)] \neq 0 \tag{8.7}$$

where $E[\ ]$ denotes an appropriate expected value operation.

Definitions and properties will now be discussed for linear, bilinear, and trilinear systems.

8.1.1 Linear Systems

By definition, as per Eq. (1.1), L_1 is a *linear system operator* with repect to inputs $\{x(t)\}$ if for all inputs $x_1(t)$, $x_2(t)$ and all constants a, b

$$L_1[ax_1(t) + bx_2(t)] = aL_1[x_1(t)] + bL_1[x_2(t)] \tag{8.8}$$

The operator L_1 has the additive and homogeneous properties of Eqs. (1.2) and (1.3).

$$\begin{aligned} \text{Additive} \quad L_1[x_1(t) + x_2(t)] &= L_1[x_1(t)] + L_1[x_2(t)] \\ \text{Homogeneous} \quad L_1[ax(t)] &= aL_1[x(t)] \end{aligned} \tag{8.9}$$

From basic theory, let $h_1(\tau)$ be the response of the linear system L_1 at time τ to a delta function input $\delta(\tau)$, namely,

$$h_1(\tau) = L_1[\delta(\tau)] \tag{8.10}$$

The quantity $h_1(\tau)$ is the *linear weighting function* of the system or the *unit impulse response function.* It is also called the *first-order time-domain kernel.*

From properties of the delta function, any arbitrary input $x(t)$ can be described by the convolution integral

$$x(t) = \int_{-\infty}^{\infty} x(t - \tau)\delta(\tau)\, d\tau \tag{8.11}$$

By letting $d\tau = \Delta\tau$ and $\tau = \tau_n = n\,\Delta\tau$ for all integers n, this can be approximated by the summation

$$x(t) \approx \sum_n x(t - \tau_n)\delta(\tau_n)\, \Delta\tau \tag{8.12}$$

Now, from properties of Eqs. (8.9) and (8.10), it follows that

$$L_1[x(t)] \approx \sum_n x(t - \tau_n)h_1(\tau_n)\, \Delta\tau$$

where

$$h_1(\tau_n) = L_1[\delta(\tau_n)]$$

Hence, in the limit as $\Delta\tau \to 0$, for any input $x(t)$, one obtains the linear system output response of Eq. (1.11):

$$y_1(t) = L_1[x(t)] = \int_{-\infty}^{\infty} x(t - \tau)h_1(\tau)\, d\tau \tag{8.13}$$

From the assumption that $x(t)$ has a zero mean value for all t where $\bar{x} = E[x(t)] = 0$, it follows that the linear output $y_1(t)$ also has a zero mean value for all t.

$$\bar{y}_1 = E[y_1(t)] = \int_{-\infty}^{\infty} h_1(\tau)E[x(t - \tau)]\, d\tau = 0 \tag{8.14}$$

As stated in Eq. (1.14), a linear system is said to be *stable* if every bounded input $x(t)$ produces a bounded output $y_1(t)$. A necessary and sufficient condition for a linear system to be stable is

$$\int_{-\infty}^{\infty} |h_1(\tau)|\, d\tau < \infty \tag{8.15}$$

The function $h_1(\tau)$ gives the response at time t by weighting inputs $x(t)$ occurring at time $(t - \tau)$. Hence, if the system L_1 is *physically realizable*, where

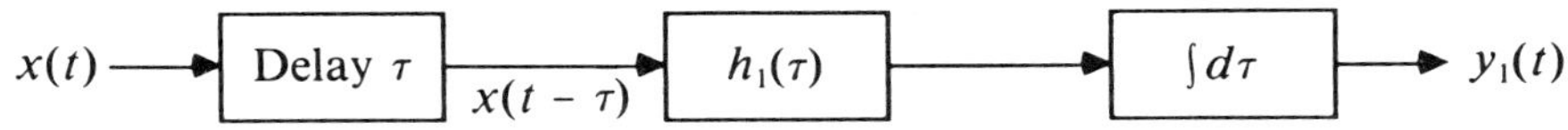

Figure 8.2. Convolution calculation of linear system output.

it responds only to past inputs that have already occurred, then $t \geqslant (t - \tau)$, which is the same as $\tau \geqslant 0$. Thus, for physical realizability, the weighting function must satisfy

$$h_1(\tau) = 0 \quad \text{for } \tau < 0 \tag{8.16}$$

For these situations, Eq. (8.13) is equivalent to Eq. (1.10), namely,

$$y_1(t) = \int_0^{\infty} x(t - \tau)h_1(\tau)\, d\tau \tag{8.17}$$

Without loss of generality, Eq. (8.13) will be used henceforth to describe linear systems, whether or not Eq. (8.16) applies. When measuring real data, only finite limits will be involved. The theoretical limits from minus infinity to plus infinity will be omitted in the following integrals to simplify the notation.

The calculation of $y_1(t)$ from $x(t)$ via the convolution integral expression of Eq. (8.13) is shown in Figure 8.2.

Consider now the Fourier transform of $h_1(\tau)$ denoted by

$$H_1(f) = \mathscr{F}[h_1(\tau)] = \int h_1(\tau)e^{-j2\pi f\tau}\, d\tau \tag{8.18}$$

This quantity $H_1(f)$ is the *linear frequency response function* of the system. It is also called the *first-order frequency-domain kernel* of the system. A sufficient condition for $H_1(f)$ to exist is that the linear system be stable. Fourier transforms of both sides of Eq. (8.13) give the simple frequency-domain result of Eq. (1.15),

$$Y_1(f) = \mathscr{F}[y_1(t)] = H_1(f)X(f) \tag{8.19}$$

from which one can express $y_1(t)$ as the inverse Fourier transform of $Y_1(f)$, namely,

$$y_1(t) = \mathscr{F}^{-1}[Y_1(f)] = \int Y_1(f)e^{j2\pi ft}\, df \tag{8.20}$$

Figure 8.3 shows the calculation of $y_1(t)$ from $x(t)$ using Fourier transform procedures, where $\mathscr{F}$ represents a suitable Fourier transform operation and

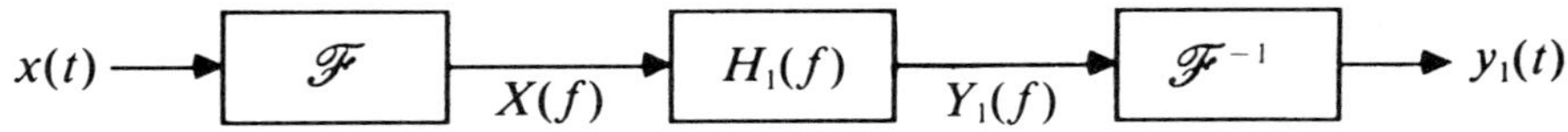

Figure 8.3. Fourier transform calculation of linear system output.

$\mathcal{F}^{-1}$ denotes the corresponding inverse operation. This diagram describes the linear system shown in Figure 8.1. Note that $y_1(t) = h_1(t)$ when $x(t) = \delta(t)$. This is the same as having $Y_1(f) = H_1(f)$ when $X(f) = 1$. When the input $x(t) = \exp(j2\pi ft)$, from Eqs. (8.13) and (8.18), the linear output $y_1(t) = H_1(f)x(t)$. This proves that sinusoidal frequencies are preserved under linear transformations and provides a simple way to determine $H_1(f)$ when $x(t)$ and $y_1(t)$ are input/output terms in a linear differential equation.

8.1.2 Bilinear Systems

By definition, L_2 is a *bilinear system operator* with respect to a pair of ordered inputs $\{x(t), y(t)\}$ if the operator L_2 is linear on $x(t)$ for a given $y(t)$, and conversely. In equation form, for any constants a, b, it is required that

$$\begin{aligned} L_2(x_1 + x_2, y) &= L_2(x_1, y) + L_2(x_2, y) \\ L_2(x, y_1 + y_2) &= L_2(x, y_1) + L_2(x, y_2) \end{aligned} \tag{8.21}$$

$$\begin{aligned} L_2(ax, y) &= aL_2(x, y) \\ L_2(x, by) &= bL_2(x, y) \\ L_2(ax, by) &= abL_2(x, y) \end{aligned} \tag{8.22}$$

where x represents $x(t)$ and y represents $y(t)$. Now

$$\begin{aligned} L_2(ax + by, ax + by) &= L_2(ax + by, ax) + L_2(ax + by, by) \\ &= L_2(ax, ax) + L_2(by, ax) + L_2(ax, by) + L_2(by, by) \\ &= a^2L_2(x, x) + abL_2(y, x) + abL_2(x, y) + b^2L_2(y, y) \end{aligned} \tag{8.23}$$

Without loss of generality, one can assume that $L_2(x, y)$ is symmetric with respect to x and y, namely,

$$L_2(x, y) = L_2(y, x) \tag{8.24}$$

Then Eq. (8.23) becomes

$$L_2(ax + by, ax + by) = a^2L_2(x, x) + 2abL_2(x, y) + b^2L_2(y, y) \tag{8.25}$$

To obtain Eq. (8.24) if $L_2(x, y) \neq L_2(y, x)$, one should replace $L_2(x, y)$ by $\tilde{L}_2(x, y)$ defined by

$$\tilde{L}_2(x, y) = \tfrac{1}{2}[L_2(x, y) + L_2(y, x)] \tag{8.26}$$

Now

$$\tilde{L}_2(x, y) = \tilde{L}_2(y, x) \tag{8.27}$$

so that $\tilde{L}_2(x, y)$ is symmetric with respect to x and y. The symmetric property of Eq. (8.24) will be assumed henceforth.

Consider the special case where $b = 0$. This gives

$$L_2(ax, ax) = a^2 L_2(x, x) \tag{8.28}$$

If one now defines a special bilinear system operator by

$$L_2(x) \equiv L_2(x, x) \tag{8.29}$$

then Eq. (8.28) shows that

$$L_2(ax) = a^2 L_2(x) \tag{8.30}$$

Hence, L_2 is a *nonlinear* operator on $x(t)$, since the homogeneous property of Eq. (8.9) is violated. Equation (8.30) produces an amplitude nonlinearity. Note also from Eqs. (8.24), (8.25), and (8.29) that

$$\begin{aligned} L_2(x_1 + x_2) &\equiv L_2(x_1 + x_2, x_1 + x_2) \\ &= L_2(x_1, x_1) + 2L_2(x_1, x_2) + L_2(x_2, x_2) \\ &= L_2(x_1) + 2L_2(x_1, x_2) + L_2(x_2) \end{aligned} \tag{8.31}$$

Thus, the additive property of Eq. (8.9) is violated whenever $L_2(x_1, x_2) \neq 0$.

The bilinear transformation $L_2(x_1, x_2)$ can generate sum and difference frequencies. To see this property, suppose $L_2(x_1, x_2) = x_1 x_2$ where $x_1 = \cos at$ and $x_2 = \cos bt$. Then

$$L_2(x_1, x_2) = \cos at \cos bt = \tfrac{1}{2}[\cos(a + b)t + \cos(a - b)t]$$

Here, the term $\cos(a - b)t$ produces an envelope nonlinearity.

From Volterra theory, let $h_2(\tau_1, \tau_2)$ be the response of the bilinear system L_2 at times τ_1, τ_2 to a pair of unit impulse function inputs $\{\delta(\tau_1), \delta(\tau_2)\}$, namely,

$$h_2(\tau_1, \tau_2) = L_2[\delta(\tau_1), \delta(\tau_2)] \tag{8.32}$$

The quantity $h_2(\tau_1, \tau_2)$ is called the *bilinear weighting function* of the system. It is also called the *second-order time-domain kernel*. From Eq. (8.24), observe that $h_2(\tau_1, \tau_2)$ is unique and satisfies

$$h_2(\tau_1, \tau_2) = h_2(\tau_2, \tau_1) \tag{8.33}$$

In words, $h_2(\tau_1, \tau_2)$ is symmetric with respect to its two variables.

As in Eq. (8.12), let $x(t)$ be approximated by

$$x(t) \approx \sum_n x(t - \tau_n)\delta(\tau_n)\,\Delta\tau = \sum_n a_n\delta(\tau_n)$$

where

$$a_n = x(t - \tau_n)\,\Delta\tau = x(t - n\,\Delta\tau)\,\Delta\tau$$

Now

$$\begin{aligned} L_2[x(t)] &\equiv L_2[x(t), x(t)] \\ &\approx L_2\left[\sum_m a_m\delta(\tau_m), \sum_n a_n\delta(\tau_n)\right] \end{aligned}$$

using different indices of summation for proper accounting. Properties of L_2 and Eq. (8.32) give

$$\begin{aligned} L_2[x(t)] &\approx \sum_m \sum_n a_m a_n L_2[\delta(\tau_m), \delta(\tau_n)] \\ &\approx \sum_m \sum_n x(t - \tau_m)x(t - \tau_n)h_2(\tau_m, \tau_m)\,\Delta\tau_1\,\Delta\tau_2 \end{aligned}$$

with $\tau_m = m\,\Delta\tau_1$ and $\tau_n = n\,\Delta\tau_2$. Hence, in the limit as $\Delta\tau_1 \to 0$ and $\Delta\tau_2 \to 0$, for any input $x(t)$, one obtains the bilinear system output response $y_2(t)$ given by

$$y_2(t) = L_2[x(t)] = \iint x(t - \tau_1)x(t - \tau_2)h_2(\tau_1, \tau_2)\,d\tau_1\,d\tau_2 \tag{8.34}$$

where the possible double limits of integration are from $-\infty$ to $+\infty$. Equation (8.34) is the bilinear extension to the linear result of Eq. (8.13).

From Eq. (8.34) it is simple to verify Eqs. (8.30) and (8.31) that

$$\begin{aligned} L_2[ax(t)] &= a^2L_2[x(t)] \\ L_2[x_1(t) + x_2(t)] &= L_2[x_1(t)] + 2L_2[x_1(t), x_2(t)] + L_2[x_2(t)] \end{aligned}$$

where

$$L_2[x_1(t), x_2(t)] = \iint x_1(t-\tau_1)x_2(t-\tau_2)h_2(\tau_1, \tau_2)\, d\tau_1\, d\tau_2 \tag{8.35}$$

Thus L_2, as defined by Eq. (8.34), is *not* a linear operator on $x(t)$. It is clear also from Eq. (8.34) that replacing $h_2(\tau_1, \tau_2)$ by $h_2(\tau_2, \tau_1)$ gives the identical result for $y_2(t)$. Finally, if the bilinear system L_2 is *physically realizable* where it responds only to past inputs, then

$$h_2(\tau_1, \tau_2) = 0 \quad \text{for } \tau_1 < 0 \text{ and/or } \tau_2 < 0 \tag{8.36}$$

so that the lower limits in Eqs. (8.34) and (8.35) become zero instead of minus infinity. In words, the quantity $h_2(\tau_1, \tau_2)$ exists only in the first quadrant for physically realizable bilinear systems.

A bilinear system is said to be *stable* if every bounded input $x(t)$ produces a bounded output $y_2(t)$. A sufficient condition for a bilinear system to be stable is

$$\iint |h_2(\tau_1, \tau_2)|\, d\tau_1\, d\tau_2 < \infty \tag{8.37}$$

Equation (8.37) should be compared with Eq. (8.15).

For a delta function input, Eq. (8.34) gives the output

$$y_2(t) = L_2[\delta(t)] = h_2(t, t) \tag{8.38}$$

This describes the bilinear weighting function $h_2(\tau_1, \tau_2)$ only along the line $\tau_1 = \tau_2 = t$. The general description of $h_2(\tau_1, \tau_2)$ requires Eq. (8.32). To obtain this result, one can use Eq. (8.31) solved for $L_2[x_1, x_2]$.

$$L_2[x_1, x_2] = \tfrac{1}{2}(L_2[x_1 + x_2] - L_2[x_1] - L_2[x_2]) \tag{8.39}$$

As a special case of Eq. (8.39), let $x_1 = \delta(\tau_1)$ and $x_2 = \delta(\tau_2)$. Now, from Eq. (8.32),

$$h_2(\tau_1, \tau_2) = \tfrac{1}{2}(L_2[\delta(\tau_1) + \delta(\tau_2)] - L_2[\delta(\tau_1)] - L_2[\delta(\tau_2)]) \tag{8.40}$$

This calculation is pictured in Figure 8.4. Note that

$$L_2(x_1, 0) = L_2(0, x_2) = 0 \tag{8.41}$$

when either $x_2 = 0$ or $x_1 = 0$.

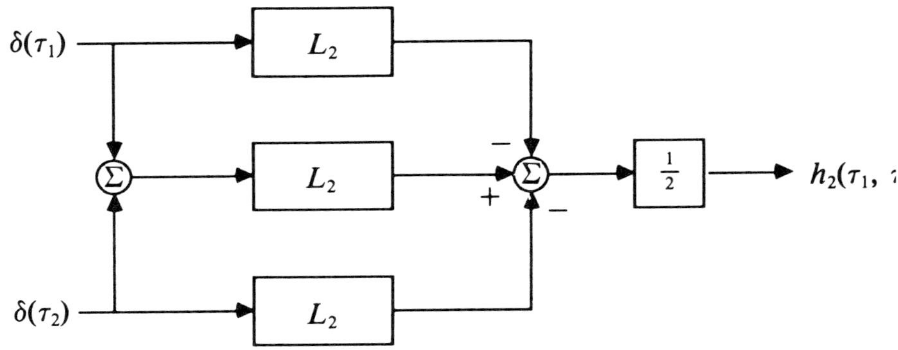

Figure 8.4. Calculation of bilinear weighting function.

The calculation of $y_2(t)$ from $x(t)$ via the double convolution integral expression of Eq. (8.34) is shown in Figure 8.5.

Let $R_{xx}(\tau)$ be the autocorrelation function of $x(t)$ as defined by

$$R_{xx}(\tau) = E[x(t)x(t + \tau)] \tag{8.42}$$

and let $S_{xx}(f)$ be its associated two-sided autospectral density function of $x(t)$ satisfying

$$S_{xx}(f) = \int R_{xx}(\tau)e^{-j2\pi f\tau}\,d\tau \tag{8.43}$$

Then, from Eq. (8.34), since $R_{xx}(\tau) = R_{xx}(-\tau)$, the bilinear output mean value for all t is

$$\bar{y}_2 = E[y_2(t)] = \iint h_2(\tau_1, \tau_2)R_{xx}(\tau_2 - \tau_1)\,d\tau_1\,d\tau_2 \tag{8.44}$$

In general, $\bar{y}_2 \neq 0$.

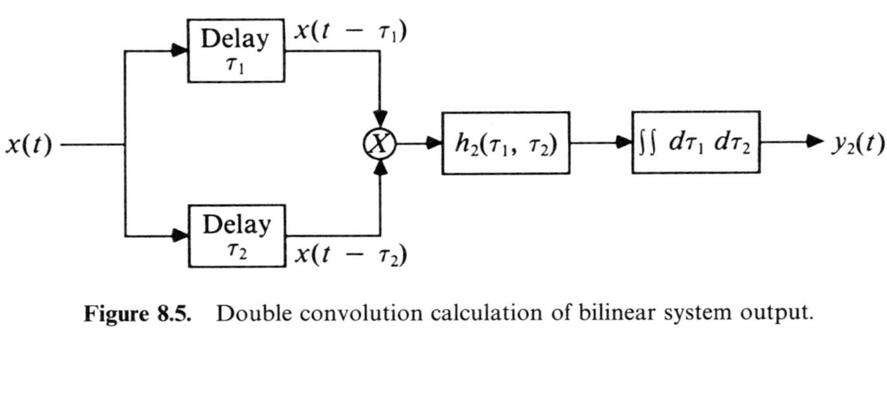

Figure 8.5. Double convolution calculation of bilinear system output.

Consider now the double Fourier transform of $h_2(\tau_1, \tau_2)$ denoted by

$$H_2(f_1, f_2) = \iint h_2(\tau_1, \tau_2)e^{-j2\pi(f_1\tau_1 + f_2\tau_2)}\, d\tau_1\, d\tau_2 \tag{8.45}$$

This quantity $H_2(f_1, f_2)$ is the *bilinear frequency response function* and is called also the *second-order frequency-domain kernel.* It is symmetric with respect to its two frequency variables because of Eq. (8.33).

$$H_2(f_1, f_2) = H_2(f_2, f_1) \tag{8.46}$$

A sufficient condition for $H_2(f_1, f_2)$ to exist is that the bilinear system be stable, Eq. (8.37). From Eq. (8.45), since $h_2(\tau_1, \tau_2)$ is real-valued, it follows that the complex conjugate quantity

$$H_2^*(f_1, f_2) = H_2(-f_1, -f_2)$$

Then, using Eq. (8.46),

$$H_2^*(f, -f) = H_2(-f, f) = H_2(f, -f) \tag{8.47}$$

This proves that $H_2(f, -f)$ is a real-valued even function of f with a phase function that is zero for all f.

The inverse Fourier transform relation to Eq. (8.43) gives

$$R_{xx}(\tau_2 - \tau_1) = \int S_{xx}(f)e^{j2\pi f(\tau_2 - \tau_1)}\, df$$

Substitution into Eq. (8.44) shows that the bilinear output mean value is

$$\begin{aligned} \bar{y}_2 &= \int S_{xx}(f)\left[\iint h_2(\tau_1, \tau_2)e^{-j2\pi(f\tau_1 - f\tau_2)}\, d\tau_1\, d\tau_2\right] df \\ &= \int H_2(f, -f)S_{xx}(f)\, df \end{aligned} \tag{8.48}$$

where $H_2(f, -f)$ can be found from Eq. (8.45). Thus, $\bar{y}_2$ can be determined by Eq. (8.48) in place of Eq. (8.44). This also shows that $H_2(f, -f)$ must be a real-valued even function of f.

Return to Eq. (8.34). Express $x(t - \tau_1)$ and $x(t - \tau_2)$ in terms of their inverse

Fourier transforms. Then, using Eq. (8.45),

$$\begin{aligned} y_2(t) &= \iint h_2(\tau_1, \tau_2)\left[\int X(\alpha)e^{j2\pi\alpha(t-\tau_1)}\, d\alpha\right]\left[\int X(\beta)e^{j2\pi\beta(t-\tau_2)}\, d\beta\right] d\tau_1\, d\tau_2 \\ &= \iint\left[\iint h_2(\tau_1, \tau_2)e^{-j2\pi(\alpha\tau_1+\beta\tau_2)}\, d\tau_1\, d\tau_2\right] X(\alpha)X(\beta)e^{j2\pi(\alpha+\beta)t}\, d\alpha\, d\beta \\ &= \iint H_2(\alpha, \beta)X(\alpha)X(\beta)e^{j2\pi(\alpha+\beta)t}\, d\alpha\, d\beta \end{aligned}$$

Change variables from (α, β) to (α, f) where $f = \alpha + \beta$, $df = d\beta$. Now

$$y_2(t) = \iint H_2(a, f - \alpha)X(\alpha)X(f - \alpha)e^{j2\pi ft}\, d\alpha\, df \tag{8.49}$$

But

$$y_2(t) = \int \mathscr{F}[y_2(t)]e^{j2\pi ft}\, df \tag{8.50}$$

Hence, one obtains the result

$$\mathscr{F}[y_2(t)] = \int H_2(\alpha, f - \alpha)X(\alpha)X(f - \alpha)\, d\alpha \tag{8.51}$$

Equation (8.51) has the form of a single convolution integral.

From Eq. (8.6), the quantity $Y_2(f)$ is defined by

$$Y_2(f) = \mathscr{F}[y_2(t) - \bar{y}_2] = \mathscr{F}[y_2(t)] - \mathscr{F}[\bar{y}_2] \tag{8.52}$$

Note that $Y_2(f) \neq \mathscr{F}[y_2(t)]$ when $\mathscr{F}[\bar{y}_2] \neq 0$. The Fourier transform of the constant $\bar{y}_2$ is given by

$$\mathscr{F}[\bar{y}_2] = \bar{y}_2\delta(f) \approx \bar{y}_2\delta_1(f) \tag{8.53}$$

where $\delta_1(f)$ is the *finite delta function* defined by

$$\delta_1(f) = \begin{cases} T, & (-1/2T) < f < (1/2T) \\ 0, & \text{otherwise} \end{cases} \tag{8.54}$$

The quantity $\delta_1(f)$ is an approximation to the usual theoretical delta function and occurs in digital computation of Fourier transforms with subrecords of

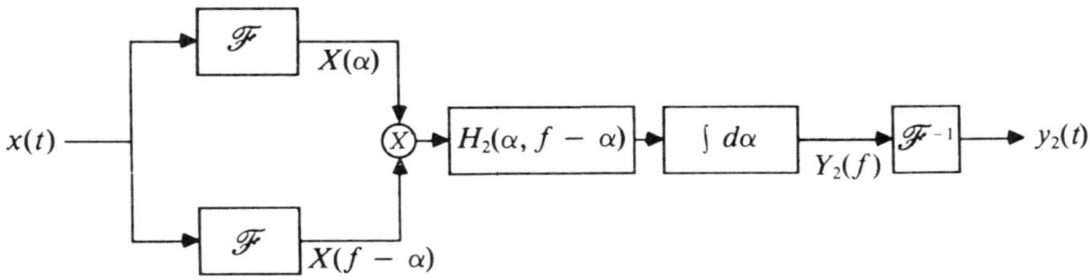

Figure 8.6. Fourier transform calculation of bilinear system output.

finite length T. Properties of $\delta_1(f)$ will often be used instead of $\delta(f)$ throughout this book to obtain more realistic results. From Eqs. (8.51)–(8.54), it follows that

$$Y_2(f) = \int H_2(\alpha, f - \alpha)X(\alpha)X(f - \alpha)\, d\alpha - \bar{y}_2\delta_1(f) \tag{8.55}$$

Note that $Y_2(f) = \mathscr{F}[y_2(t)]$ for all $f \neq 0$. Equation (8.55) should be compared with Eq. (8.19).

Figure 8.6 shows the calculation of $y_2(t)$ from $x(t)$ using Fourier transform procedures as required by Eqs. (8.50) and (8.51). This diagram describes the bilinear system shown in Figure 8.1. Note that $y_2(t)$ involves a multiplication of two quantities. Hence, the bilinear output $y_2(t)$ will contain mixing frequencies when the input contains frequencies f_1 and f_2. When $x(t) = \exp(j2\pi f_1 t) + \exp(j2\pi f_2 t)$, Eqs. (8.34) and (8.45) show that the bilinear output $y_2(t)$ contains the term $H_2(f_1, f_2)\exp[j2\pi(f_1 + f_2)t]$. This input/output pair can be used to determine $H_2(f_1, f_2)$ in a nonlinear differential equation.

8.1.3 Trilinear Systems

Previous ideas extend to trilinear systems (and to higher-order systems) as will now be carried out for trilinear systems. By definition, L_3 is a *trilinear system operator* with respect to a triplet of ordered inputs $\{x(t), y(t), z(t)\}$ if the operator L_3 is linear on $x(t)$ for given $y(t)$ and $z(t)$, and similarly for other combinations. In equation form, for any constants a, b, c, one has

$$\begin{aligned} L_3(a_1x_1 + a_2x_2, y, z) &= a_1L_3(x_1, y, z) + a_2L_3(x_2, y, z) \\ L_3(x, b_1y_1 + b_2y_2, z) &= b_1L_3(x, y_1, z) + b_2L_3(x, y_2, z) \\ L_3(x, y, c_1z_1 + c_2z_2) &= c_1L_3(x, y, z_1) + c_2L_3(x, y, z_2) \end{aligned} \tag{8.56}$$

This then leads without loss of generality to

$$\begin{aligned} L_3(ax + by + cz, ax + by + cz, ax + by + cz) \\ = a^3L_3(x, x, x) + b^3L_3(y, y, y) + c^3L_3(z, z, z) \\ + 3a^2bL_3(x, x, y) + 3ab^2L_3(x, y, y) + 3a^2cL_3(x, x, z) \\ + 3ac^2L_3(x, z, z) + 3b^2cL_3(y, y, z) + 3bc^2L_3(y, z, z) \\ + 6abcL_3(x, y, z) \end{aligned} \tag{8.57}$$

where Eq. (8.57) uses the assumption that $L_3(x, y, z)$ is symmetric with respect to interchange of variables, namely,

$$\begin{aligned} L_3(x, y, z) = L_3(x, z, y) = L_3(y, x, z) = L_3(y, z, x) \\ = L_3(z, x, y) = L_3(z, y, x) \end{aligned} \tag{8.58}$$

For the special case where $b = c = 0$, Eq. (8.57) becomes

$$L_3(ax, ax, ax) = a^3L_3(x, x, x) \tag{8.59}$$

As per Eq. (8.29), define a special trilinear system operator by

$$L_3(x) \equiv L_3(x, x, x) \tag{8.60}$$

Then, Eq. (8.59) shows that $L_3(x)$ is not homogeneous, since

$$L_3(ax) = a^3L_3(x) \tag{8.61}$$

Also, from Eqs. (8.60) and (8.57), $L_3(x)$ is not additive, since

$$\begin{aligned} L_3(x_1 + x_2) \equiv L_3(x_1 + x_2, x_1 + x_2, x_1 + x_2) \\ = L_3(x_1) + 3L_3(x_1, x_1, x_2) + 3L_3(x_1, x_2, x_2) + L_3(x_2) \end{aligned} \tag{8.62}$$

Thus, L_3 like L_2 is a *nonlinear* operator on $x(t)$.

From Volterra theory, the *trilinear weighting function* $h_3(\tau_1, \tau_2, \tau_3)$, also called the *third-order time-domain kernel*, is defined as the response of the trilinear system L_3 at times τ_1, τ_2, τ_3 to a triplet of delta function inputs $\{\delta(\tau_1), \delta(\tau_2), \delta(\tau_3)\}$, namely,

$$h_3(\tau_1, \tau_2, \tau_3) = L_3[\delta(\tau_1), \delta(\tau_2), \delta(\tau_3)] \tag{8.63}$$

From Eq. (8.58), $h_3(\tau_1, \tau_2, \tau_3)$ is unique and is symmetric with respect to its three variables. As per the derivations of Eqs. (8.13) and (8.34), one can now prove that for any input $x(t)$, the trilinear system output response $y_3(t)$ is given by

$$y_3(t) = L_3[x(t)] = \iiint x(t - \tau_1)x(t - \tau_2)x(t - \tau_3)h_3(\tau_1, \tau_2, \tau_3)\, d\tau_1\, d\tau_2\, d\tau_3 \tag{8.64}$$

where the possible triple limits of integration are from $-\infty$ to $+\infty$. Equation (8.64) is the trilinear extension to the linear result of Eq. (8.13) and to the bilinear result of Eq. (8.34).

For a delta function input, Eq. (8.64) gives the output

$$y_3(t) = L_3[\delta(t)] = h_3(t, t, t) \tag{8.65}$$

This describes the trilinear weighting function only along the line $\tau_1 = \tau_2 = \tau_3 = t$. The general description of $h_3(\tau_1, \tau_2, \tau_3)$ requires Eq. (8.63).

If the system L_3 is *physically realizable*, then

$$h_3(\tau_1, \tau_2, \tau_3) = 0$$

$$\text{for } \tau_1 < 0 \text{ and/or } \tau_2 < 0 \text{ and/or } \tau_3 < 0 \tag{8.66}$$

so that the lower limits in Eq. (8.64) become zero instead of minus infinity. In words, $h_3(\tau_1, \tau_2, \tau_3)$ exists only in the first octant for physically realizable trilinear systems. From the Gaussian assumption made on $x(t)$, it follows from Eq. (8.64) that the trilinear output mean value for all t is zero.

$$\bar{y}_3 = E[y_3(t)] = 0 \tag{8.67}$$

A trilinear system is said to be *stable* if every bounded input $x(t)$ produces a bounded output $y_3(t)$. A sufficient condition for a trilinear system to be stable is

$$\iiint |h_3(\tau_1, \tau_2, \tau_3)|\, d\tau_1\, d\tau_2\, d\tau_3 < \infty \tag{8.68}$$

Equation (8.68) should be compared with Eqs. (8.15) and (8.37).

The calculation of $y_3(t)$ from $x(t)$ via the triple convolution integral expression of Eq. (8.64) is shown in Figure 8.7.

Consider now the triple Fourier transform of $h_3(\tau_1, \tau_2, \tau_3)$ denoted by

$$H_3(f_1, f_2, f_3) = \iiint h_3(\tau_1, \tau_2, \tau_3)e^{-j2\pi(f_1\tau_1 + f_2\tau_2 + f_3\tau_3)}\, d\tau_1\, d\tau_2\, d\tau_3 \tag{8.69}$$

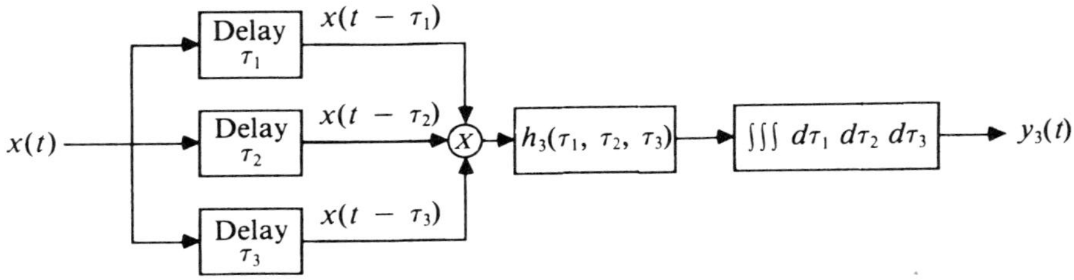

Figure 8.7. Triple convolution calculation of trilinear system output.

This quantity is the *trilinear frequency response function* and is also called the *third-order frequency-domain kernel.* From Eqs. (8.58) and (8.63) this kernel is symmetric with respect to its three variables. A sufficient condition for $H_3(f_1, f_2, f_3)$ to exist is that the trilinear system be stable, Eq. (8.68). A derivation similar to obtaining Eq. (8.51) proves the result

$$Y_3(f) = \mathscr{F}(y_3(t)) = \iint H_3(\alpha, \beta-\alpha, f-\beta) X(\alpha) X(\beta-\alpha) X(f-\beta)\, d\alpha\, d\beta \tag{8.70}$$

This should be compared with Eqs. (8.19) and (8.55). Note that Eq. (8.70) has the form of a double convolution integral.

Figure 8.8 shows the calculation of $y_3(t)$ from $x(t)$ using Fourier transform procedures as required by Eq. (8.70). This diagram describes the trilinear system shown in Figure 8.1. The quantity $y_3(t)$ involves a multiplication of three quantities. Hence, the trilinear output $y_3(t)$ will contain mixing frequencies when the input contains frequencies f_1, f_2, and f_3. When $x(t) = \exp(j2\pi f_1 t) + \exp(j2\pi f_2 t) + \exp(j2\pi f_3 t)$, Eqs. (8.64) and (8.69) show that the trilinear output $y_3(t)$ contains the term $H_3(f_1, f_2, f_3)\exp[j2\pi(f_1 + f_2 + f_3)t]$. This input/output pair can be used to determine $H_3(f_1, f_2, f_3)$ in a nonlinear differential equation.

Important special cases of bilinear and trilinear systems are treated in Section 8.2, where Figures 8.5–8.8 become relatively simple. Specifically, the bilinear systems are assumed to be squaring operations that are either followed

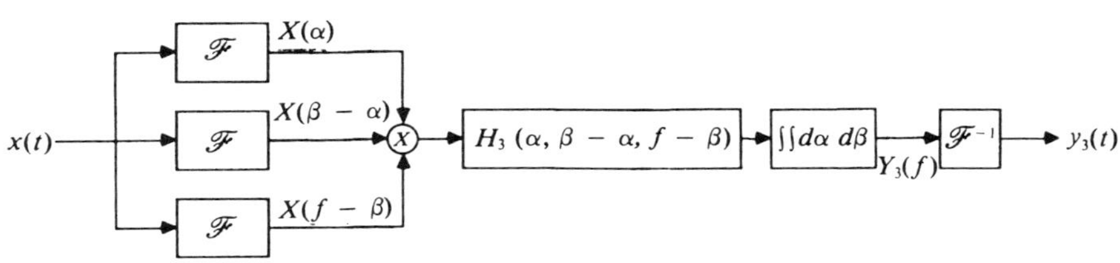

Figure 8.8. Fourier transform calculation of trilinear system output.

or preceded by linear systems, and the trilinear systems are assumed to be cubing operations that are either followed or preceded by linear systems. For these cases, the bilinear and trilinear frequency response functions can be replaced by linear frequency response functions that are the basis for the new nonlinear system analysis and identification procedures in Chapters 3–7.

8.1.4 Summary of Results

A brief summary follows of various basic quantities and formulas that indicate the inner mathematical structure as one proceeds from linear to bilinear to trilinear systems. All limits of integration are theoretically from minus infinity to plus infinity unless stated otherwise. The first three Volterra functionals in Eq. (8.1) are given by

$$y_1(t) = \int h_1(\tau)x(t - \tau)\,d\tau \tag{8.71}$$

$$y_2(t) = \iint h_2(\tau_1, \tau_2)x(t - \tau_1)x(t - \tau_2)\,d\tau_1\,d\tau_2 \tag{8.72}$$

$$y_3(t) = \iiint h_3(\tau_1, \tau_2, \tau_3)x(t - \tau_1)x(t - \tau_1)x(t - \tau_3)\,d\tau_1\,d\tau_2\,d\tau_3 \tag{8.73}$$

with $\bar{y}_1 = \bar{y}_3 = 0$ and $\bar{y}_2$ given by Eqs. (8.44) and (8.48). The time-domain quantities $h_1(\tau)$, $h_2(\tau_1, \tau_2)$, and $h_3(\tau_1, \tau_2, \tau_3)$ are called the *first-*, *second-*, and *third-order time-domain kernels*, respectively. They are also called the *linear*, *bilinear*, and *trilinear weighting functions*, respectively. Their corresponding single, double, and triple Fourier transforms are given by

$$H_1(f) = \int h_1(\tau)e^{-j2\pi f\tau}\,d\tau \tag{8.74}$$

$$H_2(f_1, f_2) = \iint h_2(\tau_1, \tau_2)e^{-j2\pi(f_1\tau_1 + f_2\tau_2)}\,d\tau_1\,d\tau_2 \tag{8.75}$$

$$H_3(f_1, f_2, f_3) = \iiint h_3(\tau_1, \tau_2, \tau_3)e^{-j2\pi(f_1\tau_1 + f_2\tau_2 + f_3\tau_3)}\,d\tau_1\,d\tau_2\,d\tau_3 \tag{8.76}$$

These frequency-domain quantities $H_1(f)$, $H_2(f_1, f_2)$, and $H_3(f_1, f_2, f_3)$ are called the *first-*, *second-*, and *third-order frequency-domain kernels*, respectively. They are also called the *linear*, *bilinear*, and *trilinear frequency response functions*, respectively.

Fourier transforms of Eqs. (8.71)–(8.73) with mean values satisfying Eq. (8.7) yield the formulas for the first three Fourier transform functionals in Eq. (8.4).

$$Y_1(f) = H_1(f)X(f) \tag{8.77}$$

$$Y_2(f) = \int H_2(\alpha, f - \alpha)X(\alpha)X(f - \alpha)\, d\alpha - \bar{y}_2\delta_1(f) \tag{8.78}$$

$$Y_3(f) = \iint H_3(\alpha, \beta - \alpha, f - \beta)X(\alpha)X(\beta - \alpha)X(f - \beta)\, d\alpha\, d\beta \tag{8.79}$$

Results from Eqs. (8.77)–(8.79) can return to Eqs. (8.71)–(8.73), respectively, by taking their inverse Fourier transforms. Thus, the same information is available in either the time domain or the frequency domain. In general, however, the analysis will be simpler when conducted in the frequency domain.

The above formulas describe the output terms in the general third-order nonlinear input/output model of Figure 8.1. Advantages of these formulas are that they are mathematically exact and widely applicable to many problems. Disadvantages are that the Volterra multidimensional bilinear and trilinear weighting functions and frequency response functions are usually very expensive and time-consuming to compute as well as very difficult to interpret.

8.2 EXAMPLES

A number of engineering examples follow of linear, bilinear, and trilinear systems based on formulas in Section 8.1. The bilinear and trilinear systems considered here are finite-memory nonlinear systems represented by zero-memory square-law systems and cubic systems that are preceded or followed by linear systems. The importance of these particular examples for many applications is shown in Chapter 4.

8.2.1 Linear Systems

Example 1. Low-Pass Filter

A low-pass filter can be defined by the linear weighting function

$$\begin{aligned} h_1(\tau) &= Ae^{-a\tau}, \quad A, a > 0, \tau \geqslant 0 \\ &= 0, \qquad \tau < 0 \end{aligned} \tag{8.80}$$

The associated linear frequency response function is

$$\begin{aligned} H_1(f) &= \int_0^\infty h_1(\tau)e^{-j2\pi f\tau}\, d\tau = \frac{A}{a + j2\pi f} \\ &= \frac{A(a - j2\pi f)}{a^2 + (2\pi f)^2} = |H_1(f)|e^{-j\phi_1(f)} \end{aligned} \tag{8.81}$$

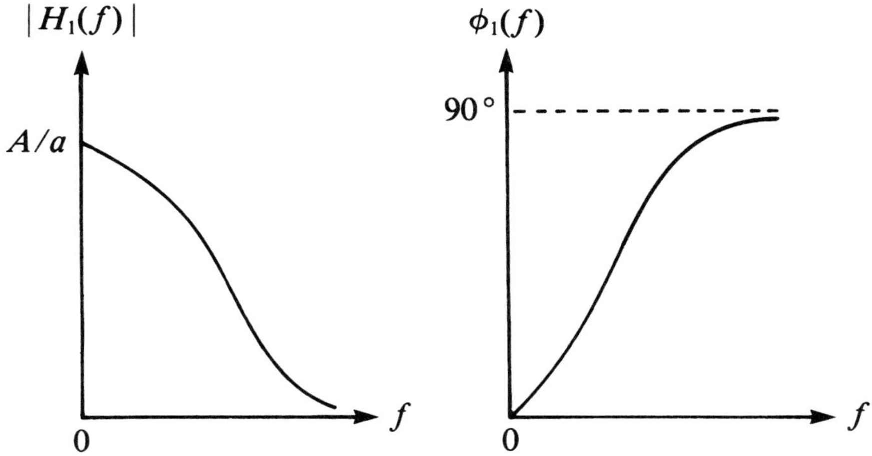

Figure 8.9. Typical gain factor and phase factor for low-pass filter.

Typical plots are shown in Figure 8.9 for the *gain factor* $|H_1(f)|$ and *phase factor* $\phi_1(f)$ given by

$$|H_1(f)| = \frac{A}{\sqrt{a^2 + (2\pi f)^2}} \tag{8.82}$$

$$\phi_1(f) = \tan^{-1}(2\pi f/a) \tag{8.83}$$

Example 2. Force Input/Displacement Output System (SDOF System)

References 1 and 2 derive the linear frequency response function $H_1(f)$ for a simple single-degree-of-freedom (SDOF) mechanical system consisting of a mass, a spring with spring constant k, and a dashpot, when a force is applied to the mass and the output is the resulting displacement of the mass, namely,

$$H_1(f) = \frac{1/k}{1 - (f/f_n)^2 + j2\zeta(f/f_n)} = |H_1(f)|e^{-j\phi_1(f)} \tag{8.84}$$

Here, ζ is a dimensionless quantity called the *damping ratio* and f_n is the *undamped natural frequency*. Typical plots are shown in Figure 8.10 for the *gain factor* $|H_1(f)|$ and *phase factor* $\phi_1(f)$ given by

$$|H_1(f)| = \frac{1/k}{\sqrt{[1 - (f/f_n)^2]^2 + [2\zeta(f/f_n)]^2}} \tag{8.85}$$

$$\phi_1(f) = \tan^{-1}\left(\frac{2\zeta(f/f_n)}{1 - (f/f_n)^2}\right) \tag{8.86}$$

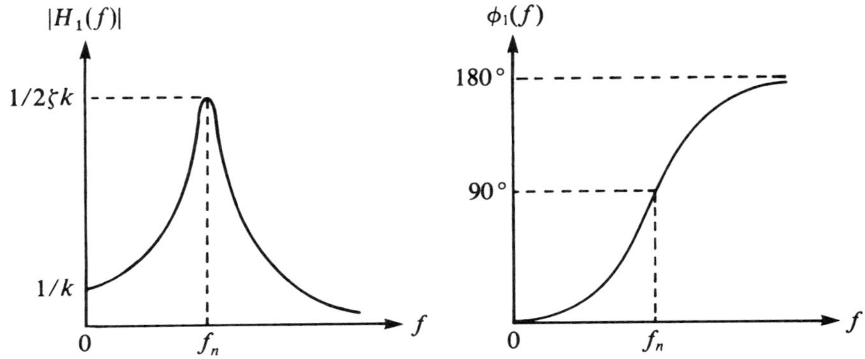

Figure 8.10. Typical gain factor and phase factor for SDOF system.

For a force input $x(t)$ with a Fourier transform $X(f)$ producing a displacement output $y_1(t)$ with Fourier transform $Y_1(f)$, one obtains

$$Y_1(f) = H_1(f)X(f) \tag{8.87}$$

using the $H_1(f)$ of Eq. (8.84). This physical linear model is sketched below.

Force input → $\boxed{H_1(f)}$ →Displacement output

Suppose the roles of $x(t)$ and $y_1(t)$ are reversed. This will give a *reverse linear model* where a displacement input $y_1(t)$ with Fourier transform $Y_1(f)$ produces a force output $x(t)$ with Fourier transform $X(f)$. Now, from Eq. (8.87),

$$X(f) = [H_1(f)]^{-1}Y_1(f) = A_1(f)Y_1(f) \tag{8.88}$$

where

$$A_1(f) = [H_1(f)]^{-1} = k[1 - (f/f_n)^2 + j2\zeta(f/f_n)] \tag{8.89}$$

using the $H_1(f)$ of Eq. (8.84). Here the inverse system $A_1(f)$ exists so that a mathematical reverse linear model is obtained as sketched below.

Displacement input → $\boxed{A_1(f)}$ →Force output

If the notation for $X(f)$ and $Y_1(f)$ is changed so that $X(f)$ represents the displacement input and $Y_1(f)$ represents the force output, then Eq. (8.88) takes the form of Eq. (8.87) with $H_1(f)$ replaced by $A_1(f)$, namely, $Y_1(f) = A_1(f)X(f)$. This transpose in roles preserves the usual meaning of keeping $x(t)$ and $X(f)$ for the input notation with $y_1(t)$ and $Y_1(f)$ for the output notation. This reverse idea was used in Section 4.9.3 in connection with nonlinear differential equations of motion.

8.2.2 Bilinear Systems

Example 3. Squarer (Square-Law) System

$$x(t) \rightarrow \boxed{\text{Squarer}} \rightarrow y_2(t) = x^2(t)$$

The input $x(t)$ can be represented by

$$x(t) = \int \delta(\tau)x(t - \tau)\, d\tau \tag{8.90}$$

where $\delta(\tau)$ is the usual delta function. The squarer output $y_2(t)$ is then given by

$$\begin{aligned} y_2(t) = x^2(t) &= \left[\int \delta(\tau_1)x(t - \tau_1)\, d\tau_1\right]\left[\int \delta(\tau_2)x(t - \tau_2)\, d\tau_2\right] \\ &= \iint \delta(\tau_1)\delta(\tau_2)x(t - \tau_1)x(t - \tau_2)\, d\tau_1\, d\tau_2 \end{aligned} \tag{8.91}$$

From Eq. (8.34), the output $y_2(t)$ is also

$$y_2(t) = \iint h_2(\tau_1, \tau_2)x(t - \tau_1)x(t - \tau_2)\, d\tau_1\, d\tau_2 \tag{8.92}$$

Hence, the bilinear weighting function must be

$$h_2(\tau_1, \tau_2) = \delta(\tau_1)\delta(\tau_2) \tag{8.93}$$

Note that this function has no memory and exists only at $\tau_1 = \tau_2 = 0$. The associated bilinear frequency response function is a constant independent of frequency, namely,

$$H_2(f_1, f_2) = 1 \quad \text{for all } f_1, f_2 \tag{8.94}$$

From Eq. (8.55), one obtains for $Y_2(f) = \mathscr{F}[y_2(t) - \bar{y}_2]$ the result

$$Y_2(f) = \int X(\alpha)X(f - \alpha)\, d\alpha - \bar{y}_2\delta_1(f) \tag{8.95}$$

with output mean value from Eq. (8.48) given by

$$\bar{y}_2 = \int S_{xx}(f)\, df = \sigma_x^2 \tag{8.96}$$

Consider a sinusoidal input at frequency f_0 where

$$x(t) = \sin(2\pi f_0 t) \tag{8.97}$$

From basic trigonometry,

$$y_2(t) = \sin^2(2\pi f_0 t) = \tfrac{1}{2} - \tfrac{1}{2}\cos 4\pi f_0 t \tag{8.98}$$

Thus, the output $y_2(t)$ contains a constant term $(\frac{1}{2})$, plus a nonlinear term at the sinusoidal $2f_0$ frequency. The output constant term shows that $y_2(t)$ will have a nonzero mean value. The output frequency $2f_0$ term shows that squaring operations always produce output frequencies that are twice the input frequencies. This property identifies a square-law system.

Example 4. Squarer Preceded by Linear System

$$x(t) \longrightarrow \boxed{h(\tau)} \xrightarrow[u(t)]{} \boxed{\text{Squarer}} \longrightarrow y_2(t) = u^2(t)$$

This example involves memory operations before the squarer. The output is

$$\begin{aligned} y_2(t) = u^2(t) &= \left[\int h(\tau)x(t-\tau)\, d\tau\right]^2 \\ &= \left[\int h(\tau_1)x(t-\tau_1)\, d\tau_1\right]\left[\int h(\tau_2)x(t-\tau_2)\, d\tau_2\right] \\ &= \iint h(\tau_1)h(\tau_2)x(t-\tau_1)x(t-\tau_2)\, d\tau_1\, d\tau_2 \end{aligned} \tag{8.99}$$

From Eq. (8.34) the bilinear weighting function is here

$$h_2(\tau_1, \tau_2) = h(\tau_1)h(\tau_2) \tag{8.100}$$

Note that this bilinear weighting function is the product of two linear weighting functions.

From Eqs. (8.45) and (8.18), the associated bilinear frequency response function is

$$H_2(f_1, f_2) = H(f_1)H(f_2) \tag{8.101}$$

Note the product form of the linear frequency response functions.

As a specific case, suppose the linear system is a low-pass filter where $h(\tau)$ is a damped exponential function $Ae^{-a\tau}$. For this case, substituting Eqs. (8.80) and (8.81) into Eqs. (8.100) and (8.101) gives

$$h_2(\tau_1, \tau_2) = A^2 e^{-a(\tau_1 + \tau_2)}$$

$$H_2(f_1, f_2) = \frac{A^2}{(a + j2\pi f_1)(a + j2\pi f_2)} \tag{8.102}$$

From Eq. (8.55), the Fourier transform $Y_2(f)$ is

$$Y_2(f) = \int H_2(\alpha, f - \alpha) X(a) X(f - \alpha)\, d\alpha - \bar{y}_2 \delta_1(f) \tag{8.103}$$

where $H_2(\alpha, f - \alpha)$ is given here by

$$\begin{aligned} H_2(\alpha, f - \alpha) &= H(\alpha) H(f - \alpha) \\ &= \frac{A^2}{(a + j2\pi f\alpha)[a + j2\pi(f - \alpha)]} \end{aligned} \tag{8.104}$$

Also,

$$Y_2(f) = \int U(\alpha) U(f - \alpha)\, d\alpha - \bar{y}_2 \delta_1(f) \tag{8.105}$$

where

$$U(\alpha) = H(\alpha) X(\alpha) = \frac{AX(\alpha)}{a + j2\pi f\alpha} \tag{8.106}$$

The output mean value $\bar{y}_2$ in this example is

$$\bar{y}_2 = E[y_2(t)] = E[u^2(t)] = \int S_{uu}(f)\, df \tag{8.107}$$

where for linear transformations defined by $H(f)$,

$$S_{uu}(f) = |H(f)|^2 S_{xx}(f) \tag{8.108}$$

Thus, for any $H(f)$ in Example 4,

$$\bar{y}_2 = \int |H(f)|^2 S_{xx}(f)\, df \tag{8.109}$$

From Eq. (8.48), a general formula for $\bar{y}_2$ is

$$\bar{y}_2 = \int H_2(f, -f) S_{xx}(f)\, df \tag{8.110}$$

This provides another way to derive Eq. (8.109) by substituting Eq. (8.101) into Eq. (8.110).

For the low-pass filter case of Eq. (8.102),

$$H_2(f, -f) = H(f)H(-f) = |H(f)|^2 = \frac{A^2}{a^2 + 4\pi^2 f^2} \tag{8.111}$$

Note that this $H_2(f, -f)$ satisfies the general property of being a real-valued even function of f. Here, for arbitrary $S_{xx}(f)$, Eq. (8.109) or (8.110) gives the output mean value

$$\bar{y}_2 = \int |H(f)|^2 S_{xx}(f)\, df = \int \frac{A^2 S_{xx}(f)}{a^2 + 4\pi^2 f^2}\, df \tag{8.112}$$

1. Suppose $S_{xx}(f) = 1$ for all f over $(-\infty, \infty)$. For this situation of a theoretical white noise input,

$$\bar{y}_2 = \int_{-\infty}^{\infty} \frac{A^2}{a^2 + 4\pi^2 f^2}\, df = \frac{A^2}{2a} \tag{8.113}$$

2. Suppose $S_{xx}(f) = 1$ for $-B \leqslant f \leqslant B$, and is zero outside. For this situation of a bandwidth-limited white noise input,

$$\bar{y}_2 = \int_{-B}^{B} \frac{A^2}{a^2 + 4\pi^2 f^2}\, df = \frac{A^2}{\pi a} \tan^{-1}\left(\frac{B}{a}\right) \tag{8.114}$$

As a check, Eq. (8.114) yields Eq. (8.113) as B approaches infinity.

The example of a squarer preceded by a linear system, Eq. (8.99), can be pictured as the product of the outputs from two identical linear transformations.

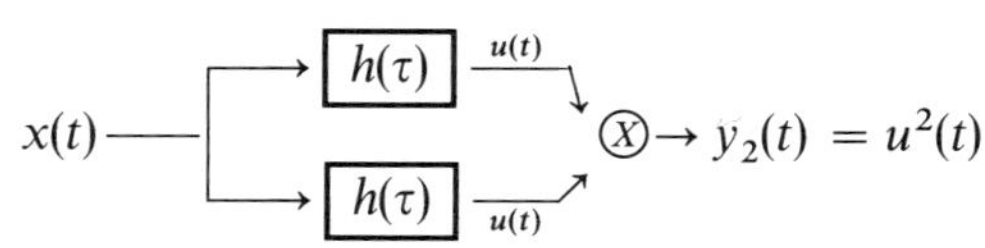

An extension of this example occurs when the two linear transformations have different weighting functions, say, $h_a(\tau)$ and $h_b(\tau)$ where $h_a(\tau) \neq h_b(\tau)$. Different outputs $u_a(t)$ and $u_b(t)$ are produced where $u_a(t) \neq u_b(t)$ with $y_2(t) = u_a(t)u_b(t)$.

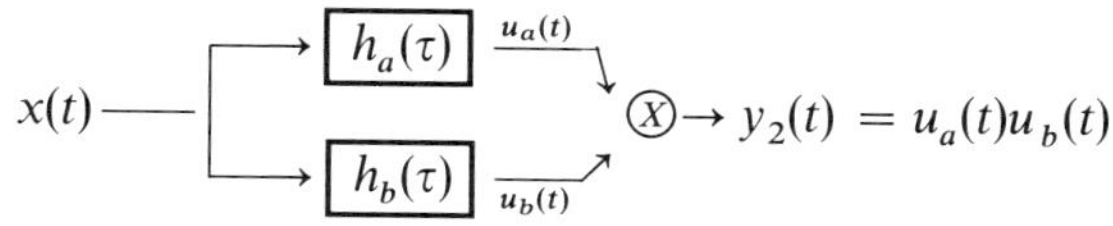

Now, instead of Eqs. (8.100) and (8.101), the results are

$$h_2(\tau_1, \tau_2) = h_a(\tau_1)h_b(\tau_2) \tag{8.115}$$

$$H_2(f_1, f_2) = H_a(f_1)H_b(f_2) \tag{8.116}$$

These bilinear functions are not symmetric. To obtain symmetric results, as shown in Eqs. (8.33) and (8.46), Eqs. (8.115) and (8.116) should be replaced by

$$h_2(\tau_1, \tau_2) = \tfrac{1}{2}[h_a(\tau_1)h_b(\tau_2) + h_a(\tau_2)h_b(\tau_1)] \tag{8.117}$$

$$H_2(f_1, f_2) = \tfrac{1}{2}[H_a(f_1)H_b(f_2) + H_a(f_2)H_b(f_1)] \tag{8.118}$$

Example 5. Squarer Followed by Linear System

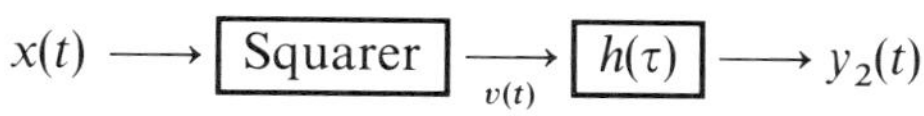

This example involves memory operations after the squarer. The output is

$$y_2(t) = \int h(\tau_1)v(t - \tau_1)\,d\tau_1 = \int h(\tau_1)x^2(t - \tau_1)\,d\tau_1$$

where

$$x^2(t - \tau_1) = \int \delta(\tau_1 - \tau_2)x(t - \tau_1)x(t - \tau_2)\,d\tau_2$$

Hence,

$$y_2(t) = \iint h(\tau_1)\delta(\tau_1 - \tau_2)x(t - \tau_1)x(t - \tau_2)\,d\tau_1\,d\tau_2 \tag{8.119}$$

From Eq. (8.34), the bilinear weighting function is here

$$h_2(\tau_1, \tau_2) = h(\tau_1)\delta(\tau_1 - \tau_2) \tag{8.120}$$

This memory result exists only on the line $\tau_1 = \tau_2$. From Eqs. (8.45) and (8.18) the associated bilinear frequency response function is

$$H_2(f_1, f_2) = H(f_1 + f_2) \tag{8.121}$$

Note the frequency additive form of the linear frequency response function.

As a specific case, consider the linear low-pass filter system where $h(r) = Ae^{-a\tau}$. For this case, substituting Eqs. (8.80) and (8.81) into Eqs. (8.120) and (8.121) gives

$$\begin{aligned} h_2(\tau_1, \tau_2) &= Ae^{-a\tau_1}\delta(\tau_1 - \tau_2) \\ H_2(f_1, f_2) &= \frac{A}{a + j2\pi(f_1 + f_2)} \end{aligned} \tag{8.122}$$

Equations (8.121) and (8.122) show here that

$$H_2(\alpha, f - \alpha) = H_2(f) = \frac{A}{a + j2\pi f} \tag{8.123}$$

Hence, from Eqs. (8.55) and (8.123), the Fourier transform

$$Y_2(f) = \frac{A}{a + j2\pi f} \int X(\alpha)X(f - \alpha)\, d\alpha - \bar{y}_2\delta_1(f) \tag{8.124}$$

Also, Eqs. (8.121) and (8.122) show that

$$H_2(f, -f) = H(0) = \frac{A}{a} \tag{8.125}$$

which is real-valued as required. It follows now from Eq. (8.48) that the output mean value

$$\bar{y}_2 = H(0) \int S_{xx}(f)\, df = H(0)\sigma_x^2 \tag{8.126}$$

This result applies to any $S_{xx}(f)$.

For a bandwidth-limited white noise input where $S_{xx}(f) = 1$ exists only for $-B \leqslant f \leqslant B$, Eq. (8.126) gives the special result

$$\bar{y}_2 = H(0) \int_{-B}^{B} S_{xx}(f)\, df = 2BH(0) \tag{8.127}$$

Example 6. Squarer Preceded and Followed by Linear Systems

Time-Domain Relations

$$x(t) \longrightarrow \boxed{h_b(\tau)} \underset{u(t)}{\longrightarrow} \boxed{\text{Squarer}} \underset{v(t)}{\longrightarrow} \boxed{h_a(\tau)} \longrightarrow y_2(t)$$

$$u(t) = \int h_b(\tau)x(t - \tau)\, dt$$

$$v(t) = u^2(t) = \iint h_b(\tau_1)h_b(\tau_2)x(t - \tau_1)x(t - \tau_2)\, d\tau_1\, d\tau_2$$

$$y_2(t) = \int h_a(\tau)v(t - \tau) + d\tau$$

$$v(t - \tau) = \iint h_b(\tau_1)h_b(\tau_2)x(t - \tau - \tau_1)x(t - \tau - \tau_2)\, d\tau_1\, d\tau_2$$

$$= \iint h_b(\sigma_1 - \tau)h_b(\sigma_2 - \tau)x(t - \sigma_1)x(t - \sigma_2)\, d\sigma_1\, d\sigma_2$$

by changing variables to $\sigma_1 = \tau + \tau_1$ and $\sigma_2 = \tau + \tau_2$. Thus,

$$y_2(t) = \iiint h_b(\sigma_1 - \tau)h_b(\sigma_2 - \tau)h_a(\tau)x(t - \sigma_1)x(t - \sigma_2)\, d\sigma_1\, d\sigma_2\, d\tau \quad (8.128)$$

Now, from Eq. (8.34),

$$h_2(\sigma_1, \sigma_2) = \int h_b(\sigma_1 - \tau)h_b(\sigma_2 - \tau)h_a(\tau)\, d\tau \quad (8.129)$$

This time-domain bilinear result is complicated to compute and interpret.

Frequency-Domain Relations

$$X(f) \longrightarrow \boxed{H_b(f)} \underset{U(f)}{\longrightarrow} \boxed{\text{Squarer}} \underset{V(f)}{\longrightarrow} \boxed{H_a(f)} \longrightarrow Y_2(f)$$

$$U(f) = H_b(f)X(f)$$

For $f \neq 0$,

$$V(f) = \int U(\alpha)U(f - \alpha)\, d\alpha$$

$$= \int H_b(\alpha)H_b(f - \alpha)X(\alpha)X(f - \alpha)\, d\alpha$$

Then

$$Y_2(f) = H_a(f)V(f) = H_a(f)\int H_b(\alpha)H_b(f-\alpha)X(\alpha)X(f-\alpha)\,d\alpha$$

Now, from Eq. (8.55), for $f \neq 0$,

$$H_2(\alpha, f-\alpha) = H_b(\alpha)H_b(f-\alpha)H_a(f) \tag{8.130}$$

Letting $\alpha = f_1$ and $(f-\alpha) = f_2$ yields the result

$$H_2(f_1, f_2) = H_b(f_1)H_b(f_2)H_a(f_1+f_2) \tag{8.131}$$

This frequency-domain bilinear result is simpler than Eq. (8.129) to compute and interpret. Equation (8.131) is the double Fourier transform of Eq. (8.129). Special cases of Example 6 are Examples 3–5. More complicated examples that require intricate three-dimensional plots are discussed in references 22, 29, and 57. These and other such Volterra results are not practical to implement.

8.2.3 Trilinear Systems

Example 7. Cuber (Cubic) System

$$x(t) \longrightarrow \boxed{\text{Cuber}} \longrightarrow y_3(t) = x^3(t)$$

The input $x(t)$ can be represented by

$$x(t) = \int \delta(\tau)x(t-\tau)\,d\tau \tag{8.132}$$

The cuber output $y_3(t)$ is then given by

$$y_3(t) = x^3(t) = \iiint \delta(\tau_1)\delta(\tau_2)\delta(\tau_3)x(t-\tau_1)x(t-\tau_2)x(t-\tau_3)\,d\tau_1\,d\tau_2\,d\tau_3 \tag{8.133}$$

Hence, from Eq. (8.64), the trilinear weighting function is

$$h_3(\tau_1, \tau_2, \tau_3) = \delta(\tau_1)\delta(\tau_2)\delta(\tau_3) \tag{8.134}$$

Note that this function has no memory and exists only at $\tau_1 = \tau_2 = \tau_3 = 0$. The associated trilinear frequency response function is a constant independent of frequency, namely,

$$H_3(f_1, f_2, f_3) = 1 \quad \text{for all } f_1, f_2, f_3 \tag{8.135}$$

From Eq. (8.70), the Fourier transform $Y_3(f)$ is

$$Y_3(f) = \iint X(\alpha)X(\beta - \alpha)X(f - \beta)\, d\alpha\, d\beta \tag{8.136}$$

Consider a sinusoidal input at frequency f_0 where

$$x(t) = \sin 2\pi f_0 t \tag{8.137}$$

From basic trigonometry,

$$y_3(t) = \sin^3(2\pi f_0 t) = \tfrac{3}{4}\sin 2\pi f_0 t - \tfrac{1}{4}\sin 6\pi f_0 t \tag{8.138}$$

Here, the output $y_3(t)$ contains a linear term at the same sinusoidal input frequency f_0 plus a nonlinear term at the sinusoidal $3f_0$ frequency. The output linear term shows that $y_3(t)$ will be correlated with $x(t)$ as well as with any other linear operations on $x(t)$. The output frequency $3f_0$ term shows that cubing operations always produce output frequencies that are triple the input frequencies. This property identifies a cubic system.

Example 8. Cuber Preceded by Linear System

$$x(t) \longrightarrow \boxed{h(\tau)} \xrightarrow[u(t)]{} \boxed{\text{Cuber}} \longrightarrow y_3(t)$$

This example involves memory operations before the cuber. The output is

$$\begin{aligned} y_3(t) &= u^3(t) = \left[\int h(\tau)x(t - \tau)\, d\tau\right]^3 \\ &= \iiint h(\tau_1)h(\tau_2)h(\tau_3)x(t - \tau_1)x(t - \tau_2)x(t - \tau_3)\, d\tau_1\, d\tau_2\, d\tau_3 \end{aligned} \tag{8.139}$$

From Eqs. (8.64) and (8.69), the trilinear weighting function and frequency response function are

$$h_3(\tau_1, \tau_2, \tau_3) = h(\tau_1)h(\tau_2)h(\tau_3) \tag{8.140}$$

$$H_3(f_1, f_2, f_3) = H(f_1)H(f_2)H(f_3) \tag{8.141}$$

Note the product form of the linear results as in Example 4. This gives

$$H_3(\alpha, \beta - \alpha, f - \beta) = H(\alpha)H(\beta - \alpha)H(f - \beta) \tag{8.142}$$

so that Eq. (8.70) becomes

$$Y_3(f) = \iint H(\alpha)H(\beta - \alpha)H(f - \beta)X(\alpha)X(\beta - \alpha)X(f - \beta)\, d\alpha\, d\beta \tag{8.143}$$

Example 9. Cuber Followed by Linear System

$$x(t) \longrightarrow \boxed{\text{Cuber}} \longrightarrow \boxed{h(\tau)} \longrightarrow y_3(t)$$

This example involves memory operations after the cuber. The output is

$$y_3(t) = \int h(\tau_1)v(t - \tau_1)\, d\tau_1 = \int h(\tau_1)x^3(t - \tau_1)\, d\tau_1$$

$$x^3(t - \tau_1) = \iint \delta(\tau_1 - \tau_2)\delta(\tau_1 - \tau_3)x(t - \tau_1)x(t - \tau_2)x(t - \tau_3)\, d\tau_2\, d\tau_3$$

Thus,

$$y_3(t) = \iiint h(\tau_1)\delta(\tau_1 - \tau_2)\delta(\tau_1 - \tau_3)x(t - \tau_1)x(t - \tau_2)x(t - \tau_3)\, d\tau_1\, d\tau_2\, d\tau_3 \tag{8.144}$$

From Eqs. (8.64) and (8.69), one obtains here the trilinear weighting function

$$h_3(\tau_1, \tau_2, \tau_3) = h(\tau_1)\delta(\tau_1 - \tau_2)\delta(\tau_1 - \tau_3) \tag{8.145}$$

This memory result exists only on the line $\tau_1 = \tau_2 = \tau_3$. The associated trilinear frequency response function is

$$H_3(f_1, f_2, f_3) = H(f_1 + f_2 + f_3) \tag{8.146}$$

Note the frequency additive form of the linear result as in Example 5. This gives

$$H_3(\alpha, \beta - \alpha, f - \beta) = H(f) \tag{8.147}$$

so that Eq. (8.70) becomes

$$x(t) = H(f)\iint X(\alpha)X(\beta - \alpha)X(f - \beta)\, d\alpha\, d\beta \tag{8.148}$$

Example 10. Cuber Preceded and Followed by Linear Systems

Time-Domain Relations

$$x(t) \longrightarrow \boxed{h_b(\tau)} \underset{u(t)}{\longrightarrow} \boxed{\text{Cuber}} \underset{v(t)}{\longrightarrow} \boxed{h_a(\tau)} \longrightarrow y_3(t)$$

$$u(t) = \int h_b(\tau)x(t - \tau)\,d\tau$$

$$v(t) = u^3(t)$$

$$= \iiint h_b(\tau_1)h_b(\tau_2)h_b(\tau_3)x(t - \tau_1)x(t - \tau_2)x(t - \tau_3)\,d\tau_1\,d\tau_2\,d\tau_3$$

$$y_3(t) = \int h_a(\tau)v(t - \tau)\,d\tau$$

One can new verify by substitutions and changes of variables that

$$y_3(t) = \iiiint h_b(\tau_1 - \tau)h_b(\tau_2 - \tau)h_b(\tau_3 - \tau)h_a(\tau) \times x(t - \tau_1)x(t - \tau_2)x(t - \tau_3)\,d\tau_1\,d\tau_2\,d\tau_3\,d\tau \tag{8.149}$$

Equation (8.64) then gives

$$h_3(\tau_1, \tau_2, \tau_3) = \int h_b(\tau_1 - \tau)h_b(\tau_2 - \tau)h_b(\tau_3 - \tau)h_a(\tau)\,d\tau \tag{8.150}$$

This time-domain trilinear result is complicated to compute and interpret.

Frequency-Domain Relations

$$X(f) \longrightarrow \boxed{H_b(f)} \underset{U(f)}{\longrightarrow} \boxed{\text{Cuber}} \underset{V(f)}{\longrightarrow} \boxed{H_a(f)} \longrightarrow Y_3(f)$$

$$U(f) = H_b(f)X(f)$$

$$V(f) = \int U(\alpha)U(\beta - \alpha)U(f - \beta)\,d\alpha\,d\beta$$

$$= \int H_b(\alpha)H_b(\beta - \alpha)H_b(f - \beta)X(\alpha)X(\beta - \alpha)X(f - \beta)\,d\alpha\,d\beta$$

$$Y_3(f) = H_a(f)V(f)$$

Substitutions then prove

$$H_3(\alpha, \beta - \alpha, f - \beta) = H_b(\alpha)H_b(\beta - \alpha)H_b(f - \beta)H_a(f) \tag{8.151}$$

Letting $\alpha = f_1$, $(\beta - \alpha) = f_2$, and $(f - \beta) = f_3$ yields the result

$$H_3(f_1, f_2, f_3) = H_b(f_1)H_b(f_2)H_b(f_2)H_a(f_1 + f_2 + f_3) \tag{8.152}$$

This frequency-domain trilinear result is simpler than Eq. (8.150) to compute and interpret. Equation (8.152) is the triple Fourier transform of Eq. (8.150). Special cases of Example 10 are Examples 7–9. Other complicated Volterra examples can require four-dimensional plots.

8.3 SYNTHESIS OF TIME-VARYING LINEAR SYSTEMS

As derived in Eq. (8.13), a *constant-parameter linear system* can be described by its weighting function $h(\tau)$, which gives the response of the system at any time to a unit impulse function applied τ time units earlier. For any input $x(t)$, the output $y(t)$ is

$$y(t) = \int h(\tau)x(t - \tau)\, d\tau \tag{8.153}$$

where the theoretical limits of integration can be from $-\infty$ to $+\infty$. For physically realizable systems, $h(\tau) = 0$ for $\tau < 0$. A *time-varying linear system* can be described by means of a time-varying weighting function $h(\tau, t)$, which gives the response at time t to a unit impulse function applied at time $(t - \tau)$. Not only is the time interval τ important, but also the particular time t at which the system is being observed. Such physical situations are governed by linear integrodifferential equations with time-varying coefficients that may be more realistic than constant-parameter coefficients. The appropriate formula for the output $y(t)$ to any input $x(t)$ follows immediately from the constant-parameter case by merely replacing $h(\tau)$ by the more general $h(\tau, t)$. Thus, the output $y(t)$ from any input $x(t)$ is now

$$y(t) = \int h(\tau, t)x(t - \tau)\, d\tau \tag{8.154}$$

For a physically realizable system, $h(\tau, t) = 0$ for $\tau < 0$.

The following theorem provides new insight on the basic nature of time-varying linear systems to constant-parameter bilinear systems.

Theorem (Schetzen) Any time-varying linear system can be synthesized by a constant-parameter bilinear system where the bilinear system will be physically realizable if and only if the time-varying linear system is physically realizable.

Proof Assume that $y(t)$ is the output of a time-varying linear system as per Eq. (8.154), where the quantity $h(\tau, t)$ is known. For the same input $x(t)$ used in Eq. (8.154), a bilinear weighting function $h_2(\tau, \beta)$ will give the output of Eq.

(8.34), namely,

$$y_2(t) = \iint h_2(\tau, \beta)x(t - \beta)x(t - \tau)\, d\beta\, d\tau \tag{8.155}$$

This $y_2(t)$ will be the same as $y(t)$ of Eq. (8.154) if

$$h(\tau, t) = \int h_2(\tau, \beta)x(t - \beta)\, d\beta \tag{8.156}$$

Equation (8.156) shows that $h_2(\tau, \beta) = 0$ for $\tau < 0$ if and only if $h(\tau, t) = 0$ for $\tau < 0$ so that the bilinear system will be physically realizable if and only if the time-varying linear system is physically realizable. It will now be shown how to obtain $h_2(\tau, \beta)$ from knowledge of $h(\tau, t)$ and $x(t)$.

Let the quantity $H(\tau, g)$ be defined as the Fourier transform of $h(\tau, t)$ with respect to t holding τ constant. Then, from Eq. (8.156),

$$\begin{aligned} H(\tau, g) &= \int h(\tau, t)e^{-j2\pi gt}\, dt \\ &= \iint h_2(\tau, \beta)x(t - \beta)e^{-j2\pi gt}\, d\beta\, dt \\ &= \int \left[\int x(t - \beta)e^{-j2\pi g(t-\beta)}\, dt\right] h_2(\tau, \beta)e^{-j2\pi g\beta}\, d\beta \end{aligned} \tag{8.157}$$

For the inside integral, let $u = t - \beta$, $du = dt$ to obtain

$$H(\tau, g) = X(g) \int h_2(\tau, \beta)e^{-j2\pi g\beta}\, d\beta \tag{8.158}$$

where $X(g)$ is the Fourier transform

$$X(g) = \int x(u)e^{-j2\pi gu}\, du \tag{8.159}$$

Now define $J(f, g)$ to be the Fourier transform of $H(\tau, g)$ with respect to τ holding g constant. Then

$$\begin{aligned} J(f, g) &= \int H(\tau, g)e^{-j2\pi f\tau}\, d\tau \\ &= X(g) \iint h_2(\tau, \beta)e^{-j2\pi(f\tau + g\beta)}\, d\tau\, d\beta \\ &= X(g)H_2(f, g) \end{aligned} \tag{8.160}$$

where, from Eq. (8.45), $H_2(f, g)$ is the bilinear frequency response function of the bilinear system. Observe that $J(f, g)$ is the double Fourier transform of the given $h(\tau, t)$ for the time-varying linear system, namely,

$$J(f, g) = \iint h(\tau, t) e^{-j2\pi(f\tau + gt)} \, d\tau \, dt \tag{8.161}$$

Thus, from calculation of $J(f, g)$ and $X(g)$ one can obtain $H_2(f, g)$ by

$$H_2(f, g) = \frac{J(f, g)}{X(g)} \tag{8.162}$$

This completes the proof and defines the appropriate constant-parameter bilinear system to synthesize the given time-varying linear system from knowledge of $h(\tau, t)$ and $x(t)$. Relations derived here are special cases of results proved by Schetzen in reference 52.

8.4 HIGHER-ORDER CORRELATION AND SPECTRA

Appropriate formulas will now be stated for higher-order cross-correlation and cross-spectral density functions where results in Section 1.3 are first-order cases. These formulas will be used in Chapter 9 to help analyze and identify the dynamic input/output properties of Gaussian stationary random data passing through bilinear and trilinear systems.

8.4.1 First-Order Functions

Consider any pair of real-valued zero mean value stationary random data $x(t)$ and $y(t)$. The *(first-order) cross-correlation function* between $x(t)$ and $y(t)$ is defined by

$$R_{xy}(\tau) = E[x(t)y(t + \tau)] = E[x(t - \tau)y(t)] \tag{8.163}$$

where $E[\ ]$ denotes an expected value operation. The quantities $R_{xx}(\tau)$ and $R_{yy}(\tau)$ are called *(first-order) autocorrelation functions.*

Assume that $x(t)$ and $y(t)$ are each of length T_{total} and that their Fourier transforms $X(f)$ and $Y(f)$ are obtained by dividing T_{total} into n_d disjoint parts, each of which is of length T. This yields discrete estimates of $X(f)$ and $Y(f)$ spaced $(1/T)$ apart. The inverse Fourier transform of $X(f)$ gives

$$x(t) = \int X(f) e^{j2\pi ft} \, df = \int X^*(f) e^{-j2\pi ft} \, df$$

where $X^*(f)$ is the complex conjugate of $X(f)$. Also,

$$y(t+\tau) = \int Y(g) e^{j2\pi g(t+\tau)}\, dg$$

Thus, Eq. (8.163) is the same as

$$\begin{aligned} R_{xy}(\tau) &= E\left[\iint X^*(f) Y(g) e^{-j2\pi f t} e^{j2\pi g(t+\tau)}\, df\, dg\right] \\ &= \iint E[X^*(f)Y(g)] e^{-j2\pi(f-g)t} e^{j2\pi g\tau}\, dg\, df \\ &= \int S_{xy}(f) e^{j2\pi f\tau}\, df \end{aligned} \tag{8.164}$$

where the double integral is a function of t and the final single integral is independent of t. This equation shows that $R_{xy}(\tau)$ is the inverse Fourier transform of $S_{xy}(f)$. It is the Wiener-Khinchin relation in reference 1.

For the last equation to eliminate t, $X(f)$ and $Y(g)$ should each be computed at discrete frequencies f and g that are spaced $\Delta f = (1/T)$ apart such that

$$E[X^*(f)Y(g)] = S_{xy}(f)\delta_1(f-g) \tag{8.165}$$

where, in agreement with Eq. (8.54), the *finite delta function*

$$\delta_1(f-g) = \begin{cases} T, & (-1/2T) < (f-g) < (1/2T) \\ 0, & \text{otherwise} \end{cases} \tag{8.166}$$

Hence,

$$E[X^*(f)Y(g)] = \begin{cases} TS_{xy}(f) & f = g \\ 0, & f \neq g \end{cases} \tag{8.167}$$

This proves that $S_{xy}(f)$ exists only when $f = g$ and gives the well-known formula of Eq. (1.41), namely,

$$S_{xy}(f) = \frac{1}{T} E[X^*(f)Y(f)] \tag{8.168}$$

Equation (8.168) is a direct Fourier transform way to compute $S_{xy}(f)$ without first computing $R_{xy}(\tau)$. The quantity $S_{xy}(f)$ is the (*first-order*) *cross-spectral density function* between $x(t)$ and $y(t)$ and is also called the *cross-spectrum*. This single-valued function is simple to compute and interpret. Equation (8.164) also shows that $S_{xy}(f)$ is the Fourier transform of $R_{xy}(\tau)$ defined by

$$S_{xy}(f) = \int R_{xy}(\tau)e^{-j2\pi f\tau}\, d\tau \tag{8.169}$$

As special cases of Eq. (8.168), (*first-order*) *autospectral density functions*, also called *autospectra*, are given by the formulas

$$\begin{aligned} S_{xx}(f) &= \frac{1}{T} E[X^*(f)X(f)] \\ S_{yy}(f) &= \frac{1}{T} E[Y^*(f)Y(f)] \end{aligned} \tag{8.170}$$

From Eqs. (8.163) and (8.164) the first-order autocorrelation function

$$R_{xx}(\tau) = E[x(t)x(t + \tau)] = \int S_{xx}(f)e^{j2\pi f\tau}\, df \tag{8.171}$$

At the value $\tau = 0$, one obtains the mean square value

$$R_{xx}(0) = E[x^2(t)] = \int S_{xx}(f)\, df \tag{8.172}$$

Thus, physically, the autospectrum $S_{xx}(f)$ gives a decomposition of the mean square value $E[x^2(t)]$ as a function of a single frequency variable. Specifically, $S_{xx}(f)$ represents the rate of change of the mean square value $E[x^2(t)]$ as a function of frequency. Similarly, the autospectrum $S_{yy}(f)$ gives a decomposition of the mean square value $E[y^2(t)]$ as a function of frequency. Many applications of these first-order results are in references 1 and 2.

8.4.2 Second-Order Functions

The *second-order cross-correlation function* between $x(t)$ and $y(t)$ is defined by the expected value operation

$$R_{xxy}(\tau_1, \tau_2) = E[x(t - \tau_1)x(t - \tau_2)y(t)] \tag{8.173}$$

The time-domain quantities can be represented by

$$\begin{aligned} x(t - \tau_1) &= \int X^*(f)e^{-j2\pi f(t - t_1)}\, df \\ x(t - \tau_2) &= \int X^*(g)e^{-j2\pi g(t - t_2)}\, dg \\ y(t) &= \int Y(h)e^{j2\pi ht}\, dh \end{aligned}$$

Thus, Eq. (8.173) is the same as

$$R_{xxy}(\tau_1, \tau_2) = \iiint E[X^*(f)X^*(g)Y(h)]e^{-j2\pi(f+g-h)t}e^{j2\pi(f\tau_1+g\tau_2)}\,dh\,df\,dg$$

$$= \iint S_{xxy}(f, g)e^{j2\pi(f\tau_1+g\tau_2)}\,df\,dg \tag{8.174}$$

where the triple integral is a function of t and the final double integral is independent of t. This equation shows that $R_{xxy}(\tau_1, \tau_2)$ is the double inverse Fourier transform of $S_{xxy}(f, g)$.

For the last equation to eliminate t, $X(f)$, $X(g)$, and $Y(h)$ should be measured at frequencies f, g, and h such that

$$E[X^*(f)X^*(g)Y(h)] = S_{xxy}(f, g)\delta_1(f + g - h) \tag{8.175}$$

with $\delta_1(f)$ defined by Eq. (8.54). Hence, $S_{xxy}(f, g)$ exists only when $h = (f + g)$ and is given by the formula

$$S_{xxy}(f, g) = \frac{1}{T}E[X^*(f)X^*(g)Y(f + g)] \tag{8.176}$$

Equation (8.176) is a direct Fourier transform way to compute $S_{xxy}(f, g)$. The quantity $S_{xxy}(f, g)$ is the *second-order cross-spectral* (*bispectral*) *density function* between $x(t)$ and $y(t)$. This double-valued function is tedious to compute and interpret. Equation (8.174) proves that $S_{xxy}(f, g)$ is the double Fourier transform of $R_{xxy}(\tau_1, \tau_2)$ defined by

$$S_{xxy}(f, g) = \iint R_{xxy}(\tau_1, \tau_2)e^{-j2\pi(f\tau_1+g\tau_2)}\,d\tau_1\,d\tau_2 \tag{8.177}$$

If $x(t)$ is a zero mean value Gaussian input to a linear system with output $y_1(t)$, then Eqs. (8.77) and (8.176) show that $S_{xxy}(f, g)$ will be identically zero for all f and g, since the expected value in Eq. (8.176) will involve a third-order moment of the Gaussian input data.

Special cases of Eq. (8.177) give the *second-order autospectral density functions*, also called *bispectra*, by the formulas

$$S_{xxx}(f, g) = \frac{1}{T}E[X^*(f)X^*(g)X(f + g)] \tag{8.178}$$

$$S_{yyy}(f, g) = \frac{1}{T}E[Y^*(f)Y^*(g)Y(f + g)] \tag{8.179}$$

From Eqs. (8.173) and (8.174), the *second-order autocorrelation functions* at

$\tau_1 = \tau_2 = 0$ are related to the *skewness* quantities

$$R_{xxx}(0, 0) = E[x^3(t)] = \iint S_{xxx}(f, g)\, df\, dg \tag{8.180}$$

$$R_{yyy}(0, 0) = E[y^3(t)] = \iint S_{yyy}(f, g)\, df\, dg \tag{8.181}$$

This shows that the bispectrum $S_{xxx}(f, g)$ gives a decomposition of the skewness $E[x^3(t)]$ as a function of two frequency variables. Similarly, the bispectrum $S_{yyy}(f, g)$ gives a decomposition of the skewness $E[y^3(t)]$ as a function of two frequency variables.

For zero mean value Gaussian random data $x(t)$, the bispectrum $S_{xxx}(f, g) = 0$ for all f and g since it represents a third-order moment and all odd-order moments of such Gaussian data are zero. If an output $y(t)$ occurs by passing real-valued Gaussian random input data through a zero-memory nonlinear system, then $y(t)$ will be non-Gaussian and the output bispectrum $S_{yyy}(f, g)$ will not be zero for arbitrary f and g. Moreover, this output bispectrum $S_{yyy}(f, g)$ will be real-valued for all f and g because the nonlinear transformation has no memory. This means that if a measured output bispectrum is complex-valued, then $y(t)$ could not have come from Gaussian random data passing through a zero-memory nonlinear system. A possible model for this result would be the passage of the Gaussian random data through a finite-memory nonlinear system. Phase estimation using the bispectrum and nonlinear system identification by observing only the output are discussed in references 37 and 49. Reference 41 reviews bispectrum matters.

8.4.3 Third-Order Functions

The *third-order cross-correlation function* between $x(t)$ and $y(t)$ is defined by the expected value operation

$$R_{xxxy}(\tau_1, \tau_2, \tau_3) = E[x(t - \tau_1)x(t - \tau_2)x(t - \tau_3)y(t)] \tag{8.182}$$

As per the derivation of Eqs. (8.174) and (8.176), one can prove after a number of steps that

$$R_{xxxy}(\tau_1, \tau_2, \tau_3) = \iiint S_{xxxy}(f, g, h)e^{j2\pi(f\tau_1 + g\tau_2 + h\tau_3)}\, df\, dg\, dh \tag{8.183}$$

and also that

$$S_{xxxy}(f, g, h) = \frac{1}{T} E[X^*(f)X^*(g)X^*(h)Y(f + g + h)] \tag{8.184}$$

Equation (8.184) is a direct Fourier transform way to compute $S_{xxxy}(f, g, h)$. Equation (8.183) shows that $S_{xxxy}(f, g, h)$ is the triple Fourier transform of $R_{xxxy}(\tau_1, \tau_2, \tau_3)$ defined by

$$S_{xxxy}(f, g, h) = \iiint R_{xxxy}(\tau_1, \tau_2, \tau_3) e^{-j2\pi(f\tau_1 + g\tau_2 + h\tau_3)} \, d\tau_1 \, d\tau_2 \, d\tau_3 \quad (8.185)$$

The quantity $S_{xxxy}(f, g, h)$ is the *third-order cross-spectral* (*trispectral*) *density function* between $x(t)$ and $y(t)$. This triple-valued function is very difficult to compute and interpret.

If $x(t)$ is a zero mean value Gaussian input to a linear system with output $y_1(t)$, then Eqs. (8.77) and (8.184) show that, in general, $S_{xxxy_1}(f, g, h) \neq 0$, since the expected value in Eq. (8.184) will involve a fourth-order moment of the Gaussian data. On the other hand, for this same input to a bilinear system with output $y_2(t)$, Eqs. (8.78) and (8.181) show that $S_{xxxy_2}(f, g, h)$ will be identically zero for all f, g, and h, since this calculation will involve odd-order moments of the Gaussian input data.

From Eq. (8.185), the *third-order autospectral density functions*, also called *trispectra*, are

$$S_{xxxx}(f, g, h) = \frac{1}{T} E[X^*(f) X^*(g) X^*(h) X(f + g + h)] \quad (8.186)$$

$$S_{yyyy}(f, g, h) = \frac{1}{T} E[Y^*(f) Y^*(g) Y^*(h) Y(f + g + h)] \quad (8.187)$$

From Eqs. (8.182) and (8.183), the *third-order autocorrelation functions* at $\tau_1 = \tau_2 = \tau_3 = 0$ are related to the *kurtosis* quantities

$$R_{xxxx}(0, 0, 0) = E[x^4(t)] = \iiint S_{xxxx}(f, g, h) \, df \, dg \, dh \quad (8.188)$$

$$R_{yyyy}(0, 0, 0) = E[y^4(t)] = \iiint S_{yyyy}(f, g, h) \, df \, dg \, dh \quad (8.189)$$

This shows that the trispectrum $S_{xxxx}(f, g, h)$ gives a decomposition of the kurtosis $E[x^4(t)]$ as a function of three frequency variables. Similarly, the trispectrum $S_{yyyy}(f, g, h)$ gives a decomposition of the kurtosis $E[y^4(t)]$ as a function of three frequency variables.

For zero mean value Gaussian random data $x(t)$, the trispectrum $S_{xxxx}(f, g, h)$, being a fourth-order moment of Gaussian data, reduces to products of second-order moments (autospectra). Thus the trispectrum $S_{xxxx}(f, g, h)$ is real-valued and nonnegative for all f, g, and h. Also, if an output $y(t)$ occurs by passing real-valued Gaussian random input data through a zero-memory nonlinear system, then the output trispectrum $S_{yyyy}(f, g, h)$ will be real-valued and nonnegative for all f, g, and h because the nonlinear transformation has no memory.

8.4.4 Special Bispectral and Trispectral Density Functions

Consider cases of Eqs. (8.176) and (8.184) when $f = g$ and when $f = g = h$. For these situations, one obtains special bispectral and trispectral density functions defined for stationary random data by

$$S_{xxy}(f) = S_{xxy}(f, f) = \frac{1}{T} E[X^*(f)X^*(f)Y(2f)] \tag{8.190}$$

$$S_{xxxy}(f) = S_{xxxy}(f, f, f) = \frac{1}{T} E[X^*(f)X^*(f)X^*(f)Y(3f)] \tag{8.191}$$

These two results are straightforward extensions of the usual cross-spectral density function $S_{xy}(f)$ of Eq. (8.168). They are functions of only one variable and are much easier to compute and interpret than the more generalized functions $S_{xxy}(f, g)$ of two variables and $S_{xxxy}(f, g, h)$ of three variables. For Gaussian input data, Section 9.4 shows how these special bispectral and trispectral density functions can be used to identify the frequency response functions in Case 1 and Case 2 square-law and cubic nonlinear models.

8.4.5 Transient Random Data

For transient random data, the stationary random data results $S_{xy}(f)$, $S_{xxy}(f, g)$, and $S_{xxxy}(f, g, h)$ should be replaced by $\mathscr{S}_{xy}(f)$, $\mathscr{S}_{xxy}(f, g)$, and $\mathscr{S}_{xxxy}(f, g, h)$, respectively, where

$$\mathscr{S}_{xy}(f) = TS_{xy}(f) = E[X^*(f)Y(f)] \tag{8.192}$$

$$\mathscr{S}_{xxy}(f, g) = T^2 S_{xxy}(f, g) = TE[X^*(f)X^*(g)Y(f + g)] \tag{8.193}$$

$$\mathscr{S}_{xxxy}(f, g, h) = T^3 S_{xxxy}(f, g, h) = T^2 E[X^*(f)X^*(g)X^*(h)Y(f + g + h)] \tag{8.194}$$

Formulas for $\mathscr{S}_{yy}(f)$, $\mathscr{S}_{yyy}(f, g)$, and $\mathscr{S}_{yyyy}(f, g, h)$ follow when $x(t) = y(t)$. Other special formulas occur when $f = g$ and when $f = g = h$.

8.4.6 Summary of Results

The main equations for stationary random data will now be compared for their inner mathematical structure as one proceeds from first- to second- to third-order functions. By time-domain quantities, the successive cross-correlation functions are defined by the expected values

$$R_{xy}(\tau) = E[x(t-\tau)y(t)] \tag{8.195}$$

$$R_{xxy}(\tau_1,\tau_2) = E[x(t-\tau_1)x(t-\tau_2)y(t)] \tag{8.196}$$

$$R_{xxxy}(\tau_1,\tau_2,\tau_3) = E[x(t-\tau_1)x(t-\tau_2)x(t-\tau_3)y(t)] \tag{8.197}$$

Appropriate Fourier transforms of these cross-correlation functions yield the successive cross-spectral density functions

$$S_{xy}(f) = \int R_{xy}(\tau)e^{-j2\pi f\tau}\,d\tau \tag{8.198}$$

$$S_{xxy}(f_1,f_2) = \iint R_{xxy}(\tau_1,\tau_2)e^{-j2\pi(f_1\tau_1+f_2\tau_2)}\,d\tau_1\,d\tau_2 \tag{8.199}$$

$$S_{xxxy}(f_1,f_2,f_3) = \iiint R_{xxxy}(\tau_1,\tau_2,\tau_3)e^{-j2\pi(f_1\tau_1+f_2\tau_2+f_3\tau_3)}\,d\tau_1\,d\tau_2\,d\tau_3 \tag{8.200}$$

Successive cross-spectral density functions are defined directly by frequency-domain quantities and expected values

$$S_{xy}(f) = \frac{1}{T}E[X^*(f)Y(f)] \tag{8.201}$$

$$S_{xxy}(f_1,f_2) = \frac{1}{T}E[X^*(f_1)X^*(f_2)Y(f_1+f_2)] \tag{8.202}$$

$$S_{xxxy}(f_1,f_2,f_3) = \frac{1}{T}E[X^*(f_1)X^*(f_2)X^*(f_3)Y(f_1+f_2+f_3)] \tag{8.203}$$

These equations show that desired spectral density functions can be computed by two distinct methods:

- *Method 1.* By first computing the various first-, second, and third-order correlation functions in the time domain, and then taking their single, double, and triple Fourier transforms to move into the frequency domain.
- *Method 2.* By computing finite Fourier transforms of the original measured data, and then taking appropriate first-, second, and third-order expected values of these quantities. Here, all quantities are in the frequency domain without computing any associated correlation functions in the time domain. This direct method is both simpler and more efficient.

This concludes Chapter 8.

9

INPUT/OUTPUT RELATIONS FOR BILINEAR AND TRILINEAR SYSTEMS

This chapter develops multidimensional input/output correlation and spectral relations for Gaussian stationary random data passing through bilinear and trilinear systems. These formulas are not valid for non-Gaussian input data. Linear and nonlinear coherence functions are defined for general third-order Volterra nonlinear models consisting of linear systems in parallel with bilinear and trilinear systems. Formulas are derived on how to identify the optimum nature of parallel linear, bilinear, and trilinear systems that minimize the output noise spectrum, based on measurements only of the Gaussian input data and the total output data. The complexity of required multidimensional bispectral and trispectral density functions shows clearly the need and importance of the practical MI/SO techniques in Chapters 3–7. The chapter concludes with formulas on how to identify Case 1 and Case 2 square-law and cubic nonlinear models by using special single-frequency bispectral and trispectral density functions.

9.1 LINEAR, BILINEAR, AND TRILINEAR SYSTEMS

A number of theoretical input/output correlation and spectral relations will now be developed for Gaussian stationary random data passing through linear, bilinear, and trilinear systems, as shown in the three parts of Figure 9.1. The separate results are discussed in Sections 9.1.1–9.1.4 and summarized in Section 9.1.5. Implementation of these relations to analyze and identify the properties of general third-order Volterra nonlinear models is detailed in Sections 9.2 and 9.3. This is useful material for nonlinear problems when techniques in the preceding chapters are not appropriate.

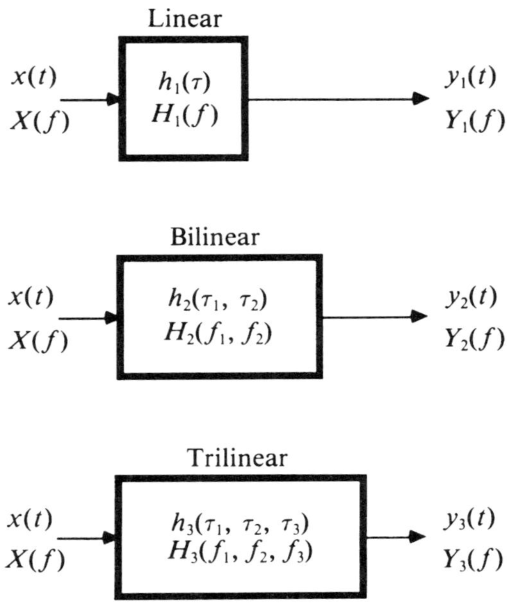

Figure 9.1. Linear, bilinear, and trilinear systems.

9.1.1 Linear Systems

For linear systems alone, from Eqs. (8.13) and (8.19), it is known that

$$y_1(t) = \int h_1(\alpha)x(t-\alpha)\,d\alpha \tag{9.1}$$

$$Y_1(f) = H_1(f)X(f) \tag{9.2}$$

where $X(f)$, $Y_1(f)$, and $H_1(f)$ are Fourier transforms of $x(t)$, $y_1(t)$, and $h_1(t)$, respectively. Hence, the input/output cross-correlation function for stationary random data is

$$\begin{aligned} R_{xy_1}(\tau) &= E[x(t)y_1(t+\tau)] = E[x(t-\tau)y_1(t)] \\ &= \int h_1(\alpha)E[x(t-\tau)x(t-\alpha)]\,d\alpha = \int h_1(\alpha)R_{xx}(\tau-\alpha)\,d\alpha \end{aligned} \tag{9.3}$$

Taking the Fourier transforms of both sides yields the linear system input/output cross-spectrum result

$$S_{xy_1}(f) = \int R_{xy_1}(\tau)e^{-j2\pi f\tau}\,d\tau = H_1(f)S_{xx}(f) \tag{9.4}$$

This simple product relation is equivalent to the convolution integral of Eq. (9.3). From Eqs. (9.2), (1.39), and (1.41), without computing correlation functions, one can prove Eq. (9.4) directly by taking expected values of Fourier transform quantities as follows:

$$S_{xy_1}(f) = \frac{1}{T} E[X^*(f)Y_1(f)] = \frac{1}{T} E[X^*(f)H_1(f)X(f)]$$
$$= H_1(f)\left(\frac{1}{T} E[X^*(f)X(f)]\right) = H_1(f)S_{xx}(f) \tag{9.5}$$

Thus, $H_1(f)$ is obtained from dividing the input/output cross-spectrum $S_{xy_1}(f)$ by the input autospectrum $S_{xx}(f)$, namely,

$$H_1(f) = \frac{S_{xy_1}(f)}{S_{xx}(f)} \tag{9.6}$$

The linear system output autocorrelation function $R_{y_1y_1}(\tau)$ is

$$R_{y_1y_1}(\tau) = E[y_1(t)y_1(t+\tau)]$$
$$= E\left[\int h_1(\alpha)x(t-\alpha)\,d\alpha \int h_1(\beta)x(t+\tau-\beta)\,d\beta\right]$$
$$= \iint h_1(\alpha)h_1(\beta)R_{xx}(\tau+\alpha-\beta)\,d\alpha\,d\beta \tag{9.7}$$

This is a double convolution integral. The Fourier transforms of both sides shows after several steps that the linear system output autospectrum is

$$S_{y_1y_1}(f) = \int R_{y_1y_1}(\tau)e^{-j2\pi f\tau}\,d\tau = |H_1(f)|^2 S_{xx}(f) \tag{9.8}$$

Equation (9.8) is equivalent to Eq. (9.7). From Eq. (9.2), without computing correlation functions, one can prove directly that

$$S_{y_1y_1}(f) = \frac{1}{T} E[Y_1^*(f)Y_1(f)] = \frac{1}{T} E[H_1^*(f)X^*(f)H_1(f)X(f)]$$
$$= |H_1(f)|^2\left(\frac{1}{T} E[X^*(f)X(f)]\right) = |H_1(f)|^2 S_{xx}(f) \tag{9.9}$$

Extensions of these input/output results will be carried out for bilinear and trilinear systems.

9.1.2 Definitions and Assumptions

Three preliminary matters are needed to derive the desired input/output relations for bilinear and trilinear systems. These involve proper understanding and application of the following items:

- Finite delta functions
- Extended spectral density functions
- Higher-order moments for Gaussian data

Finite Delta Functions. Definitions for the finite delta functions are stated in Eqs. (8.54) and (8.166). The finite delta function has the properties

$$\delta_1(-f) = \delta_1(f) = \begin{cases} T, & (-1/2T) < f < (1/2T) \\ 0, & \text{otherwise} \end{cases}$$
$$\int_{-\infty}^{\infty} \delta_1(f)\, df = \int_{(-1/2T)}^{(1/2T)} \delta_1(f)\, df = 1 \tag{9.10}$$

Also, with Fourier transforms from records of length T,

$$S_{xx}(g)\delta_1(f-g) = S_{xx}(f)\delta_1(f-g) \tag{9.11}$$

when $S_{xx}(g) = S_{xx}(f)$ for $(-1/2T) < (f-g) < (1/2T)$. This follows from computing $S_{xx}(f)$ only at the discrete frequencies $f_k = k\,\Delta f = (k/T)$ apart.

Extended Spectral Density Functions. Extended definitions of spectral density functions are stated in Eqs. (8.165) and (8.167). From these definitions, one has the expected value properties

$$E[X^*(f)Y(g)] = S_{xy}(f)\delta_1(f-g) = \begin{cases} TS_{xy}(f), & f = g \\ 0, & f \neq g \end{cases} \tag{9.12}$$

$$E[X^*(f)X(g)] = S_{xx}(f)\delta_1(f-g) = \begin{cases} TS_{xx}(f), & f = g \\ 0, & f \neq g \end{cases} \tag{9.13}$$

assuming $X(f)$ computed only at frequencies $f_k = k\,\Delta f = (k/T)$ apart. From $X(f) = X^*(-f)$, $S_{xx}(-f) = S_{xx}(f)$, $S_{xy}(-f) = S_{yx}(f)$, and $\delta_1(-f) = \delta_1(f)$, further properties are

$$\begin{aligned} E[X(f)Y(g)] &= E[X^*(-f)Y(g)] = S_{yx}(f)\delta_1(f+g) \\ E[X(f)X(g)] &= E[X^*(-f)X(g)] = S_{xx}(f)\delta_1(f+g) \end{aligned} \tag{9.14}$$

$$\begin{aligned} E[X^*(f)Y^*(g)] &= E[X^*(f)Y(-g)] = S_{xy}(f)\delta_1(f+g) \\ E[X^*(f)X^*(g)] &= E[X^*(f)X(-g)] = S_{xx}(f)\delta_1(f+g) \end{aligned} \tag{9.15}$$

Note that for data where $X(0) = 0$, one will obtain $S_{xx}(0) = 0$.

Higher-Order Moments for Gaussian Data. To obtain closed-form answers, unless stated otherwise, the real-valued input data $x(t)$ henceforth will be assumed to be a zero mean value Gaussian stationary random process with arbitrary spectral properties.

From the assumed Gaussian input properties, all odd-ordered moments of $x(t)$ are zero and all even-ordered moments reduce to products of different combinations of second-order moments. Specifically, from reference 1, if $x_i \equiv x(t_i)$ and if $R_{ij} = R_{xx}(t_i - t_j) = R_{xx}(t_j - t_i)$ for the stationary random input data, then

$$E[x_1] = E[x_1x_2x_3] = E[x_1x_2x_3x_4x_5] = 0 \tag{9.16}$$

$$E[x_1x_2x_3x_4] = R_{12}R_{34} + R_{13}R_{24} + R_{14}R_{23} \tag{9.17}$$

$$\begin{aligned} E[x_1x_2x_3x_4x_5x_6] = {} & R_{12}[R_{34}R_{56} + R_{35}R_{46} + R_{36}R_{45}] \\ & + R_{13}[R_{24}R_{56} + R_{25}R_{46} + R_{26}R_{45}] \\ & + R_{14}[R_{23}R_{56} + R_{25}R_{36} + R_{26}R_{35}] \\ & + R_{15}[R_{23}R_{46} + R_{24}R_{36} + R_{26}R_{34}] \\ & + R_{16}[R_{23}R_{45} + R_{24}R_{35} + R_{25}R_{34}] \end{aligned} \tag{9.18}$$

and so on. These results apply not only to the original $x(t)$ but to any linear transformations of these $x(t)$, since linear transformations preserve the Gaussian nature of the data.

In particular, Fourier transforms are linear operations so that these higher-order moments apply to $X(f)$ in addition to $x(t)$ where

$$X(f) = \mathscr{F}[x(t)] = \int x(t)e^{-j2\pi ft}\,dt$$

They apply also to $X^*(f)$ where $X^*(f) = X(-f)$ when $x(t)$ is real-valued. Thus, if $X_i \equiv X(f_i)$ are Fourier transforms from a zero mean value Gaussian stationary random process, then

$$E[X_1] = E[X_1X_2X_3] = E[X_1X_2X_3X_4X_5] = 0 \tag{9.19}$$

$$\begin{aligned} E[X_1X_2X_3X_4] = {} & E[X_1X_2]E[X_3X_4] + E[X_1X_2]E[X_2X_4] \\ & + E[X_1X_4]E[X_2X_3] \end{aligned} \tag{9.20}$$

$$\begin{aligned} E[X_1X_2X_3X_4X_5X_6] = {} & E[X_1X_2]E[X_3X_4X_5X_6] + E[X_1X_3]E[X_2X_4X_5X_6] \\ & + E[X_1X_4]E[X_2X_3X_5X_6] \\ & + E[X_1X_5]E[X_2X_3X_4X_6] \\ & + E[X_1X_6]E[X_2X_3X_4X_5] \end{aligned} \tag{9.21}$$

where each fourth-order moment should then be replaced by appropriate pairs of second-order moments, and so on. All of the obtained different second-order moments then reduce to various autospectral and cross-spectral density functions as per the rules in Eqs. (9.12)–(9.15).

9.1.3 Bilinear Systems

For bilinear systems alone, from Eqs. (8.34), (8.44), and (8.48), one has the bilinear system output

$$y_2(t) = \iint h_2(\alpha, \beta)x(t - \alpha)x(t - \beta)\, d\alpha\, d\beta \tag{9.22}$$

with $\bar{y}_2 = E[y_2(t)]$ given by

$$\bar{y}_2 = \iint h_2(\alpha, \beta)R_{xx}(\alpha - \beta)\, d\alpha\, d\beta = \int H_2(f, -f)S_{xx}(f)\, df \tag{9.23}$$

In general, $\bar{y}_2 \neq 0$. Also, from Eq. (8.55),

$$Y_2(f) = \mathscr{F}[y_2(t) - \bar{y}_2] = \int H_2(u, f - u)X(u)X(f - u)\, du - \bar{y}_2\delta_1(f) \tag{9.24}$$

Observe that $Y_2(f) = \mathscr{F}[y_2(t)]$ of Eq. (8.51) for all f except $f = 0$.

Define the zero mean value bilinear system output $y_b(t)$ from $y_2(t)$ and $\bar{y}_2$ by

$$y_b(t) = y_2(t) - \bar{y}_2 \tag{9.25}$$

Then $\bar{y}_b = E[y_b(t)] = 0$ and

$$y_b(t) = \iint h_2(\alpha, \beta)[x(t - \alpha)x(t - \beta) - R_{xx}(\alpha - \beta)]\, d\alpha\, d\beta \tag{9.26}$$

with

$$Y_b(f) = \mathscr{F}[y_b(t)] = \mathscr{F}[y_2(t) - \bar{y}_2] = Y_2(f) \tag{9.27}$$

for the $Y_2(f)$ of Eq. (9.24). Also, $E[Y_b(f)] = E[Y_2(f)] = 0$. Further analysis should be conducted with $y_b(t)$ instead of $y_2(t)$. The cross-correlation functions between the input $x(t)$ and output $y_b(t)$ or $y_2(t)$ involve third-order moments of $x(t)$ so that from Eq. (9.16),

$$\begin{aligned} R_{xy_b}(\tau) &= R_{xy_2}(\tau) = 0 \\ S_{xy_b}(f) &= S_{xy_2}(f) = 0 \end{aligned} \tag{9.28}$$

In words, $x(t)$ is uncorrelated with both $y_b(t)$ and $y_2(t)$.

Bilinear System Output Autocorrelation and Autospectrum. The bilinear system output autocorrelation function and output autospectral density function from Eq. (9.25) satisfy

$$\begin{aligned} R_{y_b y_b}(\tau) &= R_{y_2 y_2}(\tau) - (\bar{y}_2)^2 \\ S_{y_b y_b}(f) &= S_{y_2 y_2}(f) - (\bar{y}_2)^2 \delta_1(f) \end{aligned} \tag{9.29}$$

Starting from Eq. (9.26), after many steps one can prove in the time domain that the bilinear system output autocorrelation function is

$$R_{y_b y_b}(\tau) = 2 \iiiint h_2(\alpha, \beta) h_2(u, v) R_{xx}(\tau + \alpha - u) R_{xx}(\tau + \beta - v)\, d\alpha\, d\beta\, du\, dv \tag{9.30}$$

This complicated fourfold integral is practically impossible to compute and apply.

In the frequency domain, by expressing $R_{xx}(\tau + \alpha - u)$ and $R_{xx}(\tau + \beta - v)$ in Eq. (9.30) in terms of their Fourier transforms, one can prove after many steps that the bilinear system output autospectral density function is

$$S_{y_b y_b}(f) = 2 \int |H_2(u, f - u)|^2 S_{xx}(u) S_{xx}(f - u)\, du \tag{9.31}$$

Equation (9.31) is simpler than Eq. (9.30) but still difficult to compute. Equation (9.31) can also be derived directly from Fourier transform data without using or knowing the correlation result of Eq. (9.30).

Mathematical details will now be described to derive Eqs. (9.30) and (9.31). Similar analysis procedures can be used to derive other formulas in this chapter. Readers not concerned with details of these proofs should go to Eq. (9.32).

Derivation of the Output Autocorrelation Function. From Eq. (9.26), with different variables of integration,

$$y_b(t + \tau) = \iint h_2(u, v)[x(t + \tau - u)x(t + \tau - v) - R_{xx}(u - v)]\, du\, dv$$

By multiplying $y_b(t)$ by $y_b(t + \tau)$ and taking expected values one finds that $R_{y_b y_b}(\tau)$ involves a fourfold integral where the integrand contains the expected value

$$\begin{aligned} &E\{[x(t-\alpha)x(t-\beta) - R_{xx}(\alpha-\beta)][x(t+\tau-u)x(t+\tau-v) - R_{xx}(u-v)]\} \\ &\quad = E[x(t-\alpha)x(t-\beta)x(t+\tau-u)x(t+\tau-v)] \\ &\qquad - R_{xx}(u-v)E[x(t-\alpha)x(t-\beta)] - R_{xx}(\alpha-\beta)E[x(t+\tau-u)x(t+\tau-v)] \\ &\qquad + R_{xx}(\alpha-\beta)R_{xx}(u-v) \end{aligned}$$

The second-order moments are

$$E[x(t-\alpha)x(t-\beta)] = R_{xx}(\alpha-\beta)$$
$$E[x(t+\tau-u)x(t+\tau-v)] = R_{xx}(u-v)$$

From Eq. (9.17), the fourth-order moment

$$\begin{aligned} &E[x(t-\alpha)x(t-\beta)x(t+\tau-u)x(t+\tau-v)] \\ &\quad = R_{xx}(\alpha-\beta)R_{xx}(u-v) + R_{xx}(\tau+\alpha-u)R_{xx}(\tau+\beta-v) \\ &\quad\quad + R_{xx}(\tau+\alpha-v)R_{xx}(\tau+\beta-u) \end{aligned}$$

Substitution shows that only the last two terms remain in the integrand and that these two terms give the same fourfold integral, since $h_2(u, v) = h_2(v, u)$. Hence,

$$\begin{aligned} R_{y_by_b}(\tau) &= E[y_b(t)y_b(t+\tau)] \\ &= 2\iiiint h_2(\alpha, \beta)h_2(u, v)R_{xx}(\tau+\alpha-u)R_{xx}(\tau+\beta-v)\,d\alpha\,d\beta\,du\,dv \end{aligned}$$

This completes the proof for the output autocorrelation function of Eq. (9.30).

Indirect Derivation of the Output Autospectral Density Function. In terms of their inverse Fourier transforms, the autocorrelation functions

$$R_{xx}(\tau+\alpha-u) = \int S_{xx}(\zeta)e^{j2\pi(\tau+\alpha-u)\zeta}\,d\zeta$$

$$R_{xx}(\tau+\beta-v) = \int S_{xx}(\eta)e^{j2\pi(\tau+\beta-v)\eta}\,d\eta$$

The exponential product

$$e^{j2\pi(\tau+\alpha-u)\zeta}e^{j2\pi(\tau+\beta-v)\eta} = e^{j2\pi(\alpha\zeta+\beta\eta)}e^{-j2\pi(u\zeta+v\eta)}e^{j2\pi(\zeta+\eta)\tau}$$

Hence, Eq. (9.30) is equivalent to

$$R_{y_by_b}(\tau) = 2\iint H_2(\zeta, \eta)H_2^*(\zeta, \eta)S_{xx}(\eta)e^{j2\pi(\zeta+\eta)\tau}\,d\zeta\,d\eta$$

where

$$H_2(\zeta, \eta) = \iint h_2(\alpha, \beta)e^{j2\pi(\alpha\zeta+\beta\eta)}\,d\alpha\,d\beta$$

$$H_2^*(\zeta, \eta) = \iint h_2(u, v)e^{-j2\pi(u\zeta+v\eta)}\,du\,dv$$

Thus,

$$R_{y_by_b}(\tau) = 2 \iint |H_2(\zeta, \eta)|^2 S_{xx}(\zeta) S_{xx}(\eta) e^{j2\pi(\zeta+\eta)\tau} \, d\zeta \, d\eta$$

Change variables from (ζ, η) to (ζ, f) where $f = \zeta + \eta$, $df = d\eta$. Now

$$R_{y_by_b}(\tau) = 2 \iint |H_2(\zeta, f - \zeta)|^2 S_{xx}(\zeta) S_{xx}(f - \zeta) e^{j2\pi f\tau} \, d\zeta \, df$$

But

$$R_{y_2y_2}(\tau) = \int S_{y_by_b}(f) e^{j2\pi f\tau} \, df$$

Hence, one obtains the result

$$S_{y_by_b}(f) = 2 \int |H_2(\zeta, f - \zeta)|^2 S_{xx}(\zeta) S_{xx}(f - \zeta) \, d\zeta$$

Replacing the variable ζ by u with $d\zeta = du$ completes the indirect proof for the output autospectral density function of Eq. (9.31).

Direct Derivation of Output Autospectral Density Function. Start with Eqs. (9.24) and (9.27) where

$$Y_b(f) = Y_2(f) = \int H_2(u, f - u) X(u) X(f - u) \, du - \bar{y}_2 \delta_1(f)$$

Define

$$Z(f) = \int H_2(u, f - u) X(u) X(f - u) \, du$$

This quantity $Z(f)$ is the same as $\mathscr{F}[y_2(t)]$ in Eq. (8.51). Then

$$Y_b(f) = Z(f) - \bar{y}_2 \delta_1(f)$$

Expected values of both sides using $E[Y_b(f)] = 0$ gives

$$E[Z(f)] = \bar{y}_2 \delta_1(f)$$

Note also that

$$E[Z^*(f)] = E[Z(f)]$$

The direct derivation of Eq. (9.31) comes from computing

$$\begin{aligned}
S_{y_b y_b}(f) &= \frac{1}{T} E[Y_b^*(f) Y_b(f)] \\
&= \frac{1}{T} E\{[Z^*(f) - \bar{y}_2 \delta_1(f)][Z(f) - \bar{y}_2 \delta_1(f)]\} \\
&= S_{zz}(f) - (\bar{y}_2)^2 \delta_1(f)
\end{aligned}$$

where

$$S_{zz}(f) = \frac{1}{T} E[Z^*(f) Z(f)]$$

and from properties of $\delta_1(f)$,

$$\frac{1}{T} E[(\bar{y}_2)^2 \delta_1^2(f)] = (\bar{y}_2)^2 \delta_1(f)$$

Use variables of integration α and β to write

$$\begin{aligned}
Z(f) &= \int H_2(\alpha, f - \alpha) X(\alpha) X(f - \alpha)\, d\alpha \\
Z^*(f) &= \int H_2^*(\beta, f - \beta) X^*(\beta) X^*(f - \beta)\, d\beta
\end{aligned}$$

Now

$$S_{zz}(f) = \frac{1}{T} \iint H_2(\alpha, f - \alpha) H_2^*(\beta, f - \beta) E[X(\alpha) X(f - \alpha) X^*(\beta) X^*(f - \beta)]\, d\alpha\, d\beta$$

From Eq. (9.20) plus Eqs. (9.12)–(9.15), the fourth-order moment

$$\begin{aligned}
&E[X(\alpha) X(f - \alpha) X^*(\beta) X^*(f - \beta)] \\
&\quad = E[X(\alpha) X(f - \alpha)] E[X^*(\beta) X^*(f - \beta)] \\
&\qquad + E[X(\alpha) X^*(\beta)] E[X(f - \alpha) X^*(f - \beta)] \\
&\qquad + E[X(\alpha) X^*(f - \beta)] E[X(f - \alpha) X^*(\beta)] \\
&\quad = [S_{xx}(\alpha) \delta_1(f)][S_{xx}(\beta) \delta_1(f)] \\
&\qquad + [S_{xx}(\alpha) \delta_1(f - \alpha - \beta)][S_{xx}(f - \alpha) \delta_1(f - \alpha - \beta)] \\
&\qquad + [S_{xx}(\alpha) \delta_1(f - \alpha - \beta)][S_{xx}(f - \alpha) \delta_1(f - \alpha - \beta)]
\end{aligned}$$

Substituting into $S_{zz}(f)$ and integrating term by term gives, for the first pair of second-order moments,

$$\frac{1}{T}\iint H_2(\alpha, f-\alpha)H_2^*(\beta, f-\beta)S_{xx}(\alpha)S_{xx}(\beta)\delta_1^2(f)\,d\alpha\,d\beta$$

$$= \frac{\delta_1^2(f)}{T}\left[\int H_2(\alpha, f-\alpha)S_{xx}(\alpha)\,d\alpha\right]\left[\int H_2^*(\beta, f-\beta)S_{xx}(\beta)\,d\beta\right]$$

$$= (\bar{y}_2)^2\delta_1(f)$$

The second pair of second-order moments gives

$$\frac{1}{T}\iint H_2(\alpha, f-\alpha)H_2^*(\beta, f-\beta)S_{xx}(\alpha)S_{xx}(f-\alpha)\delta_1^2(\alpha-\beta)\,d\alpha\,d\beta$$

$$= \frac{\delta_1(0)}{T}\int H_2(\alpha, f-\alpha)H_2^*(\alpha, f-\alpha)S_{xx}(\alpha)S_{xx}(f-\alpha)\,d\alpha$$

$$= \int |H_2(\alpha, f-\alpha)|^2 S_{xx}(\alpha)S_{xx}(f-\alpha)\,d\alpha$$

The third pair of second-order moments gives

$$\frac{1}{T}\iint H_2(\alpha, f-\alpha)H_2^*(\beta, f-\beta)S_{xx}(\alpha)S_{xx}(f-\alpha)\delta_1^2(f-\alpha-\beta)\,d\alpha\,d\beta$$

$$= \frac{\delta_1(0)}{T}\int H_2(\alpha, f-\alpha)H_2^*(f-\alpha, \alpha)S_{xx}(\alpha)S_{xx}(f-\alpha)\,d\alpha$$

$$= \int |H_2(\alpha, f-\alpha)|^2 S_{xx}(\alpha)S_{xx}(f-\alpha)\,d\alpha$$

since $H_2^*(f-\alpha, \alpha) = H_2^*(\alpha, f-\alpha)$. Thus,

$$S_{zz}(f = (\bar{y}_2)^2\delta_1(f) + 2\int |H_2(\alpha, f-\alpha)|^2 S_{xx}(\alpha)S_{xx}(f-\alpha)\,d\alpha$$

Finally,

$$S_{y_b y_b}(f) = S_{zz}(f) - (\bar{y}_2)^2\delta_1(f) = 2\int |H_2(\alpha, f-\alpha)|^2 S_{xx}(\alpha)S_{xx}(f-\alpha)\,d\alpha$$

Replacing α with u gives the result of Eq. (9.31). This completes the direct proof for the output autospectral density function.

Second-Order Cross-Correlation and Cross-Spectrum. Consider now the second-order cross-correlation function $R_{xxy_b}(\tau_1, \tau_2)$. From Eqs. (8.173), (9.26), and (9.17),

$$\begin{aligned} R_{xxy_b}(\tau_1, \tau_2) &= E[x(t-\tau_1)x(t-\tau_2)y_b(t)] \\ &= \iint h_2(\alpha, \beta)E\left[x(t-\tau_1)x(t-\tau_2)\left\{x(t-\alpha)x(t-\beta)\right.\right. \\ &\quad \left.\left. - R_{xx}(\alpha-\beta)\right\}\right] d\alpha\, d\beta \\ &= \iint h_2(\alpha, \beta)[R_{xx}(\tau_1-\alpha)R_{xx}(\tau_2-\beta) \\ &\quad + R_{xx}(\tau_1-\beta)R_{xx}(\tau_2-\alpha)]\, d\alpha\, d\beta \\ &= 2\iint h_2(\alpha, \beta)R_{xx}(\tau_1-\alpha)R_{xx}(\tau_2-\beta)\, d\alpha\, d\beta \end{aligned} \tag{9.32}$$

The last step uses the symmetric property that $h_2(\alpha, \beta) = h_2(\beta, \alpha)$. Equation (9.32) is difficult to compute and interpret.

From Eq. (8.177), the second-order cross-spectrum (bispectrum) is defined by

$$S_{xxy_b}(f, g) = \iint R_{xxy_b}(\tau_1, \tau_2)e^{-j2\pi(f\tau_1+g\tau_2)}\, d\tau_1\, d\tau_2 \tag{9.33}$$

Substitution of Eq. (9.32) into Eq. (9.33) after many steps gives the second-order bispectrum result in the frequency domain that is equivalent to Eq. (9.32), namely,

$$S_{xxy_b}(f, g) = 2H_2(f, g)S_{xx}(f)S_{xx}(g) \tag{9.34}$$

Here, $H_2(f, g)$ is the double Fourier transform of $h_2(\alpha, \beta)$ as per Eq. (8.45). Equation (9.34) can also be derived directly without computing correlation functions. Note the simplicity of Eq. (9.34) compared with Eq. (9.32).

First-order correlation and spectral density functions are identically zero between the output $y_1(t)$ from the linear system and the outputs $y_b(t)$ or $y_2(t)$ from the bilinear system. This follows from Eq. (9.28), since $y_1(t)$ is linearly related to $x(t)$. Thus,

$$\begin{aligned} R_{y_1y_b}(\tau) &= R_{y_1y_2}(\tau) = 0 \\ S_{y_1y_b}(f) &= S_{y_1y_2}(f) = 0 \end{aligned} \tag{9.35}$$

Finally, it should be noted here that the second-order correlation and spectral density functions between the input $x(t)$ and the linear output $y_1(t)$ are also identically zero,

$$R_{xxy_1}(\tau_1, \tau_2) = 0$$
$$S_{xxy_1}(f_1, f_2) = 0 \tag{9.36}$$

since they involve third-order moments of $x(t)$ that are zero from the Gaussian assumptions on $x(t)$.

The direct derivation of Eq. (9.34) will now be detailed. Readers not concerned with this proof should go to Section 9.1.4.

Direct Derivation of the Second-Order Cross-Spectrum

From Eqs. (8.176), (9.24), and (9.27),

$$S_{xxy_b}(f, g) = \frac{1}{T} E[X^*(f)X^*(g)Y_b(f + g)]$$

where

$$Y_b(f + g) = \int H_2(u, f + g - u)X(u)X(f + g - u)\, du - \bar{y}_2\delta_1(f + g)$$

with

$$\bar{y}_2 = \int H_2(u, -u)S_{xx}(u)\, du$$

Now,

$$S_{xxy_b}(f, g) = \frac{1}{T}\int H_2(u, f + g - u)E[X^*(f)X^*(g)X(u)X(f + g - u)]\, du$$
$$- \bar{y}_2\delta_1(f + g)\left[\frac{1}{T}E[X^*(f)X^*(g)\right]$$

The second-order moment

$$E[X^*(f)X^*(g)] = S_{xx}(f)\delta_1(f + g) = \begin{cases} TS_{xx}(f), & f + g = 0 \\ 0, & \text{otherwise} \end{cases}$$

Hence, using Eq. (9.23), the last term in $S_{xxy_b}(f, g)$ is

$$-S_{xx}(f)\delta_1(f + g)\int H_2(u, -u)S_{xx}(u)\, du$$

The fourth-order moment in the first term of $S_{xxy_b}(f, g)$ reduces to

$$\begin{aligned} &E[X^*(f)X^*(g)X(u)X(f+g-u)] \\ &\quad = E[X^*(f)X^*(g)]E[X(u)X(f+g-u)] \\ &\qquad + E[X^*(f)X(u)]E[X^*(g)X(f+g-u)] \\ &\qquad + E[X^*(f)X(f+g-u)]E[X^*(g)X(u)] \\ &\quad = [S_{xx}(f)\delta_1(f+g)][S_{xx}(u)\delta_1(f+g)] + [S_{xx}(f)\delta_1(f-u)][S_{xx}(g)\delta_1(f-u)] \\ &\qquad + [S_{xx}(f)\delta_1(g-u)][S_{xx}(g)\delta_1(g-u)] \end{aligned}$$

Substituting into the first term in $S_{xxy_b}(f, g)$ and integrating term by term gives, for the first pair of second-order moments,

$$\begin{aligned} \frac{1}{T}\int H_2(u, f+g-u)S_{xx}(f)S_{xx}(u)\delta_1^2(f+g)\,du \\ = S_{xx}(f)\delta_1(f+g)\int H_2(u, -u)S_{xx}(u)\,du \end{aligned}$$

This cancels out with the last term in $S_{xxy_b}(f, g)$. The second pair of second-order moments gives

$$\frac{1}{T}\int H_2(u, f+g-u)S_{xx}(f)S_{xx}(g)\delta_1^2(f-u)\,du = H_2(f, g)S_{xx}(f)S_{xx}(g)$$

using the fact that

$$\frac{1}{T}\iint \delta_1^2(f-u)\,du = \int \delta_1(f-u)\,du = 1$$

The third pair of second-order moments gives

$$\frac{1}{T}\int H_2(u, f+g-u)S_{xx}(f)S_{xx}(g)\delta_1^2(g-u)\,du = H_2(g, f)S_{xx}(f)S_{xx}(g)$$

where $H_2(g, f) = H_2(f, g)$ by the symmetry property. Hence,

$$S_{xxy_b}(f, g) = 2H_2(f, g)S_{xx}(f)S_{xx}(g)$$

This completes the direct proof for the second-order bispectrum result of Eq. (9.34).

9.1.4 Trilinear Systems

For trilinear systems alone, from Eqs. (8.64) and (8.70), the trilinear system outputs are

$$y_3(t) = \iiint h_3(\alpha, \beta, \gamma)x(t - \alpha)x(t - \beta)x(t - \gamma)\, d\alpha\, d\beta\, d\gamma \tag{9.37}$$

$$Y_3(f) = \iint H_3(u, v - u, f - v)X(u)X(v - u)X(f - v)\, du\, dv \tag{9.38}$$

From the assumed properties of the zero mean value Gaussian input $x(t)$, and the symmetric nature of $h_3(\alpha, \beta, \gamma)$ and $H_3(f_1, f_2, f_3)$, one can prove that $\bar{y}_3 = E[y_3(t)] = 0$, $E[Y_3(f)] = 0$, and that the input/output cross-correlation function is

$$R_{xy_3}(\tau) = \int c(\alpha)R_{xx}(\tau - \alpha)\, d\alpha \tag{9.39}$$

where

$$c(\alpha) = 3\iint h_3(\alpha, \beta, \gamma)R_{xx}(\beta - \gamma)\, d\beta\, d\gamma \tag{9.40}$$

Taking the Fourier transforms of both sides of Eq. (9.39) shows that the corresponding input/output cross-spectral density function is

$$S_{xy_3}(f) = C(f)S_{xx}(f) \tag{9.41}$$

with $C(f)$, the Fourier transform of $c(\alpha)$, given by

$$C(f) = [S_{xy_3}(f)/S_{xx}(f)] = 3\int H_3(u, -u, f)S_{xx}(u)\, du \tag{9.42}$$

This result for $C(f)$ varies as a function of $S_{xx}(f)$. One can also show, using Eq. (9.1), that the cross-correlation function

$$R_{y_1y_3}(\tau) = \int h_1(\beta)R_{xy_3}(\tau + \beta)\, d\beta = \iint h_1(\beta)c(\alpha)R_{xx}(\tau + \beta - \alpha)\, d\alpha\, d\beta \tag{9.43}$$

with the associated cross-spectral density function

$$S_{y_1y_3}(f) = H_1^*(f)S_{xy_3}(f) = H_1^*(f)C(f)S_{xx}(f) \tag{9.44}$$

Note that $y_3(t)$ is correlated with both $x(t)$ and $y_1(t)$. Detailed derivations for the above equations are straightforward exercises along the lines of similar derivations for bilinear systems in Section 9.1.3. All frequency-domain results can be derived directly without going through correlation functions.

To obtain uncorrelated trilinear results with $x(t)$ and $y_1(t)$, one can define the trilinear system output $y_c(t)$ from $y_3(t)$ by

$$y_c(t) = y_3(t) - \int c(\alpha)x(t - \alpha)\, d\alpha \tag{9.45}$$

Then $\bar{y}_c = E[y_c(t)] = 0$ with

$$Y_c(f) = Y_3(f) - C(f)X(f) \tag{9.46}$$

In place of Eqs. (9.39)–(9.44), one obtains

$$\begin{aligned} R_{xy_c}(\tau) &= R_{y_1y_c}(\tau) = 0 \\ S_{xy_c}(f) &= S_{y_1y_c}(f) = 0 \end{aligned} \tag{9.47}$$

Thus $y_c(t)$, unlike $y_3(t)$, will be uncorrelated with $x(t)$ and $y_1(t)$. Further analysis should be conducted with $y_c(t)$ instead of $y_3(t)$.

This quantity $y_c(t)$ is uncorrelated with the $y_b(t)$ of Eq. (9.26), since odd-order moments occur.

$$\begin{aligned} R_{y_by_c}(\tau) &= R_{y_2y_3}(\tau) = 0 \\ S_{y_by_c}(f) &= S_{y_1y_3}(f) = 0 \end{aligned} \tag{9.48}$$

One can also verify that the following second-order correlation and spectral density functions are identically zero:

$$\begin{aligned} R_{xxy_c}(\tau_1, \tau_2) &= R_{xxy_3}(\tau_1, \tau_2) = 0 \\ S_{xxy_c}(f, g) &= S_{xxy_3}(f, g) = 0 \end{aligned} \tag{9.49}$$

Trilinear System Output Autocorrelation Function. The trilinear system output autocorrelation function $R_{y_cy_c}(\tau)$ is given from Eq. (9.45) by

$$\begin{aligned} R_{y_cy_c}(\tau) &= E[y_c(t)y_c(t + \tau)] \\ &= R_{y_3y_3}(\tau) - \int c(\alpha)R_{xy_3}(\tau + \alpha)\, d\alpha - \int c(\beta)R_{xy_3}(\beta - \tau)\, d\beta \\ &\quad + \iint c(\alpha)c(\beta)R_{xx}(\tau + \alpha - \beta)\, d\alpha\, d\beta \end{aligned}$$

From Eq. (9.39),

$$R_{xy_3}(\tau + \alpha) = \int c(\beta)R_{xx}(\tau + \alpha - \beta)\, d\beta$$

$$R_{xy_3}(\beta - \tau) = \int c(\alpha)R_{xx}(\beta - \tau - \alpha)\, d\alpha = \int c(\alpha)R_{xx}(\tau + \alpha - \beta)\, d\alpha$$

Hence,

$$R_{y_cy_c}(\tau) = R_{y_3y_3}(\tau) - \iint c(\alpha)c(\beta)R_{xx}(\tau + \alpha - \beta)\, d\alpha\, d\beta \tag{9.50}$$

An expression will now be derived for $R_{y_3y_3}(\tau)$. From Eq. (9.37),

$$R_{y_3y_3}(\tau) = \iiiint\!\!\iint h_3(\alpha, \beta, \gamma)h_3(u, v, w)J(\alpha, \beta, \gamma, u, v, w, \tau)\, d\alpha\, d\beta\, d\gamma\, du\, dv\, dw \tag{9.51}$$

where the integrand quantity J is defined by

$$J(\alpha, \beta, \gamma, u, v, w, \tau) = E[x(t-\alpha)x(t-\beta)x(t-\gamma)x(t+\tau-u)x(t+\tau-v)x(t+\tau-w)] \tag{9.52}$$

This breaks down into 15 terms from which symmetry properties of $h_3(\alpha, \beta, \gamma)$ and $h_3(u, v, w)$ show, after many detailed steps, that

$$\begin{aligned} J &= 6R_{xx}(\tau + \alpha - u)R_{xx}(\tau + \beta - v)R_{xx}(\tau + \gamma - w) \\ &\quad + 9R_{xx}(\alpha - \beta)R_{xx}(u - v)R_{xx}(\tau + \gamma - w) \end{aligned}$$

The second term above, when substituted into Eq. (9.51), is the same from Eq. (9.40) as the second term in Eq. (9.50) except for sign. Hence, one obtains

$$R_{y_3y_3}(\tau) = R_{y_cy_c}(\tau) + \iint c(\alpha)c(\beta)R_{xx}(\tau + \alpha - \beta)\, d\alpha\, d\beta \tag{9.53}$$

where

$$\begin{aligned} R_{y_cy_c}(\tau) = 6 \iiiint\!\!\iint & h_3(\alpha, \beta, \gamma)h_3(u, v, w)R_{xx}(\tau + \alpha - u) \\ & \times R_{xx}(\tau + \beta - v)R_{xx}(\tau + \gamma - w)\, da\, d\beta\, d\gamma\, du\, dv\, dw \end{aligned} \tag{9.54}$$

This complicated sixfold integral is practically impossible to compute and apply.

Trilinear System Output Autospectral Density Function. As per the indirect derivation of Eq. (9.31) from Eq. (9.30), one can prove after many steps that the corresponding trilinear system output autospectral density function

$$S_{y_cy_c}(f) = 6 \iint |H_3(u, v-u, f-v)|^2 S_{xx}(u) S_{xx}(v-u) S_{xx}(f-v)\, du\, dv \quad (9.55)$$

Equation (9.55) is simpler than Eq. (9.54) but still very difficult to compute.

Equation (9.55) can also be derived directly from Eqs. (8.170) and (9.46) without computing correlation functions. From Eq. (9.53) one obtains, by taking Fourier transforms of both sides,

$$S_{y_3y_3}(f) = S_{y_cy_c}(f) + |C(f)|^2 S_{xx}(f) \quad (9.56)$$

where $C(f)$ is given by Eq. (9.42) and $S_{y_cy_c}(f)$ by Eq. (9.55).

Third-Order Cross-Correlation Function. Consider now the third-order cross-correlation function $R_{xxxy_c}(\tau_1, \tau_2, \tau_3)$. From Eqs. (8.182) and (9.45),

$$\begin{aligned} R_{xxxy_c}&(\tau_1, \tau_2, \tau_3) \\ &= E[x(t-\tau_1)x(t-\tau_2)x(t-\tau_3)y_c(t)] \\ &= R_{xxxy_3}(\tau_1, \tau_2, \tau_3) - \int c(\alpha) E[x(t-\tau_1)x(t-\tau_2)x(t-\tau_3)x(t-\alpha)]\, d\alpha \quad (9.57) \end{aligned}$$

where the fourth-order moment

$$\begin{aligned} E[x(t&-\tau_1)x(t-\tau_2)x(t-\tau_3)x(t-\alpha)] \\ &= R_{xx}(\tau_1-\tau_2)R_{xx}(\tau_3-\alpha) + R_{xx}(\tau_1-\tau_3)R_{xx}(\tau_2-\alpha) \\ &\quad + R_{xx}(\tau_2-\tau_3)R_{xx}(\tau_1-\alpha) \end{aligned}$$

Hence, from Eq. (9.39), it follows that

$$\begin{aligned} R_{xxxy_c}(\tau_1, \tau_2, \tau_3) &= R_{xxxy_3}(\tau_1, \tau_2, \tau_3) - R_{xx}(\tau_1-\tau_2)R_{xy_3}(\tau_3) \\ &\quad - R_{xx}(\tau_1-\tau_3)R_{xy_3}(\tau_2) - R_{xx}(\tau_2-\tau_3)R_{xy_3}(\tau_1) \quad (9.58) \end{aligned}$$

Separate calculation can also be made of $R_{xxxy_3}(\tau_1, \tau_2, \tau_3)$ from

$$\begin{aligned} R_{xxxy_3}&(\tau_1, \tau_2, \tau_3) \\ &= E[x(t-\tau_1)x(t-\tau_2)x(t-\tau_3)y_3(t)] \quad (9.95) \\ &= \iiint h_3(\alpha, \beta, \gamma) E[x(t-\tau_1)x(t-\tau_2)x(t-\tau_3)x(t-\alpha)x(t-\beta)x(t-\gamma)]\, d\alpha\, d\beta\, d\gamma \end{aligned}$$

The sixth-order moment here breaks down into 15 terms from which symmetry properties of $h_3(\alpha, \beta, \gamma)$ show that

$$R_{xxxy_3}(\tau_1, \tau_2, \tau_3) = 6 \iiint h_3(\alpha, \beta, \gamma) R_{xx}(\tau_1 - \alpha) R_{xx}(\tau_2 - \beta) R_{xx}(\tau_3 - \gamma)\, d\alpha\, d\beta\, d\gamma$$
$$+ R_{xx}(\tau_1 - \tau_2) R_{xy_3}(\tau_3) + R_{xx}(\tau_1 - \tau_3) R_{xy_3}(\tau_2)$$
$$+ R_{xx}(\tau_2 - \tau_3) R_{xy_3}(\tau_1) \tag{9.60}$$

This proves that the third-order cross-correlation function is

$$R_{xxxy_c}(\tau_1, \tau_2, \tau_3)$$
$$= 6 \iiint h_3(\alpha, \beta, \gamma) R_{xx}(\tau_1 - \alpha) R_{xx}(\tau_2 - \beta) R_{xx}(\tau_3 - \gamma)\, d\alpha\, d\beta\, d\gamma \tag{9.61}$$

Third-Order Cross-Spectrum. From Eq. (8.185), the third-order cross-spectrum (trispectrum) is defined by

$$S_{xxxy_c}(f, g, h) = \iiint R_{xxxy_c}(\tau_1, \tau_2, \tau_3) e^{-j2\pi(f\tau_1 + g\tau_2 + h\tau_3)}\, d\tau_1\, d\tau_2\, d\tau_3 \tag{9.62}$$

Substitution of Eq. (9.61) into Eq. (9.62) after many steps gives the important trispectrum result in the frequency domain that is equivalent to Eq. (9.61), namely,

$$S_{xxxy_c}(f, g, h) = 6H_3(f, g, h) S_{xx}(f) S_{xx}(g) S_{xx}(h) \tag{9.63}$$

Here, $H_3(f, g, h)$ is the triple Fourier transform of $h_3(\tau_1, \tau_2, \tau_3)$ as per Eq. (8.69). Equation (9.63) can also be derived directly from Eq. (8.184) without computing correlation functions. Note the simplicity of Eq. (9.63) compared with Eq. (9.61).

9.1.5 Summary of Results

As in Section 8.4.6, the main equations should be compared for their inner mathematical structure as one proceeds from linear to bilinear to trilinear systems.

Time-domain cross-correlation results are

$$R_{xy_1}(\tau) = \int h_1(\alpha) R_{xx}(\tau - \alpha)\, d\alpha \tag{9.64}$$

$$R_{xxy_b}(\tau_1, \tau_2) = 2 \iint h_2(\alpha, \beta) R_{xx}(\tau_1 - \alpha) R_{xx}(\tau_2 - \beta)\, d\alpha\, d\beta \tag{9.65}$$

$$R_{xxxy_c}(\tau_1, \tau_2, \tau_3) = 6 \iiint h_3(\alpha, \beta, \gamma) R_{xx}(\tau_1 - \alpha) R_{xx}(\tau_2 - \beta) R_{xx}(\tau_3 - \gamma)\, d\alpha\, d\beta\, d\gamma \tag{9.66}$$

The quantities $y_b(t)$ and $y_c(t)$ are defined by Eqs. (9.25) and (9.45), respectively. Corresponding frequency-domain cross-spectrum results are

$$S_{xy_1}(f) = H_1(f)S_{xx}(f) \tag{9.67}$$

$$S_{xxy_b}(f, g) = 2H_2(f, g)S_{xx}(f)S_{xx}(g) \tag{9.68}$$

$$S_{xxxy_c}(f, g, h) = 6H_3(f, g, h)S_{xx}(f)S_{xx}(g)S_{xx}(h) \tag{9.69}$$

The output autocorrelation functions are

$$R_{y_1y_1}(\tau) = \iint h_1(\alpha)h_1(\beta)R_{xx}(\tau + \alpha - \beta)\, d\alpha\, d\beta \tag{9.70}$$

$$R_{y_by_b}(\tau) = 2 \iiiint h_2(\alpha, \beta)h_2(u, v)R_{xx}(\tau + \alpha - u)R_{xx}(\tau + \beta - v)\, d\alpha\, d\beta\, du\, dv \tag{9.71}$$

$$R_{y_cy_c}(\tau) = 6 \int\!\!\int\!\!\int\!\!\int\!\!\int\!\!\int h_3(\alpha, \beta, \gamma)h_3(u, v, w)R_{xx}(\tau + \alpha - u) \times R_{xx}(\tau + \beta - v)R_{xx}(\tau + \gamma - w)\, d\alpha\, d\beta\, d\gamma\, du\, dv\, dw \tag{9.72}$$

Corresponding output autospectral density functions are

$$S_{y_1y_1}(f) = |H_1(f)|^2 S_{xx}(f) \tag{9.73}$$

$$S_{y_by_b}(f) = 2 \int |H_2(u, f - u)|^2 S_{xx}(u)S_{xx}(f - u)\, du \tag{9.74}$$

$$S_{y_cy_c}(f) = 6 \iint |H_3(u, v - u, f - v)|^2 S_{xx}(u)S_{xx}(v - u)S_{xx}(f - v)\, du\, dv \tag{9.75}$$

The benefits of frequency-domain results compared to time-domain results are clear from these equations.

9.2 GENERAL THIRD-ORDER NONLINEAR MODELS

9.2.1 Nonlinear Model with Correlated Outputs

Some of the preceding results will now be extended so as to analyze various desired nonlinear models excited by zero mean value Gaussian stationary random data. Consider the general third-order Volterra nonlinear time-domain model of Figure 9.2, where the total output is given by

$$y(t) = y_1(t) + y_2(t) + y_3(t) \tag{9.76}$$

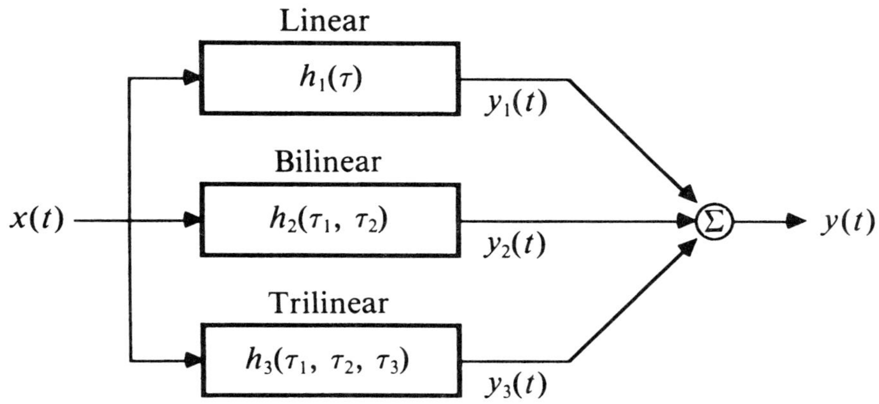

Figure 9.2. Third-order nonlinear time-domain model with correlated outputs.

Here, $y_1(t)$, $y_2(t)$, and $y_3(t)$ satisfy Eqs. (9.1), (9.22), and (9.37), respectively. The output autocorrelation function is

$$R_{yy}(\tau) = R_{y_1y_1}(\tau) + R_{y_2y_2}(\tau) + R_{y_3y_3}(\tau) + R_{y_1y_3}(\tau) + R_{y_3y_1}(\tau) \tag{9.77}$$

where $R_{y_1y_1}(\tau)$, $R_{y_2y_2}(\tau)$, $R_{y_1y_3}(\tau)$ and $R_{y_3y_3}(\tau)$ satisfy Eqs. (9.7), (9.29), (9.43), and (9.53), respectively. These equations are complicated to apply. Also, correlation occurs between the outputs $y_1(t)$ and $y_3(t)$. One has here the mean values $\bar{y}_1 = \bar{y}_3 = 0$ with

$$\bar{y} = \bar{y}_2 = E[y(t)] = \int H_2(f, -f)S_{xx}(f)\,df \tag{9.78}$$

First-order cross-correlation functions are

$$R_{xy_1}(\tau) = \int h_1(\alpha)R_{xx}(\tau - \alpha)\,d\alpha \tag{9.79}$$

$$R_{xy_2}(\tau) = 0 \tag{9.80}$$

$$R_{xy_3}(\tau) = \int c(\alpha)R_{xx}(\tau - \alpha)\,d\alpha \tag{9.81}$$

as shown in Eqs. (9.3), (9.28), and (9.39), respectively. One can also write appropriate equations from Figure 9.2 for the second- and third-order cross-correlation functions. However, it is sufficient here to note that all of these equations are complicated and that for the assumed zero mean value Gaussian input data, the second-order functions are

$$R_{xxy_1}(\tau_1, \tau_2) = R_{xxy_3}(\tau_1, \tau_2) = 0, \quad R_{xxy_2}(\tau_1, \tau_2) \neq 0 \tag{9.82}$$

while the third-order functions are

$$\begin{gathered} R_{xxxy_1}(\tau_1, \tau_2, \tau_3) \neq 0, \quad R_{xxxy_3}(\tau_1, \tau_2, \tau_3) \neq 0 \\ R_{xxxy_2}(\tau_1, \tau_2, \tau_3) = 0 \end{gathered} \tag{9.83}$$

With respect to the total output $y(t)$, one has

$$R_{xy}(\tau) = R_{xy_1}(\tau) + R_{xy_3}(\tau) \tag{9.84}$$

$$R_{xxy}(\tau_1, \tau_2) = R_{xxy_2}(\tau_1, \tau_2) \tag{9.85}$$

$$R_{xxxy}(\tau_1, \tau_2, \tau_3) = R_{xxxy_1}(\tau_1, \tau_2, \tau_3) + R_{xxxy_3}(\tau_1, \tau_2, \tau_3) \tag{9.86}$$

Taking the Fourier transforms of Eq. (9.76) gives the third-order Volterra nonlinear frequency-domain model of Figure 9.3 where

$$Y(f) = \mathscr{F}[y(t) - \bar{y}] = Y_1(f) + Y_2(f) + Y_3(f) \tag{9.87}$$

The terms $Y_1(f)$, $Y_2(f)$, and $Y_3(f)$ satisfy Eqs. (9.2), (9.24), and (9.38), respectively.

The output autospectral density function for the general third-order nonlinear model of Figure 9.3 is as follows. From Eq. (9.77), by taking Fourier transforms,

$$S_{yy}(f) = S_{y_1y_1}(f) + S_{y_2y_2}(f) + S_{y_3y_3}(f) + S_{y_1y_3}(f) + S^*_{y_1y_3}(f) \tag{9.88}$$

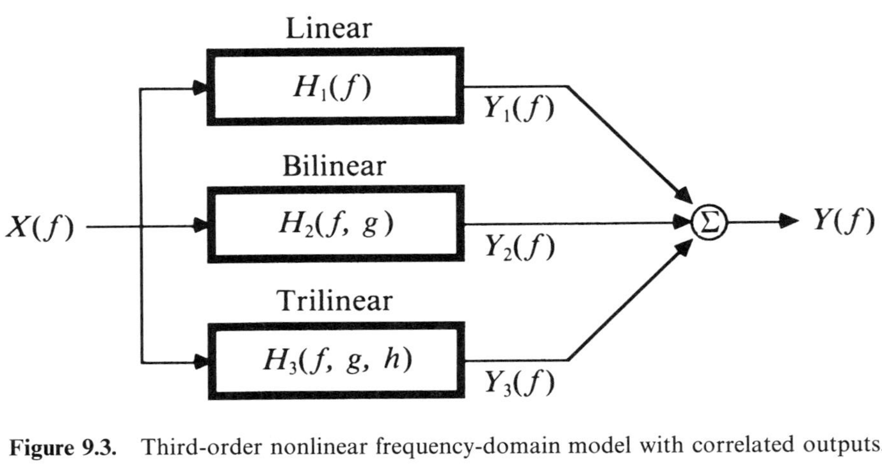

Figure 9.3. Third-order nonlinear frequency-domain model with correlated outputs.

where the linear system output spectrum $S_{y_1y_1}(f)$ is given by Eq. (9.2). The bilinear system output spectrum from Eq. (9.29) is

$$S_{y_2y_2}(f) = S_{y_by_b}(f) + (\bar{y}_2)^2\delta_1(f) \tag{9.89}$$

with $S_{y_by_b}(f)$ computed by Eq. (9.31) and $\bar{y}_2$ computed by Eq. (9.23). The trilinear system output spectrum from Eq. (9.56) is

$$S_{y_3y_3}(f) = S_{y_cy_c}(f) + |C(f)|^2 S_{xx}(f) \tag{9.90}$$

with $S_{y_cy_c}(f)$ computed by Eq. (9.55) and $C(f)$ computed by Eq. (9.42).

$$C(f) = 3\int H_3(u, -u, f)S_{xx}(u)\,du \tag{9.91}$$

The cross-spectrum term from Eq. (9.44) is

$$S_{y_1y_3}(f) = H_1^*(f)C(f)S_{xx}(f) \tag{9.92}$$

These correlation and spectral density relations can be replaced by simpler expressions using revised Volterra nonlinear models equivalent to Figures 9.2 and 9.3 where the outputs are uncorrelated. This will now be developed.

9.2.2 Nonlinear Model with Uncorrelated Outputs

It is desired to decompose the zero mean value output $[y(t) - \bar{y}]$ into parts that will have zero mean values and be mutually uncorrelated. Hence, one should replace $y_1(t)$, $y_2(t)$, and $y_3(t)$ in Eq. (9.76) by $y_a(t)$, $y_b(t)$, and $y_c(t)$, respectively, defined by

$$y_a(t) = \int [h_1(\alpha) + c(\alpha)]x(t - \alpha)\,d\alpha \tag{9.93}$$

$$y_b(t) = y_2(t) - \bar{y}_2 \tag{9.94}$$

$$y_c(t) = y_3(t) - \int c(\alpha)x(t - \alpha)\,d\alpha \tag{9.95}$$

Now, Eq. (9.76) becomes, with $\bar{y} = \bar{y}_2$,

$$y(t) = \bar{y} + y_a(t) + y_b(t) + y_c(t) \tag{9.96}$$

Figure 9.2 can thus be replaced by the equivalent Figure 9.4. It will be shown later that the linear output term $y_a(t)$ is the result of passing $x(t)$ through the optimum linear system $[h_1(\alpha) + c(\alpha)]$ to predict $y(t)$ from $x(t)$.

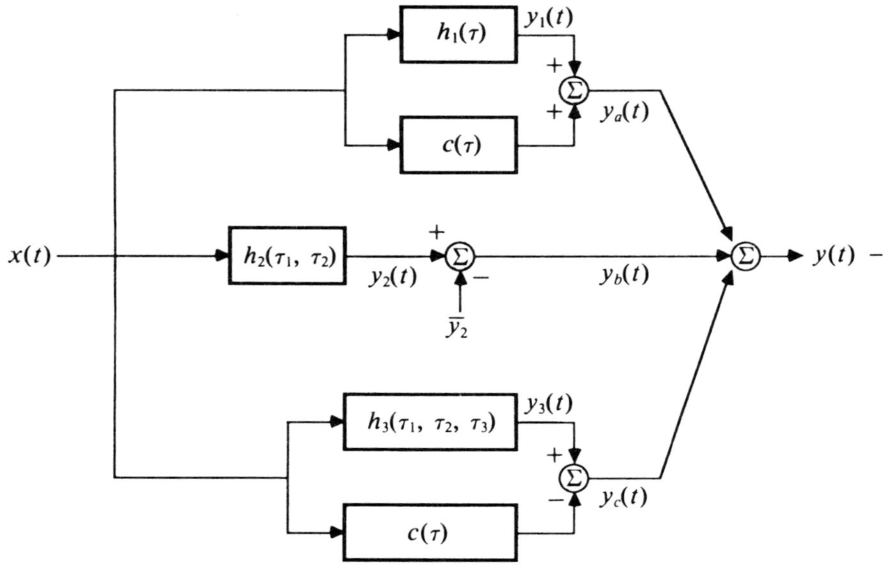

Figure 9.4. Revised third-order nonlinear time-domain model with uncorrelated outputs.

In the revised Volterra nonlinear model of Figure 9.4, the output autocorrelation function

$$R_{yy}(\tau) = (\bar{y})^2 + R_{y_a y_a}(\tau) + R_{y_b y_b}(\tau) + R_{y_c y_c}(\tau) \tag{9.97}$$

where $R_{y_a y_a}(\tau)$ is given by

$$R_{y_a y_a}(\tau) = \iint [h_1(\alpha) + c(\alpha)][h_1(\beta) + c(\beta)]R_{xx}(\tau + \alpha - \beta)\, d\alpha\, d\beta \tag{9.98}$$

and $R_{y_b y_b}(\tau)$ and $R_{y_c y_c}(\tau)$ are given by Eqs. (9.30) and (9.54), respectively. The various output cross-correlation functions and mean values are now

$$R_{y_a y_b}(\tau) = R_{y_a y_c}(\tau) = R_{y_b y_c}(\tau) = 0 \tag{9.99}$$

$$\bar{y}_a = \bar{y}_b = \bar{y}_c = 0 \tag{9.100}$$

The input/output cross-correlation functions become

$$R_{xy}(\tau) = R_{xy_a}(\tau) = \int [h_1(\alpha) + c(\alpha)]R_{xx}(\tau - \alpha)\, d\alpha \tag{9.101}$$

$$R_{xy_b}(\tau) = R_{xy_c}(\tau) = 0 \tag{9.102}$$

In words, $y_a(t)$, $y_b(t)$, and $y_c(t)$ are mutually uncorrelated, and $x(t)$ is uncorrelated with both $y_b(t)$ and $y_c(t)$.

Results for $y_a(t)$, $y_b(t)$, and $y_c(t)$ will now be expressed in the frequency domain using Fourier transforms and spectral density functions. The Fourier transform of Eq. (9.96) gives

$$Y(f) = \mathscr{F}[y(t) - \bar{y}] = Y_a(f) + Y_b(f) + Y_c(f) \tag{9.103}$$

where

$$Y_a(f) = [H_1(f) + C(f)]X(f) \tag{9.104}$$

$$Y_b(f) = Y_2(f) \tag{9.105}$$

$$Y_c(f) = Y_3(f) - C(f)X(f) \tag{9.106}$$

with the term $C(f)$ computed by Eq. (9.42), while $Y_2(f)$ and $Y_3(f)$ satisfy Eqs. (9.24) and (9.38), respectively. The result of Eq. (9.103) is displayed in the revised nonlinear model of Figure 9.5, which is equivalent to Figure 9.3.

From Eq. (9.99), the various output cross-spectral density functions will be identically zero for the mutually uncorrelated outputs, namely,

$$S_{y_a y_b}(f) = S_{y_a y_c}(f) = S_{y_b y_c}(f) = 0 \tag{9.107}$$

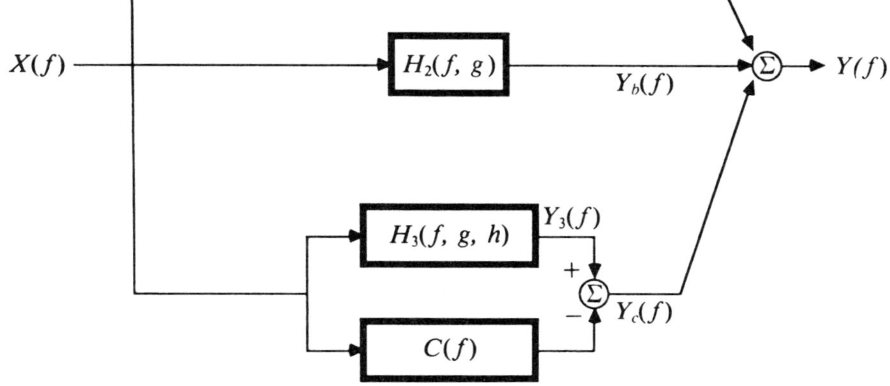

Figure 9.5. Revised third-order nonlinear frequency-domain model with uncorrelated outputs.

The total output autospectral density function in Figure 9.5 becomes

$$S_{yy}(f) = S_{y_a y_a}(f) + S_{y_b y_b}(f) + S_{y_c y_c}(f) \tag{9.108}$$

where from Eqs. (9.104), (9.31), and (9.55),

$$S_{y_a y_a}(f) = |H_1(f) + C(f)|^2 S_{xx}(f) \tag{9.109}$$

$$S_{y_b y_b}(f) = 2 \int |H_2(u, f-u)|^2 S_{xx}(u) S_{xx}(f-u)\, du \tag{9.110}$$

$$S_{y_c y_c}(f) = 6 \iint |H_3(u, v-u, f-v)|^2 S_{xx}(u) S_{xx}(v-u) S_{xx}(f-v)\, du\, dv \tag{9.111}$$

These spectral formulas are much simpler than corresponding correlation formulas. Equations (9.109)–(9.111) solve the *Output Prediction Problem* on how to predict output data properties in Figure 9.5 from knowledge of input data properties and system properties. The *Input Determination Problem* of $S_{xx}(f)$ cannot be solved here from knowledge of the system properties and the total output data properties of $y(t)$. From Eq. (9.109), one requires knowledge of the system $[H_1(f) + C(f)]$ and the linear output term $y_a(t)$.

Fourier transforms of Eqs. (9.101) and (9.102) prove that the first-order cross-spectrum relations in Figure 9.5 are

$$S_{xy}(f) = S_{xy_a}(f) = [H_1(f) + C(f)] S_{xx}(f) \tag{9.112}$$

$$S_{xy_b}(f) = S_{xy_c}(f) = 0 \tag{9.113}$$

Second-order cross-spectrum relation from Eq. (9.34) and uncorrelated outputs are

$$S_{xxy}(f, g) = S_{xxy_b}(f, g) = 2H_2(f, g) S_{xx}(f) S_{xx}(g) \tag{9.114}$$

$$S_{xxy_a}(f, g) = S_{xxy_c}(f, g) = 0 \tag{9.115}$$

Third-order cross-spectrum relations from Eq. (9.63) and uncorrelated outputs are

$$S_{xxxy}(f, g, h) = S_{xxxy_c}(f, g, h) = 6H_3(f, g, h) S_{xx}(f) S_{xx}(g) S_{xx}(h) \tag{9.116}$$

$$S_{xxxy_a}(f, g, h) = S_{xxxy_b}(f, g, h) = 0 \tag{9.117}$$

Equations (9.42), (9.112), (9.114), and (9.116) solve the *System Identification Problem* to determine all of the systems in Figure 9.5 from the measurements only of the input data $x(t)$ and the total output data $y(t)$. Furthermore, these calculations can be done using their Fourier transform quantities $X(f)$ and $Y(f)$ without ever computing any first- or higher-order correlation functions.

9.2.3 Identification of Linear, Bilinear, and Trilinear Systems

Useful procedures will now be developed to determine linear, bilinear, and trilinear system properties from measured data. Assume that only $x(t)$ and $y(t)$ are measured and that the Volterra models of Figures 9.3 and 9.5 apply except, perhaps, for extraneous zero mean value Gaussian output noise $n(t)$, which is assumed here to be uncorrelated with the mutually uncorrelated outputs $y_a(t)$, $y_b(t)$, and $y_c(t)$. This noise term represents all possible deviations from the ideal nonlinear models of Figures 9.3 and 9.5. With such output noise, they become Figures 9.6 and 9.7, where $N(f)$ is the Fourier transform of $n(t)$. In place of Eq. (9.103), one now obtains for the Volterra model of Figure 9.7 the result

$$Y(f) = Y_a(f) + Y_b(f) + Y_c(f) + N(f) \tag{9.118}$$

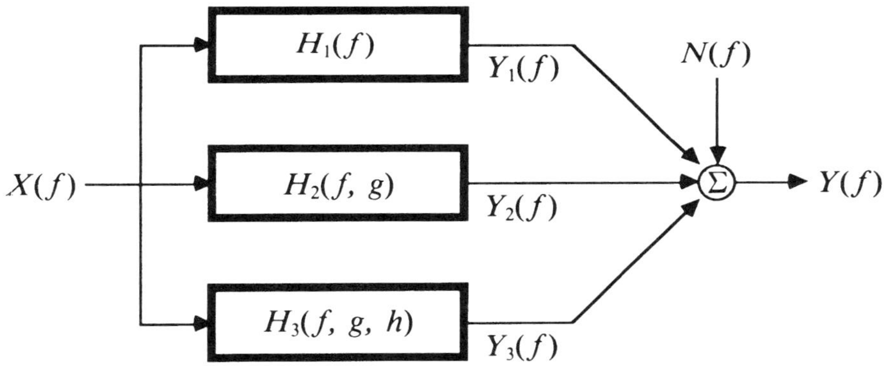

Figure 9.6. General Volterra third-order nonlinear frequency-domain model with correlated outputs.

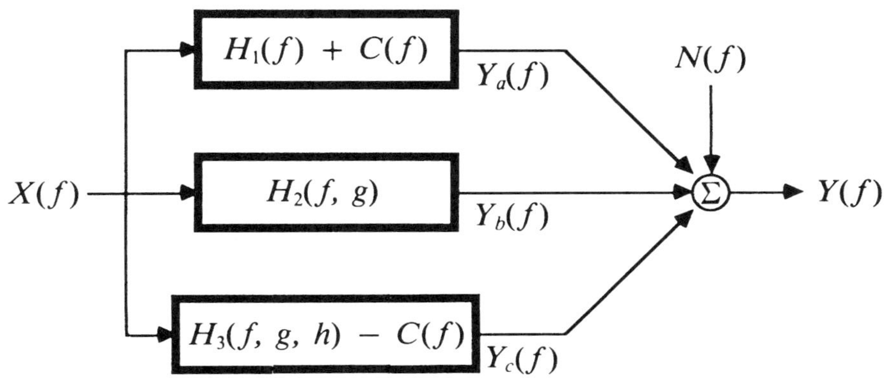

Figure 9.7. General Volterra third-order nonlinear frequency-domain model with uncorrelated outputs.

and in place of Eq. (9.108),

$$S_{yy}(f) = S_{y_a y_a}(f) + S_{y_b y_b}(f) + S_{y_c y_c}(f) + S_{nn}(f) \tag{9.119}$$

where $S_{nn}(f)$ is the autospectral density function of the output noise.

First-, second-, and third-order spectral density functions can be computed as outlined in Section 8.4 by dividing $x(t)$ and $y(t)$ into n_d subrecords of length T each, determining $X(f)$ and $Y(f)$ for each subrecord, and performing n_d available averages so as to estimate the following functions:

$$S_{xx}(f) = \frac{1}{T} E[X^*(f)X(f)] \tag{9.120}$$

$$S_{xy}(f) = \frac{1}{T} E[X^*(f)Y(f)] \tag{9.121}$$

$$S_{xxy}(f, g) = \frac{1}{T} E[X^*(f)X^*(g)Y(f + g)] \tag{9.122}$$

$$S_{xxxy}(f, g, h) = \frac{1}{T} E[X^*(f)X^*(g)X^*(h)Y(f + g + h)] \tag{9.123}$$

Computation of the multidimensional bispectral and trispectral density functions will generally be quite difficult.

The ratio of $S_{xy}(f)$ divided by $S_{xx}(f)$ gives the overall *optimum* linear system frequency response function $H_o(f)$ to estimate $y(t)$ from $x(t)$. Equation (9.112) then shows that

$$H_o(f) = \frac{S_{xy}(f)}{S_{xx}(f)} = H_1(f) + C(f) \tag{9.124}$$

This result for $H_o(f)$ varies as a function of $S_{xx}(f)$ when $C(f) \neq 0$. Note that $H_o(f) = H_1(f)$ if and only if $C(f) = 0$. The first-order linear system $H_1(f)$ is given by

$$H_1(f) = H_o(f) - C(f) \tag{9.125}$$

where $H_o(f)$ is known from Eq. (9.124) and $C(f)$ is still to be determined. In general, $H_1(f) \neq H_o(f)$.

The second-order bilinear system $H_2(f, g)$ is found from Eq. (9.114) by

$$H_2(f, g) = \frac{S_{xxy}(f, g)}{2S_{xx}(f)S_{xx}(g)} \tag{9.126}$$

When $f = u$ and $g = -u$, since $S_{xx}(-u) = S_{xx}(u)$, one obtains

$$H_2(u, -u) = \frac{S_{xxy}(u, -u)}{2S_{xx}^2(u)} \tag{9.127}$$

Here, from Eq. (9.122), using $X^*(-u) = X(u)$,

$$S_{xxy}(u, -u) = \frac{1}{T} E[X^*(u)X(u)Y(0)] \tag{9.128}$$

When $f = u$ and $g = f - u$,

$$H_2(u, f - u) = \frac{S_{xxy}(u, f - u)}{2S_{xx}(u)S_{xx}(f - u)} \tag{9.129}$$

where from Eq. (9.122),

$$S_{xxy}(u, f - u) = \frac{1}{T} E[X^*(u)X^*(f - u)Y(f)] \tag{9.130}$$

The third-order trilinear system $H_3(f, g, h)$ is found from Eq. (9.116) by

$$H_3(f, g, h) = \frac{S_{xxxy}(f, g, h)}{6S_{xx}(f)S_{xx}(g)S_{xx}(h)} \tag{9.131}$$

When $f = u, g = -u$, and $h = f$, one obtains

$$H_3(u, -u, f) = \frac{S_{xxxy}(u, -u, f)}{6S_{xx}(u)S_{xx}(u)S_{xx}(f)} \tag{9.132}$$

where from Eq. (9.123), using $X^*(-u) = X(u)$,

$$S_{xxxy}(u, -u, f) = \frac{1}{T} E[X^*(u)X(u)X^*(f)Y(f)] \tag{9.133}$$

When $f = u, g = v - u$, and $h = f - v$,

$$H_3(u, v - u, f - v) = \frac{S_{xxxy}(u, v - u, f - v)}{6S_{xx}(u)S_{xx}(v - u)S_{xx}(f - v)} \tag{9.134}$$

Here, from Eq. (9.123),

$$S_{xxxy}(u, v - u, f - v) = \frac{1}{T} E[X^*(u)X^*(v - u)X^*(f - v)Y(f)] \tag{9.135}$$

Substitution of Eq. (9.132) into Eq. (9.91) shows that

$$C(f) = 3\int H_3(u, -u, f)S_{xx}(u)\, du = \int \frac{S_{xxxy}(u, -u, f)}{2S_{xx}(u)S_{xx}(f)}\, du \tag{9.136}$$

The value of $C(f)$ together with $H_o(f)$ now gives $H_1(f)$ by Eq. (9.125). Thus, Eqs. (9.125), (9.126), and (9.131) solve the *System Identification Problem* on how to identify Volterra system properties from knowledge of input data properties and total output data properties when input data are Gaussian.

From Eqs. (9.124), (9.129), and (9.134), it follows by substitution that the uncorrelated output autospectra terms of Eqs. (9.109)–(9.111) in Figures 9.5 and 9.7 can be computed by the formulas

$$S_{y_a y_a}(f) = \frac{|S_{xy}(f)|^2}{S_{xx}(f)} \tag{9.137}$$

$$S_{y_b y_b}(f) = \frac{1}{2}\int \frac{|S_{xxy}(u, f-u)|^2}{S_{xx}(u)S_{xx}(f-u)}\, du \tag{9.138}$$

$$S_{y_c y_c}(f) = \frac{1}{6}\iint \frac{|S_{xxxy}(u, v-u, f-v)|^2}{S_{xx}(u)S_{xx}(v-u)S_{xx}(f-v)}\, du\, dv \tag{9.139}$$

One can also calculate the correlated output autospectra terms in Figure 9.3. All of these results can be obtained solely from measurement and analysis of the Gaussian input data $x(t)$ and the total output data $y(t)$ that includes the extraneous output noise.

9.2.4 Linear and Nonlinear Coherence Functions

Linear and nonlinear coherence functions will now be stated for the linear and nonlinear parts of Figure 9.7. The *linear coherence function* between $x(t)$ and $y(t)$ is defined for $f \neq 0$ by

$$\gamma_{xy}^2(f) = \frac{|S_{xy}(f)|^2}{S_{xx}(f)S_{yy}(f)} = \frac{S_{y_a y_a}(f)}{S_{yy}(f)} \tag{9.140}$$

where $0 \leqslant \gamma_{xy}^2(f) \leqslant 1$ for all f. The linear output spectrum $S_{y_a y_a}(f)$ is computed by Eq. (9.109) or (9.137). This gives the proportion of $S_{yy}(f)$ due to $x(t)$ passing through the optimum linear system $H_o(f) = [H_1(f) + C(f)]$. From Eqs. (9.109), (9.124), and (9.140), the linear output spectrum

$$S_{y_a y_a}(f) = |H_o(f)|^2 S_{xx}(f) = \gamma_{xy}^2(f)S_{yy}(f) \tag{9.141}$$

The *bilinear coherence function* between $x(t)$ and $y(t)$ via the bilinear system

$H_2(f, g)$ is defined for $f \neq 0$ by

$$q_{xy_b}^2(f) = \frac{S_{y_b y_b}(f)}{S_{yy}(f)} \tag{9.142}$$

where $0 \leqslant q_{xy_b}^2(f) \leqslant 1$ for all f. The bilinear output spectrum $S_{y_b y_b}(f)$ is computed by Eq. (9.110) or (9.138). This gives the proportion of $S_{yy}(f)$ due to $x(t)$ via the bilinear system $H_2(f, g)$. From Eq. (9.142), the bilinear output spectrum satisfies

$$S_{y_b y_b}(f) = q_{xy_b}^2(f) S_{yy}(f) \tag{9.143}$$

The *trilinear coherence function* between $x(t)$ and $y(t)$ via the trilinear system $[H_3(f, g, h) - C(f)]$ is defined for $f \neq 0$ by

$$q_{xy_c}^2(f) = \frac{S_{y_c y_c}(f)}{S_{yy}(f)} \tag{9.144}$$

where $0 \leqslant q_{xy_c}^2(f) \leqslant 1$ for all f. The trilinear output spectrum $S_{y_c y_c}(f)$ is computed by Eq. (9.111) or (9.139). This gives the proportion of $S_{yy}(f)$ due to $x(t)$ via the trilinear system $[H_3(f, g, h) - C(f)]$. From Eq. (9.144), the trilinear output spectrum satisfies

$$S_{y_c y_c}(f) = q_{xy_c}^2(f) S_{yy}(f) \tag{9.145}$$

Substitution of Eqs. (9.141), (9.143), and (9.145) into Eq. (9.119) now shows that the *output noise spectrum* in Figure 9.7 is given by

$$S_{nn}(f) = [1 - \gamma_{xy}^2(f) - q_{xy_b}^2(f) - q_{xy_c}^2(f)] S_{yy}(f) \tag{9.146}$$

The ratio of $S_{nn}(f)$ to $S_{yy}(f)$ gives the proportion of $S_{yy}(f)$ that is *not* due to $x(t)$ passing through any of the systems in Figure 9.7. Specifically, $S_{nn}(f)$ determines that part of the total output spectrum $S_{yy}(f)$ due to uncorrelated noise at the output, or due to $x(t)$ passing through higher-order nonlinear systems than those included in Figure 9.7.

A *goodness-of-fit* measure for the validity of this revised nonlinear model can be defined by seeing how close the sum $[\gamma_{xy}^2(f) + q_{xy_b}^2 + q_{xy_c}^2(f)]$ is to unity. This produces greatly improved results over previous work restricted to only linear system analysis where the bilinear and trilinear coherence functions are not computed and are assumed to be zero. In previous work, all of the possible bilinear and trilinear effects are included in the extraneous output noise. Thus, in general, previous work has computed an output noise spectrum that is much larger than the smaller result of Eq. (9.146). Also, previous work has computed an erroneous $H_1(f)$ when assuming incorrectly that $C(f) = 0$ in Eq. (9.125). The conclusion is that even if only linear system analysis and identification are desired, it can be made more accurate by following these procedures that allow for possible bilinear and trilinear effects. Of course, good engineering work now requires that these nonlinear effects be evaluated properly by following the procedures developed in this book, especially the simple, practical procedures in Chapters 3–7. Also, since results are functions of frequency, emphasis should be placed on those frequencies where there are peaks in $S_{yy}(f)$ as these peak values are of major physical interest where most of the energy or power occurs.

9.2.5 Summary of Results

A summary follows of some of the main results. Fourier transform relations for the third-order nonlinear frequency-domain model with correlated outputs of Figure 9.6 where $Y(f) = \mathscr{F}[y(t) - \bar{y}]$ are

$$Y(f) = Y_1(f) + Y_2(f) + Y_3(f) + N(f) \tag{9.147}$$

$$Y_1(f) = H_1(f)X(f) \tag{9.148}$$

$$Y_2(f) = \int H_2(u, f-u)X(u)X(f-u)\,du - \bar{y}_2\delta_1(f) \tag{9.149}$$

$$Y_3(f) = \iint H_3(u, v-u, f-v)X(u)X(v-u)X(f-v)\,du\,dv \tag{9.150}$$

$$\bar{y}_2 = \int H_2(f, -f)S_{xx}(f)\,df \tag{9.151}$$

These output results can all be computed from measurements of $x(t)$ and knowledge of the systems $H_1(f)$, $H_2(f, g)$, and $H_3(f, g, h)$. Note that for $f \neq 0$, the terms $\bar{y}_2$ and $\delta_1(f)$ can be omitted.

Spectral results for the third-order nonlinear frequency-domain model with correlated outputs of Figure 9.6 are

$$S_{yy}(f) = S_{y_1y_1}(f) + S_{y_2y_2}(f) + S_{y_3y_3}(f) + S_{y_1y_3}(f) + S^*_{y_1y_3}(f) + S_{nn}(f) \tag{9.152}$$

$$S_{y_1y_1}(f) = |H_1(f)|^2 S_{xx}(f) \tag{9.153}$$

$$S_{y_2y_2}(f) = S_{y_by_b}(f) + (\bar{y}_2)^2\delta_1(f) \tag{9.154}$$

$$S_{y_by_b}(f) = 2\int |H_2(u, f-u)|^2 S_{xx}(u)S_{xx}(f-u)\,du \tag{9.155}$$

$$S_{y_3y_3}(f) = S_{y_cy_c}(f) + |C(f)|^2 S_{xx}(f) \tag{9.156}$$

$$S_{y_cy_c}(f) = 6\iint |H_3(u, v-u, f-v)|^2 S_{xx}(u)S_{xx}(v-u)S_{xx}(f-v)\,du\,dv \tag{9.157}$$

$$S_{y_1y_3}(f) = H_1^*(f)C(f)S_{xx}(f) \tag{9.158}$$

$$C(f) = 3\int H_3(u, -u, f)S_{xx}(u)\,du \tag{9.159}$$

Fourier transform relations for the third-order nonlinear frequency-domain model with uncorrelated outputs of Figure 9.7 are

$$Y(f) = Y_a(f) + Y_b(f) + Y_c(f) + N(f) \tag{9.160}$$

$$Y_a(f) = [H_1(f) + C(f)]X(f) \tag{9.161}$$

$$Y_b(f) = \int H_2(u, f - u)X(u)X(f - u)\, du - \bar{y}_2\delta_1(f) \tag{9.162}$$

$$Y_c(f) = Y_3(f) - C(f)X(f) \tag{9.163}$$

$$Y_3(f) = \iint H_3(u, v - u, f - v)X(u)X(v - u)X(f - v)\, du\, dv \tag{9.164}$$

$$C(f) = 3\int H_3(u, -u, f)S_{xx}(u)\, du \tag{9.165}$$

$$\bar{y}_2 = \int H_2(f, -f)S_{xx}(f)\, df \tag{9.166}$$

These output results can all be computed from measurements of $x(t)$ and knowledge of the systems $H_1(f)$, $H_2(f, g)$, and $H_3(f, g, h)$. Note that for $f \neq 0$, the terms $\bar{y}_2$ and $\delta_1(f)$ can be omitted.

Spectral results for the third-order nonlinear frequency-domain model with uncorrelated outputs of Figure 9.7 are

$$S_{yy}(f) = S_{y_a y_a}(f) + S_{y_b y_b}(f) + S_{y_c y_c}(f) + S_{nn}(f) \tag{9.167}$$

$$S_{y_a y_a}(f) = |H_1(f) + C(f)|^2 S_{xx}(f) \tag{9.168}$$

$$S_{y_b y_b}(f) = 2\int |H_2(u, f - u)|^2 S_{xx}(u)S_{xx}(f - u)\, du \tag{9.169}$$

$$S_{y_c y_c}(f) = 6\iint |H_3(u, v - u, f - v)|^2 S_{xx}(u)S_{xx}(v - u)S_{xx}(f - v)\, du\, dv \tag{9.170}$$

Identification of the linear, bilinear, and trilinear systems from simultaneous measurements of $x(t)$ and $y(t)$ in Figures 9.6 and 9.7 can be carried out as follows. The first-order linear system $H_1(f)$ is given by

$$H_1(f) = H_o(f) - C(f) \tag{9.171}$$

where the optimum linear system is

$$H_o(f) = \frac{S_{xy}(f)}{S_{xx}(f)} \tag{9.172}$$

and where $C(f)$ from Eq. (9.136) is

$$C(f) = \int \frac{S_{xxxy}(u, -u, f)}{2S_{xx}(u)S_{xx}(f)} du \tag{9.173}$$

The second-order bilinear system $H_2(f, g)$ is given by

$$H_2(f, g) = \frac{S_{xxy}(f, g)}{2S_{xx}(f)S_{xx}(g)} \tag{9.174}$$

The third-order trilinear system $H_3(f, g, h)$ is given by

$$H_3(f, g, h) = \frac{S_{xxxy}(f, g, h)}{6S_{xx}(f)S_{xx}(g)S_{xx}(h)} \tag{9.175}$$

Sections 9.2.1–9.2.4 contain other useful relations.

Advantages of the above procedures are that they are mathematically exact for Gaussian input data. Disadvantages are that the multidimensional functions are usually very complicated, difficult to interpret, and cannot be converted to equations of motion.

9.3 OPTIMUM LINEAR AND NONLINEAR SYSTEMS

Formulas for identifying the optimum forms for the linear, bilinear, and trilinear systems $H_1(f)$, $H_2(f, g)$, and $H_3(f, g, h)$ will now be derived. These formulas minimize the output noise spectrum $S_{nn}(f)$ in the third-order Volterra nonlinear model of Figure 9.7 with respect to all possible choices of H_1, H_2, and H_3. Unlike Section 9.2.3, no assumption is made here that $n(t)$ is uncorrelated with $y_a(t)$, $y_b(t)$, and $y_c(t)$. Instead, it will be shown that this is automatically true when $S_{nn}(f)$ is minimized.

For arbitrary choices of the systems H_1, H_2, and H_3, the quantity $N(f)$ in Figure 9.7 is given by

$$N(f) = Y(f) - Y_a(f) - Y_b(f) - Y_c(f) \tag{9.176}$$

The autospectrum $S_{nn}(f)$ follows directly by multiplying $N^*(f)$ by $N(f)$, taking expected values, and multiplying by the factor $(1/T)$, namely,

$$S_{nn}(f) = \frac{1}{T} E[N^*(f)N(f)] \tag{9.177}$$

From Eq. (9.176), omitting f to simplify notation,

$$
\begin{aligned}
E[N^*N] &= E[(Y^* - Y_a^* - Y_b^* - Y_c^*)(Y - Y_a - Y_b - Y_c)] \\
&= E[Y^*(Y - Y_a - Y_b - Y_c) - Y_a^*(Y - Y_a) - Y_b^*(Y - Y_b) - Y_c^*(Y - Y_c)]
\end{aligned}
$$

plus other terms whose expected values are zero because $y_a(t)$, $y_b(t)$, and $y_c(t)$ are mutually uncorrelated in Figure 9.7. Thus,

$$
\begin{aligned}
E[N^*N] = E[Y^*Y] - E[Y^*Y_a] - E[Y^*Y_b] - E[Y^*Y_c] - E[Y_a^*Y] \\
+ E[Y_a^*Y_a] - E[Y_b^*Y] + E[Y_b^*Y_b] - E[Y_c^*Y] + E[Y_c^*Y_c]
\end{aligned}
$$

This proves the general result applicable to any systems $H_1(f)$, $H_2(f, g)$, and $H_3(f, g, h)$ in Figure 9.7 that

$$
\begin{aligned}
S_{nn}(f) &= S_{yy}(f) + [S_{y_ay_a}(f) - S_{yy_a}(f) - S_{y_ay}(f)] \\
&\quad + [S_{y_by_b}(f) - S_{yy_b}(f) - S_{y_by}(f)] + [S_{y_cy_c}(f) - S_{yy_c}(f) - S_{y_cy}(f)]
\end{aligned}
\tag{9.178}
$$

The system H_1 appears only in the first brackets, the system H_2 appears only in the second brackets, and the system H_3 appears only in the third brackets of Eq. (9.178). Hence, the minimization of $S_{nn}(f)$ with respect to H_1, H_2, and H_3 will not involve any overlapping terms and can be carried out separately.

9.3.1 Optimum Linear System

From Eq. (9.104),

$$
Y_a(f) = [H_1(f) + C(f)]X(f) \tag{9.179}
$$

It follows that the three spectral terms in the *first* brackets of Eq. (9.178) involving $H_1(f)$ are

$$
\begin{aligned}
S_{y_ay_a}(f) &= [H_1^*(f) + C^*(f)][H_1(f) + C(f)]S_{xx}(f) \\
S_{yy_a}(f) &= [H_1(f) + C(f)]S_{yx}(f) \\
S_{y_ay}(f) &= [H_1^*(f) + C^*(f)]S_{xy}(f)
\end{aligned}
\tag{9.180}
$$

Hence, omitting f to simplify notation, the first brackets in Eq. (9.178) become

$$
[S_{y_ay_a} - S_{yy_a} - S_{y_ay}] = (H_1^* + C^*)(H_1 + C)S_{xx} - (H_1 + C)S_{yx} - (H_1^* + C^*)S_{xy} \tag{9.181}
$$

The optimum linear system H_1 is defined as that particular H_1 over all possible H_1 that gives a minimum value for S_{nn}. This can be determined by setting the partial derivative of S_{nn} of Eq. (9.178) with respect to H_1^* equal to zero, holding H_1 fixed. As shown in reference 1, this procedure is equivalent to the classical operations of separately setting the partial derivatives of S_{nn} with respect to the real and imaginary parts of H_1 equal to zero. The desired partial derivative of S_{nn} with respect to H_1^* is the partial derivative of Eq. (9.181) with respect to H_1^*. When set equal to zero, this gives the result

$$\frac{\partial S_{nn}}{\partial H_1^*} = (H_1 + C)S_{xx} - S_{xy} = 0 \tag{9.182}$$

Hence, the optimum linear system $H_1 = H_1(f)$ must satisfy the equation

$$H_1(f) = \frac{S_{xy}(f)}{S_{xx}(f)} - C(f) = H_o(f) - C(f) \tag{9.183}$$

This is the same as Eq. (9.124) where $H_o(f)$, the overall optimum linear system, is not the same as $H_1(f)$ when $C(f) \neq 0$.

For this optimum choice of H_1, Eq. (9.180) becomes

$$S_{y_a y_a} = (H_1^* + C^*)S_{xy} = (H_1 + C)S_{yx} \tag{9.184}$$

since $S_{y_a y_a}$ is real-valued. Thus,

$$S_{y_a y_a} = S_{y y_a} = S_{y_a y} \tag{9.185}$$

and Eq. (9.181) is the same as

$$[S_{y_a y_a} - S_{y y_a} - S_{y_a y}] = -S_{y_a y_a} \tag{9.186}$$

where $S_{y_a y_a}$ satisfies Eq. (9.141). Also, from Eq. (9.176),

$$S_{n y_a} = S_{y y_a} - S_{y_a y_a} = 0 \tag{9.187}$$

so that $n(t)$ is uncorrelated with $y_a(t)$.

9.3.2 Optimum Bilinear System

From Eqs. (9.24) and (9.27),

$$Y_b(f) = \int H_2(u, f - u)X(u)X(f - u)\, du - \bar{y}_2\delta_1(f) \tag{9.188}$$

It follows here by a direct proof that for Gaussian input data the three spectral terms in the *second* brackets of Eq. (9.178) involving $H_2(f, g)$ are

$$S_{y_b y_b}(f) = 2\int H_2^*(u, f-u)H_2(u, f-u)S_{xx}(u)S_{xx}(f-u)\,du$$

$$S_{yy_b}(f) = \int H_2(u, f-u)S_{xxy}^*(u, f-u)\,du \tag{9.189}$$

$$S_{y_b y}(f) = \int H_2^*(u, f-u)S_{xxy}(u, f-u)\,du$$

Now the partial derivative of S_{nn} of Eq. (9.178) with respect to H_2^* (holding H_2 fixed) when set equal to zero gives the result

$$\frac{\partial S_{nn}}{\partial H_2^*} = 2H_2(u, f-u)S_{xx}(u)S_{xx}(f-u) - S_{xxy}(u, f-u) = 0$$

Hence, the optimum bilinear system $H_2(u, f-u)$ must satisfy the equation

$$H_2(u, f-u) = \frac{S_{xxy}(u, f-u)}{2S_{xx}(u)S_{xx}(f-u)} \tag{9.190}$$

This is the same as Eqs. (9.129) and (9.126). For this choice of H_2 it follows that

$$S_{y_b y_b} = S_{yy_b} = S_{y_b y} \tag{9.191}$$

The sum of the three terms in the second brackets is now

$$[S_{y_b y_b} - S_{yy_b} - S_{y_b y}] = -S_{y_b y_b} \tag{9.192}$$

where $S_{y_b y_b}(f)$ satisfies Eq. (9.143). Also, from Eq. (9.176),

$$S_{ny_b}(f) = 0 \tag{9.193}$$

so that $n(t)$ is uncorrelated with $y_b(t)$.

9.3.3 Optimum Trilinear System

From Eqs. (9.38) and (9.46),

$$Y_c(f) = \iint H_3(u, v-u, f-v)X(u)X(v-u)X(f-v)\,du\,dv - C(f)X(f) \tag{9.194}$$

Here it follows by a direct proof that for Gaussian input data the three spectral terms in the *third* brackets of Eq. (9.178) involving $H_3(f, g, h)$ are

$$S_{y_c y_c}(f) = 6 \iint |H_3(u, v-u, f-v)|^2 S_{xx}(u) S_{xx}(v-u) S_{xx}(f-v)\, du\, dv \tag{4.195}$$

$$S_{yy_c}(f) = \iint H_3(u, v-u, f-v) S^*_{xxxy}(u, v-u, f-v)\, du\, dv$$

$$S_{y_c y}(f) = \iint H^*_3(u, v-u, f-v) S_{xxxy}(u, v-u, f-v)\, du\, dv$$

Now the partial derivative of S_{nn} of Eq. (9.178) with respect to H^*_3 (holding H_3 fixed) when set equal to zero gives the result

$$\frac{\partial S_{nn}}{\partial H^*_3} = 6H_3(u, v-u, f-v) S_{xx}(u) S_{xx}(v-u) S_{xx}(f-v) - S_{xxxy}(u, v-u, f-v) = 0$$

Hence, the optimum trilinear system $H_3(u, v-u, f-v)$ must satisfy the equation

$$H_3(u, v-u, f-v) = \frac{S_{xxxy}(u, v-u, f-v)}{6S_{xx}(u) S_{xx}(v-u) S_{xx}(f-v)} \tag{9.196}$$

This is the same as Eqs. (9.134) and (9.131). For this choice of H_3, it follows that

$$S_{y_c y_c} = S_{yy_c} = S_{y_c y} \tag{9.197}$$

The sum of the three terms in the third brackets is now

$$[S_{y_c y_c} - S_{yy_c} - S_{y_c y}] = -S_{y_c y_c} \tag{9.198}$$

where $S_{y_c y_c}(f)$ satisfies Eq. (9.145). Also, from Eq. (9.176),

$$S_{ny_c} = 0 \tag{9.199}$$

so that $n(t)$ is uncorrelated with $y_c(t)$.

9.3.4 Minimum Output Noise Spectrum

Substitution of Eqs. (9.186), (9.192), and (9.198) into Eq. (9.178) proves that the minimum output noise spectrum is given by

$$\begin{aligned} S_{nn}(f) &= S_{yy}(f) - S_{y_a y_a}(f) - S_{y_b y_b}(f) - S_{y_c y_c}(f) \\ &= [1 - \gamma^2_{xy}(f) - q^2_{xy_b}(f) - q^2_{xy_c}(f)] S_{yy}(f) \end{aligned} \tag{9.200}$$

This is the same as Eq. (9.146) and provides validity for the nonlinear model of Figure 9.7, as discussed below Eq. (9.146).

In summary, Eqs. (9.183), (9.190), and (9.196) give the optimum forms for linear, bilinear, and trilinear systems, respectively, that minimize the output noise spectrum $S_{nn}(f)$ in the general third-order nonlinear model of Figure 9.7. Equations (9.187), (9.193), and (9.199) prove that the resulting output noise $n(t)$ will be automatically uncorrelated with the linear output $y_a(t)$, the bilinear output $y_b(t)$, and the trilinear output $y_c(t)$. Finally, in terms of linear and nonlinear coherence functions, Eq. (9.200) gives the minimum value for the output noise spectrum. These results are all calculated solely from measurements of only the input data $x(t)$ and the total output data $y(t)$ under the assumption that the input data represent a zero mean value Gaussian stationary random process with arbitrary spectral properties.

9.4 CASE 1 AND CASE 2 NONLINEAR MODELS

The total output Fourier transform $Y(f)$ for the general third-order Volterra nonlinear model in Figure 9.6 is

$$Y(f) = Y_1(f) + Y_2(f) + Y_3(f) + N(f) \tag{9.201}$$

where $Y_1(f)$, $Y_2(f)$, and $Y_3(f)$ are the Fourier transform outputs from the linear, bilinear, and trilinear systems, respectively. The Case 1 and Case 2 single-input/single-output nonlinear models of Figures 4.13 and 4.16 are special cases of Figure 9.6 that involve square-law and cubic nonlinear systems. For Gaussian input data, formulas will now be developed for these Case 1 and Case 2 nonlinear models that follow from using the general multidimensional relations in this chapter. Techniques for arbitrary data are in Chapter 4.

9.4.1 Case 1 Square-Law and Cubic Models

Case 1 single-input/single-output nonlinear models from Section 4.6.1 occur when higher-order frequency response functions can be represented by additive first-order frequency response functions such that the linear, bilinear, and trilinear systems in Figure 9.6 are

$$H_1(f) = A_1(f) \tag{9.202}$$

$$H_2(f_1, f_2) = A_2(f_1 + f_2) \tag{9.203}$$

$$H_3(f_1, f_2, f_3) = A_3(f_1 + f_2 + f_3) \tag{9.204}$$

Here,

$$H_2(f, f) = A_2(2f) \tag{9.205}$$

$$H_2(f, f, f) = A_3(3f) \tag{9.206}$$

$$H_2(f, -f) = A_2(0) \tag{9.207}$$

$$H_2(u, f - u) = A_2(f)$$

$$H_3(u, v - u, f - v) = A_3(f) \tag{9.208}$$

The $A_i(f)$ are frequency response functions of constant-parameter linear systems either known or to be determined. For this case, Eqs. (9.2), (9.24), and (9.38) become

$$Y_1(f) = A_1(f)X(f) \tag{9.209}$$

$$Y_2(f) = A_2(f)\int X(u)X(f-u)\,du - \bar{y}_2\delta_1(f) = Y_b(f) \tag{9.210}$$

$$Y_3(f) = A_3(f)\iint X(u)X(v-u)X(f-v)\,du\,dv \tag{9.211}$$

with $\bar{y}_2$ given from Eq. (9.23) by

$$\bar{y}_2 = A_2(0)\int S_{xx}(f)\,df = A_2(0)\sigma_x^2 \tag{9.212}$$

Equation (9.210) shows that $\bar{y}_2$ occurs only at $f = 0$.

These results have a simple physical interpretation as pictured in Figure 9.8, where zero-memory squaring and cubing operations are followed by the A for "after" linear systems. The single integral in Eq. (9.210) is the Fourier transform of $x^2(t)$, denoted by $X_2(f)$, and the double integral in Eq. (9.211) is the Fourier transform of $x^3(t)$, denoted by $X_3(f)$. The total output Fourier transform $Y(f)$ of Eq. (9.201) uses here the terms in Eqs. (9.209)–(9.212).

The Case 1 single-input/single-output nonlinear model of Figure 9.8 is equivalent to the three-input/single-output linear model of Figure 9.9 where the inputs $X(f)$, $X_2(f)$, and $X_3(f)$ are correlated. Note that Figures 9.8 and 9.9 are exactly the same as Figures 4.13 and 4.14.

From knowledge of $A_1(f)$, $A_2(f)$, and $A_3(f)$, Eqs. (9.202)–(9.208) together with Eqs. (9.152)–(9.159) show that the total output autospectral density function in Figures 9.8 and 9.9 is given by

$$S_{yy}(f) = S_{y_1y_1}(f) + S_{y_2y_2}(f) + S_{y_3y_3}(f) + S_{y_1y_3}(f) + S_{y_1y_3}^*(f) + S_{nn}(f) \tag{9.213}$$

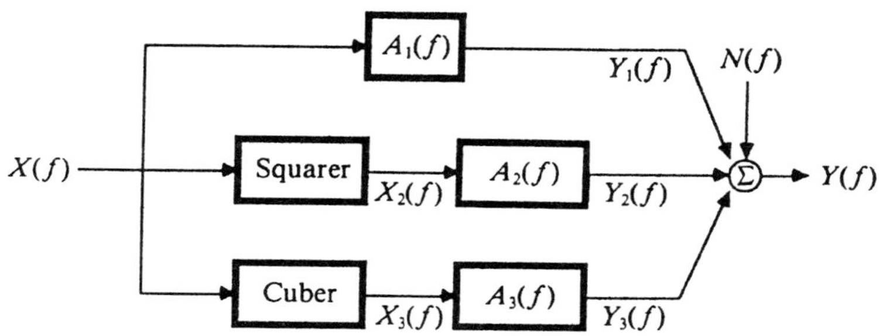

Figure 9.8. Case 1 nonlinear model with correlated outputs.

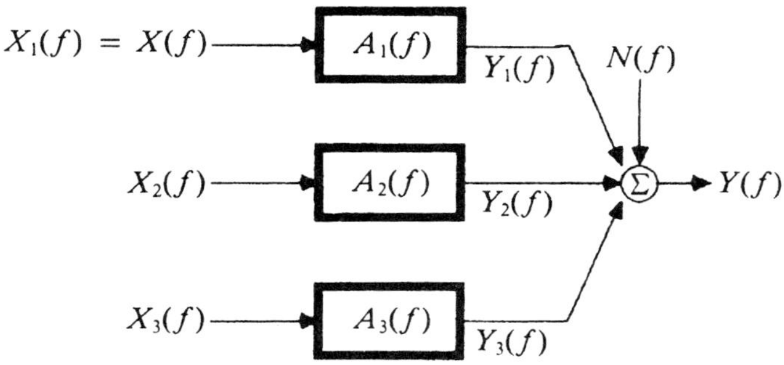

Figure 9.9. Three-input/single-output linear model with correlated inputs equivalent to Figure 9.8.

where

$$S_{y_1y_1}(f) = |A_1(f)|^2 S_{xx}(f) \tag{9.214}$$

$$S_{y_2y_2}(f) = S_{y_by_b}(f) + (\bar{y}_2)^2 \delta_1(f) \tag{9.215}$$

$$S_{y_by_b}(f) = 2|A_2(f)|^2 \int S_{xx}(u) S_{xx}(v - u)\, du \tag{9.216}$$

$$S_{y_3y_3}(f) = S_{y_cy_c}(f) + |C(f)|^2 S_{xx}(f) \tag{9.217}$$

$$S_{y_cy_c}(f) = 6|A_3(f)|^2 \iint S_{xx}(u) S_{xx}(v - u) S_{xx}(f - v)\, du\, dv \tag{9.218}$$

$$S_{y_1y_3}(f) = A_1^*(f) C(f) S_{xx}(f) = A_1^*(f) S_{xy_3}(f) \tag{9.219}$$

$$C(f) = 3A_3(f) \int S_{xx}(u)\, du = 3\sigma_x^2 A_3(f) \tag{9.220}$$

Equation (9.213) shows when $S_{nn}(f) = 0$ that the total output spectrum $S_{yy}(f)$ does not equal the sum of the terms for $S_{y_1y_1}(f)$ plus $S_{y_2y_2}(f)$ plus $S_{y_3y_3}(f)$ because the output terms $Y_1(f)$ and $Y_3(f)$ are correlated. To achieve a more desirable decomposition, the model of Figure 9.8 should be revised as shown in Figure 9.10, where the mutually uincorrelated outputs $Y_a(f)$, $Y_b(f)$, and $Y_c(f)$ replace the previous correlated outputs $Y_1(f)$, $Y_2(f)$, and $Y_3(f)$. The steps required are the same as when Figure 9.6 is replaced by Figure 9.7.

The revised Case 1 nonlinear model of Figure 9.10 is equivalent to the three-input/single-output linear model of Figure 9.11, where the three inputs $X(f)$, $X_2(f)$, and $Z(f) = X_3(f) - 3\sigma_x^2 X(f)$ are uncorrelated. Thus, it is simple here to identify the linear systems $H_o(f)$, $A_2(f)$, and $A_3(f)$ from measurements of $X(f)$ and $Y(f)$ since Figure 9.11 represents three separate single-input/single-output linear systems. Note in Figures 9.10 and 9.11 that $H_o(f)$ is a

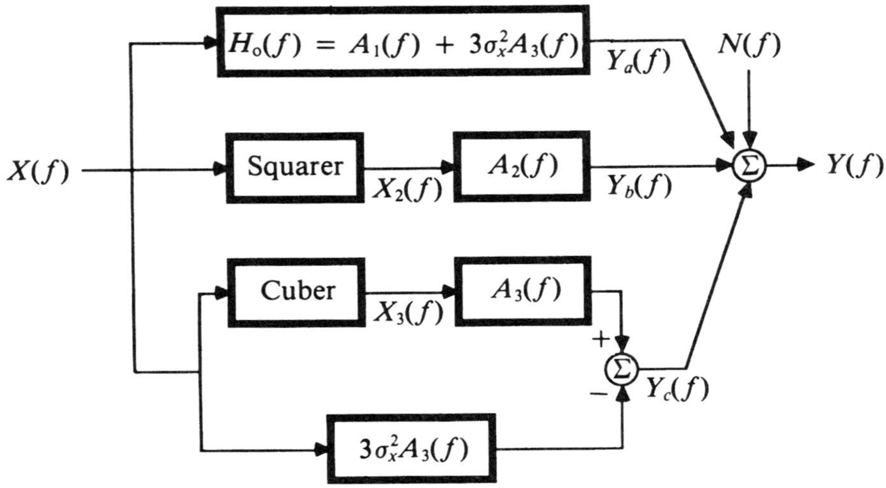

Figure 9.10. Revised Case 1 nonlinear model with uncorrelated outputs.

function of the input variance σ_x^2 and so will vary with different input spectra $S_{xx}(f)$. However, $A_1(f)$, $A_2(f)$, and $A_3(f)$ will be independent of $S_{xx}(f)$ when Figures 9.10 and 9.11 are appropriate nonlinear models.

For the models in Figures 9.10 and 9.11, the total output Fourier transform is given by

$$Y(f) = Y_a(f) + Y_b(f) + Y_c(f) + N(f) \tag{9.221}$$

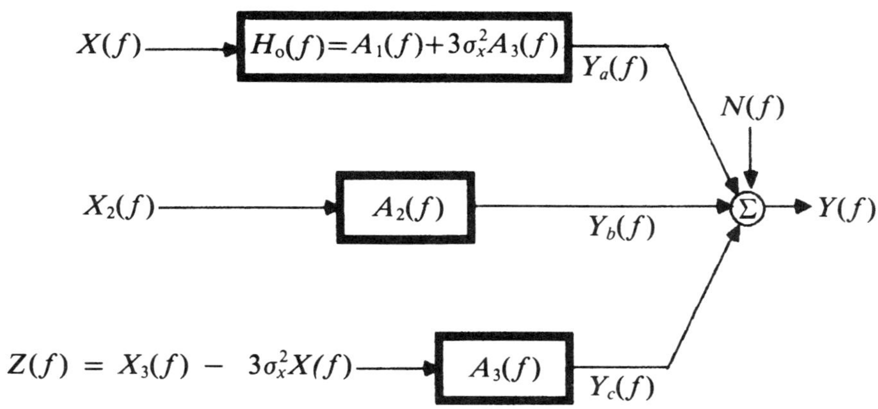

Figure 9.11. Three-input/single-output linear model with uncorrelated inputs equivalent to Figure 9.10.

The total output autospectral density function is

$$S_{yy}(f) = S_{y_a y_a}(f) + S_{y_b y_b}(f) + S_{y_c y_c}(f) + S_{nn}(f) \tag{9.222}$$

where from Eqs. (9.109)–(9.111), the separate spectral terms are

$$S_{y_a y_a}(f) = |H_o(f)|^2 S_{xx}(f) \tag{9.223}$$

$$S_{y_b y_b}(f) = 2|A_2(f)|^2 \int S_{xx}(u) S_{xx}(f-u)\, du \tag{9.224}$$

$$S_{y_c y_c}(f) = 6|A_3(f)|^2 \iint S_{xx}(u) S_{xx}(v-u) S_{xx}(f-v)\, du\, dv \tag{9.225}$$

From Eqs. (9.124), (9.202), and (9.220), the optimum linear system $H_o(f)$ satisfies here the relation

$$H_o(f) = \frac{S_{xy}(f)}{S_{xx}(f)} = A_1(f) + C(f) = A_1(f) + 3\sigma_x^2 A_3(f) \tag{9.226}$$

Equation (9.223) represents the output autospectral density function for the uncorrelated output term $y_a(t)$ in the optimum linear system path. Equation (9.224) represents the output autospectral density function for the uncorrelated output term $y_b(t)$ in the finite-memory nonlinear square-law path, and Eq. (9.225) represents the output autospectral density function for the uncorrelated output term $y_c(t)$ in the finite-memory nonlinear cubic path.

Equation (9.226) shows that the optimum linear system $H_o(f) \neq A_1(f)$ when $C(f) \neq 0$. Thus, conventional linear spectral procedures that consider only single-input/single-output linear models and compute only $H_o(f)$ can give erroneous results for the desired $A_1(f)$, and fail to identify the systems $A_2(f)$ and $A_3(f)$ in Figures 9.8–9.11.

Useful properties of Figures 9.10 and 9.11 can be assessed by computing the linear and nonlinear coherence functions that determine the proportion of the output spectrum that comes from the three paths. The linear coherence function from Eq. (9.140) is

$$\gamma_{xy}^2(f) = \frac{S_{y_a y_a}(f)}{S_{yy}(f)} \tag{9.227}$$

where $0 \leqslant \gamma_{xy}^2(f) \leqslant 1$ for all f states the proportion via the optimum linear path. From Eq. 9.142), the bilinear coherence function is

$$q_{xy_b}^2(f) = \frac{S_{y_b y_b}(f)}{S_{yy}(f)} \tag{9.228}$$

where $0 \leqslant q_{xy_b}^2(f) \leqslant 1$ for all f states the proportion via the finite-memory square-law path. Finally, by Eq. (9.144), the trilinear coherence function is

$$q_{xy_c}^2(f) = \frac{S_{y_c y_c}(f)}{S_{yy}(f)} \tag{9.229}$$

where $0 \leqslant q_{xy_c}^2(f) \leqslant 1$ for all f states the proportion via the finite-memory cubic path. The models of Figures 9.10 and 9.11 will be valid at frequencies where the sum of these three coherence functions is close to unity. The output noise spectrum $S_{nn}(f)$ is given by Eq. (9.146) using the linear and nonlinear coherence functions. Computation of $S_{nn}(f)$ provides a quantitative evaluation of Figures 9.10 and 9.11.

The linear coherence function $\gamma_{xy}^2(f)$ of Eq. (9.227) is the same as the ordinary coherence function between the input $X(f)$ and the total output $Y(f)$. The bilinear coherence function $q_{xy_b}^2(f)$ of Eq. (9.228) is the same as the ordinary coherence function between the second mutually uncorrelated input, that is, $X_2(f)$ in Figure 9.11 and the total output $Y(f)$. The trilinear coherence function $q_{xy_c}^2(f)$ of Eq. (9.229) is the same as the ordinary coherence function between the third mutually uncorrelated input, that is, $Z(f)$ in Figure 9.11 and the total output $Y(f)$.

From Eqs. (9.114) and (9.116), it follows for Case 1 models of Figure 9.8 that the special bispectral density function $S_{xxy}(f)$ defined in Eq. (8.190) is given by

$$S_{xxy}(f) = S_{xxy}(f, f) = 2A_2(2f)S_{xx}^2(f) \tag{9.230}$$

and the special trispectral density function $S_{xxxy}(f)$ defined in Eq. (8.191) is given by

$$S_{xxxy}(f) = S_{xxxy}(f, f, f) = 6A_3(3f)S_{xx}^3(f) \tag{9.231}$$

These special functions of only one frequency variable are much simpler to compute than the general functions $S_{xxy}(f, g)$ and $S_{xxxy}(f, g, h)$ of two and three variables. Note that Eq. (9.230) has $A_2(2f)$ instead of $A_2(f)$ because of the preceding squaring operation, and Eq. (9.231) has $A_3(3f)$ instead of $A_3(f)$ because of the preceding cubing operation.

9.4.2 System Identification in Case 1 Models

To identify the A systems in Case 1 models from measurements of $x(t)$ and $y(t)$ only, one can proceed as outlined in Section 9.2.3. Here, in place of Eqs. (9.122) and (9.123), one should compute the special bispectral and trispectral density functions of single variables

$$S_{xxy}(f) = S_{xxy}(f, f) = \frac{1}{T} E[X^*(f)X^*(f)Y(2f)] \tag{9.232}$$

$$S_{xxxy}(f) = S_{xxxy}(f, f, f) = \frac{1}{T} E[X^*(f)X^*(f)X^*(f)Y(3f)] \tag{9.233}$$

These functions defined earlier in Eqs. (8.190) and (8.191) are direct extensions of the ordinary cross-spectral density functions $S_{xy}(f)$ of Eq. (8.168) or (9.121). Both $S_{xy}(f)$ and $S_{xx}(f)$ should also be computed.

From Eq. (9.226), at frequency f,

$$A_1(f) = H_o(f) - 3\sigma_x^2 A_3(f) \tag{9.234}$$

where the optimum linear system $H_o(f)$ is known from $S_{xx}(f)$ and $S_{xy}(f)$. Note that computation of $A_1(f)$ requires knowledge of σ_x^2 and $A_3(f)$. From Eq. (9.230), at frequency $2f$,

$$A_2(2f) = \frac{S_{xxy}(f)}{2S_{xx}^2(f)} \tag{9.235}$$

To determine $A_2(f)$ instead of $A_2(2f)$, one should replace f by $(f/2)$ above. From Eq. (9.231), at frequency $3f$,

$$A_3(3f) = \frac{S_{xxxy}(f)}{6S_{xx}^3(f)} \tag{9.236}$$

To determine $A_3(f)$ instead of $A_3(3f)$, one should replace f by $(f/3)$ above. Thus, to obtain $A_2(f)$ and $A_3(f)$, one should compute the following quantities:

$$A_2(f) = \frac{S_{xxy}(f/2)}{2S_{xx}^2(f/2)} \tag{9.237}$$

$$A_3(f) = \frac{S_{xxxy}(f/3)}{6S_{xx}^3(f/3)} \tag{9.238}$$

where $S_{xxy}(f/2)$ and $S_{xxy}(f/3)$ are calculated from Eqs. (9.232) and (9.233) with f replaced as shown by $(f/2)$ and $(f/3)$, respectively. Substitution of $A_3(f)$ into Eq. (9.234) gives $A_1(f)$.

This completes the procedure for identifying the A systems in Case 1 models based on computing the special bispectral and trispectral density functions. From the work in Section 9.3, these A systems represent the optimum systems for minimizing the output noise spectrum in Case 1 models.

9.4.3 Case 2 Square-Law and Cubic Models

Case 2 single-input/single-output nonlinear models from Section 4.6.3 occur when higher-order frequency response functions can be represented by the product of first-order frequency response functions such that the linear, bilinear, and trilinear systems in Figure 9.6 are

$$H_1(f) = B_1(f) \tag{9.239}$$

$$H_2(f, f_2) = B_2(f_1)B_2(f_2) \tag{9.240}$$

$$H_3(f_1, f_2, f_3) = B_3(f_1)B_3(f_2)B_3(f_3) \tag{9.241}$$

Here,

$$H_2(f, f) = B_2^2(f) \tag{9.242}$$

$$H_3(f, f, f) = B_3^3(f) \tag{9.243}$$

$$H_2(f, -f) = |B_2(f)|^2 \tag{9.244}$$

$$H_2(u, f-u) = B_2(u)B_2(f-u) \tag{9.244}$$

$$H_3(u, v-u, f-v) = B_3(u)B_3(v-u)B_3(f-v) \tag{9.245}$$

The $B_i(f)$ are frequency response functions of constant-parameter linear systems either known or to be determined. For this case, Eqs. (9.2), (9.24), and (9.38) become

$$Y_1(f) = B_1(f)X(f) \tag{9.246}$$

$$Y_2(f) = \int U_2(\alpha)U_2(f-\alpha)\,d\alpha - \bar{y}_2\delta_1(f) = Y_b(f) \tag{9.247}$$

$$Y_3(f) = \iint U_3(\alpha)U_3(\beta-\alpha)U_3(f-\beta)\,d\alpha\,d\beta \tag{9.248}$$

where

$$U_2(f) = B_2(f)X(f) \tag{9.249}$$

$$U_3(f) = B_3(f)X(f) \tag{9.250}$$

$$\bar{y}_2 = \sigma_2^2 = \int S_{u_2u_2}(f)\,df = \int |B_2(f)|^2 S_{xx}(f)\,df \tag{9.251}$$

$$\sigma_3^2 = \int S_{u_3u_3}(f)\,df = \int |B_3(f)|^2 S_{xx}(f)\,df \tag{9.252}$$

These results have a simple physical interpretation as pictured in Figure 9.12, where zero-memory squaring and cubing operations are preceded by the

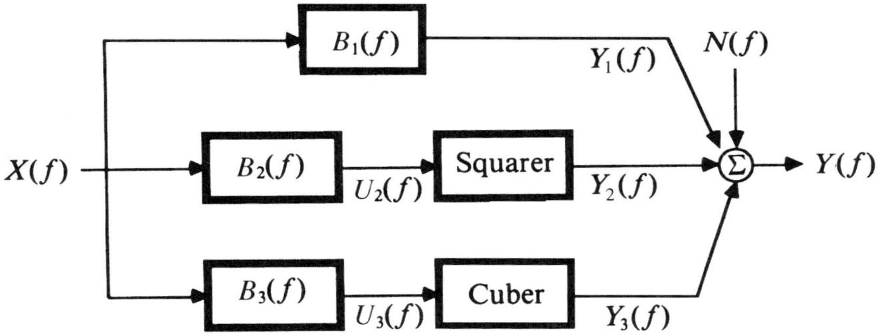

Figure 9.12. Case 2 nonlinear model with correlated outputs.

B for "before" linear systems. The quantity $u_2(t)$ with Fourier transform $U_2(f)$ occurs by passing $x(t)$ through $B_2(f)$, and the single integral in Eq. (9.247) is the Fourier transform of $u_2^2(t)$. The quantity $u_3(t)$ with Fourier transform $U_3(f)$ occurs by passing $x(t)$ through $B_3(f)$, and the double integral in Eq. (9.248) is the Fourier transform of $u_3^3(t)$. The total output Fourier transform $Y(f)$ of Eq. (9.201) uses here the terms in Eqs. (9.246)–(9.248). Note that Figure 9.12 is exactly the same as Figure 4.16.

From knowledge of $B_1(f)$, $B_2(f)$, and $B_3(f)$, Eqs. (9.239)–(9.245) together with Eqs. (9.152)–(9.159) show that the total output spectral density function in Figure 9.12 is

$$S_{yy}(f) = S_{y_1y_1}(f) + S_{y_2y_2}(f) + S_{y_3y_3}(f) + S_{y_1y_3}(f) + S^*_{y_1y_3}(f) + S_{nn}(f) \qquad (9.253)$$

where

$$S_{y_1y_1}(f) = |B_1(f)|^2 S_{xx}(f) \qquad (9.254)$$

$$S_{y_2y_2}(f) = S_{y_by_b}(f) + (\bar{y}_2)^2 \delta_1(f) \qquad (9.255)$$

$$S_{y_3y_3}(f) = S_{y_cy_c}(f) + |C(f)|^2 S_{xx}(f) \qquad (9.256)$$

$$S_{y_1y_3}(f) = B_1^*(f) C(f) S_{xx}(f) \qquad (9.257)$$

$$C(f) = 3B_3(f) \int |B_3(u)|^2 \, du = 3\sigma_3^2 B_3(f) \qquad (9.258)$$

with further formulas for $S_{y_by_b}(f)$ and $S_{y_cy_c}(f)$ as stated in Eqs. (9.261) and (9.262). Note that $C(f)$ in Eq. (9.258) for Case 2 models is quite different from the $C(f)$ in Eq. (9.220) for Case 1 models.

Equation (9.257) shows that the output terms $Y_1(f)$ and $Y_3(f)$ are correlated so that it is difficult to decompose the total output spectrum of Eq. (9.253)

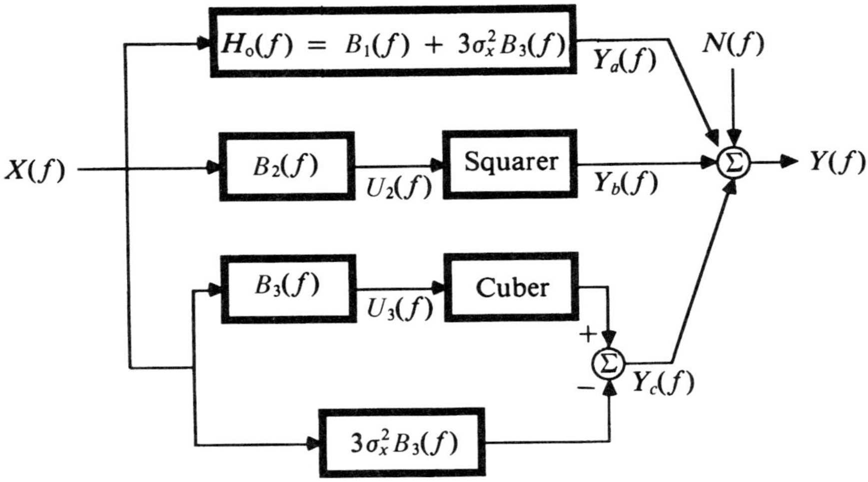

Figure 9.13. Revised Case 2 nonlinear model with uncorrelated outputs.

into distinct parts. To achieve this goal, the model of Figure 9.12 should be revised as shown in Figure 9.13, where the mutually uncorrelated outputs $Y_a(f)$, $Y_b(f)$, and $Y_c(f)$ replace the previous correlated outputs $Y_1(f)$, $Y_2(f)$, and $Y_3(f)$. The steps required are the same as when Figure 9.8 is replaced by Figure 9.10, and when Figure 9.3 is replacd by Figure 9.5.

For the model in Figure 9.13, the total output autospectral density function is

$$S_{yy}(f) = S_{y_a y_a}(f) + S_{y_b y_b}(f) + S_{y_c y_c}(f) + S_{nn}(f) \tag{9.259}$$

where from Eqs. (9.109)–(9.111),

$$S_{y_a y_a}(f) = |H_o(f)|^2 S_{xx}(f) \tag{9.260}$$

$$S_{y_b y_b}(f) = 2\int S_{u_2 u_2}(\alpha) S_{u_2 u_2}(f - \alpha)\, d\alpha \tag{9.261}$$

$$S_{y_c y_c}(f) = 6\iint S_{u_3 u_3}(\alpha) S_{u_3 u_3}(\beta - \alpha) S_{u_3 u_3}(f - \beta)\, d\alpha\, d\beta \tag{9.262}$$

with

$$H_o(f) = B_1(f) + C(f) = B_1(f) + 3\sigma_3^2 B_3(f) \tag{9.263}$$

$$S_{u_2 u_2}(f) = |B_2(f)|^2 S_{xx}(f) \tag{9.264}$$

$$S_{u_3 u_3}(f) = |B_3(f)|^2 S_{xx}(f) \tag{9.265}$$

Linear and nonlinear coherence functions can now be computed exactly as per Eqs. (9.227)–(9.229) to assess the validity of the revised Case 2 model of Figure 9.13. These results can be compared with similar results from the revised Case 1 model of Figure 9.10 to help decide which of these nonlinear models is more appropriate to fit the measured data in physical problems.

From Eqs. (9.114) and (9.116), it follows for Case 2 models of Figure 9.13 that the special bispectral density function $S_{xxy}(f)$ of Eq. (8.190) is

$$S_{xxy}(f) = S_{xxy}(f, f) = 2B_2^2(f)S_{xx}^2(f) \tag{9.266}$$

and the special trispectral density function $S_{xxxy}(f)$ of Eq. (8.191) is

$$S_{xxxy}(f) = S_{xxxy}(f, f, f) = 6B_3^3(f)S_{xx}^3(f) \tag{9.267}$$

These results should be compared with Eqs. (9.230) and (9.231). Note that Eq. (9.266) has $B_2^2(f)$ instead of $B_2(f)$ because of the following squaring operation, and Eq. (9.267) has $B_3^3(f)$ instead of $B_3(f)$ because of the following cubing operation.

9.4.4 System Identification in Case 2 Models

To identify the B systems in Case 2 models from measurements of $x(t)$ and $y(t)$ only, one should follow the procedure outlined in Section 9.2.3. As before, begin by computing the spectral quantities

$$S_{xx}(f) = \frac{1}{T} E[|X(f)|^2] \tag{9.268}$$

$$S_{xy}(f) = \frac{1}{T} E[X^*(f)Y(f)] \tag{9.269}$$

$$S_{xxy}(f) = \frac{1}{T} E[X^*(f)X^*(f)Y(2f)] \tag{9.270}$$

$$S_{xxxy}(f) = \frac{1}{T} E[X^*(f)X^*(f)X^*(f)Y(3f)] \tag{9.271}$$

From Eq. (9.263), one obtains

$$B_1(f) = H_o(f) - 3\sigma_3^2 B_3(f) \tag{9.272}$$

where $H_o(f)$ is found by the usual formula

$$H_o(f) = \frac{S_{xy}(f)}{S_{xx}(f)} \tag{9.273}$$

Note that computation of $B_1(f)$ requires knowledge of $B_3(f)$. In general, $H_o(f) \neq B_1(f)$.

From Eq. (9.266) involving the special bispectral density function,

$$B_2^2(f) = \frac{S_{xxy}(f)}{2S_{xx}^2(f)} = |B_2(f)|^2 e^{-2j\phi_2(f)} \tag{9.274}$$

This gives $B_2(f)$ by the formula

$$B_2(f) = \frac{[S_{xxy}(f)/2]^{1/2}}{S_{xx}(f)} = |B_2(f)| e^{-j\phi_2(f)} \tag{9.275}$$

From Eq. (9.267) involving the special trispectral density function,

$$B_3^3(f) = \frac{S_{xxxy}(f)}{6S_{xx}^3(f)} = |B_3(f)|^3 e^{-3j\phi_3(f)} \tag{9.276}$$

This gives $B_3(f)$ by the formula

$$B_3(f) = \frac{[S_{xxxy}(f)/6]^{1/3}}{S_{xx}(f)} = |B_3(f)| e^{-j\phi_3(f)} \tag{9.277}$$

This completes the procedure for identifying the B systems in Case 2 models. From the work in Section 9.3, these B systems represent the optimum systems for minimizing the output noise spectrum in Case 2 models.

This concludes Chapter 9.

REFERENCES

1. Bendat, J. S., and Piersol, A. G., *Random Data: Analysis and Measurement Procedures*, 2nd ed., Wiley-Interscience, New York, 1986.
2. Bendat, J. S., and Piersol, A. G., *Engineering Applications of Correlation and Spectral Analysis*, 2nd ed., Wiley-Interscience, New York, 1993.
3. Bendat, J. S., and Piersol, A. G., Spectral analysis of nonlinear systems involving square-law operations, Journal of Sound and Vibration, **81**(2), 199 (1982).
4. Bendat, J. S., and Piersol, A. G., Decomposition of wave forces into linear and nonlinear components, *Journal of Sound and Vibration*, **106**(3), 391 (1986).
5. Bendat, J. S., and Palo, P. A., Practical techniques for nonlinear system analysis and identification, *Sound and Vibration*, Bay Village, Ohio, June 1990.
6. Bendat, J. S., Palo, P. A., and Coppolino, R. N., A general identification technique for nonlinear differential equations of motion, *Probabilistic Engineering Mechanics*, **7**(1), 43 (1992).
7. Bendat, J. S., Coppolino, R. N., and Palo, P. A., Identification of physical parameters with memory in nonlinear systems, *International Journal of Nonlinear Mechanics*, **30**(6), 841 (1995).
8. Bendat, J. S., Spectral techniques for nonlinear system analysis and identification, *Shock and Vibration*, **1**(1), 21 (1993).
9. Bendat, J. S., Statistical errors for nonlinear system measurements involving square-law operations, *Journal of Sound and Vibration*, **90**(2), 275 (1983).
10. Bendat, J. S., Solutions for the multiple input/output problem, *Journal of Sound and Vibration*, **44**(3), 311 (1976).
11. Bendat, J. S., System identification from multiple input/output data, *Journal of Sound and Vibration*, **49**(3), 293 (1976).
12. Bendat, J. S., Statistical errors in measurement of coherence functions and input/output quantities, *Journal of Sound and Vibration*, **59**(3), 405 (1978).
13. Bendat, J. S., Modern analysis procedures for multiple input/output problems, *Journal of the Acoustical Society of America*, **68**(2), 498 (1980).
14. Bendat, J. S., *The Hilbert Transform and Applications to Correlation Measurements*, Bruel & Kjaer, Naerum, Denmark, 1986.
15. Braun, S., *Mechanical Signature Analysis*, Academic Press, New York, 1986.

16. Broch, J. T., *Nonlinear Systems and Random Vibration*, Bruel & Kjaer, Naerum, Denmark, 1975.

17. Bussgang, J. J., Cross-correlation functions of amplitude distorted Gaussian inputs, in *Nonlinear Systems*, A. H. Haddad, Ed., Dowden, Hutchinson & Ross, Stroudsburg, PA, 1975.

18. Chakrabarti, S. K., *Nonlinear Methods in Offshore Engineering*, Elsevier, Amsterdam, 1990.

19. Chertoff, M. E., Steele, T., Ator, G. A., and Bian, L., Characterizing cochlear mechano-electric transduction using a nonlinear systems identification procedure, *Journal of the Acoustical Society of America*, **100**(6), 3741 (1996).

20. Choi, D., Miksad, R. W., and Powers, E. J., Application of digital cross-bispectral analysis techniques to model the nonlinear response of a moored vessel system in random seas, *Journal of Sound and Vibration*, **99**(3), 309 (1985).

21. Crandall, S. H., and Mark, W. D., *Random Vibration in Mechanical Systems*, Academic Press, New York, 1963.

22. Dalzell, J. F., Cross-bispectral analysis: application to ship resistance in waves, *Journal of Ship Research*, **18**, 62 (1974).

23. Deutsch, R., *Nonlinear Transformations of Random Processes*, Prentice-Hall, Englewood Cliffs, NJ, 1962.

24. Dodds, C. J., and Robson, J. D., Partial coherence in multivariate random processes, *Journal of Sound and Vibration*, **42**(2), 243 (1975).

25. Ewins, D. J., *Modal Testing: Theory and Practice*, Wiley, New York, 1985.

26. Fitzpatrick, J. A., and Rice, H. J., Simplified coherence functions for multiple input/output analysis, *Journal of Sound and Vibraion*, **122**(1), 171 (1988).

27. Fitzpatrick, J. A., and Rice, H. J., An alternative to the bispectral approach for the analysis of flow induced vibration problems, *Proceedings of the International Symposium on Flow Induced Vibration and Noise*, American Society of Mechanical Engineers, Chicago, **2**, 197 (1988).

28. French, A. S., Practical nonlinear system analysis by Wiener kernal estimation in the frequency domain, *Biological Cybernetics*, **24**, 111 (1976).

29. Hasselman, K., Munk, W., and MacDonald, G., Bispectra of ocean waves, in *Time Series Analysis*, W. Rosenblatt, Ed., Wiley, New York, 1963.

30. Hildebrand, F. B., *Advanced Calculus for Applications*, Prentice Hall, Englewood Cliffs, NJ, 1962.

31. Hooft, J. P., *Advanced Dynamics of Marine Structures*, Wiley-Interscience, New York, 1982.

32. Kim, C. H., and Breslin, J. P., Prediction of slow drift oscillations of a moored ship in head seas, *Proceedings Behavior of Offshore Structures*, BOSS '76, Trondheim, Norway, 320 (1976).

33. Kim, K. S., and Shin, Y. S., Application of a new Hilbert transform method to nonlinearity identification, *Proceedings of the 7th International Modal Analysis Conference*, Las Vegas, Nevada, 1386 (1989).

34. Marmelis, P. Z., and Marmelis, V. Z., *Analysis of Physiological Systems: The White-Noise Approach*, Plenum Press, New York, 1978.

35. *MATLAB Reference Guide*, The Mathworks Inc., Natick, MA, 1992.

36. *MATLAB, Signal Processing Toolbox User's Guide*, The Mathworks Inc., Natick, MA, 1994.

37. Matsuoka, T., and Ulrych, T. J., Phase estimation using the bispectrum, *Proceedings of the IEEE*, **72**(10), 1403 (1988).

38. Morison, J. R., et al., The force exerted by surface waves on piles, *Journal of Petroleum Techniques*, **189**, 149 (1950).

39. Newland, D. E., *Random Vibrations and Spectral Analysis*, 2nd ed., Longman, New York, 1984.

40. Newman, J. N., *Marine Hydrodynamics*, MIT Press, Cambridge, MA, 1977.

41. Nikias, C. L., and Raghuveer, M. R., Bispectrum estimation: a digital signal processing framework, *Proceedings of the IEEE*, **75**(7), 869 (1987).

42. Price, R., A useful theorem for nonlinear devices having Gaussian inputs, in *Nonlinear Systems*, A. H. Haddad, Ed., Dowden, Hutchinson & Ross, Stroudsburg, PA, 1975.

43. Price, W. G., and Biship, R., *Probabilistic Theory of Ship Dynamics*, Chapman and Hall, London, 1974.

44. Rice, H. J., and Fitzpatrick, J. A., A generalized technique for spectral analysis of nonlinear systems, *Mechanical Systems and Signal Processing*, **2**(2), 195 (1988).

45. Rice, H. J., Esmonde, H., and Fitzpatrick, J. A., A spectral method for identifying quadratic damping in single degree of freedom systems, *Proceedings of the International Symposium on Flow Induced Vibration and Noise*, American Society of Mechanical Engineers, Chicago, **3**, 269 (1988).

46. Rice, S. O., Mathematical analysis of random noise, in *Selected Papers on Noise and Stochastic Processes*, N. Wax, Ed., Dover, New York, 1954.

47. Rice, S. O., and Bedrosian, E., The output properties of Volterra systems (nonlinear systems with memory) driven by harmonic and Gaussian inputs, *Proceedings of the IEEE*, **59**(12), 1688 (1971).

48. Roberts, J. B., Nonlinear analysis of slow drift oscillations of moored vessels in random seas, *Journal of Ship Research*, **25**, 130 (1981).

49. Rozario, N., and Papoulis, A., The identification of certain nonlinear systems by only observing the output, *Workshop on Higher-Order Spectral Analysis*, Office of Naval Research and National Science Foundation, Vail, Colorado, 78 (1989).

50. Rugh, W. J., *Nonlinear System Theory: The Volterra/Wiener Approach*, Johns Hopkins University Press, Baltimore, MD, 1981.

51. Sarpkaya, T., and Isaacson, M., *Mechanics of Wave Forces on Offshore Structures*, Van Nostrand Reinhold, New York, 1981.

52. Schetzen, M., *The Volterra and Wiener Theories of Nonlinear Systems*, Wiley-Interscience, New York, 1980.

53. Schueller, G. I., and Bucher, C. G., Non-Gaussian response of systems under dynamic excitation, in *Stochastic Structural Dynamics*, S. T. Ariaratnam, G. I. Schueller, and I. Elishakoff, Eds., Elsevier Applied Science, New York, 1988.

54. Smallwood, D. O., Matrix methods for estimating the coherence functions from estimates of the cross-spectral density matrix, *Shock and Vibration*, **3**(4), 237 (1996).

55. Spanos, P. D., and Lu, R., Nonlinear system identification in offshore structural reliability, *Journal of Offshore Mechanics and Arctic Engineering* **117**, 171 (1995).

56. Tick, L. J., The estimation of transfer functions of quadratic systems, *Technometrics*, **3**, 563 (1961).

57. Vinh, T., Liu, H., and Djouder, M., Second order transfer function: computation and physical interpretation, *Proceedings 5th International Modal Analysis Conference*, London, 587 (1987).

58. Vugts, J. H., and Bouquet, A. G., A nonlinear frequency domain description of wave forces on an element of a vertical pile, *Proceedings Behavior of Offshore Structures*, BOSS '85, Delft, Netherlands, 239 (1985).

59. Waters, J. K., Model test of extreme wave loading on a moored rectangular barge, *Department of Naval Architecture, Ocean and Marine Engineering*, United States Naval Academy, Annapolis, Maryland, November 1995.

60. Yang, C. Y., *Random Vibration of Structures*, Wiley-Interscience, New York, 1986.

61. Zseleczky, J. J., and Wallendorf, L. A., Extreme roll motions of a navy frigate model in large beam seas, *Hydromechanics Laboratory*, United States Naval Academy, Annapolis, Maryland, October 1994.

INDEX

GLOSSARY OF SYMBOLS

a, b	Arbitrary constants
$A(f, B(f)$	Linear system frequency response functions
$E[\ ]$	Expected value of [ ]
f	Cyclical frequency
$\mathscr{F}[\ ]$	Fourier transform of [ ]
$G_{xx}(f)$	Autospectral density function (one-sided)
$G_{xy}(f)$	Cross-spectral density function (one-sided)
$\mathscr{G}(f)$	Energy spectral density function (one-sided)
$h(\tau)$	Linear weighting function (first-order)
$h(\tau, t)$	Time-varying linear weighting function
$h(\tau_1, \tau_2)$	Bilinear weighting function (second-order)
$h(\tau_1, \tau_2, \tau_3)$	Trilinear weighting function (third-order)
$H(f)$	Linear frequency response function (first-order)
$H_o(f)$	Optimum linear frequency response function
$\|H(f\|$	System gain factor
$H(f_1,f_2)$	Bilinear frequency response function (second-order)
$H(f_1, f_2, f_3)$	Trilinear frequency response function (third-order)
j	$\sqrt{-1}$
$J(f, g)$	Double Fourier transform of $h(\tau, t)$
$L_1(x)$	Linear system operator
$L_2(x)$	Special bilinear system operator
$L_3(x)$	Special trilinear system operator
$L_2(x, y)$	Bilinear system operator
$L_3(x, y, z)$	Trilinear system operator
n_d	Number of distinct averages
$p(x)$	Probability density function (first-order)
$p_2(y)$	Nonlinear system output probability density function
$p(x_1, x_2)$	Joint probability density function (second-order)
Prob [ ]	Probability of [ ]

$q_{xy}^2(f)$	Nonlinear coherence function
$R_{xx}(\tau)$	Autocorrelation function (first-order)
$R_{xy}(\tau)$	Cross-correlation function (first-order)
$R_{xxx}(\tau_1, \tau_2)$	Second-order autocorrelation function
$R_{xxy}(\tau_1, \tau_2)$	Second-order cross-correlation function
$R_{xxxx}(\tau_1, \tau_2, \tau_3)$	Third-order autocorrelation function
$R_{xxxy}(\tau_1, \tau_2, \tau_3)$	Third-order cross-correlation function
$S_{xx}(f)$	Autospectral density function (two-sided)
$S_{xy}(f$	Cross-spectral density function (two-sided)
$S_{xxy}(f)$	Special bispectral density function
$S_{xxxy}(f)$	Special trispectral density function
$S_{xxx}(f_1, f_2)$	Second-order autospectral density function (bispectrum)
$S_{xxy}(f_1, f_2)$	Second-order cross-spectral density function
$S_{xxxx}(f_1, f_2, f_3)$	Third-order autospectral density function (trispectrum)
$S_{xxxy}(f_1, f_2, f_3)$	Third-order cross-spectral density function
$\mathscr{S}(f)$	Energy spectral density function (two-sided)
t	Time variable
T	Record length
$x(t)$, $y(t)$	Time-dependent real-valued records
$X(f)$, $Y(f)$	Fourier transforms of $x(t)$, $y(t)$
$y_1(t)$, $y_a(t)$	Linear system output
$y_2(t)$, $y_b(t)$	Bilinear or squarer system output
$y_3(t)$, $y_c(t)$	Trilinear or cuber system output
$\|[\ ]\|$	Absolute value of [ ]
$[\tilde{\ }]$	Raw estimate of [ ]; Hilbert transform of [ ]
$[\hat{\ }]$	Smooth estimate of [ ]
α	A probability value
$\gamma_{xy}^2(f)$	Linear (ordinary) coherence function
$\delta(\ )$	Delta function
$\delta_1(\ )$	Finite delta function
ε	Normalized random error
$\theta_{xy}(f)$	Argument of $G_{xy}(f)$
μ	Mean value
$\rho(\tau)$	Correlation coefficient function
σ	Standard deviation
σ^2	Variance
τ	Time displacement
$\phi(f)$	System phase factor
ψ^2	Mean square value

Books of related interest by Julius S. Bendat and Allan G. Piersol.

RANDOM DATA

ANALYSIS AND MEASUREMENT PROCEDURES

Second Edition

With more than 100,000 copies in use throughout the world from successive English editions and four translations, this is the definitive book for the treatment of random processes in linear systems. This Second Edition has been substantially revised and expanded to serve as a convenient reference for practicing engineers and scientists, and as a specialized textbook for upper division and graduate students to help them and their teachers conduct research work. It promotes random data analysis in many areas of application, and contains updated discussions of well-established recommended data analysis procedures for optimum linear system identification from measured data. This book also includes complete derivations of both older and newer practical formulas for statistical error analysis of computed estimates that can be used to design experiments and evaluate the results. Problems sets are given at the end of each chapter. Readers will find applications in digital data analysis, dynamic test programs, input/output modeling, noise source location, structural response prediction, random vibration, nonstationary data analysis techniques, Hilbert transform technques, and more.

1986 (0-471-04000-2) 566 pp.

ENGINEERING APPLICATIONS OF CORRELATION AND SPECTRAL ANALYSIS

Second Edition

This Second Edition engineering book with many applications and case studies from different fields supplements the theory and data processing procedures in *Random Data: Analysis and Measurement Procedures, Second Edition*, by discussing how to solve various engineering problems based upon correlation and spectral analysis of measured data. Illustrations presented in detail include: (1) predicting the response properties of random excitations passing through linear systems, (2) estimating time delays and propagation velocities, (3) determining the location of random excitations which produce random responses, and (4) identifying optimum linear system frequency response function properties from measured excitations and responses. Linear system problems are solved for single-input/output and multiple-input/output physical data where arbitrary probability, correlation and spectral properties can exist among the records. Statistical error analysis formulas are used to design experiments and evaluate computed results. Practical techniques for nonstationary data analysis and for nonlinear system analysis are also discussed and illustrated with examples.

1993 (0-471-57055-9) 458 pp.